Chemisorption and Reactivity on Supported Clusters and Thin Films

NATO ASI Series

Advanced Science Institutes Series

A Series presenting the results of activities sponsored by the NATO Science Committee, which aims at the dissemination of advanced scientific and technological knowledge, with a view to strengthening links between scientific communities.

The Series is published by an international board of publishers in conjunction with the NATO Scientific Affairs Division

A Life Sciences **B Physics**	Plenum Publishing Corporation London and New York
C Mathematical and Physical Sciences **D Behavioural and Social Sciences** **E Applied Sciences**	Kluwer Academic Publishers Dordrecht, Boston and London
F Computer and Systems Sciences **G Ecological Sciences** **H Cell Biology** **I Global Environmental Change**	Springer-Verlag Berlin, Heidelberg, New York, London, Paris and Tokyo

PARTNERSHIP SUB-SERIES

1. Disarmament Technologies	Kluwer Academic Publishers
2. Environment	Springer-Verlag / Kluwer Academic Publishers
3. High Technology	Kluwer Academic Publishers
4. Science and Technology Policy	Kluwer Academic Publishers
5. Computer Networking	Kluwer Academic Publishers

The Partnership Sub-Series incorporates activities undertaken in collaboration with NATO's Cooperation Partners, the countries of the CIS and Central and Eastern Europe, in Priority Areas of concern to those countries.

NATO-PCO-DATA BASE

The electronic index to the NATO ASI Series provides full bibliographical references (with keywords and/or abstracts) to more than 50000 contributions from international scientists published in all sections of the NATO ASI Series.
Access to the NATO-PCO-DATA BASE is possible in two ways:

– via online FILE 128 (NATO-PCO-DATA BASE) hosted by ESRIN, Via Galileo Galilei, I-00044 Frascati, Italy.

– via CD-ROM "NATO-PCO-DATA BASE" with user-friendly retrieval software in English, French and German (© WTV GmbH and DATAWARE Technologies Inc. 1989).

The CD-ROM can be ordered through any member of the Board of Publishers or through NATO-PCO, Overijse, Belgium.

Series E: Applied Sciences - Vol. 331

Chemisorption and Reactivity on Supported Clusters and Thin Films

Towards an Understanding of Microscopic Processes in Catalysis

edited by

Richard M. Lambert

Department of Chemistry,
University of Cambridge,
Cambridge, U.K.

and

Gianfranco Pacchioni

Department of Inorganic, Organometallic and Analytical Chemistry,
University of Milan,
Milan, Italy

Kluwer Academic Publishers

Dordrecht / Boston / London

Published in cooperation with NATO Scientific Affairs Division

Proceedings of the NATO Advanced Study Institute on
Chemisorption and Reactivity on Supported Clusters and Thin Films:
Towards an Understanding of Microscopic Processes in Catalysis
Erice, Trapani, Sicily
July 15–26, 1996

A C.I.P. Catalogue record for this book is available from the Library of Congress.

ISBN 0-7923-4448-0

Published by Kluwer Academic Publishers,
P.O. Box 17, 3300 AA Dordrecht, The Netherlands.

Kluwer Academic Publishers incorporates the publishing programmes of
D. Reidel, Martinus Nijhoff, Dr W. Junk and MTP Press.

Sold and distributed in the U.S.A. and Canada
by Kluwer Academic Publishers,
101 Philip Drive, Norwell, MA 02061, U.S.A.

In all other countries, sold and distributed
by Kluwer Academic Publishers Group,
P.O. Box 322, 3300 AH Dordrecht, The Netherlands.

Printed on acid-free paper

Printed in the Netherlands

CONTENTS

PREFACE

Heterogeneous catalysis is of immense technological and economic importance on a global scale because it provides the backbone of the world's chemical and oil industries. For this reason, among others, the subject is intensively researched in both academic and industrial laboratories. The innate complexity of practical catalytic systems suggests that useful progress should be achievable by investigating key aspects of catalysis by experimental studies on idealised model systems. Thin films and supported clusters are two promising types of model systems that can be used for this purpose because they mimic important aspects of the properties of practical dispersed catalysts. Similarly, appropriate theoretical studies of chemisorption and surface reactions using cluster models or extended slab systems can provide valuable information about the factors that underlie bonding and catalytic activity. This book is based on the lectures given at the 32nd course of the International School of Material Science at the Ettore Majorana Centre, Erice Sicily between July the 15th and 26th 1996. The title of the course was "Chemisorption and reactivity on supported Clusters and thin films: towards an understanding of microscopic processes in catalysis". The Institute dealt with the fundamentals of heterogeneous catalysis from phenomenological, experimental and theoretical points of view. An important theme of the course was its emphasis on the design and use of suitable model systems whose properties could be expected to mimic key aspects of the behaviour of practical catalysts which generally are much more complex and poorly defined materials.

This volume describes experimental and theoretical approaches to the surface chemistry and catalytic behaviour of metals, metal oxides and metal/metal oxide systems. It should be of value to advanced undergraduates taking courses in chemisorption and catalysis and to graduate students undertaking research in this field. An introduction to the principles and main themes of heterogeneous catalysis is followed by detailed accounts of the application of modern experimental and theoretical techniques to fundamental problems in catalytic science. The description of a variety of advanced experimental methods to problems of structure, bonding and reactivity at well defined surfaces is complemented by an account of theoretical procedures including Hartree-Fock, density functional and computer simulation techniques.

The editors gratefully acknowledge the support provided by NATO, the sole sponsor, to whom thanks are also due. Without this support, neither the course nor the production of this book would have been possible. We are indebted to Professor A. Zichichi, Director of the Majorana Centre, whose help was essential for the planning of the school. Finally, it a pleasure to acknowledge the hospitality and generous help provided by all the staff at the Ettore Majorana Centre, whose efficiency, attention to detail, and unfailing good humour made the school the successful and highly enjoyable occasion that it was.

Richard Lambert
Cambridge, England
Gianfranco Pacchioni
Milan, Italy
October 1996

INTRODUCTION TO HETEROGENEOUS CATALYSIS

R. M. LAMBERT
University of Cambridge
Department of Chemistry
Lensfield Road
Cambridge
CB2 1EW

INTRODUCTION

Chemisorption is a necessary precursor to heterogeneous catalysis. Thin films and clusters provide a means of studying chemisorption experimentally, and there are theoretical approaches related to each of these avenues. Our purpose is to achieve a fundamental understanding of heterogeneous catalysis, therefore let us begin by asking why this is considered to be an important and worthwhile goal. Few would disagree that most research on catalysis is at least partly driven by the desire to improve existing technology or to invent new technology. A principal purpose of this introductory chapter is to put the subject in context, to indicate what is known and what remains to be done, and to give an indication the most important avenues for future development. Another is to introduce basic ideas about the nature of solid catalysts, to show how they may be approximated by idealised thin film systems and to indicate what may be learned by this approach. A third objective is to give a simple overview of where we stand with respect to theories of catalysis. Finally, these themes are illustrated with some specific examples. The reader will be aware that many of the themes introduced here are developed in much greater detail in the chapters that follow this one.

In terms of the world economy and with respect environmental protection, heterogeneous catalysis, especially by metals, and especially by the platinum metals, is of global strategic importance. To some degree, this is reflected in the fact that the most academic research is focussed on adsorption and catalysis by metal surfaces - although it must be admitted that the relative ease of studying metals plays a very significant part here. To appreciate the scope of our subject, consider first some urgent environmental issues. Reactions occurring on the surfaces of polar stratospheric clouds1 or volcanic inorganic aerosols[2] are implicated in ozone destruction. Much research in electrocatalysis[3] and photocatalysis[4] is aimed at the development of fuel cells, solar energy conversion or CO_2 conversion - issues that are of importance with respect to minimising the greenhouse effect. Control of toxic emissions from automotive and stationary combustion engines is achieved entirely by the application of catalytic systems. The economic impact of heterogeneous catalysis is no less striking. More than 90% of the output of world chemical industry depends on the operation of heterogeneously catalysed reactions. More than 25% of global industrial output (not just the production of chemicals) depends on heterogeneous catalysis by the Pt metals alone. 20% of world GNP is derived through the operation of heterogeneously catalysed processes, equivalent to about $5,000 billion/year. Governments and industry constantly demand more efficient and cleaner processes, cheaper

R.M. Lambert and G. Pacchioni (eds.),
Chemisorption and Reactivity on Supported Clusters and Thin Films, 1-26.

feedstocks and cheaper chemical plants - and better performing products. Catalysis enables all of these. This is exemplified by the huge sales of catalysts themselves (as opposed to the products made by employing them) amounting to $9 billion/year and growing at 5%/year. The principal industrial applications of catalytic technology lie in the in following areas.
Oil refining (cracking, isomerisation, reforming, hydrocracking). The industry is undergoing major consolidation and developments in the Far East are likely to be crucial.
Petrochemicals production (polymers, syngas, selective oxidation and reduction). Most research is driven by the desire to reduce costs by discovering new routes or by improving the selectivity of existing routes.
Environmental protection (80% autocatalysts; 20% stationary sources). Suppression of diesel exhaust and volatile organic compounds emission probably hold the key to future growth in this area.

The majority of solid catalysts are broadly divisible into two classes: metals (including alloys) and oxides (including binary or ternary mixed oxides). We shall consider only these classes and neglect all others - for example sulphides of the transition metals which are also industrially important. Many catalytic processes are operated at the solid gas interface, and it is this interface that we shall concentrate on here. However, it should be borne in mind that many important catalytic processes, including electrocatalysis, are operated in the liquid phase. Such systems are difficult to address experimentally: the presence of the solvent introduces great complexity, both in the acquisition and in the interpretation of data. Therefore we shall not consider reactions at the solid liquid interface.

Experimental studies of catalytically relevant phenomena have advanced enormously in the last two decades or so, thanks to the use of highly characterised surfaces and the development of a range of sophisticated techniques. These so-called "surface science" methods are to a large extent based on electron spectroscopic techniques which exploit the energy dependence of the inelastic mean free path of electrons in solids[5] . By working with ejected electrons whose kinetic energies lie in the range ~ 50 to 500 eV, one may achieve the required surface sensitivity. Such methods enable us to study the atomic arrangement and electronic properties of surfaces, especially extended surfaces, in great detail. With well defined surfaces, working under ultra high vacuum conditions,very detailed information about the dynamics of chemisorption and the structure and bonding of adsorbed species can be obtained. To extrapolate from this information to an understanding of the observed activity and selectivity at elevated pressures of highly dispersed working catalysts is a formidable task, because of the complexity of the latter. Typically, practical heterogeneous catalysts are relatively poorly characterised. The catalytically active surface phase may include a variety of different crystal planes of unknown chemical composition interacting to an unknown degree with the support phase. It is in this connection that experiments on well characterised thin films and clusters can be especially valuable. Most such work has tended to focus on metals as catalysts, albeit in some cases addressing their interaction with oxide supports. However, corresponding work on oxides in their own right, though much more scarce, is gaining momentum. Although experiments on oxide surfaces involve a higher degree of experimental difficulty, this is now an important growth area[6] and is likely to remain so - many industrially important catalysts are oxides and most industrially important catalyst supports are oxides.

Advances in theory have provided vital fundamental understanding, and the most important *ab initio* theoretical techniques are discussed in detail elsewhere in this volume. Suffice to say here that in recent years density functional calculations have come to play an increasingly important role in chemisorption calculations, in regard to both energetics and dynamics[7], especially with respect to metal surfaces. At the simplest level of approximation (jellium model) density functional theory has been used to address a variety of chemisorption problems, in cluding promoter effects, and a number of important conclusions have emerged[8,9] Such calculations are now starting to make an important contribution to our understanding of adsorption on both perfect and defective oxide surfaces[10]. Standard density functional calculations treat the exchange-correlation energy in the local density approximation, i.e. the electron density is treated as locally uniform. However, this gives rise to systematic "overbinding" which can introduce significant errors in the results. The accuracy of such calculations can be improved by application of generalised gradient approximations in which the exchange-correlation energy is represented in a form that explicitly contains gradients of the electron density[11].

Despite impressive advances in both experiment and theory, it remains true that valid generalisations are few and it is difficult to make specific predictions that are of direct practical utility. This reflects the sheer chemical and structural complexity of real catalytic systems. A great proportion of current research on catalysis concentrates on finding the link between structure and function. This link is one of the great unifying themes of chemical theory which has found triumphant expression in the field of organic chemistry where it has culminated in the ability to predict accurately the outcome of chemical reactions taking place between very complicated molecules *in the solution phase*. Many of these *homogeneous* reactions are operated catalytically. Let us take a step closer to our present subject. The structure, spectra and catalytic behaviour of metal cluster compounds in solution are known and understood with a high degree of certainty. Related to this is the fact that homogeneous catalysts are far more selective than their heterogeneous counterparts. Why? Because in solution reactions take place between well defined molecular entities that can be studied in very great detail by methods that on the whole yield unambiguous information. Usually, we need consider only one type of reactive encounter in each elementary step: A collides with B; we need not worry about A', A" ... B',B" and so on. Not only do we know the molecular and electronic structure of the reactants and products very well, in many cases it is possible to isolate and study the intermediates that are formed in some overall process.

With a heterogeneous catalyst the corresponding problem is much more difficult. A whole range of very different adsorption sites and surface species may be present and one cannot be certain that the measurement technique provides information about the chemically significant fraction of the adsorbed layer - a point that we shall return to below. Because the heterogeneous system is characterised by a spread of adsorption sites and adsorbed species, chemical specificity, usually termed *selectivity*, is often rather poor. Many reaction channels may be accessible and a distribution of products is produced. Why then bother with heterogeneous catalysis at all? The answer is simple. Compared with their homogeneous counterparts, heterogenous catalysts are capable of delivering very high productivity because intrinsic turnover rates are high and because they can be operated in a continuous mode (as opposed to batch mode). Also, again in contrast with homogeneous catalysts, one is not faced with the problem of separating the catalyst from the products.

Therefore heterogeneous catalysts dominate modern chemical technology and they will continue to do so for the forseeable future. To a large extent, the most important (and most interesting) academic and industrial questions involve the issue of catalyst selectivity, as opposed to catalyst activity. The more complex the species undergoing reaction, the greater the number of possible reaction pathways and the more critical the question of selectivity. The ability to assemble complex organic molecules on a solid surface rapidly and with precision would transform catalytic science and industry. If such control could be extended to produce enantioselective catalysts - capable of producing a chirally specific product from an achiral reactant - the transformation would become a revolution. In academic laboratories, a start has already been made[12,13].

THIN FILMS AND CLUSTERS

Vacuum-deposited polycrystalline thin films of metals and alloys provide a convenient means of exploring a range of metal adsorbents with the intention of isolating key phenomena thought to be of importance in some overall catalytic process. Such films are quite well suited to investigation by spectroscopic techniques [see for example Chapters 5,6,7 in reference 14]. Moreover, their surface area is sufficiently large so as to make the measurement of catalytic rates a relatively straightforward matter. Early work on thin films employed deposition on a variety of polycrystalline or amorphous supports, some of which are also employed as supports for practical catalysts. One could therefore argue that such studies are useful for obtaining information about the catalyst/support interaction, a phenomenon that is undoubtedly of critical importance in the development of technologically useful systems. With such samples, the application of electron microscopy permits investigation of key aspects of film growth such as nucleation, diffusion and coalescence. By working at sufficiently low metal loadings, one may obtain valuable information about the evolution of electronic structure with cluster size[15,16].

However, large area thin films seldom consist of epitaxial single crystals. They are highly defective, energetically and structurally heterogeneous, and not completely free of contamination. Although this means that in some respects they must approach the behaviour of practical dispersed catalysts, detailed interpretation of chemisorption and reaction data is correspondingly more difficult. The early popularity of the large area thin film approach derived from the fact that "clean" metal surfaces could be conveniently and simply produced by deposition of a fresh film, achieved by evaporating the metal from a suitable source. More recently, the ready availability of sufficiently large single crystals has revolutionised the study of chemisorption and catalysis on metal surfaces[17]. Single crystal surfaces can be prepared in chemically defined states with well defined and controllable structure. They are ideally suited to the application of both momentum and energy resolved spectroscopies. Thirty years of detailed study have taught us that a full understanding of surface chemistry and catalysis at the gas/metal interface requires the kind of atomic scale characterisation which only single crystal measurements can provide. Nevertheless, we have a long way to go. Despite decades of intensive research in academic and industrial laboratories, not one new catalyst nor a single new catalytic process have been invented as a result of the application of "surface science" methods. This situation contrasts very strongly with that in the semiconductor and solid state device industries where the routine use of modern surface analytical techniques and the application of thin film technology are of crucial importance in both production and development[18].

The same point about the relative lack of impact of fundamental surface science on heterogeneous catalysis may be made in another way. Many years ago, Fritz Haber showed that iron was the best metal with which to catalyse the synthesis of ammonia from nitrogen and hydrogen. This process is vital for the manufacture of fertilisers and hence for world agriculture - the "green revolution" depends critically on our ability to make atmospheric nitrogen available to crops. Today, despite a great deal of sophisticated research on this apparently simple chemical conversion, we still use iron to make ammonia by the process that bears Haber's name. It is hard to over-emphasise the point that developments in heterogeneous catalysis depend on empirical discoveries: technology advances incrementally as a result of traditional trial-and-error methodology. Thus far, experimental surface scientists have helped us to optimise or rationalise existing processes; this is of course an important advance, but the design of heterogeneous catalysts from first principles remains a rather distant goal.

THIN FILM CALORIMETRY

Accurate knowledge of adsorption enthalpies (ΔH_a)is fundamental to the detailed interpretation of catalytic data. The construction of plausible working hypotheses is impossible in the absence of such information. The observed reaction kinetics are governed by the adsorption enthalpies of reactants and products, therefore proposed reaction mechanisms built purely on kinetic data lack credibility in the absence of information about the relevant ΔH_a values. Nevertheless, the framing of quite elaborate reaction mechanisms in the complete absence of relevant thermodynamic data continues to be a popular pursuit, largely because the required adsorption enthalpies are unknown. In this connection, it is important to recognise that for a given metal ΔH_a depends on both the crystal face and on the surface coverage. The consequences of this for catalysis are readily grasped if one considers the implications for the surface lifetime of adsorbed species as discussed in more detail below. Thus knowedge of the appropriate ΔH_a values is vital if one seeks to rationalise the dependence of catalytic rates on adsorbate coverage (i.e. partial pressure of reactants and products) and on crystal face (i.e. particle size). Despite this, there is a paucity of reliable data, which reflects the difficulty of making appropriate measurements. When adsorption/desorption is reversible, i.e. the adsorbate can be desorbed without decomposition by simply heating the sample, temperature programmed desorption can in principle provide the required ΔH_a, and in favourable cases multiple adsorption states and their associated ΔH_as can be identified [19]. However, extracting numerical values from the desorption data is not usually free from assumptions. Broadly speaking, the more complex the adsorbate, the more complex the desorption spectra and the less reliable the derived information.

When adsorption/desorption is not reversible, which is generally the case for those systems that are of most interest in heterogeneous catalysis - organic molecules on metal surfaces - recourse must be made to adsorption calorimetry. Calorimetry has two huge advantages. It can be applied to any system. It provides the required information in a direct way and extraction of $\Delta H_a(\theta)$ from the experimental heats is free from unverifiable assumptions. Calorimetric measurements on thin films have therefore played an important part: the techniques employed and the quality of the attainable results are exemplified well by the work of Wedler and co-workers, conveniently summarised in Chapter 9 of reference 6 13. Large area thin films are employed to maximise the amount of energy released (~ 10^{-3} J) in each adsorption step during which a small dose of adsorbate is administered to the film surface. The equipment is sensitive to temperature changes of ~ 10^{-4} K so that differential heats in the order of 20 kJ/mol resulting from the adsorption of ~ 0.05 monolyers of gas are measurable. Despite the elegant simplicity of this approach, it suffers from significant shortcomings due to the intrinsic nature of polycrystalline thin films that are not subjected to any surface characterisation. The method of film preparation necessarily generates an unknown number of extended and point imperfections such as edges, steps, kinks and adatoms. These special sites will generally exhibit the highest ΔH_a values and are probably the first to be occupied; subsequent adsorption takes place on the (unknown) distribution of exposed crystal planes. This leads to obvious difficulties of interpretation, exacerbated by the fact that successive measurements on the same film can lead to (unknown) changes in film structure. Though such measurements on polycrystalline films can be very useful in the interpretation of catalytic data, despite the inherent uncertainties, matters are complicated by the lack of direct transferability of information. Thin film calorimeters are generally not suitable as catalytic reactors: ΔH_a values measured on one film in one apparatus must be applied with caution in interpreting reaction data obtained with the same dispersed metal in another apparatus. The way forward is experimentally demanding and makes use of fragile thin film single crystals dosed with adsorbate under ultra high vacuum conditions, as described below.

An important recent development is that of the ultra high vacuum *thin film single crystal* microcalorimeter developed relatively recently by King and co-workers[20]. In this ingenious device, an oriented single crystal ~ 20 nm thick is dosed repetitively with ~50 ms pulses of the adsorbate from a chopped molecular beam. Each pulse contains about 10^{12} molecules which is equivalent to ~2% of a monolayer. The heat capacity of the active sample area is only ~1μJ/K which permits the measurement of differential heats of adsorption with a precision of ~5 kJ/mol. The heat released in the sample is found by focusing the infra-red radiation from its back face onto a sensitive detector and the system is calibrated by using a chopped laser beam to deliver a known heat increment. The beauty of this approach is that it provides calorimetric information about adsorption and reaction processes taking place on a well defined single crystal surface whose structure and purity are controlled by LEED and Auger spectroscopy. Furthermore, it measures *differential* heats of adsorption, so that one obtains thermodynamic information about the coverage dependence of the adsorption or reaction process under study. A few examples will serve to illustrate the power of the method. It has been used to measure (a) the difference in energy between the Pt{100} (1x1) and hex reconstructed surfaces[21] (b) the heat of reaction of CO oxidation on Pt{110}[22] (c) the potassium-promoted change in heat of adsorption of CO on Ni{100}[23] (d) the coverage dependence of the adsorption enthalpy of ethylene on Pt{110}-(1x2)[24]. These data yield clear and relatively unambiguous information about surface processes. Thus case (b) shows that when CO is dosed on to an oxygen-precovered surface, the desorbing CO_2 product molecules are almost thermally accommodated with the metal surface. However, the reverse sequence leads to CO_2 molecules which carry away ~ 50 kJ/mol of excess energy - these are thought to result from the reaction of "hot" (i.e. non-accommodated) adatoms resulting from the dissociative adsorption of O_2. Case (c) tells us quantitatively and directly how potassium adatoms affect the binding strength of co-adsorbed CO molecules (alkalis are commonly used as promoters in catalysis). The strongly increased heat of adsorption of CO at high K coverages indicates that the presence of CO leads to an *increased* ionisation of the K and formation of ordered $K^+CO^- K^+$ islands. Case (d) presents yet another example that is directly relevant to a central issue in catalysis - the evolution of adsorption heat with coverage and what this implies in terms of the surface species that are formed. In the case of ethylene on Pt{110}, the authors found an essentially stepwise decreases in ΔH_a with increasing adsorbate coverage. They argue, convincingly, that these correspond to the successive formation of ethylylidyne, ethylidyne, di σ-adsorbed ethylene and π-adsorbed ethylene. An average Pt-C bond energy is extracted for the three most stable of these. Other laboratories are now taking up this technique: we may therefore anticipate the prospect of a reliable catalogue of adsorption and reaction heats, and their coverage dependence, with which to interpret catalytic data.

Thin film microcalorimetry is a demanding technique that will probably remain the province of a few specialised laboratories. Much more commonly, the application of thin films in modern research on chemisorption and catalysis involves using single crystal substrates for the preparation of catlytically significant thin film structures. Such systems have reproducible and well controlled properties; atomistic interpretation of the data becomes possible and contact with theory is facilitated. A number of variations are possible, all of which produce catalytically relevant structures. These include the growth of an oxide film on the parent metal (e.g. Ni(111)/NiO(111)[25]; growth of the oxide of one element on a substrate consisting of another (e.g. Cu(111)/α-Nd_2O_3)[26]; growth of a metal film on an oxide (e.g. TiO_2/Rh[27]); growth of more complex structures (e.g Mo(110)/SiO_2/Cu clusters[28]). Oxide/oxide systems are important in catalysis and this has already stimulated fundamental studies of supported oxides[29]: we can expect this to be a growth area.

CATALYSIS BY METALS

We begin by defining some straightforward and commonly used terms.

Activity: refers to the total rate of conversion of reactants to all products

$$A \rightarrow B + C + D ; \qquad -dA/dt = \text{activity}$$

This quantity is most usefully expressed in normalised units as the specific rate (molecules converted/area.s) or as the *turnover frequency* (molecules converted/surface atom.s). Clearly, it provides a means of making comparisons between different catalysts, provided that a measure of the actual number of active surface metal atoms is available. This latter is not always easy to obtain; normally, quantitative measurements of CO or hydrogen chemisorption capacity are used to assess the metal surface area available to reactant molecules.

Selectivity: refers to the rate for a particular channel relative to total rate; in many cases it is the property of principal interest

$$A \rightarrow B, C \text{ ; selectivity} = \frac{dB/dt}{-dA/dt}$$

A simple calculation enables one to appreciate the effectiveness of heterogeneous catalysts. Consider a relatively simple reaction such as an isomerisation which can take place either homogeneously in the gas phase or at a surface. We may use absolute rate theory[30]to obtain an expression for the ratio of the heterogeneous and homogeneous rates

$$\frac{R(\text{hetero})}{R(\text{homo})} = n_s \exp(\Delta E/RT)/f_{gas}$$
$$= 10^{-12} \exp(\Delta E/RT)$$

where $\Delta E = E_a(\text{homo}) - E_a(\text{hetero})$, f_{gas} is the partition function of the gaseous species and n_s is the number density of surface sites.

This implies that at ~ 500 K and one atmosphere pressure we would obtain equal rates in a molar volume in the gas phase and over ~10^{-4} m^2 catalyst when ΔE ~ 70 kJ/mol. Such a lowering of the activation energy by the catalyst is not difficult to achieve in principle when one considers that chemisorption leads to breaking of bonds and formation of new bonds. Given that supported metal catalysts with active surface areas in the order on 10^2 m^2 g^{-1} are readily prepared, it is easy to see how very high catalytic rates are achievable with even a simple reactor.

However, for the catalyst to be effective, chemisorption bond strengths must lie between fairly narrow limits. Thus if the adsorption bond is weak so that surface lifetime of a chemisorbed reactant is very short, the probability of a reactive encounter with another species will be negligible. On the other hand, if the adsorption is too strong, the long-lived adsorbate may undergo many encounters with other species, without reacting, because it is too stable - i.e. the activation energy required to access the relevant transition state is too high. A simple numerical estimate serves to illustrate the point. To a reasonable approximation, the surface lifetime (τ) is given by $\tau = \tau_0 \exp(\Delta H_a/RT)$ where τ_0 (~ 10^{-13} s) is the vibrational period of the chemisorption bond. At ~ 500 K, the temperature at which many catalytic processes are operated, and with ΔH_a = 40 kJ/mol, the surface lifetime is ~ one microsecond ... for ΔH_a = 120 kJ/mol, it is ~ one century.

This has major consequences because ΔH_a can be a strong a function of both coverage and surface crystallography. Thus on practical catalysts, which are polycrystalline materials, there can be a wide distribution of adsorption lifetimes and chemical reactivity for a given adsorbate on a given metal. The interpretation of experimental data must therefore be approached with care: although some particular measurement may refer to all or most surface species that are present, only a small fraction of these may actually be catalytically active. Thus in the catalytic hydrogenation of ethylene over a polycrystalline Ni surface, straightforward experiments using ordinary ethene and ^{14}C labelled ethylene showed that only a small fraction of the adsorbed molecules participated in catalytic turnover. The same conclusion was reached by making use of selective poisoning: a fractional monolayer of S quenched the hydrogenation rate almost completely, but the total uptake of ethylene was hardly affected.

Early attempts to rationalise the catalytic behaviour of metals in terms of solid state properties resulted in an approach which divided the issue into two supposedly independent parts. (i) the so-called "geometric factor", according to which attention was focused on the crystallography of the metal surface with a view to rationalising patterns of activity in terms of the bonding geometry and stability of the adsorbed reactant. (ii) the "electronic factor" which concentrated on changes in catalytic behaviour as a function of metal surface electronic structure at constant crystallography, a goal that was supposed to be attained by investigating binary alloys. This attempt to separate two solid state properties that are in fact inextricably convolved could not really be expected to meet with great success - one only needs to recall that the work function of a metal depends on the crystallography of the surface to appreciate the crudity of the "geometric" and electronic" separation. The point is illustrated in a valuable way by the results of a calculation carried out in the free electron by limit in which one considers the magnitude of the electric field in the vicinity of a surface step of atomic dimensions on an otherwise flat metal surface. This is a useful way of obtaining insight into the effect of surface structure on catalytic reactivity. It was shown that the step field was proportional to $N(E)_{Ef}$ - the density of states at the Fermi energy. The implication is that the electric field associated such atomic scale features will be about ten times stronger for a metal such as Pt (for which E_f lies in the d-band) than for Au (for which E_f lies in the s-band). In the former case, the field strength is sufficiently high that adsorbate dissociation can facilitated by the molecular polarisation, i.e. the activation energy for intramolecular bond breaking is lowered. One is therefore led to the idea that on transition metals surface chemistry and catalysis should be sensitive to the details of surface structure (steps make a difference), whereas on Cu, Ag, Au, for example, such structure-sensitive behaviour should be much less apparent. These conclusions are supported by the reulsts of more sophisticated calculations[31] To a large degree, these expectations borne out in practice; a nice illustration is provided by the results of single crystal/molecular beam studies of the H_2+D_2 scrambling reaction carried out over Cu and Pt surfaces. In the first case, the reaction has an appreciable activation energy on the smooth metal surface and the introduction of steps makes no difference. With Pt however, the presence of steps accelerates the rate[32].

Although, as the preceding example shows, the geometric/electronic separation is oversimplified, it was nevertheless a good place to start: correlations were made that were a useful guide to thought, and in the case of electronic properties in particular, many workers in catalysis were encouraged to start thinking about the subject in terms of the framework provided by solid state physics. This latter was an important advance and owes much to the stimulus provided by Dowden[33]; his ideas stimulated much work on alloy catalysis, as outlined in the following section.

CATALYSIS BY ALLOYS

Chemisorption and catalysis at alloy surfaces are important in a variety of ways. First, as noted above, alloys provide a testing ground for theories of catalysis, themselves dependent on the degree to which alloy electronic structure can be reliably calculated and verified by experiment. Second, alloy catalysts are very important in actual large scale technology, as we shall see. Early studies of chemisorption and catalysis on alloy surfaces were stimulated by the notion that the methods of solid state physics could be applied straightforwardly in such cases, the hope being that a general theory linking electronic structure would arise. Despite some successes, it is now apparent that these initial attempts were too simplistic - nevertheless, it is instructive to examine what was done. A key idea was to investigate the catalytic properties of binary alloys formed between Group VIII and Group IB metals, the object being to vary the electronic structure of the adsorbate by controlled variations in composition. The central theoretical concept was provided by the Rigid Band Model which has its origins in theories of alloy electronic structure advanced many years ago by Hume-Rothery. It is supposed that when a Group IB metal is added substitutionally to a Group VIII metal the Brillouin zone structure of the solid may be regarded as having been preserved, the net effect being to add one extra electron per added IB atom to a common bands structure. With increasing IB metal content, the Fermi energy then sweeps through a rigidly fixed density of states, a process which is of course accompanied by changes in $N(E)_{Ef}$. This quantity is taken to play a critical role in the catalytic behaviour of the alloy because it should determine the ease with which electrons can be transferred to or from an adsorbate from or to the alloy. Therefore it should be possible to calculate the alloy composition at which large changes in $N(E)_{Ef}$ occur and these should be accompanied by pronounced changes in catalytic behaviour. The Rigid Band Model of alloy catalysis is appealing in its simplicity, but it depends on the erroneous (in principle) assumption that the two types of atom in the alloy lose their chemical identity. In practice, it is also found to be seriously in error (see below). Moreover, interpretation of experimental data can be further confounded by segregation effects whose importance was not appreciated until modern methods of surface analysis became available - even though such effects are to be expected on thermodynamic grounds. For example, Cu-Ni alloys show strong surface segregation of Cu over almost the entire range of compositions: catalytic data interpretation based on the assumption that the surface composition is the same as that of the bulk are bound to lead to incorrect conclusions. A further complication is that the extent of segregation can depend on the composition of the gas phase: the presence of CO can cause reverse segregation in a Pd-Ag alloy because chemisorption occurs only on the Pd component.

Despite all these complications, the Rigid Band Model played a valuable role in systematising thinking; it also had some apparently surprising successes. Two examples will suffice. When Au-Pd alloys were used in the catalytic conversion of ortho to para hydrogen, it was found that there was a distinct increase in activation energy at exactly the composition at which magnetic susceptibility measurements indicated completion of filling of the d-band by addition of Au to Pd. Furthermore, this composition was approximately that predicted by the Rigid Band Model. The results were rationalised by arguing that the rate limiting step (hydrogen dissociative chemisorption) involved electron transfer *to* the alloy (confirmed by work function measurements). Therefore the rate should be high when this process was energetically favourable ($N(E)_{Ef}$ high) and disfavoured when $N(E)_{Ef}$ was low. A converse and apparently confirmatory example was provided by studies of hydrogen peroxide decomposition on Cu-Ni alloys. Arguments based on kinetic studies indicated that the rate determining step was $OH_a + e^- \rightarrow OH^-$, i.e. electron transfer *from the solid.* The Rigid band prediction would be that this process should be favoured at low $N(E)_{Ef}$ (pure copper) and disfavoured upon addition of Ni (which was expected to increase $N(E)_{Ef}$). This is just what the experiments showed.

However, the advent of photoelectron spectroscopy coupled with rather accurate calculations of alloy electronic structure made it impossible to sustain this theory. The crucial events were the early measurements of Spicer *et al* on Cu-Ni alloys[34] and the development of the Coherent Potential Approximation for accurate calculations of alloy electronic structure[35]. The theory was formulated in terms of electron multiple scattering in a random alloy of A and B with the two types of atoms explicitly being treated as different. One scatterer is treated exactly, and the rest are approximated by a medium through which the electrons travel unscattered - the coherent potential. Among other things, it permitted calculation of the LDOS in the vicinity of either A or B and the results compared very favourably with experiment. It became clear that the two types of atom retain their chemical identity to a large degree and are therefore far from being indistinguishable, as the Rigid Band Model would claim. Adsorbates should therefore be able to "tell" whether they are in the vicinity of an A atom or a B atom - an unsurprising conclusion for any chemist, one would suppose, but one which some catalytic chemists appeared unable or unwilling to recognise, at least for a time.

EFFECTS DUE TO THE OXIDE SUPPORT

To a first approximation, the oxide materials commonly used to support metal catalyst particles may be regarded as chemically inert and not directly involved in the catalytic process. This is broadly true of the so-called non-reducible oxides (silica, alumina, magnesia) at the temperatures required for many metal-catalysed reactions to occur (~ 500 K), although it should be recognised that these materials can themselves be very effective catalysts at higher temperatures. Thus at temperatures around 1000 K, silica and magnesia are themselves active catalysts for the partial oxidation36 and oxidative coupling37 of methane, respectively. There are however cases in which the oxide support plays a key part in producing the desired properties. The factors thought to be involved may be illustrated with reference to methanol synthesis and ammonia synthesis, both of which are very important large scale industrial processes. The former involves use of a Cu/ZnO catalyst and has attracted a huge amount of research because of the controversial role of the ZnO support. Some authors consider the ZnO to be the most effective support precisely because it is directly involved in the chemistry. Others take the view that the activity is support independent. The question remains open and the debate continues. In the case of ammonia synthesis, the catalyst consists of potassium promoted Fe on an alumina support. The latter is thought to act as a so-called "structural promoter", that is, it acts to preferentially stabilise certain crystal planes at the exposed surfaces of the catalytically active iron particles. Studies of ammonia synthesis on well defined single crystal surfaces lend credence to this view because they show that Fe{111} is substantially more active than other crystal planes of iron for the dissociative chemisorption of nitrogen. What is the likely role of the potassium component? This provides an example of a so-called "chemical promoter", an important topic that will be addressed later.

NON-REDUCIBLE SUPPORTS: DETRIMENTAL AND BENEFICIAL EFFECTS

Silver is a uniquely efficacious catalyst for the production of ethylene oxide (C_2H_4O), a strategically important chemical intermediate that is manufactured on a huge scale by the Ag-catalysed selective oxidation of ethylene. (Every other metal simply catalyses the burning of ethylene to carbon dioxide and water.) The catalyst consists of Ag particles supported on crystalline α-alumina which is required purely as an inert support for the catalytically active metal. In practice, the alumina phase always contains some silica impurity, and the resulting silica-alumina material behaves as a solid acid (see below). The result is that ethylene oxide molecules formed on the Ag surface undergo further conversion at acid sites on the support: isomerisation to acetaldehyde (CH_3CHO) is followed by rapid combustion[38] to CO_2 and H_2O. The consequent loss in selectivity can easily be in the order of a few percentage points - which can make the difference between commercial success and failure.

However, in other cases, the presence of acid sites on the support surface can actually be actively exploited to produce highly effective catalysts for hydrocarbon conversion. These bi-functional catalysts are ingeniously devised materials consisting of platinum particles in contact with a silica-alumina phase. The surfaces of the oxide and the metal work in concert to carry out complex chemical conversions upon which the whole of the world's hydrocarbon processing industry depends. They perform a variety of industrially important hydrocarbon conversions, thanks to chemical reactions that occur on both the metal and oxide phases. Their mode of action can be illustrated with reference to the following, admittedly simplified, scheme for alkane isomerisation, an important process in petroleum technology for converting linear alkanes to branched alkanes. First, note that the silica alumina surface contains both Lewis and Brönsted acid sites whose strength and number density depend on the exact composition of the material.

Figure 1 shows a simplified scheme which illustrates the principal processes involved in alkane isomerisation. Dissociative alkane adsorption on the Pt component generates an adsorbed alkene and hydrogen. The alkene diffuses to the Pt/silica alumina interface and undergoes *spillover* onto the oxide phase where it acquires a proton and is thereby converted to a primary carbocation. This carbocation then rearranges to produce successively more stable secondary and tertiary carbocations in a manner analogous to that which is well established for such species in solution. Deprotonation followed by reverse spillover produces an isomerised alkene on the Pt surface; finally, this is hydrogenated to yield the isomerised alkane, which desorbs.

ISOMERISATION OF ALKANES : AN IMPORTANT EXAMPLE OF BIFUNCTIONAL CATALYSIS

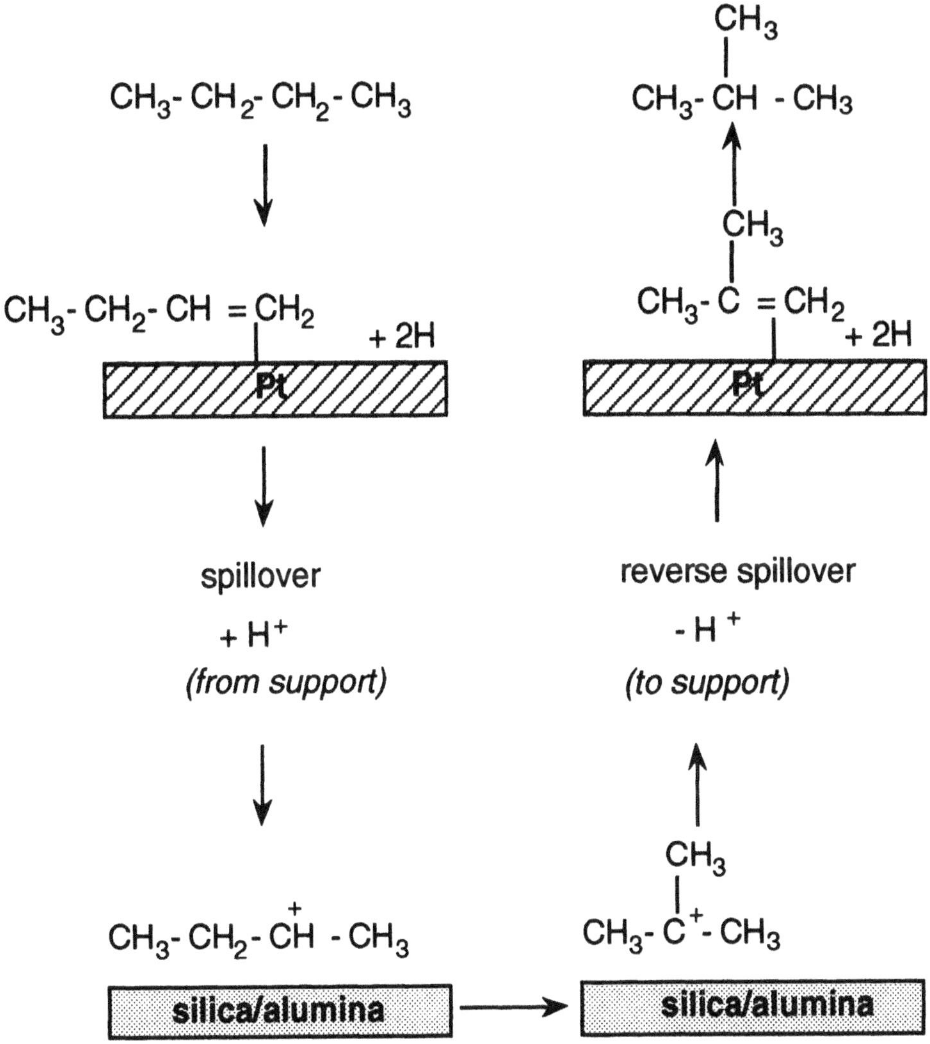

Figure 1. Schematic representation of mode of action of a bifunctional catalyst

There are also cases in which the interaction between support and metal phases is so strong that striking new phenomena appear and a new name is justified : the strong metal-support interaction (SMSI). This is an interesting and controversial topic that is the subject of much current research, including work with thin films and clusters.

THE STRONG METAL-SUPPORT INTERACTION

The idea that in certain circumstances a strong interaction could be induced between the metal and support phases of a catalyst was first introduced by Tauster, Fung And Garten[39] specifically with respect to Group VIII transition metals supported on TiO_2. In order to appreciate how the effect arises, a brief description of catalyst preparation and characterisation is called for. Typically, supported metal catalysts are prepared by thermally decomposing a precursor compound of the metal (say palladium chloride) in contact with the support; this produces a material consisting of the metal oxide in contact with the support. The active working catalyst is then produced from this intermediate material by reducing it in hydrogen at ~ 700 K to produce, in this case, a Pd/SiO_2 catalyst. The active metal area of this catalyst may be evaluated by hydrogen or CO chemisorption, and one generally finds that the metal particles adsorb ~ 1 H atom or CO molecule per surface metal atom. Metal/TiO_2 catalysts prepared in this way behave normally if low temperature reduction (LTR) is carried out, as above. However if high temperature reduction (HTR, T > 700 K) is performed, the SMSI effect appears. The H_2 and CO chemisorptive capacity of the metal particles is drastically attenuated while their catalytic behaviour is strongly, sometimes dramatically, altered. Changes in catalytic activity and very pronounced changes in selectivity can occur[40], and, perhaps most striking of all, there can be an actual increase in activity for the so-called methanation reaction between CO and hydrogen. Methanation involves co-adsorption of two species, both of whose chemisorption is strongly suppressed when the catalyst is in the SMSI mode. Yet the reaction rate is greatly accelerated.

Since the initial discovery, many other metal/metal oxide systems have been shown to exhibit this interesting phenomenon and a relatively recent review by Tauster[41] provides a concise summary of work up to 1987. Although our understanding of SMSI behaviour has advanced considerably since 1978, not least as a result of thin film studies, the subject is still controversial and a number of unresolved issues remain. This is exemplified by some recent work[42,43] on TiO_2/Au catalysts which can exhibit extraordinarily high activity for CO oxidation, among other reactions. This is remarkable, because by itself gold is catalytically totally inert. It would appear that in this case one is dealing with an extreme case of SMSI. There is no consensus about the origin of catalytic activity in this interesting system which therefore appears to be an excellent case for fundamental investigation by thin film methods. Our understanding of the SMSI effect is far from complete - some earlier views have been substantially modified, but much more work remains to be done. Some recent areas of investigation described below indicate the diversity of this subject.

Frost[44] made an interesting and controversial suggestion in connection with the mode of operation of methanol synthesis catalysts. (Reference has already been made above to the pre-existing debate surrounding the roles of the Cu and ZnO phases.) Although these catalysts consist of a $Cu/ZnO/Al_2O_3$ composite, whereas Frost's own data refer to Cu/thoria catalysts, his proposals are interesting and thought provoking. The essential idea is that a Schottky junction is formed between the metal and oxide phases - they key point about this is that it implies electronic communication between the two solid phases.

First note that a large contribution to the enthalpy of formation of oxygen vacancies in an oxide comes from the energy required to raise two electrons to the conduction band. If a Schottky junction is formed between the oxide and a metal phase, the enthalpy for oxygen vacancy creation in the oxide is lowered by twice the Schottky barrier height.

This should enhance the concentation of oxygen vacancies, and Frost goes on to argue that the catalytic chemistry takes place entirely on the oxide phase, at these vacancy sites. In other words, promotion of the oxide by the metal occurs, and not the other way around. These ideas have been vigorously challenged. Moreover, some experimental findings appear to be in conflict with them, as the following example serves to illustrate.

First, recall that in order to form a Schottky barrier, ϕ(metal) must be sufficiently high, i.e. E_F must lie below the oxide conduction band edge. We have shown[45] that alloy-derived Cu/ceria catalysts are extraordinarily active for methanol synthesis. These materials

- hugely outperform the best industrial catalysts
- contain a large amount of ultra-dispersed Cu (< 1 nm diameter)
- have a very high density of metal/oxide junctions

If the Junction Effect Theory is correct, the metal plays no chemically significant role - it merely acts as a sufficiently deep sink for electrons from the oxide. Therefore ultra-dispersed Au/ceria catalysts should be *even better* than their Cu counterparts for methanol synthesis, because $\phi(Au) > \phi(Cu)$ by ~ 0.4 eV. In fact, these Au/ceria materials are totally inert[46].

On the other hand, recent thin film studies of Pt/ceria provide compelling evidence that very strong metal-support interaction can lead to a situation in which the *metal* appears to induce a high degree of catalytic activity on the surface of an otherwise inert support material[47]. Specifically, we have shown that a platinum single crystal, fully encapsulated by a thin film of cerium oxide, is a better oxidation catalyst than the clean metal surface itself. This is reminiscent of Frost's proposal. The key point appears to be that the metal should be encapsulated by no more than a very thin film of the support material. Thus in the case of Pt{111}/ceria, the observed behaviour can be rationalised in terms of electron transfer between the two solid phases, leading to creation of oxygen vacancies in the ceria, thus inducing catalytic activity in the latter. The principal findings may be summarised as follows.

The structure and reactivity of a series Pt(111)/ceria model catalysts were studied as a function of ceria coverage and morphology. XPS and LEED data showed that the method of preparation resulted in growth of disordered CeO_2 which converted to epitaxial (111) - oriented ceria upon annealing to temperatures exceeding 700 K. Submonolayer ceria coverages strongly promoted the oxidation of CO. XPS data indicated that the chemical state of the oxide film corresponded essentially to CeO_2. In this regime, coverage dependence of the apparent activation energy suggested that new sites become available for reaction, perhaps at the metal/oxide interface.

Most strikingly, the fully encapsulated Pt(111) wafer was a much more effective catalyst than the bare metal itself. Specifically, under conditions where no bare Pt sites were detectable by CO chemisorption, the sample was a much better CO oxidation catalyst than the clean metal surface itself. Thus when the Pt crystal was fully encapsulated by an ultra thin ceria film, the CeO_2 surface itself becomes a very active catalyst. LEED data showed that pre-annealing of the ceria-encapsulated sample converted disordered films to an ordered (111)- oriented CeO_2 films. The encapsulating ceria film was a much better catalyst when it was disordered compared with when it was ordered. Entirely analogous behaviour has been found for the CO+NO reaction on ceria-encapsulated Rh{111}[48].

The high catalytic activity of CeO_2 in these special circumstances may be rationalised in terms of Frost's hypothesis[44] according to which electron transfer from a metal phase to an oxide phase can reduce the enthalpy for oxygen vacancy formation in the latter. It is known that such vacancies can play a role in CO oxidation[49,50] and that their formation can be induced by the presence of a platinum metal[51,52,53,54]. Sayle *et al*[55] have carried out a detailed theoretical investigation of the role of oxygen vacancies on ceria surfaces in relation to CO oxidation. Their results indicate that the (111), (110) and (310) faces of ceria will dominate the morphology of the material. They also find that oxygen vacancies are more stable at the surface of the crystal than at the bulk and that the segregation of such vacancies to the surface is particularly favoured energetically for the more open (110) and (310) surfaces: most importantly, abstraction of lattice oxygen by CO to form CO_2 is exothermic at the (110) and (310) surfaces.

Our results for Pt{111}/ceria may therefore be understood as follows. Given the observed marked changes in CO oxidation activity we observe as a function of clearly detectable changes in the state of disorder of the multilayer ceria film, it seems reasonable to associate this catalysis with the oxide phase. Schottky barrier formation between Pt and CeO_2 leads to electron transfer from metal to oxide with concomitant decrease in the energy required for oxygen vacancy creation in the latter. Vacancy formation is particularly favoured at more open planes such as (110) and (310) which therefore become active for CO oxidation by exothermic consumption of lattice oxygen. The resulting oxygen vacancies are filled by adsorption of oxygen from the gas phase. The ceria film needs to be very thin in order for this mechanism to operate so that effective electronic communication is maintained between the Pt and the surface of the oxide. In such circumstances, disordered ceria films will be much more active than ordered ceria films because surfaces of the former will contain open crystal planes at which oxygen vacancy creation is favoured.

A mechanism of this type could account for part or all of the catalytic activity exhibited by certain highly dispersed heterogeneous catalysts in which the support material consists of an oxide which is capable of undergoing processes such as those discussed here.

Interestingly, recent results[56] obtained with highly dispersed Pt/ceria catalysts are consistent with our findings in that they indicate strong promotion of the oxide's activity by the metal. Our own results indicate that very good contact between the metal and oxide phases is a necessary pre-requisite, suggesting that in the case of practical dispersed catalysts one should seek to achieve a very high degree of metal dispersion in order to benefit from this phenomenon. With such practical materials, it seems possible that partial (as opposed to complete) encapsulation of the metal by the oxide might actually be beneficial for the following reasons: (i) in the case of CO oxidation, the metal-catalysed rate may make a significant contribution to the overall rate. (ii) efficient dissociative chemisorption of O_2 by the metal, followed by reverse spillover may constitute an efficient path for re-filling oxygen vacancies in the oxide phase created by CO oxidation.

AN EXAMPLE OF CATALYSIS BY ALLOYS

Catalysis with bimetallic clusters and thin films

The use of alloy systems

- enables us to learn more about how catalysts work
- ideally, should permit the rational design of better catalysts; or, at least enable their improvement by empirical means.

A common strategy is to work with bimetallic systems in which one component is catalytically inert so that its presence at the surface should strongly influence the catalytic behaviour of the active component.

Acetylene coupling reactions over Pd surfaces can lead to a variety of products, depending on the conditions [57,58,59].

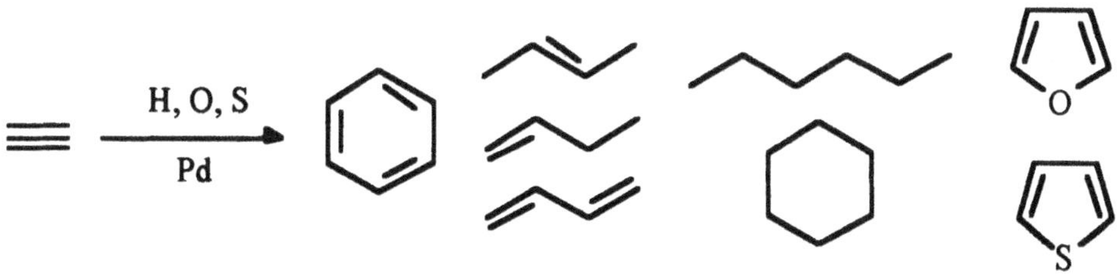

In addition, they can be operated over a wide range of pressure conditions from atmospheric pressure to UHV. This network of reactions therefore provides a useful system for examining some important effects, particularly with respect to selectivity. Pure Pd{111} is an efficient catalyst for these processes, whereas pure Au is totally inert. The catalysis is structure sensitive in that other crystal planes of Pd are much less effective than the {111} plane. With dispersed catalysts, adding (inert) Au to (active) Pd produces large increases in activity and substantial changes in selectivity, as described below. Modelling these phenomena with Au{111}/Pd and Pd{111}/Au thin film systems provides a good understanding of the behaviour of practical catalysts. In particular, such studies provide information about the bonding and reactivity of the reactant and product molecules, and about the elementary steps involved in the overall catalytic process. They also reveal how surface structure, composition and morphology determine the activity and selectivity of practical Au/Pd catalysts.

Au/Pd core/shell colloidal particles (Figure 2) are useful bimetallic catalyst precursors that can be used to study systematically the effect of changes in surface composition brought about by pre-annealing the bimetallic system to successively higher temperatures.

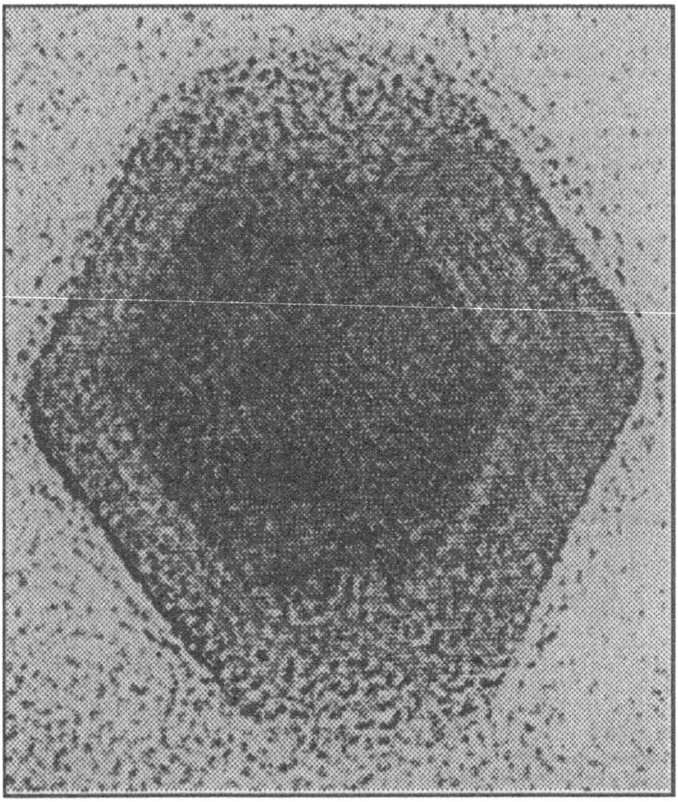

Figure 2. High resolution micrograph of Au/Pd core/shell colloidal particle. For clarity, a thick-skinned particle is shown. The catalytic experiments were conducted with thin-skinned particles, i.e. with about 2 ML Pd.

Such annealing leads to progressive intermixing of the Pd skin with the Au core. As can be seen from Figure 3, the pretreatment leads to Au/Pd intermixing (300 - 600 K) and the corresponding changes in catalytic activity and selectivity are also shown (Figures 3, 4 respectively).. Clearly, there are large enhancements in both activity and benzene selectivity as Au is added to the initially pure Pd surface - in fact, the alloy catalysts are far more active than pure Pd itself.

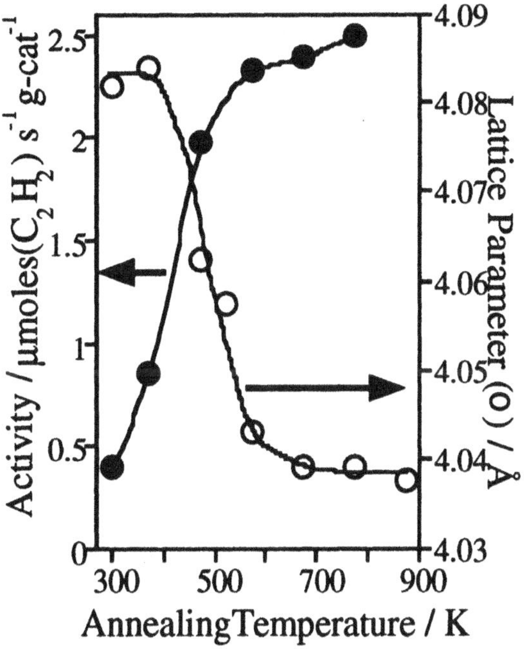

Figure 3.
Dependence of activity and lattice parameter of thin Pd-skin/Au-core colloid as a function of annealing temperature

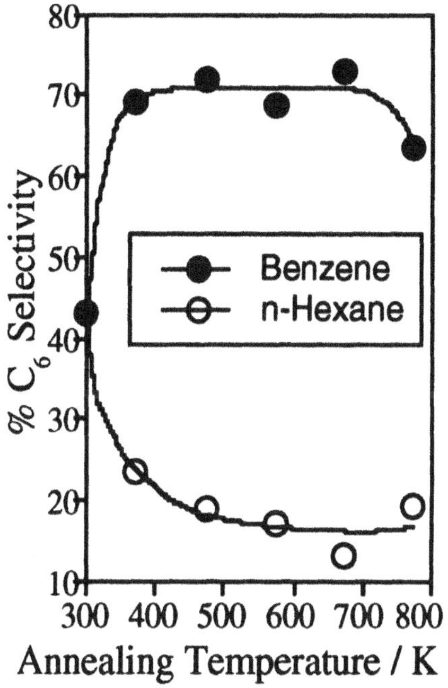

Figure 4.
C_6 selectivity of thin Pd-skin/Au-core colloid as a function of annealing temperature

Results obtained with Au{111}/Pd and Pd{111}/Au are complementary and shed useful light on the behaviour of the practical catalysts. The former system is characterised by ready formation of a stable crystalline √3 surface alloy of composition Pd_2Au which we have characterised in detail by LEED and STM. The latter gives rise to a series of random surface alloys whose composition may be varied by suitable thermal treatment. The data in Figure 5 were obtained with the Au{111}/Pd system.

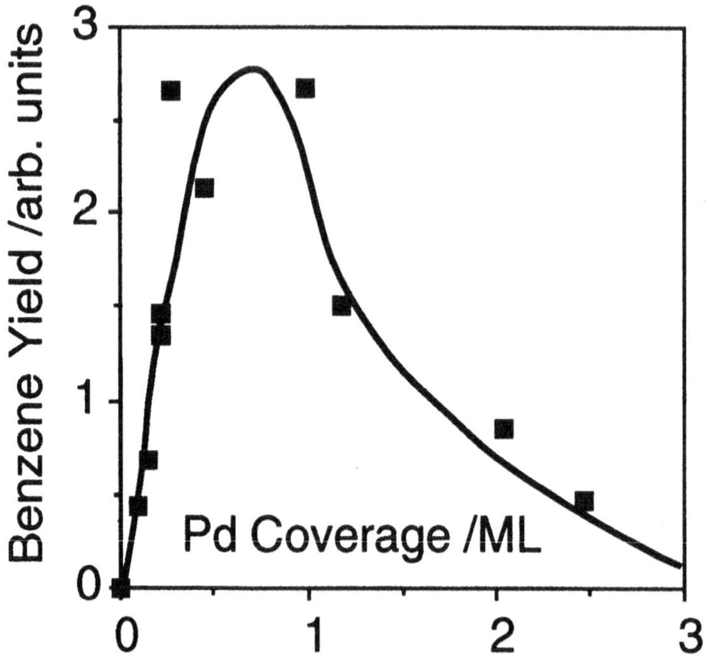

Figure 5. Au{111}/Pd. Benzene yield as a function of Pd overlayer thickness. Temperature programmed reaction aftr saturating surface with acetylene.

It can be seen that the yield of catalytically produced benzene from Pd overlayers on Au{111} scales essentially linearly with coverage in the regime 0 - 1 ML. Since benzene synthesis depends critically on the availability of {111} like ensembles of Pd atoms, this implies that even from the lowest coverages, the deposited Pd must grow as well defined islands. The STM data confirm that this is indeed the case, revealing that pseudomorphic islands of Pd nucleate and grow in the "elbows" of the herringbone reconstruction of the clean Au{111} surface[59]. The strong decrease in activity at higher Pd coverages is at first sight surprising - on a simple basis, one would expect Pd multilayers to be at least as effective as the first monolayer. Again, STM provides the answer. As-deposited multilayers are morphologically rough and contain a high proportion of non-{111} Pd sites. They are therefore ineffective, as expected on the basis of the known structure sensitivity of this reaction. The activity of a monolayer Pd film on Au{111} is comparable to that of the {111} surface of a bulk Pd sample. Annealing such a film leads to Pd_2Au √3 surface alloy formation and a very large enhancement in catalytic activity, as shown in Figure 6.

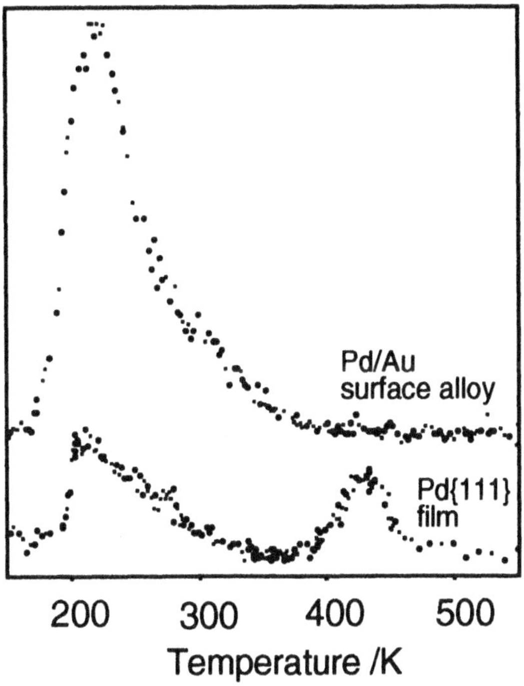

Figure 6. Au{111}/Pd. Temperature programmed reaction, showing effcct on yield of benzene of converting Pd overlayer to √3 Pd_2Au surface alloy film.

This is strongly reminiscent of the response of the practical catalysts: adding inert Au to the active Pd surface makes a much better catalyst. This behaviour can be investigated in more detail by studying the range of random surface alloys accessible in the "inverse" Pd{111}/Au. Preannealing ultra thin Au films on Pd{111} can be used to generate a series of random alloys of known composition. Figure 7 shows the experimentally observed yield of benzene in temperature programmed reaction experiments as a function of surface composition. Also shown are the results of a simple statistical calculations in which it is assumed that various ensembles of Au surface atoms are catalytically effective.

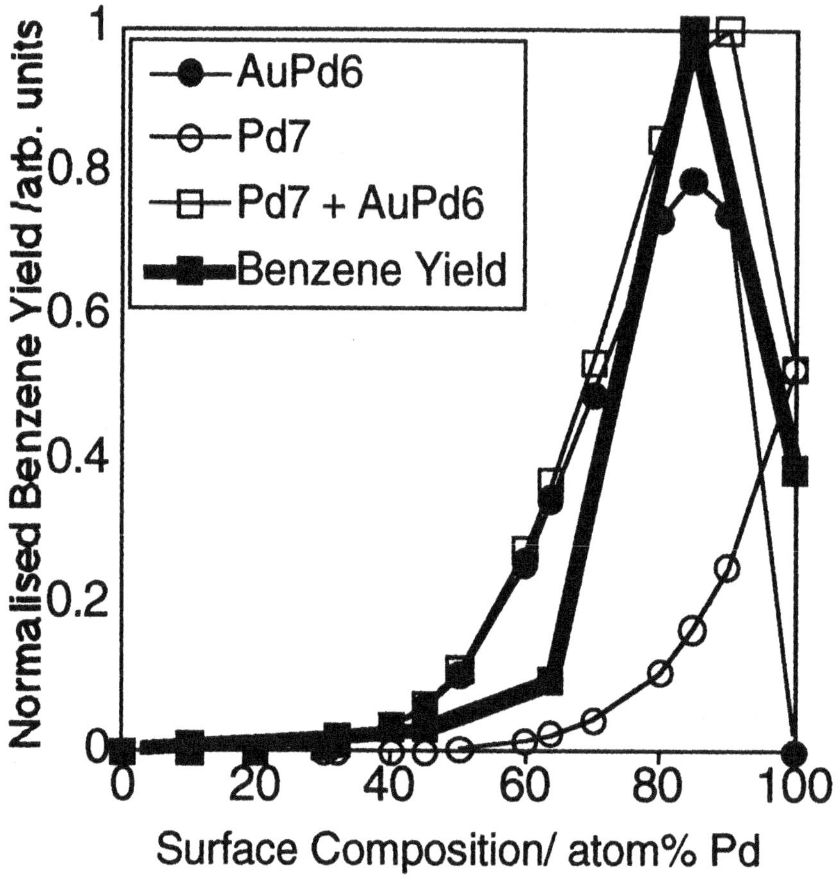

Figure 7 2-dimensional surface alloys made from Pd{111}/Au. Effect of surface composition on benzene yield. The bold line shows the experimental data, the best fit to this data is shown by the open squares.

It can be seen that the best fit is found if it assumed that essentially all the catalytic activity is due to pure Pd ensembles (Pd_7) and Pd_6Au ensembles. (Note that the highly active √3 surface alloy formed using Au{111}/Pd consists entirely of Pd_6Au ensembles.) The behaviour of the practical catalysts is now understandable. Adding Au to the Pd surface decreases the extent of decomposition of *both* the acetylene reactant and the benzene primary product: an overall increase in acetylene → benzene conversion is the result. The effect is likely to be particularly dramatic in regard to reduction of benzene decomposition because, as is apparent from Figure 6, the presence of Au greatly reduces the desorption temperature of the reactively formed benzene molecules. This reduction in surface lifetime favours benzene desorption over the competing decomposition reaction. The changes in selectivity are also understandable (Figure 4). Benzene is certainly the precursor for n-hexane, probably *via* a sequence of the type benzene → cyclohexane → n-hexane. The third step involves hydrogenolysis and therefore multi-site adsorption of the reactant hydrocarbon; this should be strongly inhibited by breaking up the large Pd ensembles with Au, as is indeed observed.

ELECTROCHEMICAL PROMOTION OF METAL FILM CATALYSTS

Electrochemical promotion (EP) has been described for many catalytic reactions on Pt, Pd, Rh, Au, Ag and IrO_2 surfaces and the subject has been extensively reviewed (see reference [60] and references therein). The phenomenon occurs when catalytic reactions take place on a variety of metal films in contact with a solid electrolyte, where the latter acts as a source of electrochemically controlled promoter species that directly modify the behaviour of the catalyst film (the working electrode). These EP systems are characterised by a very large value of the ratio between the change in turnover rate and the rate of charge transport through the electrolyte (up to 10^5). Reported gains in reaction rate with respect to the unpromoted rate lie in the range 2-100. Vayenas *et al* have advanced a theoretical model which provides a satisfactory quantitative account of the EP effect. Spillover ions pumped from the solid electrolyte modify the catalyst surface work function, with concomitant changes in adsorption enthalpies and reaction activation energies. This in turn can lead to major changes in catalytic activity and selectivity. Since the surface coverage of promoter species is controlled by varying the electrochemical potential (V_{wr}) of the catalytically active thin film, and because the effect is reversible with V_{wr}, one has *true in situ control of the working catalyst* .

The effectiveness of this approach may illustrated by reference to recent work on catalytic reactions that are important with respect to pollution abatement[61,62,63,64,65]. Large enhancement in the performance of the thin film catalysts are achieved and fundamental information is obtained about the reaction mechanism. Figure 8 shows some of the results obtained for the reduction of NO by propene using a Pt metal film catalyst in contact with β'' alumina, a solid electrolyte which acts to supply/remove Na to/from the working catalysts.

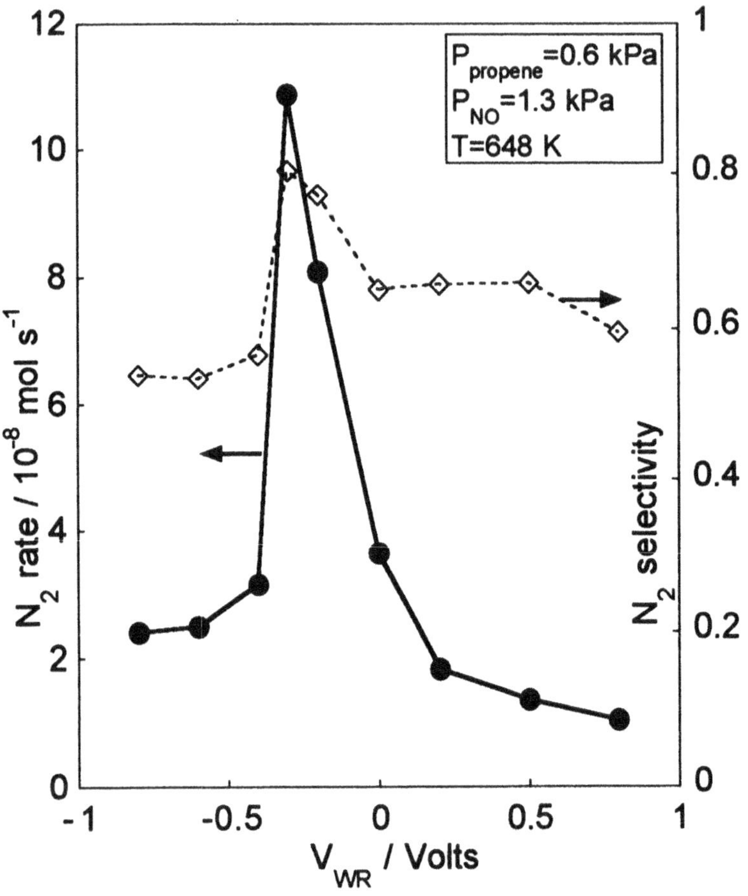

Figure 8.
Electrochemical promotion of NO reduction by propene catalysed by a Pt film elctrochemically supplied with Na from β'' alumina. Showing effect of Na pumping on N_2 rate and selectivity.

The products of reaction are N_2, N_2O, CO_2 and H_2O. Application of negative potential supplies Na to the surface of the catalyst film and leads to a large enhancement in total catalytic rate (the figure shows the N_2 rate) and significant improvements in N_2 selectivity. At relatively low Na loadings the system exhibits strong promotion: rate increases exponentially with catalyst potential, as predicted theoretically[60]. Eventually, overloading with Na leads to poisoning. It is emphasised that this behaviour is fully reversible - the system could be cycled repeatedly and reproducibly between the promoted and poisoned regimes. We propose that the reaction initiating step is dissociation of chemisorbed NO, a process for which clean Pt is not efficient. The local electric field due to the Na promoter ions acts to dissociate the chemisorbed NO by increasing the population of the π^* orbital[9], thus triggering the catalytic reaction. Of course one should not lose sight of the fact that under reaction conditions the Na promoter is present as a surface compound whose composition depends on the prevailing conditions of temperature and partial pressures of reactive gases. (This does not affect the argument given above.) In the present case, our XPS results show that the Na is present as a mixture of nitrate and nitrite, typical of the kind of promoter compounds one would expect to find at the surfaces of conventional metal catalysts. At very high Na loadings the system becomes poisoned, presumably due to site blocking by excessive amounts of the promoter surface compounds. Interestingly, we have also shown that the electrochemical promotion of NO reduction by propene is also effective in an oxygen-containing environment - solving this problem is of considerable technological interest with respect to the implementation of fuel efficient lean burn gasoline engines for which conventional catalytic control systems are inadequate.

REFERENCES

1 Golden D.M. and Williams L.H in *Low Temperature Chemistry of the Atmosphere,* Eds G.K. Moortgat, A.J. Barnes, G. Le Bras and J.R. Sodeau, NATO ASI Series I, Vol 21 p 236, Springer-Verlag, Berlin 1994.
2 Crutzen, P.J., (1995) *Faraday Discussions* 100, 1-23.
3 Trasatti S. (1995) *International Journal of Hydrogen Eneregy,* 20, 835-844.
4 Zamaraev, K.I., Khramov, M.I and Parmon, V.N. (1994) *Catalysis Reviews Science and Engineering* 36, 617-644.
5 Briggs, D., and Seah, M.P., (1990)*Practical Surfcae Analysis* Volume 1 Wiley, New York.
6 Heinrich V.E., and Cox P.A. (1994)*The Surface Science of Metal Oxides* Cambridge University Press, Cambridge.
7 Gunderson, K., Jacobsen, K.W., Nørskov, J.K., and Hammer, B. (1994) *Surface Science* 304, 131.
8 Holloway, S., and Nørskov, J., (1991) *Bonding at Surfaces,* Liverpool University Press, Liverpool.
9 Nørskov, J.K., Holloway, S., and Lang, N.D. (1984) *Surface Science* 137 65.
10 Szymansky, M., and Gillan, M.J. (1996) *Surface Science* in press.
11 Perdew, J.P., and Wang, Y. (1986) *Physical Review B* 33, 8800.
12 Blaser, H.U., Jalett, H.P., and Spindler, R.F. (1996) *Journal of Molecular Catalysis A* : Chemical 107, 85.
13 Simons, K.E., Meheux, P.A., Griffiths, S.P., Sutherland, I.M., Johnston, P., Wells, P.B., Carley, A.F., Rajumon, M.K., Roberts, M.W., Ibbotson, A. (1994), *Recueil Des Travaux Chimiques Des Pays-Bas* 113, 465.
14 Studies in Surface Science and Catalysis, Volume 32.*Thin Metal Films and Gas Chemisorption*, P.Wissman (Ed.), Elesvier, Amsterdam 1987.
15 Wertheim G.K., DiCenzo, S.B., Buchanan D.N.E and Bennett P.A. (1985) *Solid State Communications* 53 377.
16 DiCenzo S.B., Wertheim G.K., and S.E. Youngquist S.E. (1983) *Physical Review Letters* 51 2310.
17 Somorjai G.A. (1994) *Introduction to Surface Chemistry and Catalysis,* Wiley, New York.
18 Vankatesan, T. (1992) *Thin Solid Films* 216, 52.
19 Woodruff, D.P., and Delchar, T.A. (1986) *Modern Techniques of Surface Science,* Cambridge University Press, Cambridge.
20 Stuck, A., Wartnaby, C.E., Teo, Y.Y., Stuckless, J.T., Al-Sarraf, N. and King, D.A. (1996), *Surface Science* 349, 229.
21 Yeo, Y.Y., Wartnaby, C.E. and King, D.A. (1995) *Science* 268, 1731.
22 Wartnaby C.E., Stuck A., Yeo Y.Y. and King, D.A. (1995) *Journal of Chemical Physics* 102, 1855.
23 Al-Srraf N., Stuckless J.T., and King D.A. (1992), *Nature* 360, 245.
24 Stuck, A., Wartnaby, C.E., Teo, Y.Y., and King, D.A. (1995) *Physical Review Letters* 74, 578.
25 Lambert, R.M., Badyal, J.P.S., and Zhang, X. (1990) *Surface Science* , 225 L15.
26 Lambert, R.M., Nix, R.M., and Judd, R.W. (1988) *Surface Science* 205 59.
27 Evans, J., Hayden, B.E., Mosselmans, F., and Murray, A. (1994) *Surface Science,* 301 61.
28 Vesecky, S.M., and Goodman, D.W. (1992) *Science* 258, 788.
29 Jehng, J.M., and Wachs, I.E. (1991) *Journal of Molecular Catalysis* 67, 369.
30 Atkins, P.W. (1990) *Physical Chemistry* Chapter 28, Oxford University Press.
31 Tersoff, J., and Falicov, L.M., (1981) Electronic structure and local atomic configuration of flat and stepped surfaces of Ni and Cu *Physical Review B* 24, 754-764.
32 Lin, T.H., and Somorjai, G.A. (1984) *Journal of Chemical Physics* 81 704.
33 Dowden, D.A. (1950) *Journal of the Chemical Society* 242.

34 Seib, D.H. and Spicer, W.E. (1970) *Physical Review B* 2, 1676.
35 Soven, P. (1969) *Physical Review* 178, 1136.
36 Liu, H.-F., Liu, R.S., Liew, K.Y., Johnson, R.E., and Lunsford, J.H. (1984) *Journal of the American Chemical Society* 106, 4117.
37 Krylov, O.V. (1993) *Catalysis Today* 18, 209.
38 Lambert, R.M., Grant, R.B., and Tan, S.A. (1987) *Journal of Catalysis*, 106 54.
39 Tauster, S.J., Fung, S.C., and Garten, R.L. (1978) *Journal of theAmerican Chemical Society*, 100, 170.
40 Maeda, A., Kunimori, K., and Uchijima, T. (1992), *Nippon Kagaku Kaishi* 267.
41 Tauster, S.J. (1987) *Accounts of Chemical Research* 20, 389.
42 Bollinger, M.A., and Vannice, M.A. (1996) *Applied Catalysis B:Environmental* 8, 417.
43 Haruta, M., Kobayashi, T., Sano, H., and Yamada, N. (1987) *Chemistry Letters* 405.
44 Frost, J. C. (1988) *Nature* 334, 577.
45 Lambert, R.M., Nix, R.M., Rayment T, Jennings, J.R and Owen, G. (1987) *Journal of Catalysis* 106 216.
46 Lambert, R.M., Shaw, E.A., Walker, A.P., and Rayment, T. (1992) *Journal of Catalysis*, 134 747.
47 Hardacre, C., and Lambert, R.M. (1994) *Journal of Physical Chemistry* 98 10901.
48 Warren, J.P., Hardacre, C and Lambert, R.M. (1996) *submitted to Journal of Physical Chemistry*
49 Jin, T.; Okuhara, T.; Mains, G. J.; White, J. M. *J. Phys. Chem.* 1987, 91, 3310.
50 Sanchez, M. G.; Gazquez, J. L. *J. Catal.* 1987, 104, 120.
51 Shyu, J. Z.; Otto, K.; Watlins, W. L. H.; Graham, G. W.; Belitz, R. K.; Ghandi, H. S. *J. Catal.* 1988, 114, 23.
52 Hicks, R. F.; Rigano, C.; Pang, B. *Catal. Lett.* 1990, 6, 271.
53 Syhu, J. Z.; Otto, K. *J. Catal.* 1989, *115*, 16.
54 Yao, Y. F. *J. Catal.* 1984, *87*, 152.
55 Sayle, T. X. T.; Catlow, C. R. A.; Parker S. C. (1994)*Surface Science* 316, 329.
56 Golunski, S.E., Hatcher, H.A., Rajaram, R.R., and TRuex, T.J. (1995) *Applied Catalysis B: Environmental* 5, 367.
57 Lee, A.F., Baddeley, C.J., Hardacre, C., Ormerod, R.M., Lambert, R.M. Schmid G. and West, H. (1995) *Journal Of Physical Chemistry* 99 6096.
58 Baddeley, C.J.,, Tikhov, M., Hardacre, C., Lomas, J.R., and Lambert R.M. (1996) *Journal of Physical Chemistry* 100 2189.
59 Baddeley, C.J., Ormerod, R.M, Stephenson, A.W., and Lambert, R.M. (1995)*Journal of Physical Chemistry* 99 5146.
60 Vayenas, C. G., Jaksic, M. M., Bebelis S. and Neophytides S. in *"Modern Aspects of Electrochemistry"* J. O'M. Bockris, B. E. Conway and R.E. White (Eds.)., Vol 29, pp 57 - 202, Plenum Press, New York (1996).
61 Yentekakis,I.V., Moggridge, G. D., Vayenas C. G., and Lambert, R. M. (1994) *Journal of Catalysis* 146 292.
62 Harkness, I.R., and Lambert, R.M. (1995) *Journal of Catalysis* 152 211.
63 Lambert, R.M., Harkness, I.R., Hardacre, C., Yentekakis, I.V., andVayenas, C.G.(1996) *Journal of Catalysis* 160 19.
64 Palermo, A., Tikhov, M.S., Filkin, N.C., Yentekakis, I.V. ,Vayenas C.G. and Lambert, R.M., (1996) *Studies in Surface Science and Catalysis* Elsevier, Amsterdam 101, 513.
65 Palermo, A., Harkness, I., Yentekakis, I.V., Marina, O. Vayenas, C.G. and Lambert, R.M. (1996) *Journal of Catalysis* 161 471.

THIN FILMS AS MODEL CATALYSTS

DARRELL R. RAINER and D. WAYNE GOODMAN
Texas A&M University
College Station, TX, USA

1. Introduction

One of the primary ambitions driving heterogeneous catalytic research is to establish the relationship between the atomic level physical and chemical properties of the catalyst with the observed catalytic behavior. In the last few decades, a diversity of ultrahigh vacuum (UHV) based surface sensitive characterization methods have been developed that allow for the detailed study of stuctural, compositional, and electronic properties at the atomic level. While these types of techniques offer unique and exciting opportunities in the area of catalysis science, there are some inherent obstacles associated with their employment, not the least of which being that most high surface area industrial catalysts are not suitable for study using modern surface science techniques.

A common solution to this problem has been to utilize well-defined metal single crystals (by their nature imminently suited for UHV surface science study), using particular crystal planes to model surface sites that are expected to be present in the high surface area supported metal analogs. The development in recent years of a combined apparatus that couples the UHV surface analysis chamber with an elevated pressure reactor has afforded the opportunity for direct comparisons between reaction rates measured on single crystals with those recorded for more realistic catalysts [1-4]. This approach, by exploiting the opportunity for detailed post-reaction surface analysis of the single crystal, has yielded valuable insight into structure sensitivities, the identification and catalytic effects of certain inhibitors and promoters, and the identification of reaction intermediates [5-8].

While these successes have established the viability of metal single crystals as relevant model catalysts, there are still certain shortcomings associated with their use in this context. For supported metal catalysts, the role of the metal-support interaction and intrinsic particle size effects are important concerns that

R.M. Lambert and G. Pacchioni (eds.),
Chemisorption and Reactivity on Supported Clusters and Thin Films, 27-59.

must be considered in any comprehensive treatment [8-10]. Obviously, single crystal studies alone cannot address these issues. Clearly, a need exists to develop model catalysts of higher complexity capable of more fully simulating all the elements that contribute to the catalytic properties of a particular system, without wholly sacrificing the features that have made single crystal studies so unique and fruitful. Also, oxides themselves serve as catalysts in a variety of processes, and are characterized by an array of interesting properties, but are not suited for surface science investigation due to their insulating properties and poor thermal conductivities [11-14].

One approach to bridging this so-called "material gap" is through the employment of single crystal supported oxide thin films as model catalysts. Studies of several different oxide thin film systems (ranging in thickness from < 5 to 100 Å) have indicated that these types of model oxides exhibit roughly the same physical and chemical properties as their bulk oxide analogs with notable exceptions [13,14]. Because the films are so thin and in intimate contact with the grounded substrate single crystal, they are electrically conductive via defects. This feature alleviates the charging and referencing problems normally associated with performing the various electronic spectroscopies of modern surface science on insulating samples. Because the films are flat as well as conductive, they are suited to atomic level imaging using scanning tunneling microscopy (STM) and atomic force microscopy (AFM) [10]. Also, the poor thermal conductivities that thwart attempts at uniform sample heating and cooling in bulk oxide studies have no impact in the thin film experiments, allowing for the exercise of sufficient surface temperature control for temperature programmed desorption (TPD) and reaction kinetics experiments [13,14].

These thin films are interesting both in terms of their intrinsic physical, chemical, and catalytic properties and in their function as supports for metal particles. Model metal supported catalysts obtained by the vapor deposition in UHV of catalytic metals onto oxide thin film supports provide systems that serve as useful intermediates between single crystals and high surface area supported catalysts. These supported models do in fact address important issues such as particle size and the metal-support interaction, but are wholly amenable to UHV based surface science investigation [9,10]. Additionally, by varying deposition rates and substrate deposition temperatures, a degree of control can be exercised over particle growth for the model catalysts that is typically unachievable for supported powders prepared by standard incipient wetness impregnation techniques, resulting in markedly narrower particle size distributions for the models [9,15-18]. This last feature is obviously very noteworthy in the context of experimentally resolving particle size effects.

This review focuses on the utility of thin film oxides and oxide supported metal particles as vehicles for detailed investigations of heterogeneous catalysts and catalytic processes. Several diverse and representative case studies are discussed that collectively broadly illustrate some of the interesting properties and applications of these thin film systems for heterogeneous catalytic research.

2. Experimental

These experiments have been performed utilizing several different combined UHV-microreactor systems (with base pressures of ~ 5 x 10^{-9} Torr) described in detail elsewhere [19,20]. These systems are variously equipped for performing Auger electron spectroscopy (AES), low-energy electron diffraction (LEED), ion scattering spectroscopy (ISS), X-ray and ultraviolet photoelectron spectroscopy (XPS and UPS), high resolution electron energy loss spectroscopy (HREELS), infrared reflection absorption spectroscopy (IRAS), UHV-STM and AFM, and temperature programmed desorption (TPD). The UHV-STM work was performed with an Omicron UHV-STM.

Film preparation on the single crystal supports is described briefly in the appropriate sections below and in the references therein. The film deposition rates were typically on the order of 0.5 to 1.0 ML per minute. Metal dosing was accomplished by evaporation from resistively heated high purity sources. The sample temperatures were monitored using a type "C" thermocouple spot-welded to the edge of the single crystal.

The kinetics experiments over the single crystal and supported thin film model catalysts have been performed in the batch reactor mode using gas phase IR spectroscopy or gas chromatography to analyze the product gases. The CO + NO reactions over the Pd/Al_2O_3 powder catalysts were performed in a conventional atmospheric flow reactor described elsewhere [21].

3. Preparation and Characterization of Thin Film Model Catalysts

3.1 OXIDE THIN FILMS

3.1.1 *SiO_2*

SiO_2 is commonly employed as an inert high surface area support in dispersed metal catalysts. Amorphous SiO_2 thin films have been prepared on Mo(110) and Mo(100) substrates by evaporating Si in O_2 backgrounds [22-23]. The stoichiometry of the films as a function of oxygen background pressure has been monitored using AES. A comparison of the Auger spectra for bare Mo(110), and for SiO_2 thin films prepared in 1 x 10^{-6} and 4 x 10^{-6} Torr O_2 backgrounds is presented in Figure 1. For the oxygen pressure of $1x10^{-6}$ Torr, features associated with both Si and SiO_2 are evident in the spectrum. For the higher oxygen background of $4x10^{-6}$ Torr, the spectrum is consistent with a film comprised exclusively of fully oxidized SiO_2; no further changes in the AES were apparent with increasing oxygen pressures . These evaluations are based on the characteristic lineshapes and energies of the Auger transitions. SiO_2 displays characteristic LVV Auger transitions at 76, 63, and 59 eV, while a peak at 91 eV is observed for Si [24,25].

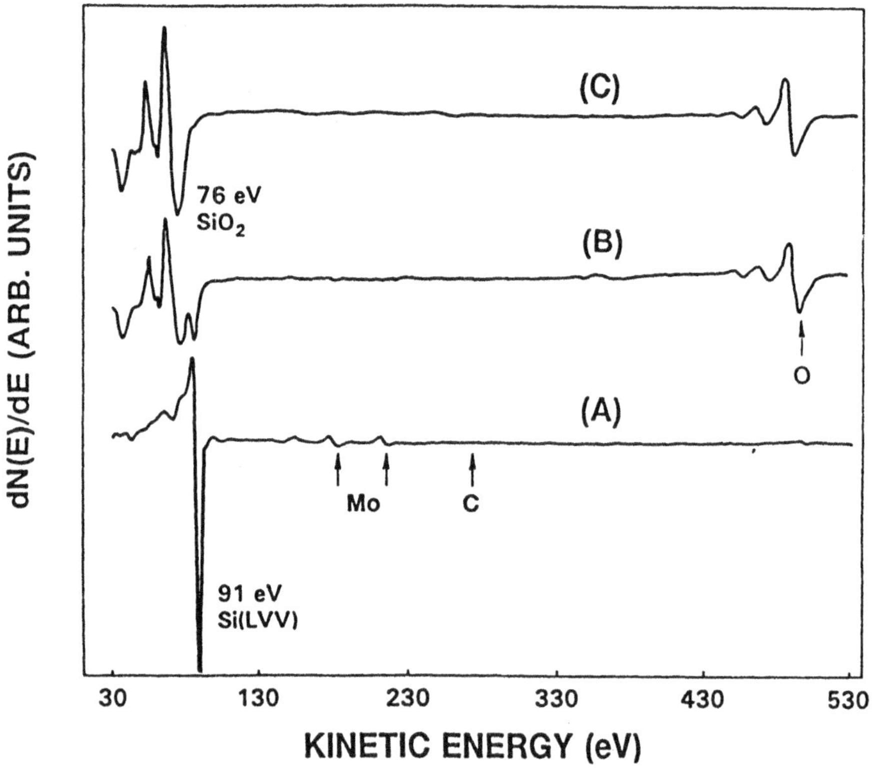

Figure 1. Auger electron spectra for silicon and silica thin films on Mo(110) prepared by depositing Si at O_2 partial pressures of (a) 0 Torr, (b) 1 x 10^{-6} Torr, and (c) 4 x 10^{-6} Torr.

XPS can also be used to characterize the stoichiometric purity of the film. Figure 2 shows XPS results for films grown on Mo(100) in three different background pressures of oxygen at 350K. The films are approximately 36 Å thick based on attenuation of the Mo(100) substrate XPS intensity. The peak assignments for the Si 2p transitions are as follows: pure Si at 99.4 eV and Si in SiO_2 at 103.6 eV, with SiO_x suboxides ($x < 2$) falling between those two values. For the films grown in oxygen pressures of 1 x 10^{-8} and 5 x 10^{-8} Torr, the spectra reveal a dominant contribution from unoxidized Si and SiO_x suboxides, while the spectrum for the 2 x 10^{-5} Torr oxygen growth pressure indicates a film comprised of over 80% SiO_2. Figure 3 shows the effects of annealing upon this film; as the annealing temperature is increased, the contribution from suboxides and unoxidized Si is greatly diminished. For the sample annealed to 1373K, the relative XPS peak areas are indicative of a film that is more than 99% pure SiO_2. The films are not stable above 1400K [26].

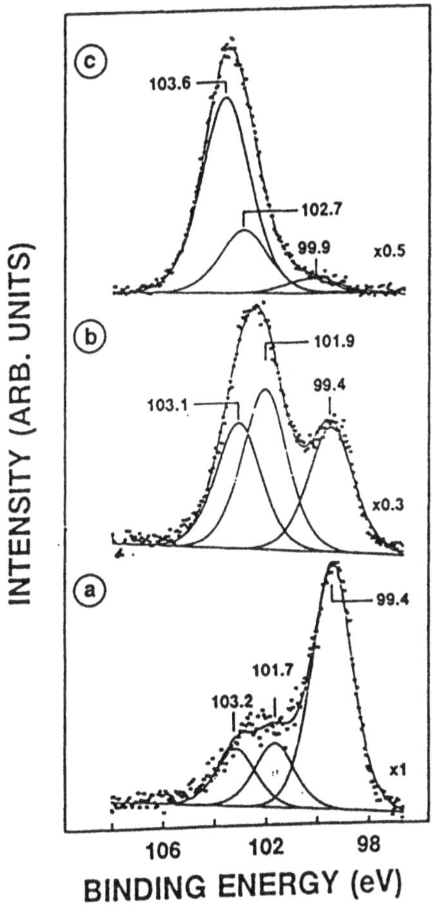

Figure 2. Si 2p XPS spectra for 36 Å silicon oxide films on Mo(100) prepared in O_2 ambient pressures of (a) 1 x 10^{-8}(b) 5 x 10^{-8}, and (c) 2 x 10^{-5} Torr.

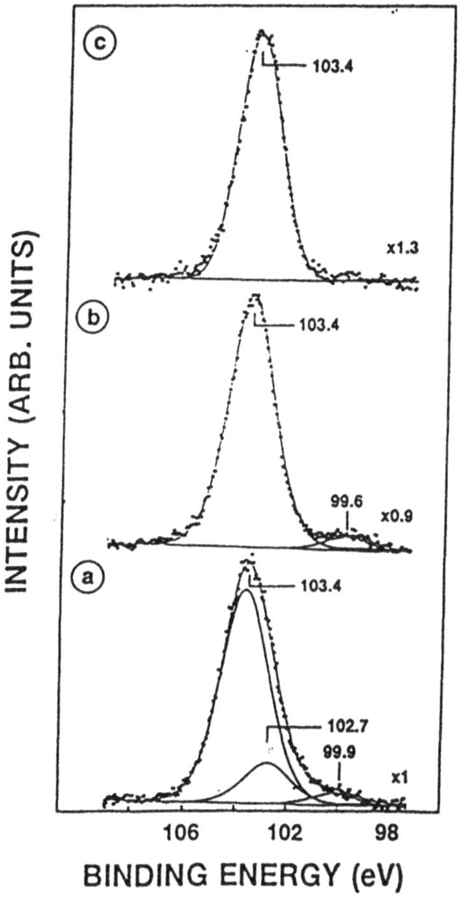

Figure 3. Si 2p XPS spectra for 36 Å silicon oxide films on Mo(100) prepared in 2 x 10^{-5} Torr O_2 and and annealed to (a) 393K, (b) 1023K, and (c) 1373K.

SiO_2 thin films have also been characterized electronically using HREELS. Figure 4 shows HREELS spectra for three different annealing temperatures; for the sample annealed to 900K, the spectrum is virtually identical to that observed for vitreous SiO_2, which exhibits electron energy loss features at 10.3, 11.8, 14.0, 17.0, and 22-23 eV [27]. These results are also quite consistent with SiO_2 films prepared by the thermal oxidation of Si substrate surfaces [28-29].

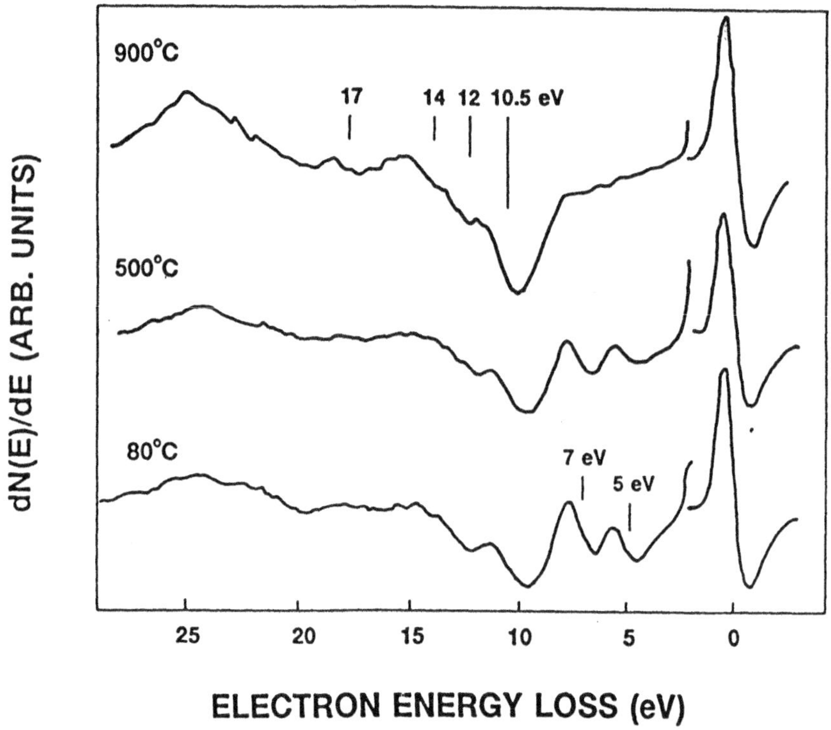

Figure 4. HREELS for a ~ 50 Å SiO2 film on Mo(110) annealed to 80, 500, and 900 °C.

The films prepared at room temperature also exhibit loss features at 5 and 7 eV, which have been associated with SiO_4 tetrahedra with broken Si-O bonds [30]. These features are significantly attenuated as a function of increasing annealing temperature, and have disappeared completely after annealing to 1173K. The removal of these defect peaks by annealing indicates a thermally induced reorganization of the film that results in the formation of long range 3D SiO_4 networks, as occur in vitreous silica.

Thermally induced structural modifications in SiO_2 films are also apparent from IRAS studies (Figure 5). A broad feature associated with the Si-O asymmetric stretch appears at 1178 cm^{-1} in the IRAS spectrum for the film deposited at a substrate temperature of 323K. As the sample is annealed to progressively higher temperatures, this peak shifts to higher frequency, reaching a maximum of 1252 cm^{-1} at 1100K. This feature is attributed primarily to a convolution of the longitudinal optical modes of the asymetric stretches (AS_1 at 1256 cm^{-1} and AS_2 at 1160 cm^{-1}) of SiO_2. It has been shown that these two modes couple via disorder in the SiO_2 film [31,32]; in highly ordered crystalline alpha-quartz, the AS_2 mode is optically weaker than in disordered amorphous SiO_2, while this trend is reversed for the AS_1 mode. Based on these observations, the increase with increasing temperature in the intensity of the feature at 1250 cm^{-1} in Figure 5, taken with the concomitant decrease in the intensity at 1160 cm^{-1} indicates that thermally induced ordering is occuring in the film, consistent with the HREELS data.

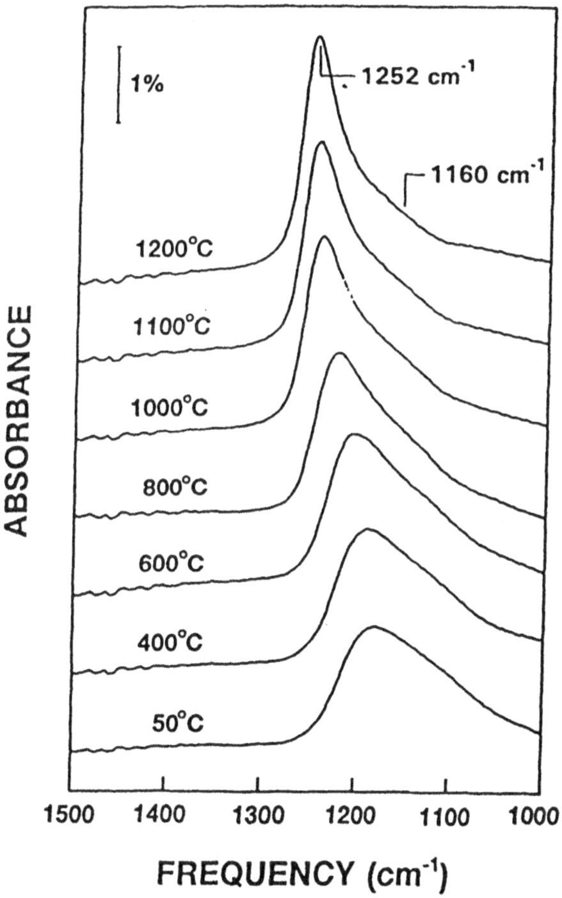

Figure 5. IRAS for a ~ 50 Å SiO_2 film on Mo(110) annealed to 50, 400, 600, 800, 1000, 1100, 1200 °C, acquired under UHV conditions.

These characterization studies indicate that the SiO_2 thin films prepared under the prescribed conditions yield chemically inert surfaces that serve as excellent models of the vitreous silica supports so frequently employed in dispersed metal catalysts.

3.1.2 *MgO*

MgO has been shown to be catalytically active in a variety of processes, including the H_2/D_2 exchange reaction and methanol dehydrogenation [33,34]. Also, Li/MgO systems have exhibited significant catalytic properties for the oxidative coupling of methane to ethane [35,36], as is discussed in greater detail in section 4.1.

Stoichiometric, epitaxial MgO films have been prepared on Mo(100) substrates by vapor deposition of Mg in a 1 x 10^{-6} Torr oxygen background pressure at a substrate temperature of 300K. These films have been characterized by a variety of techniques, including AES, LEED, XPS, HREELS, and TPD [37-40], and have been shown to be stable up to >1400K, though subject to thermally generated defects above 1100K. LEED studies for MgO deposited on the Mo(100) substrate indicate epitaxial film growth with a (100) orientation. This (100) termination has been shown to very closely approximate the truncated rocksalt structure of bulk MgO, exhibiting a very limited degree of surface relaxation [41,42]. The results from the XPS and HREELS studies reveal electronic and vibrational structure that is very similar to bulk MgO [37-38].

HREELS spectra acquired at 80K for a 20 ML thick MgO(100) film supported on Mo(100) are presented in Figure 6. The prominent loss feature at 6.3 eV represents a band gap that is significantly smaller than the 7.9 eV optical band gap associated with bulk MgO [43]. This feature has been assigned to a surface related interband transition originating with surface atoms of five-fold coordination. Henrich et al. have demonstrated that lowering the primary electron beam energy and increasing the angle of incidence enhances the intensity of this transition, consistent with the feature originating in the near surface region [44]. The spectra in Figure 6 indicate an MgO film that is largely free of point defects, up until oxygen pretreatment temperatures of $\geq$ 1100K.

At the elevated pretreatment temperatures, spectral features begin to appear in the band gap that are associated with the evolution of certain defects (so called "color centers"). Specifically, the features appearing at 5.3 and 1.1 eV are attributed to F centers (oxygen vacancies occupied by two electrons) while the peak at 3.6 eV is assigned to F aggregates (adjacent cation and anion vacancies [37,45-49]). The importance of these defects in terms of the catalytic properties will be examined in detail in the section that appears below discussing methane coupling over MgO and Li/MgO.

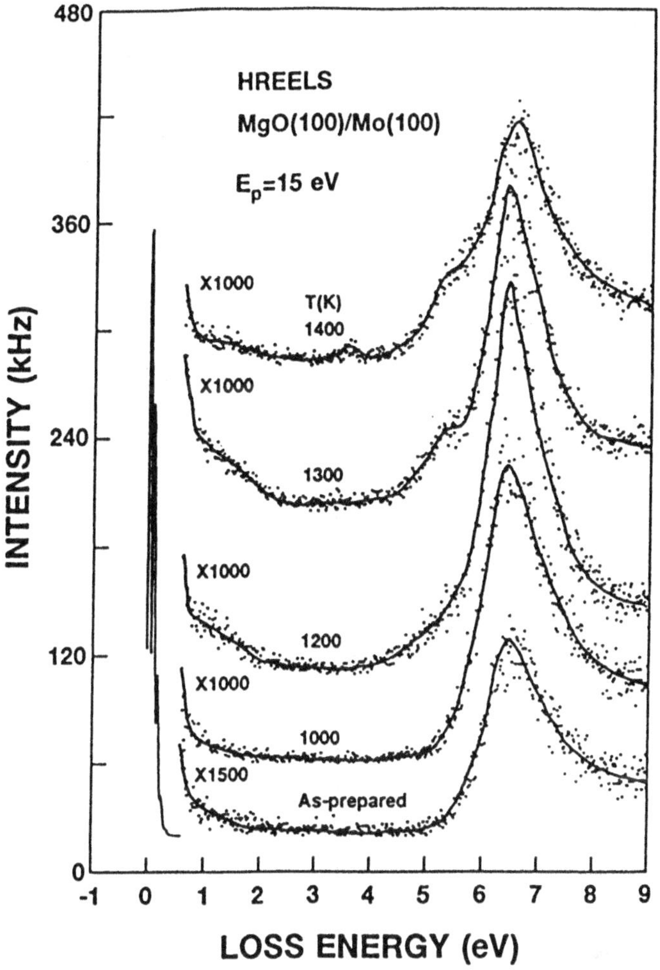

Figure 6. HREELS for a ~ 20 ML MgO film acquired after annealing in 5 x 10-7 Torr O_2 ambient for 1 min. to the indicated temperatures.

The acid-base properties of these films have been examined by HREELS and TPD in conjunction with a variety of probe molecules spanning a range of acidities. Among other findings, these studies have indicated that the thermally generated surface defect sites exhibit an enhanced basicity relative to the ordered film surface [50].

3.1.3 Al_2O_3

Thin Al_2O_3 films have been prepared on Ta(110), Mo(110), and Re(0001) substrates. Film preparation entails vapor depositing Al from a resistively heated source onto the preferred substrate at 900K in a backround pressure of O_2 in the range 5×10^{-7} to 1×10^{-6} Torr [51,52]. The films are subseqently annealed to 1200K in the O_2 background to improve their crystalline quality. AES analysis reveals a stoichiometric Al_2O_3 film devoid of metallic Al within the detection limits. ISS and LEED studies of the Al_2O_3/Ta(110) system have indicated that the film growth is largely two dimensional and proceeds heteroepitaxially, with the formation of a long range ordered epitaxial Al_2O_3 film with a slightly distorted (Beta = 117.9°) hexagonal lattice [51]. In fact, similar ordered LEED patterns have been observed for Al_2O_3 growth on each of these substrates.

LEED studies performed on very thin Al_2O_3 films (≤ 8 Å) grown on Mo(110) have yielded complex diffraction patterns with most of the spots originating from multiple scattering at the interface; with increasing film thickness, the multiple scattering features are attenuated. For films thicker than 15 Å, a simple hexagonal LEED pattern is observed, but the spots are more diffuse, indicating disorder in the thicker films. Because the lattice periodicity derived from this LEED pattern is similar to the O-O distance in bulk Al_2O_3, it has been interpreted to represent an ordered, close-packed O^{2-} anion layer associated with either the (001) face of alpha-Al_2O_3 or the (111) face of gamma-Al_2O_3 [52].

HREELS has also been used to monitor the growth of Al_2O_3 on Mo(110), as shown in Figure 7. Loss features associated with the fundamental modes of surface optical phonons are clearly distinguishable in the spectra. The 4.4 Å film exhibits surface optical phonon losses characterized by a two-mode pattern, while the spectrum for the 20 Å film displays three modes associated with these losses, consistent with previous examples in the literature [53,54]. It has been suggested that the appearance of of only the two phonon modes is characteristic of very thin Al_2O_3 films [55].

HREELS studies of the thicker Al_2O_3 films (> 20 Å) have indicated a band gap of 6.5 eV, significantly lower than the optical band gap of 9.5 eV observed for bulk Al_2O_3 [57]. Again, as in the case for MgO, this band gap is associated with the surface. Chemically, these surfaces have been shown to be relatively inert based on studies involving several different adsorbate gases, indicating an absence of unsaturated surface bonds. These properties are indicative of the inherent suitability of these films for simulating bulk Al_2O_3 supports.

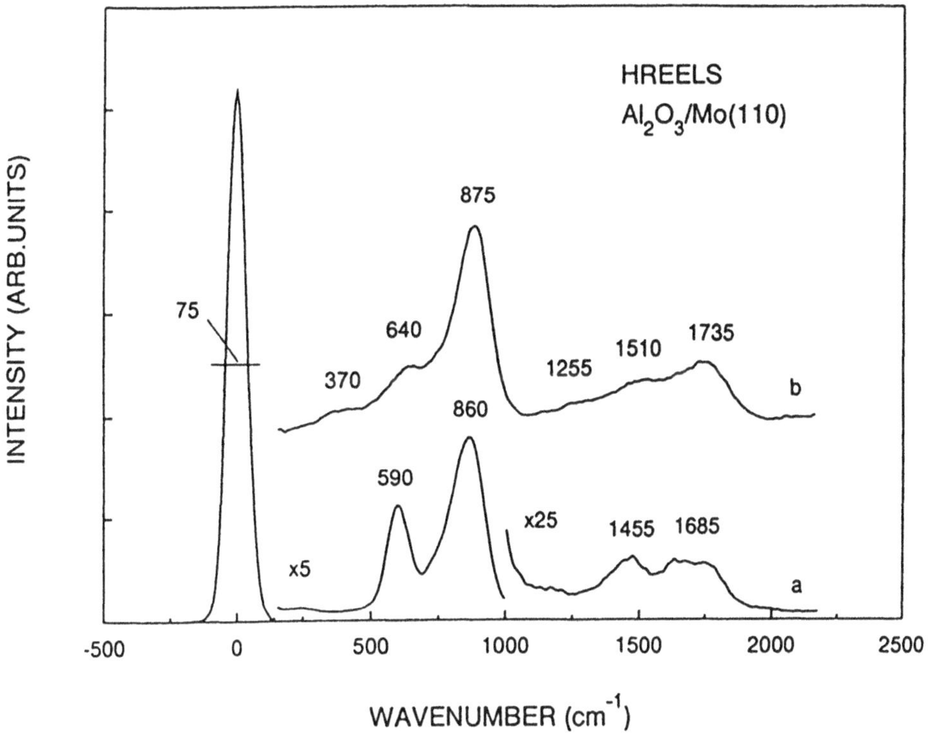

Figure 7. HREELS for Al_2O_3 films on Mo(110) of thickness (a) 4.4 Å and (b) 20 Å.

3.2 MODEL SUPPORTED CATALYSTS

3.2.1 *Cu/SiO*$_2$

Cu particles vapor deposited onto SiO_2 thin film supports have been characterized by a variety of surface science methods. The dispersions and particle size distributions have been studied using CO chemisorption and ex-situ STM and AFM. These experiments have demonstrated that the average supported metal particle size in the model catalyst can be influenced as a function of Cu coverage, deposition temperature, and annealing temperature [58-61]. Atomic resolution has been achieved in the ex-situ STM studies, identifying lattice spacings consistent with the Cu(311) face and CuO. Additionally, IRAS studies using CO as a probe have revealed that the morphologies of the particles (ranging from < 20 Å - 500 Å) are characterized by a high proportion of low index crystal planes with an increasing contribution from low coordinated defect sites with decreasing average particle size [59,60].

An interesting aspect of supported metal particles involves size dependent changes in the intrinsic electronic properties. UPS and XPS studies over a variety of supported metal systems have demonstrated pronounced size dependent modifications in the electronic structure in the small particle limit [62-66]. Specifically, XPS studies have revealed a virtually universal shift toward higher binding energies with decreasing particle size, which has variably been attributed to changes in d electron densities [65] or final state screening effects [62,63].

Copper was deposited at 300K onto a thin film SiO_2(80Å)/Mo(110) support and studied by TPD and XPS [67]. IRAS and STM studies have indicated that the growth mode for Cu on SiO_2 at 300K is three dimensional, and no significant changes in the core-level binding energies (CLBE's) for a given Cu coverage were apparent upon annealing to 700K.

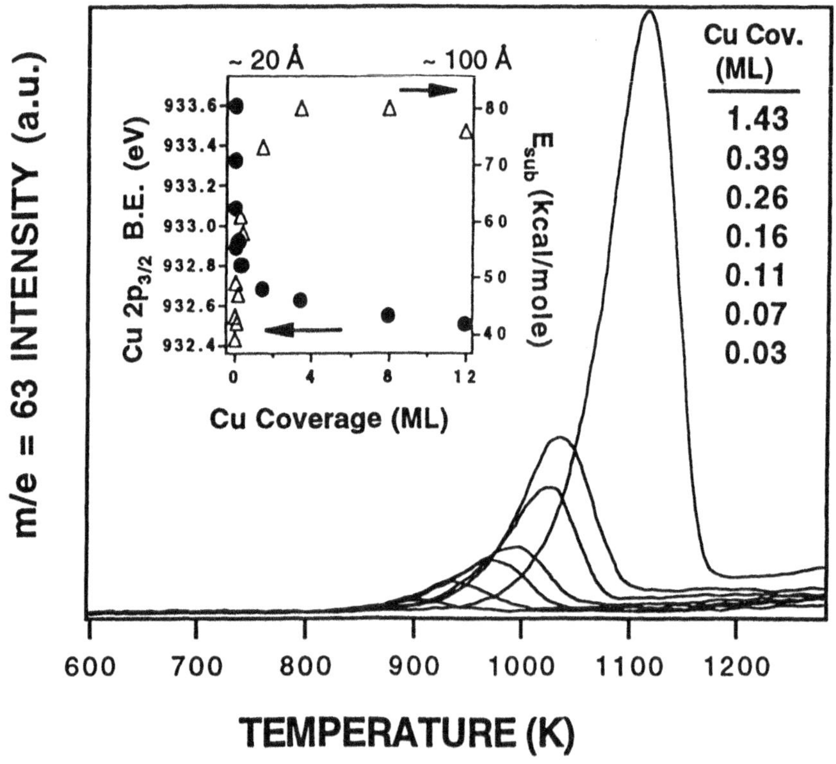

Figure 8. TPD spectra of several different Cu coverages in a Cu/SiO_2 (80 Å)/ Mo(110) model catalyst. The inset compares the CLBE's for the Cu $2p_{3/2}$ transition with the Cu sublimation energies as a function of Cu coverage. The corresponding approximate particle sizes are indicated above the inset.

TPD spectra for several different Cu loadings in the model $Cu/SiO_2/Mo(110)$ catalyst are presented in Figure 8. The inset in the figure compares the energy of sublimation E_{sub} for the Cu particles, determined by leading edge analysis of the TPD spectra, with the observed core-level shift (CLS) in the Cu $2p_{3/2}$ CLBE as a function of Cu coverage (particle size). The sublimation energy ranges from ~40 kcal/mole in the small particle limit, to a bulk value of ~ 80 kcal/mole, consistent with the 80.4 kcal/mole cohesive energy recorded for metallic bulk Cu [68]. The decrease in the cohesive energy with diminishing particle size is attributed to decreasing average Cu coordination number. The sublimation energies for the smallest particles in Figure 8 are comparable to the bond energy of a single Cu-Cu bond, 48.2 kcal/mole [69].

An interesting correlation emerges from the comparison of these two properties; each converges to the bulk value at a Cu coverage of approximately 2 ML (corresponding to an average particle size of ~ 25 Å based on CO chemisorption measurements). A similar convergence has been observed for the FWHM in the XPS transition (not shown), with the peak width broadening significantly below 2 ML [67]. This correlation between the XPS data and the sublimation energy suggests that the particle size dependent CLS's may be significantly influenced by an initial state effect, though contributions from final state effects are by no means precluded. These shifts in the XPS data may represent intrinsic modifications to the electronic band structure of the particles with changing average Cu coordination number, or alternatively, an increasing contribution from the metal support interaction as the fraction of Cu atoms in direct contact with the SiO_2 surface increases. Similarly, the rapid convergence to the bulk values could represent the crossover between metallic and non-metallic behavior at a critical particle size or perhaps a change in the growth mode at a critical Cu coverage.

These results provide an example of how photoelectronic studies can be utilized in concert with other characterization techniques to probe the relationship between electronic and physical (or chemical) properties as a function of particle size in these model catalysts. Another method of probing the electronic structure in supported particles is discussed in the following section.

3.2.2 *Ni/Al$_2$O$_3$*

The $Ni/Al_2O_3/Re(0001)$ system has been studied using UHV-STM [70]. The micrographs in Figures 9a and 9b are for 1.0 ML Ni deposited at 300K onto a 4.0 ML thick Al_2O_3 thin film supported on Re(0001), before and after a 900K anneal. The larger particles and lower particle number density for the annealed sample are indicative of the increased mobility of Ni on the surface at the elevated temperature. The quality of these micrographs illustrate the suitability of the model supported catalysts for nanoscale imaging using STM, which would be impossible on the bulk oxide analogs due to their insulating properties. In addition to the very detailed structural information regarding particle size and shape provided by this technique, Figure 9a demonstrates that a broad enough area is accessible to provide a statistically meaningful survey in terms of the overall particle size distribution.

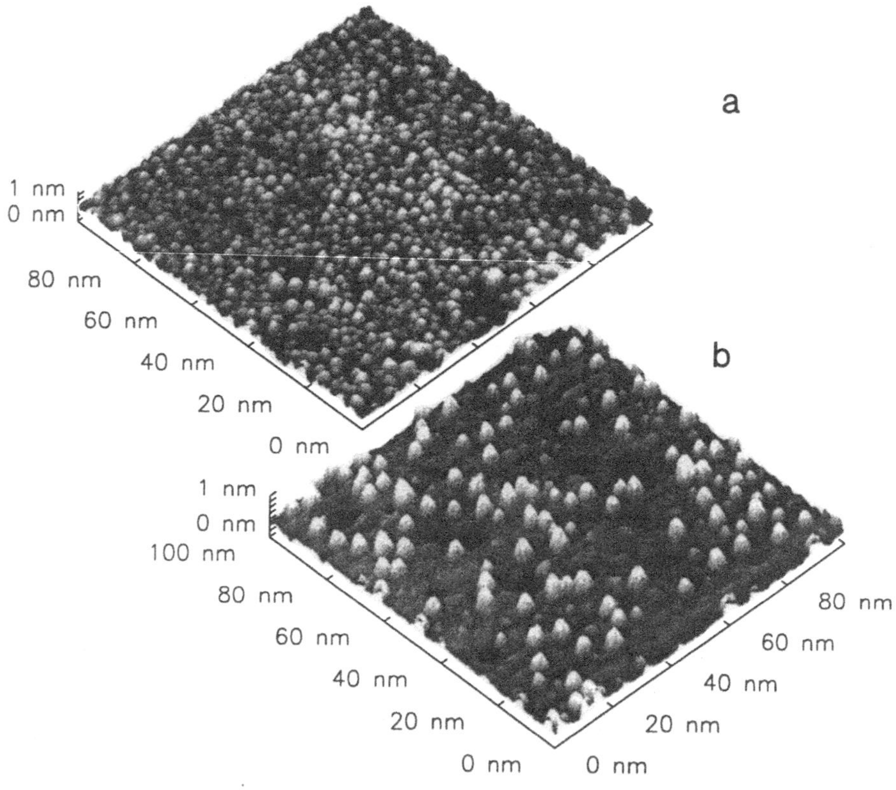

Figure 9. UHV-STM micrographs for 1.0 ML Ni on Al_2O_3(4.0 ML)/ Re(0001) after annealing to (a) 300K and (b) 900K. The bias is 5 V.

Another interesting aspect of this technique is the ability to investigate the electronic structure in a very specific and localized manner by identifying a particular feature with the STM and varying the tip bias voltage to generate an I-V curve to measure the density of states. This method, referred to as scanning tunneling spectroscopy (STS), offers a unique opportunity to unambiguously investigate the local electronic structure in a single, well defined supported metal particle of known size and shape.

A micrograph is presented in Figure 10 for the Ni(0.1 ML)/Al_2O_3(3.6 ML)/Re(0001) system, along with I-V curves corresponding to the indicated areas in the image. Regions 2 and 3 represent areas of the uncovered Al_2O_3 film; the associated I-V curves display distinct band gaps of ~ 6 eV, comparable to the surface band gap measured for Al_2O_3 using HREELS [56]. Region 4, a Ni particle exhibiting a base diameter of ~ 15 Å and a height of ~ 6 Å, produces an I-V curve indicative of a 0.6 eV band gap (a comparison with the I-V curve for region 1, bare metallic ReO_3, makes this more apparent). The origins of this band gap have not been conclusively identified. The magnitude is consistent with what might be expected from an intrinsic quantum size effect associated with the onset of non-metallic behavior, but more prosaic explanations such as the formation of a coulomb blockade or the formation of an oxide on the Ni surface cannot be discounted. Regardless of the origin, this result clearly demonstrates how STS can be applied to distinguish between the metal particle and the oxide support by measuring the density of states in the Fermi region [71].

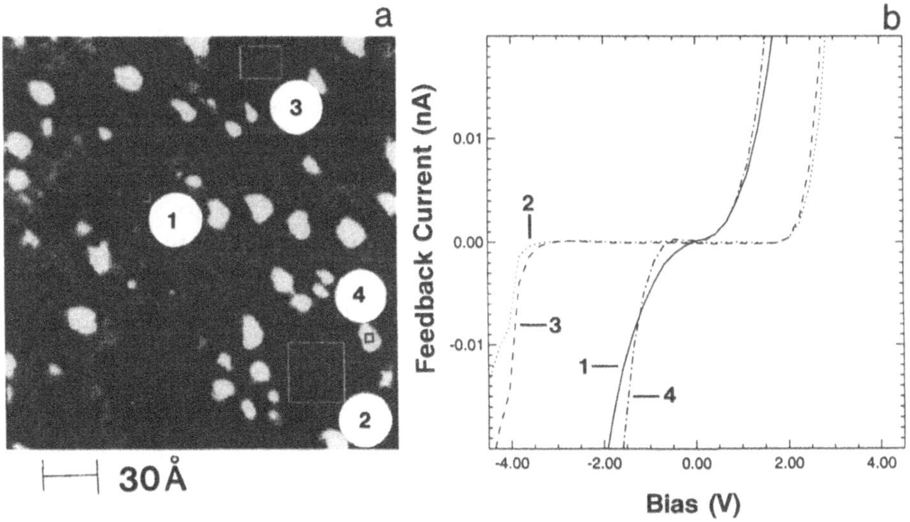

Figure 10. (a) UHV-STM micrograph for a Ni(0.1 ML)/Al_2O_3(3.6 ML)/Re(0001) model catalyst. (b) Four STS spectra acquired for the corresponding areas indicated in the micrograph (numbered boxes).

3.2.3 *Pd/Al_2O_3*

Model supported Pd/Al_2O_3 catalysts have been prepared and characterized using transmission electron microscopy (TEM) and IRAS to investigate the particle morphology as a function of particle size. TEM images for several different Pd loadings in a Pd/Al_2O_3/C catalyst are presented in Figure 11. The Pd loadings are reported in equivalent monolayer coverages, based on comparisons with Pd supported on bare Ta(110), which grows layer by layer. The Pd was deposited at a substrate temperature of 500K, and subsequently annealed to 800K. As the Pd coverage is increased, the average particle size increases and the particle number density goes down. Comparing the inset particle size histograms in Figure 11 with the histogram for the 5% Pd/gamma-Al_2O_3 catalyst imaged in Figure 12 illustrates the improved particle size control that can be exercised in the model systems relative to the high surface area supported catalysts. The histograms for the model particles are narrow and uniform in shape, while the histogram for the powder catalyst is very irregular and spans a large range (< 50 Å - ~ 600 Å) of particle sizes [21].

The particle size dependent surface morphologies for Pd in a Pd/Al_2O_3/Ta(110) catalyst were investigated using IRAS with CO as a probe molecule. The results of an IRAS annealing series in a background pressure of CO for Pd particles of 70 Å average diameter in a Pd(5.0 ML)/Al_2O_3/Ta(110) catalyst are presented in Figure 13a, with an analogous series for the Pd(111) single crystal shown in Figure 13b. Striking similarities are observed between the two sets of spectra [72,73].

For the Pd(111) single crystal at 550K, a single peak appears at ~1850 cm^{-1} that is attributed to CO adsorbed in the 3-fold hollow site. As the temperature is decreased, increasing the equilibrium surface coverage of CO, the peak shifts into the region associated with CO bound in the two fold bridging site. As the sample is further cooled, this feature narrows and shifts to higher frequency due to increased lateral CO-CO interactions. At about 300K , a small peak at ~ 2075 cm^{-1} is observed that is attributed to CO bound in the linear atop site, which increases as the sample is cooled. Between 200 and 150K, a fairly dramatic phase transition in the ordered CO overlayer occurs, shifting the CO from the bridging sites at 1962 cm^{-1} to CO bound in the linear atop position (2110 cm^{-1}) and the 3-fold hollow site (1895 cm^{-1}) [73].

The IRAS for the 70 Å particles in Figure 13a reflects this behavior. At 100K features at 2108 cm^{-1}, 1960 cm^{-1}, and 1890 cm^{-1} are evident in the spectrum, consistent with CO bound in the linear atop, bridging, and 3-fold hollow sites on the (111) facet. The additional peak at 1998 cm^{-1} is consistent with CO bound in the bridging site on the (100) facet; CO binds exclusively on

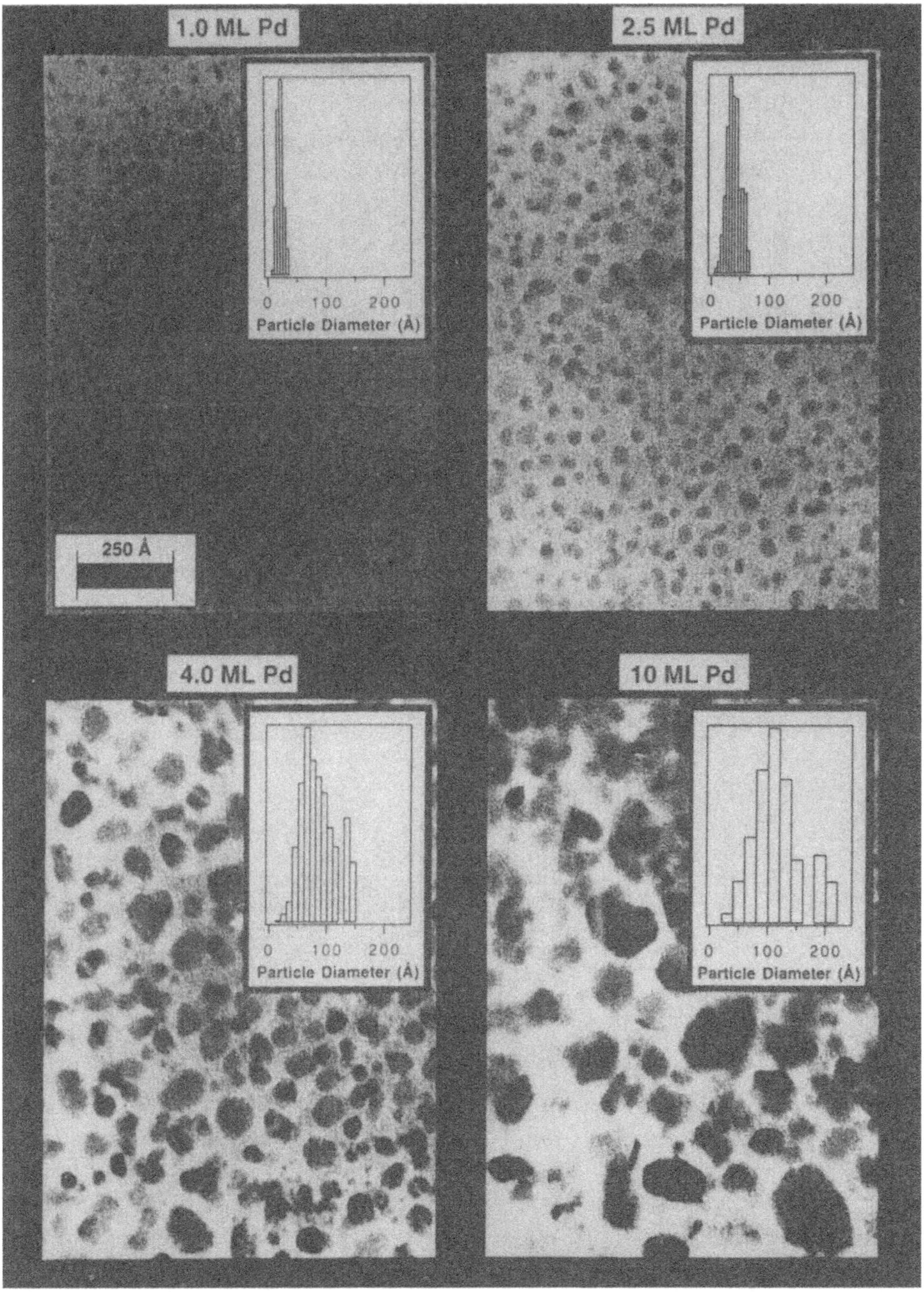

Figure 11. TEM micrographs for Pd/Al_2O_3(30 Å)/C samples for the indicated Pd coverages deposited at 500K and subsequently annealed to 800K.

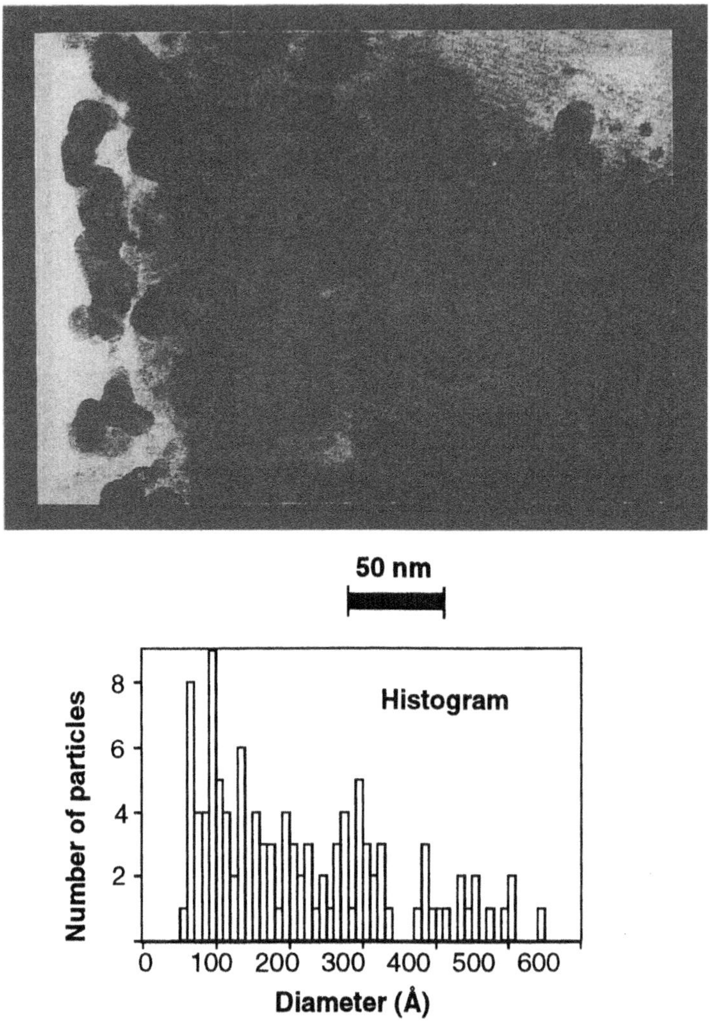

Figure 12. TEM micrograph and particle size histogram for a 5% Pd/gamma-Al_2O_3 powder catalyst prepared by incipient wetness impregnation.

these sites on the Pd(100) single crystal, saturating at 1995 cm^{-1} at 100K[73]. Based on these comparisons, the morphology of the 70 Å particles seems to be predominantly comprised of low index (111) and (100) facets. This is particularly evident in that even the very facet specific phase transition associated with the Pd(111) surface described above is observed on the particles, although to less degree. Between 200K and 100K, the linear atop and 3-fold hollow peaks are increasing in intensity, while the bridging feature at 1960 cm^{-1}

is decreasing. The transition is not as complete as that observed on the Pd(111) single crystal. At 100K on the (111) single crystal, no bridging CO is evident in the spectrum, while a peak at 1960 cm^{-1} still appears at this temperature for the 70 Å particles. A possible explanation is that the curvature in the supported particle precludes the degree of compression in the CO overlayer necessary to drive this transition to completion [72].

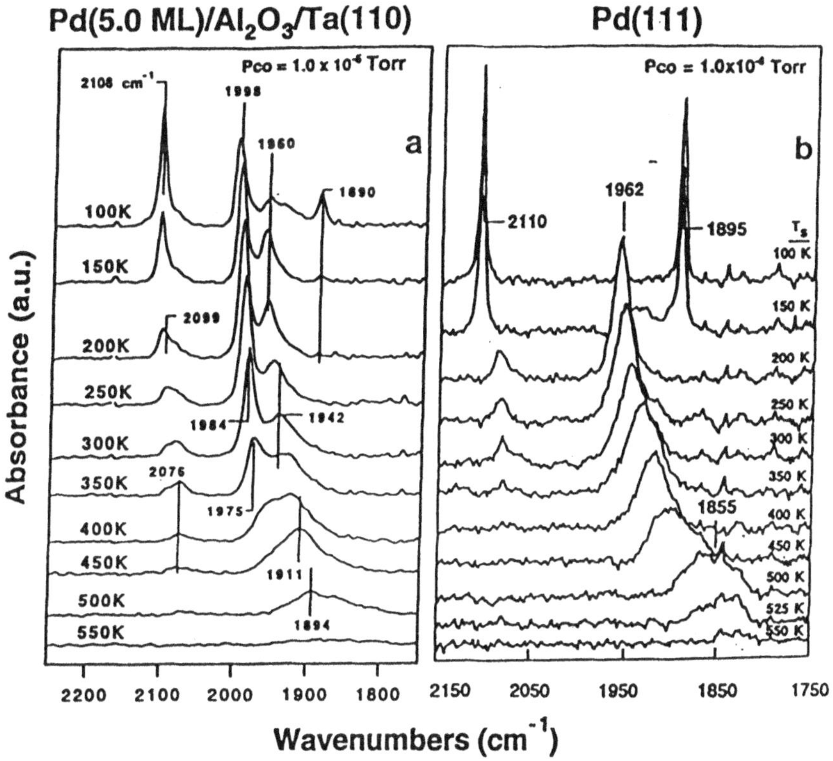

Figure 13. IRAS annealing series for (a) CO/Pd(5.0 ML)/Al_2O_3/Ta(110) in a 1 x 10^{-5} Torr CO ambient and (b) CO/Pd(111) in a 1 x 10^{-6} Torr CO ambient. The average particle diameter in the supported model is ~ 70 Å.

The same type of IRAS annealing series has also been performed on smaller Pd particles (~25 - 30 Å average diameter) in a Pd(1.0 ML)/Al_2O_3/Ta(110) model catalyst (Figure 14) [72]. Clear differences are evident. At 100K, the ratio of the linear atop to multiply bound CO stretching intensity is higher on the smaller particles relative to the larger (Figure 13a). Also, the broad peak in the region associated with bridging CO is never resolved into two peaks associated with the (111) and (100) facets, and no 3-fold hollow bound features appear. These observations are consistent with an increased concentration of low-coordinated step/edge defect sites and less well defined faceting on the smaller particles. These conclusions are supported by similar results from model $Pd/SiO_2/Mo(110)$ catalyst studies [23].

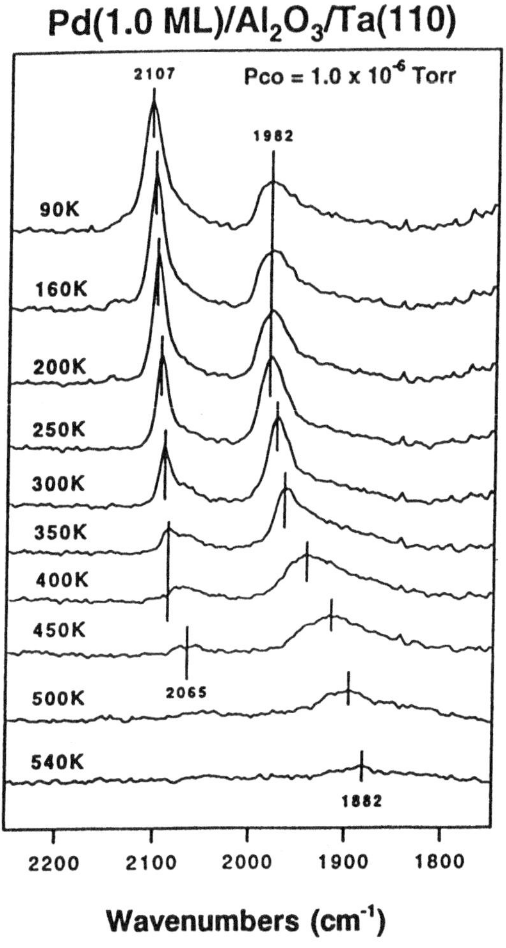

Figure 14. IRAS annealing series for CO adsorbed on a $Pd(1.0\ ML)/Al_2O_3/$ $Ta(110)$ model catalyst in a 1×10^{-5} Torr CO background.The average particle diameter is ~ 30 Å.

These morphological differences are accompanied by differences in the chemistry of CO adsorption. The 70 Å particles display an isosteric heat of adsorption in the zero coverage limit for CO (measured using IRAS intensity as an indicator of surface coverage) of 38.9 kcal/mole, consistent with previous studies on several low index Pd single crystal planes [74], and exhibit no significant activity for dissociating CO. Conversely, at sufficiently high temperatures, CO background pressures, and exposure times, the 30 Å particles demonstrated a pronounced propensity for dissociating CO [72]. These types of particle size dependent chemical and structural differences can have significant implications regarding catalytic behavior, as is well illustrated in the discussion of the CO + NO reaction over Pd presented in section 4.2.

4. Reaction Kinetics Studies Utilizing Thin Film Model Catalysts

4.1 OXIDATIVE COUPLING OF METHANE TO ETHANE OVER Li/MgO

The oxidative coupling of methane to ethane over Li promoted MgO has been studied utilizing thin film MgO and Li/MgO catalysts supported on Mo(100). This reaction has important implications in the effort to more efficiently utilize methane as an alternative fuel source. Arrhenius plots for the formation of methane are presented in Figure 15 for a pure MgO film and a Li promoted

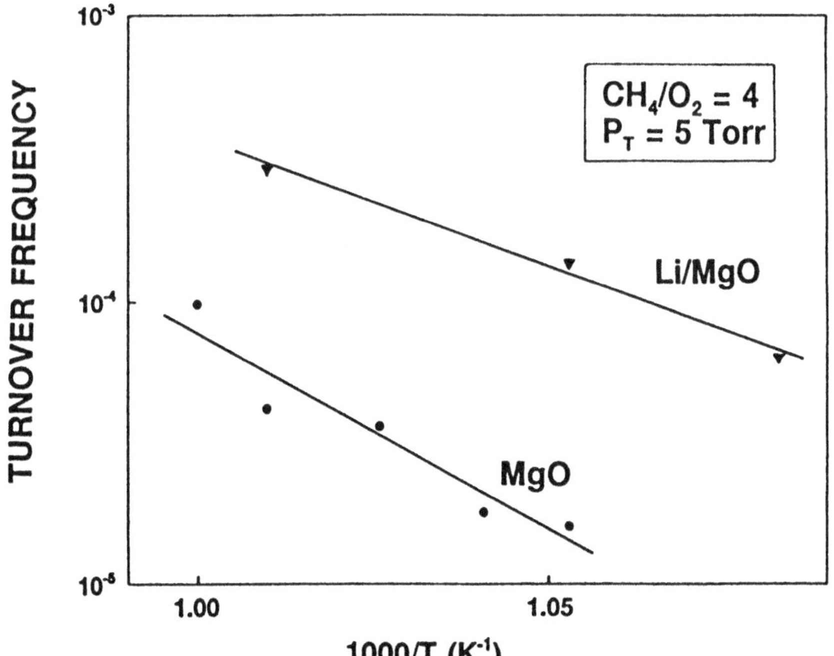

Figure 15. Arrhenius plots for the oxidative coupling of methane over MgO and Li/MgO thin films on Pt(100) at the indicated reactant partial pressures.

MgO film. These results are consistent with those observed for MgO and Li/MgO powder catalysts by Lunsford et al. [75], both in terms of the tenfold activity enhancement for the Li promoted catalyst and the absolute specific activities [76].

Based on several experiments involving powder catalysts that have demonstrated that methyl radical and ethane formation rates are dependent on the Li concentration, it has been suggested that the active site for this reaction in the Li/MgO catalyst is the [Li^+O^-] center. In this model, a hydrogen atom is abstracted at the active site, generating methyl radicals that subsequently desorb and recombine in the gas phase [75,77].

The model studies described here suggest a different active site. As discussed previously in section 3.1.2, HREELS can be used to identify specific structural defects in the MgO film. Figure 16 presents HREELS spectra for a 10 at. % Li

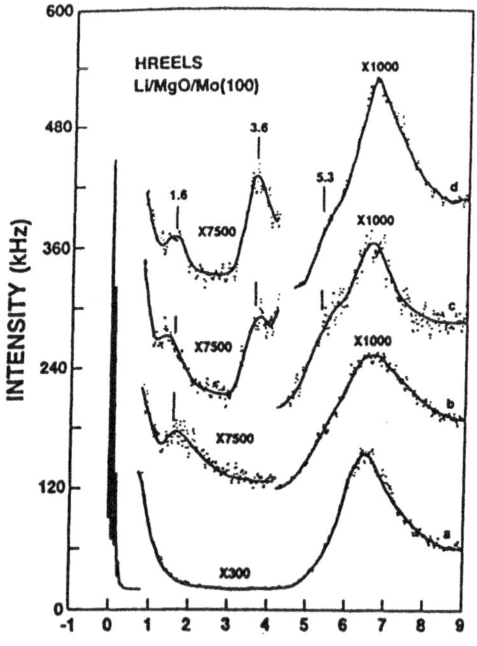

Figure 16. HREELS spectra of a 20 ML thick 10 at. % Li/MgO film after annealing to (a) 300, (b) 1000, (c) 1200, and (d) 1300K in a 5 x 10^{-7} Torr O2 background. The primary electron beam energy is 15 eV.

doped MgO film annealed in a 5 x 10^{-7} Torr oxygen background to several different pretreatment temperatures. The loss feature at 1.6 eV which appears in the Li doped sample, but not in the pure MgO film (Figure 6), is attributed to the [Li^+O^-] center, consistent with work by Abraham et al. [78-80]. The peaks at 5.3, and 3.6 eV are attributed to F centers (again, oxygen vacancies occupied by two electrons) and F aggregates respectively, consistent with their appearance in the pure MgO spectrum upon annealing to 1100K. Identical features have been observed previously in bulk MgO samples, and have been attributed to these defect centers. Furthermore, studies including electron beam energies and incidence angle as adjustable parameters have indicated that the species giving rise to these features (including the [Li^+O^-] centers in the Li/MgO films) exist in the near surface region [37,81].

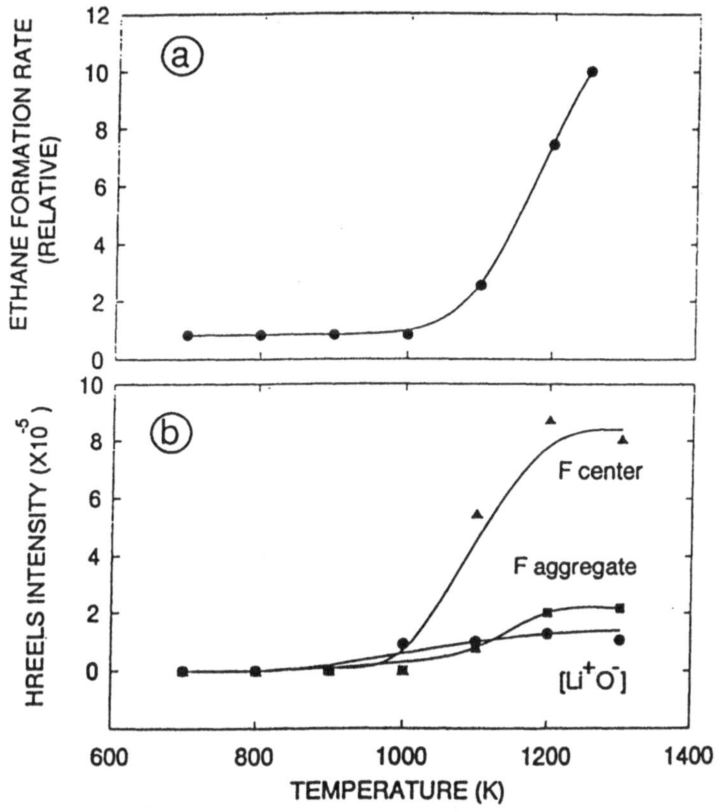

Figure 17. Correlations between (a) ethane formation rate at 990K and (b) HREELS intensities of the various defects in a 10 at. % Li/MgO thin film catalyst as a function of O_2 pretreatment temperature.

Figure 17 compares ethane formation rates as a function of sample oxygen pretreatment temperature with the corresponding concentrations of $[Li^{+}O^{-}]$, F aggregates, and F centers as determined using HREELS. This comparison shows that the substantial rate increase in the activity for ethane formation which begins to occur at a pretreatment temperature of slightly over 1000K, does not correspond to an increase in the concentration of $[Li^{+}O^{-}]$ centers so much as to a dramatic increase in the number of F centers and F aggregates that occurs with increasing Li concentration. This suggests that the key active site is not in fact the $[Li^{+}O^{-}]$ center as previously believed, but rather the F center defect [37,81].

This interpretation is supported by the similar activation energies observed for the pure MgO catalyst and the Li promoted MgO. The F center is the active site for both catalysts, giving rise to the same reaction mechanism, but the concentration of F centers in the Li doped sample is much higher, resulting in the observed activity enhancement.

4.2 CO + NO OVER PALLADIUM

The CO + NO reaction over Pd has been studied through a combined approach utilizing single crystals, model planar supported $Pd/Al_2O_3/Ta(110)$ catalysts of the type described above, and conventional high surface area Pd/Al_2O_3 powder catalysts [21]. Figure 18 compares Arrhenius plots for the reaction for each type of catalyst. The single crystal and planar model supported data were acquired in the batch mode at 1 Torr partial pressures in each reactant while the powder catalyst experiments were conducted in a conventional atmospheric flow reactor at partial pressures of 4.4 and 5.2 Torr of CO and NO respectively in a He carrier gas.

A pronounced structural dependence is evident in the data. For the single crystals, the close-packed Pd(111) surface is approximately five times more reactive than the more open (100) plane [82,83]. Among the particles in the powder catalysts, the catalyst exhibiting the largest average Pd particle size, ~1200 Å, displays a specific activity approximately 30 times higher than the catalyst supporting 60 Å particles. This enhancement in the rate over larger particles is also manifested in the planar model catalysts, although to a more limited degree [21]. A similar structure sensitivity has been reported for supported Rh catalysts, with 600 Å particles displaying a 45-fold rate enhancement relative to 25 Å particles [84].

There are clear differences among the three types of systems. The single crystals exhibit activation energies (~ 19-20 kcal/mole) that are significantly higher than those observed for the supported particles (~30-38 kcal/mole).

Additionally, the absolute activities are markedly higher on the single crystals. On the other hand there are some important correlations among the three catalyst types. It has been shown by IRAS (as previously described) that the smaller Pd particles exhibit surface morphologies that are characterized by a higher concentration of low-coordinated step/edge defect sites relative to the larger ones, and consequently of more open character [72]. In this regard, the trends observed over each type of catalyst are consistent; open faces are less active than more close-packed ones, and smaller particles are less active than large ones.

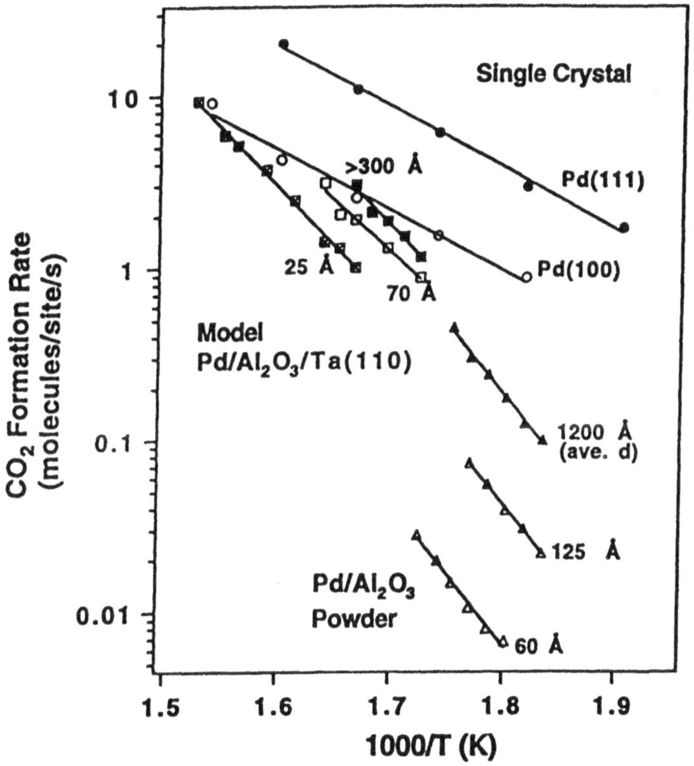

Figure 18. Arrhenius plots for the CO + NO reaction over Pd single crystals, model planar supported $Pd/Al_2O_3/Ta(110)$ catalysts, and Pd/Al_2O_3 powder catalysts.

Oh et al. have proposed that the differences in activity and activation energy between Rh single crystals and supported catalysts, as well as the observed particle size dependence, are related to the formation and stabilization of an inactive atomic N species that occurs preferentially at low-coordinated, reactive defect sites [84]. The single crystal and planar model supported studies described here do indeed provide some indication that such an inactive N_a species plays an important role in determining the activity over Pd.

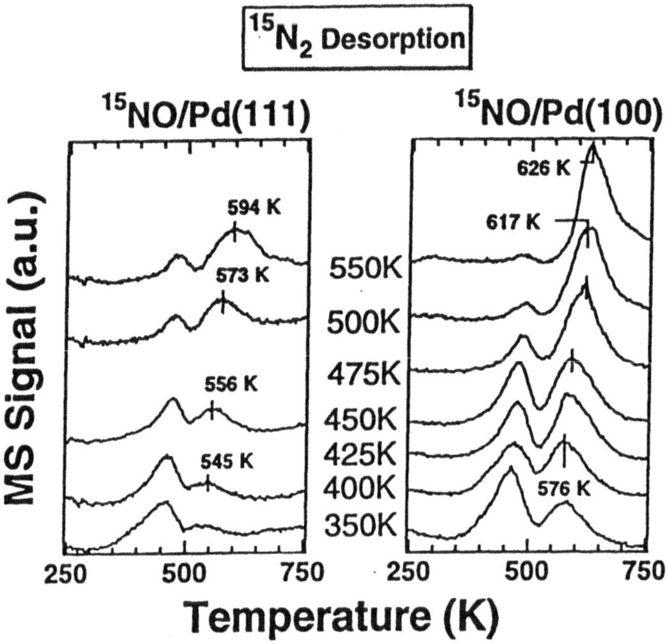

Figure 19. TPD spectra for $^{15}N2$ recombinative desorption after a 500 L exposure to ^{15}NO at the indicated substrate temperatures on Pd(111) and (100).

Figure 19 compares TPD for the recombination of N_a on Pd(111) and Pd(100) after a 500 L exposure to ^{15}NO in a background pressure of 1 x 10^{-6} Torr at the indicated substrate temperatures. Following this exposure, the samples were cooled in the presence of the ^{15}NO background; the background ^{15}NO was subsequently pumped away and the TPD experiment was performed. On both surfaces, N_2 recombinative desorption occurs in two distinct phases, giving rise to a low temperature and a high temperature feature in the TPD spectrum. However, clear differences between the two surfaces are apparent. The total combined peak $^{15}N_2$ area at each substrate exposure temperature is greater on the (100) surface, consistent with more facile NO dissociation on the more open face. Also, the ratio of the high temperature to the low temperature feature (based on TPD peak area) is significantly higher for the (100). That the desorption temperature of the more strongly bound N_a species is at the upper limit or higher than the typical reaction temperatures employed in these experiments, suggests that this species is inactive under these reaction conditions [83].

An analogous TPD experiment using a 1:1 mixture of CO + ^{15}NO yields similar results in terms of the high temperature to low temperature ratio of the $^{15}N_2$ recombinative features on each surface. By calibrating the $^{15}N_2$ peak areas against that acquired from the desorption of a monolayer coverage of N_a on Pd (100) (obtained from the reaction of ^{15}NO with H_2 at 550K), it was determined that the (100) surface is ~80% covered with the inactive form of atomic ^{15}N, while the coverage on the Pd(111) is only ~ 20% for this exposure. These observations are consistent with poisoning from inactive atomic N accounting for the structure sensitivity in the rates. The Pd(100) surface is about 5 times less active for the CO + NO reaction, and is observed to be approximately 4 times more active for the formation of the stable N_a species on the surface [83].

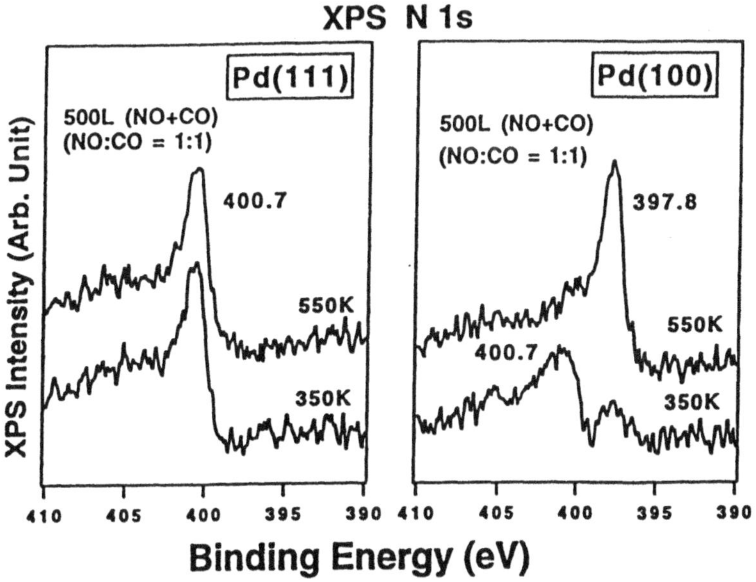

Figure 20. XPS spectra of the N 1s transition for Pd(111) and (100) after 500 L exposure to CO + NO at a 1:1 ratio at 350 and 550K.

Another illustration of the structure dependent NO dissociative behavior is provided by XPS. After 500 L exposures at 350 and 550K to a 1:1 ratio of a CO + NO mixture at a background pressure of 1 x 10^{-6} Torr, the XPS of the N1s region is significantly different on the two single crystal surfaces (Figure 20). For the Pd(111) surface, a single peak at 400.7 eV appears in the spectrum for each substrate exposure temperature, attributed to N in molecularly adsorbed NO. On the Pd(100) single crystal, an additional peak appears at 397.8 eV that is assigned to atomic N; for the 550K exposure temperature, this peak in fact dominates the spectrum. This species has been shown to be stable on the surface up to about 650K by an XPS annealing series. It is interesting to note that this temperature corresponds to the "light off" temperature (the critical threshold temperature where the exothermicity of the reaction takes over and the rate increases exponentially) for this reaction over the Pd(100) surface under similar reaction conditions [21].

These results strongly suggest that the observed structure sensitivity in the reaction rates on the single crystals is related to the enhanced formation and stabilization of an inactive N_a species on the more open surface. In an effort to determine if these findings extend to the supported particles, a TPD study analogous to the one described above was performed on several different loadings of Pd in the Pd/Al_2O_3/Ta(110) model catalyst (Figure 21), corresponding to the indicated average particle sizes. The N_2 and N_2O TPD features were monitored after a 500 L exposure to ^{15}NO at 550K.

The N_2 desorption behavior is very similar to that observed on the single crystals, exhibiting a low and a high temperature recombination feature at ~520 and 630K respectively. Furthermore, the spectra in Figure 21a exhibit the trend of an increase in the ratio of the high to low temperature peak area with decreasing particle size, with the peak maximum of the former shifting slightly to higher temperature. These results are not surprising in that one might anticipate the smaller particles, with their higher concentrations of reactive step/edge defect sites, to be more efficient in dissociating NO and stabilizing the resulting N_a species. According to the model invoking inactive N_a as an inhibitor, these results are consistent with the particle size dependent activity trends presented in Figure 18.

Another indication of these differences in NO adsorption properties with changing particle size is provided in the N_2O desorption spectra (Figure 21b). For the 25 Å particles no desorbing N_2O is detectable; however, as the particle size increases a feature at ~530K begins to appear. At the 300 Å average Pd particle size, the catalyst displays a marked activity for N_2O production in the TPD experiment. These results represent another indicator that the proportion of molecularly adsorbed NO to atomic N on the surface is strongly linked to particle size; taken with the activity data this implies that the structure sensitive formation of a stable atomic N species exerts a significant influence in determining the kinetics of the CO + NO reaction over Pd.

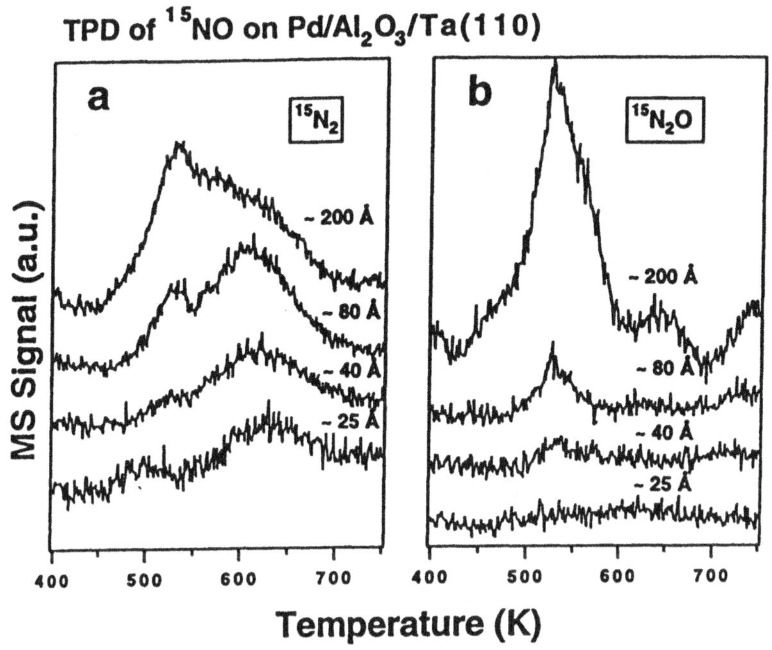

Figure 21. TPD spectra for (a) $^{15}N_2$ recomninative desorption and (b) $^{15}N_2O$ desorption over several different Pd coverages in a $Pd/Al_2O_3/Ta(110)$ catalyst after 5 min exposure to 1 x 10^{-6} Torr ^{15}NO at 550K.

Summarizing, correlations regarding structure sensitivies have emerged in the combined study of the CO + NO reaction over Pd single crystals and supported Pd/Al_2O_3 catalysts, indicating that close-packed Pd surfaces are the most active for this reaction. Evidence has been presented that suggests that an enhancement in the formation and stabilization of an inactive atomic N species on the more open planes plays a significant role in determining the relative rates [21,84]. The reasons for the differences observed in the E_a's and the absolute rates between the single crystals and the supported catalysts may well be related to the high concentration of reactive defect sites associated with small supported particles.

5. Conclusions

Because they are accessible to a wide range of UHV based surface characterization techniques for which their high surface area counterparts are poorly suited, oxide thin films supported on refractory metal substrates provide unique opportunities in the area of heterogeneous catalytic research. The ability to subject these samples to the array of charged particle spectroscopies that comprise the core of modern surface science, without the attendent charging problems associated with the application of these techniques to insulating materials, allows for structural and electronic characterizations to be performed at a level of detail unlikely to be achieved for the bulk analogs. These films are interesting both in terms of their intrinsic physical, chemical, and catalytic properties, and in their function as supports for small metal particles.

The experiments presented here illustrate some of the interesting properties associated with these systems and establish thin film model catalyst studies as a promising approach in catalysis science, particularly when employed in concert with more traditional methods.

Acknowledgments

We acknowledge with pleasure the support of this work by the Department of Energy, Office of Basic Energy Sciences, Division of Chemical Sciences and the Robert A. Welch Foundation.

6. References

1. Sexton, B.A., and Somorjai, G.A. (1977) *J. Catal.* **46**, 167.
2. Bonzel, H.P., and Krebs, H.J. (1980) *Surf. Sci.* **91**, 499.
3. Goodman, D.W., Kelley, R.D., Madey, T.E., and Yates, J.T., Jr. (1980) *J. Catal.* **63**, 226.
4. Campbell, C.T., and Paffett, M.T. (1984) *Surf. Sci.* **139**, 396.
5. Rodriguez, J.A., and Goodman, D.W. (1991) *Surf. Sci.* Rep. **14**, 1.
6. Goodman, D.W. (1986) *Annu. Rev. Phys.* **37**, 425.
7. Goodman, D.W. (1994) *Surf. Sci.* **299/300**, 837.
8. Vannice, M.A., and Garten, R.L. (1979) *J. Catal.* **56**, 536.
9. Rainer, D.R., Xu, C., and Goodman, D.W. (1996) *J. Mol. Catal.* submitted for publication.
10. Xu, X., and Goodman, D.W. (1995) in A.T. Hubbard (ed.), *Handbook of Surface Imaging and Visualization*, CRC Press, Boca raton, FL.
11. Kung, H.H. (1989) *Transition Metal Oxides*, Elsevier, Amsterdam.

12. Davydov, A.A. (1990) *Infrared Spectroscopy of Adsorbed Species on the Surface of Transition Metal Oxides*, John Wiley & Sons, New York.
13. Xu, C., and Goodman, D.W. (1996) *Handbook of Nanophase Materials*, submitted for publication.
14. Street, S.C., and Goodman, D.W. (1996) *The Chemical Physics of Solid Surfaces*, submitted for publication.
15. Poppa, H. (1993) *Sci. Eng.* **35**, 359.
16. Poppa, H. (1984) *Vacuum* **34**, 1081.
17. Marks, F.A., Lindau, I., and Browning, R. (1990) *J. Vac. Sci. Technol. A* **8**, 3437.
18. Xu, X., Szanyi, J., Xu, Q., and Goodman, D.W. (1994) *Catal. Today* **21**, 57.
19. Campbell, R.A., and Goodman, D.W. (1992) *Rev. Sci. Instrum.* **63(1)**, 172.
20. Szanyi, J., Goodman, D.W. (1993) *Rev. Sci. Instrum.* **64(8)**, 2350.
21. Rainer, D.R., Vesecky, S.M., Koranne, M., Oh, W.S., and Goodman, D.W. (1996) to be published.
22. Xu, X., and Goodman, D.W. (1992) *Appl. Phys. Lett.* **61(7)**, 774.
23. Xu, X., and Goodman, D.W. (1993) *Surf. Sci.* **282**, 323.
24. Davis, L.E., MacDonald, N.C., Palmberg, P.W., Riach, G.E., and Weber, R.E. (1976) *Handbook of Auger Electron Spectroscopy*, Perkin-Elmer, Eden Prairie, MN.
25. Joyce, B.A., and Neave, J.H. (1971) *Surf. Sci.* **27**, 499.
26. He, J.-W., Xu, X., Corneille, and Goodman, D.W. (1992) *Surf. Sci.* **279**, 119.
27. Phillip, H.R. (1971) *J. Phys. Chem. Solids* **32**, 1935.
28. Ibach, H. and Rowe, J.E. (1974) *Phys. Rev. B.* **9**, 1951.
29. Adachi, T., and Helms, C.R. (1979) *Appl. Phys. Lett.* **35**, 199.
30. Goodman, D.W. (1982) *Surf. Sci.* **123**, L679.
31. Kirk, C.T. (1988) *Phys. Rev. B.* **38**, 1255.
32. Lange, P. (1989) *J. Appl. Phys.* **66**, 201.
33. Wu, M.-C., Truong, C., and Goodman, D.W. (1993) in E. Umbach (ed.) *Adsorption on Ordered Surfaces of Ionic Solids and Thin Films*, Springer-Verlag, Berlin.
34. Mars, P., Scholten, J.J.F., and Zwietering, P. (1963) *Advan. Catal.* **14**, 35.
35. Ito, T., and Lunsford, J.H. (1985) *Nature*, **314**, 721.
36. Ito, T., Wang, J.X., Lin, C.H., and Lunsford, J.H. *J. Am. Chem. Soc.* **107**, 5062.

37. Wu, M.-C., Truong, C.M., and Goodman, D.W. (1992) *Phys. Rev. B.* **46(19)**, 12688.
38. Corneille, J.S., He. J.-W., and Goodman, D.W. (1993) *Surf. Sci.* **306**, 269.
39. Wu, M.-C., Estrada, C. A., Corneille, J.S., and Goodman, D.W. (1992) *J. Chem. Phys.* **96(5)**, 3892.
40. Wu, M.-C., Corneille, J.S., He, J.-W., Estrada, C.A., and Goodman, D.W. (1991) *J. Vac. Sci. Technol. A.* **10(4)**, 1467.
41. Urano, T., Kanaji, T., and Kaburagi, M. (1983) *Surf. Sci.* **134**, 109.
42. Prutton, M., Walker, J.A., Welton-Cook, M.R., and Felton, R.C. (1979) *Surf. Sci.* **89**, 95.
43. Roessler, D.M., and Walker, W.C. (1967) *Phys. Rev. B* **159**, 733.
44. Henrich, V.E., Dresselhaus, G., and Zeiger, H.J. (1980) *Phys. Rev. B* **22**, 4764.
45. Weber, H. (1951) *Z. Phys.* **30**, 392.
46. Hensley, E.B., and Kroes, R.L. (1968) *Bull. Am. Phys. Soc.* **13**, 420.
47. Chen, Y., Sibley, W.A., Srygley, F.D., Weeks, R.A., Hensley, E.B., and Kroes, R.L. (1968) *J. Phys. Chem. Solids* **29**, 863.
48. Henderson, B., and Bowen, D.H. (1971) *J. Phys.* C **4**, 1487.
49. Henderson, B. and Wertz, J.E. (1977) *Defects in Alkaline Earth Oxides* Taylor and francis, London.
50. Wu, M.-C., and Goodman, D.W. (1992) *Catal. Lett.* **15**, 1.
51. Chen, P.J., and Goodman, D.W. (1994) *Surf. Sci. Lett.* **312**, L767.
52. Wu, M.-C., and Goodman, D.W. (1994) *J. Phys. Chem.* **98**, 9874.
53. Crowell, J.E., Chen, J.G., and Yates, J. T., Jr. (1986) *Surf. Sci.* **165**,37.
54. Chen, P.J., Colaianni, M.L., and Yates, J.T., Jr. (1990) *Phys. Rev. B.* **41**, 8025.
55. Frederick, B.G., Apai, G., and Rhodin, T.N. (1990) *J. Electron. Spectrosc. Relat. Phenom.* **54/55**,415.
56. Street, S.C., and Goodman, D.W. (1996) to be published.
57. Duffy, J.A. (1986) *J. Solid State Chem.* **62**, 145.
58. Xu, X., He, J.-W., and Goodman, D.W. (1993) *Surf. Sci.* **284**, 103.
59. Xu, X., Vesecky, S.M., and Goodman, D.W. (1992) *Science* **258**, 788.
60. Xu, X., and Goodman, D.W. (1993) *J. Phys. Chem.* **97**, 683.
61. Xu, X., Vesecky, S.M., He, J.-W, and Goodman, D.W. (1993) *J. Vac. Sci. Technol. A* **11(4)**, 1930.
62. DiCenzo, S.B., and Wertheim, G.K. (1985) *Comm. Solid State Phys.* 11, 203.
63. Cheung, T.T.P. (1984) *Surf. Sci.* **140**, 151.
64. Bastl, Z. (1986) *Vacuum* **36**, 447.

65. Mason, M.G. (1983) *Phys. Rev. B* **27**, 748.
66. Egelhoff, W.F., Jr., and Tibbets, G.G. (1979) *Phys. Rev. B* **19**, 5028.
67. Rainer, D.R. (1996) Ph.D. Dissertation, Texas A&M.
68. Kittel, C. (1986) *Introduction to Solid State Physics*, Wiley, New York.
69. CRC Handbook of Chemistry and Physics, 72nd edition, D.R. Lide (ed), CRC Press, Boca Raton, FL.
70. Xu, C., and Goodman, D.W. (1996) to be published.
71. Xu, C., and Goodman, D.W. (1996) *Science*, sumbitted for publication.
72. Rainer, D.R., Wu, M.-C., Mahon, D., and Goodman, D.W. (1996) *J. Vac. Sci. Technol. A* in press.
73. Szanyi, J., Kuhn, W.K., and Goodman, D.W. (1993) *J. Vac. Sci. Technol. A* **11**, 1969.
74. Chou, P., and Vannice, M.A. (1987) *J. Catal.* **104**, 17.
75. Lin, C.H., Ito, T., Wang, J.X., and Lunsford, J.H. (1987) *J. Am. Chem. Soc.* **109**, 4808.
76. Coulter, K., and Goodman, D.W. (1992) *Catal. Lett.* **16**, 191.
77. Driscoll, D.J., Martir, W., Wang, J.-X., and Lunsford, J.H. (1985) *J. Am. Chem. Soc.* **107**, 58.
78. Abraham, M.M., Chen, Y., Boatner, L.A., and Reynolds, R.W. (1976) *Phys. Rev. Lett.* **37**, 849.
79. Chen, Y., Tohver, H.T., Narayan, J., and Abraham, M.M. (1977) *Phys. Rev. B* **16**, 5535.
80. Olson, D.N., Orerea, V.M., Chen, Y., and Abraham, M.M. (1980) *Phys. Rev. B* **21**, 1258.
81. Wu, M.-C., Truong, C.M., Coulter, K., and Goodman, D.W. (1992) *J. Am. Chem. Soc.* **114**, 7565.
82. Vesecky, S.M., Chen, P., Xu, X., and Goodman, D.W. (1995) *J. Vac. Sci. Technol. A* **13(3)**, 1539.
83. Vesecky, S.M., Rainer, D.R., and Goodman, D.W. (1995) *J. Vac. Sci. Technol. A* **13(3)**, 1539.
84. Oh, S.H., and Eickel, C.C. (1991) *J. Catal.* **128**, 526.

METAL DEPOSITS ON THIN WELL ORDERED OXIDE FILMS: MORPHOLOGY, ADSORPTION AND REACTIVITY

M. BÄUMER, J. LIBUDA, and H.-J. FREUND
Fritz-Haber-Institut der Max-Planck-Gesellschaft
Faradayweg 4-6, 14195 Berlin, Germany

Work on clean oxide surfaces, such as NiO, CoO, FeO, Cr_2O_3 and Al_2O_3, where adsorption and reaction of small molecules has been studied, is reviewed. The studies on the clean surfaces are the basis for investigations of transition metal particles deposited onto these oxide surfaces: Ag, Pd, Rh and Pt have been deposited on thin alumina films and the morphology has been determined via STM and SPA-LEED. Photoelectron spectroscopy was applied to investigate the electronic structure of the deposits as a function of particle size. The bonding between molecules and the metal aggregates has been characterized via thermal desorption spectroscopy and x-ray absorption spectroscopy. We find a considerable variation of adsorption strength as a function of particle size and the kind of metal. Additionally, examples for size dependent reactivity are presented.

1. Introduction

Disperse metal catalysts play an important role in several technically important reactions [1-3]. Sometimes the yield and selectivity of such reactions depend on the average size of the metal particles involved, i. e. the reaction is size selective. Examples are the hydrogenolysis of ethane and the oxidation of carbon monoxide [4]. Very often the catalysts are prepared by choosing an appropriate support material, such as SiO_2, Al_2O_3 or MgO, and depositing a transition metal compound onto the support surface from solution or by chemical vapour deposition. Via calcination and subsequent reduction, metal deposits are formed on the support surface.

Several groups have recently started to model such systems by depositing a metal from the vapour onto single crystal oxide surfaces for investigations under ultra-high vacuum conditions [5-10]. Most of them have used bulk single crystals as support materials [11-44]. However, thin films of corresponding oxides have also been employed which were prepared on top of a metal substrate [45-73]. Such systems have several experimental advantages over bulk single crystal surfaces. For instance, charging of the sample under electron bombardment or during electron spectroscopic measurements are completely avoided. Also, cooling of the surface to low temperatures in rather short times is achieved via the intimate contact between the oxide film and the metallic

R.M. Lambert and G. Pacchioni (eds.),
Chemisorption and Reactivity on Supported Clusters and Thin Films, 61-104.

substrate, whereas the often found low heat conductances hamper effective cooling of bulk materials. By controlling the thickness of the film, an electronic structure of the formed oxide may be established which is representative of its bulk electronic structure [47]. In addition, thin films may be prepared with different surface crystallographic orientations, a task which is considerably more difficult for bulk single crystals [8,9]. In particular, the preparation of polar surfaces of ionic systems is very difficult via cleavage of bulk crystals due to the very high surface energies of these systems [74].

In the following, we review some results from our laboratory on adsorption and reaction on clean metal oxide surfaces in an introductory chapter, and then dwell more specifically on results of metal deposits on a model alumina surface. Especially, we address the morphology of the metal deposits for different transition metals and then discuss the electronic structure as studied by photoelectron and other electron spectroscopies. Finally, we report on the interaction of the metal deposits and discuss associative and dissociative adsorption. Specific emphasis will be placed onto those properties which are strongly dependent on the size of the deposited metal particle.

2. Analytical Tools and Oxide Preparation

The experiments were performed in several ultrahigh-vacuum systems which are equipped with various surface science tools. The methods available are: photoemission (XPS, ARUPS), electron energy loss spectroscopy (ELS, HREELS), thermal desorption spectroscopy (TDS), Auger electron spectroscopy (AES), infra-red reflection absorption spectroscopy (IRAS), and electron spin resonance spectroscopy (ESR) [75-77]. We have investigated both, bulk single crystals cleaved in vacuo as well as in situ prepared thin oxide films. The typical preparation methods for the thin films are summarized in Fig. 1. The first one is the simple oxidation of the corresponding metal single crystal surface, applying high oxygen dosages coupled with heating cycles. If there is a large lattice mismatch between the metal and the native oxide, as it is the case of NiO/Ni, the film exhibits a rather pronounced defect density. While this may be of interest in itself if one wants to study the influence of surface defects on adsorption, it is also desirable to prepare films with a very small defect density. This is possible via the choice of an appropriate metal substrate with a lattice spacing that fits the oxide lattice spacings [21,60]. The metal to be oxidized is then evaporated onto the inert substrate either in an oxygen atmosphere or it is oxidized after deposition. In the case of a NiO(100) film, for example, a Ag(100) surface may be used as substrate. For the preparation of a NiO(111) film, Au(111) is an appropriate substrate [59]. Another attractive technique, as presented in the third panel, is the oxidation of alloy single crystals. We have used this alternative to prepare a well-ordered Al_2O_3 film [78-83]. The advantage here is the high melting point of the alloy which allows heating of the film to high temperatures and in turn leads to ordering within the film. In contrast to this situation, oxidation of an Al metal substrate would not lead to an ordered film, because Al would melt before

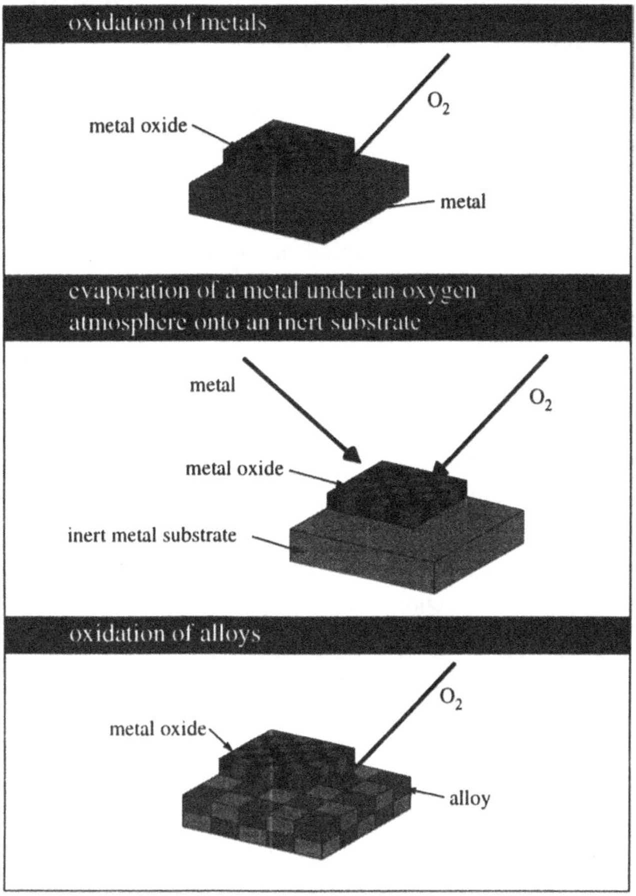

Figure 1. Methods for thin oxide film preparation.

Al_2O_3 orders. Choosing NiAl(110) as a substrate for Al_2O_3 also assures that the Ni, set free in the oxidation process, is dissolved in the bulk of the alloy because Ni_3Al is the thermodynamically favourable alloy stoichiometry so that the bulk of the alloy tries to enrich with Ni.

3. Structure and Reactivity of Clean and Modified Oxide Surfaces

Structure and reactivity of oxide surfaces may vary considerably depending on the system. We have studied rock salt type surfaces as well as corundum type surfaces, nonpolar and polar surfaces, oxides that are reactive, like NiO or Cr_2O_3 for example, and oxides which are inert, such as Al_2O_3 [74]. The clean oxide surfaces as well as those modified by deposition of alkali [63, 84, 85] have been investigated. In the following, we shall present results on a chemically reactive polar $Cr_2O_3(0001)$ surface and compare with a chemically inert but isostructural Al_2O_3 surface.

Before we turn to the explicite examples, we briefly summarize our results on other oxide surfaces. We have investigated a series of oxides with rock salt structure, i.e. NiO, CoO, and FeO [49, 74]. The question around which our research centred was the electronic and structural stability of the polar (111) faces. Vibrational spectroscopy (HREELS) was used to show that the apparently non-reconstructed (111) surfaces were hydroxyl terminated [55]. The OH termination effectively reduces the surface charge and stabilizes the surface. Upon removal of the hydroxyls as water by a thermal treatment, the surface becomes unstable and undergoes reconstruction [86, 89]. The particular reconstruction found on the ionic surfaces has been termed the octopolar reconstruction because the formation of octopolar units is found at the surface exposing microfacets with (100) orientation. Only the topmost atom of each of the octopolar units is still in a coordination typical for a (111) surface, while all other atoms are in the typical fivefold (100) coordination. The surface energy of such a reconstructed surface is still considerably higher (4.38 Jm^{-2}) compared to the (100) surface (1.38 Jm^{-2}) and the hydroxyl terminated non-reconstructed surface [90]. This energy content may be used to drive chemical reactions. In fact, the reconstructed surface does interact much stronger with adsorbed molecules as compared with the hydroxyl terminated or the non-reconstructed (100) surfaces. Papp and coworkers [91] have recently demonstrated that topotactically prepared NiO powder, preferentially exposing (111) surface planes, indeed exhibits higher activity in deNOx-reactions and that its reactivity is correlated with the water content of the powder.

Parallel to the work on polar surfaces, we have studied the adsorption properties of the non-polar (100) surface [52]. The ideal surface, containing no defects, adsorbs CO with rather small binding energy of about 0.2...0.3 eV [51]. The bond is almost exclusively electrostatic in nature and due to the interaction of the multipolar moment of the (100) surface with the quadrupol moment of CO, whose dipole moment is very small [3, 92-98]. However, the small interaction is large enough to orient the molecule on the surface in such a way that the carbon atom points towards the Ni cation and the axis is oriented perpendicularly with respect to the surface plane. For NO the binding energy is slightly larger, i.e. 0.4...0.5 eV, and there are indications from vibrational spectra [53] as well as from other observations that a small covalent contribution has to be considered [3, 92-98]. It is this small covalent contribution that leads to an orientation of the molecule with the nitrogen atom towards a Ni ion in the surface and an inclined axis with respect to the surface normal (45°). The NO stretching vibration is red shifted [53] with respect to the gasphase indicating a slight additional population of the 2π-antibonding NO orbital via interaction with the surface. In the case of CO, on the other hand, there is a very small blue shift of the vibrational stretching frequency indicative of the so called wall-effect [92], where a non chemically bonded molecule vibrates towards a heavy wall. Water only physisorbs on the NiO(100) surface as long as the surface is perfect. Upon creation of defects, however, the behaviour becomes more similar to the polar surface in the sense that water spontaneously dissociates and hydroxyl groups are formed on the surface [50]. These hydroxyl groups are readily observed in vibrational spectra and proved via isotopic labelling with heavy water [50]. Spectroscopic identification of terrace sites on the surface is possible with electron energy loss spectroscopy. We have shown [74, 99] that the coordinatively unsaturated Ni ions in the surface exhibit typical electronic d-d excitations which are energetically shifted with respect to those of the octahedrally coordinated Ni ions in the second and deeper layers. Since the ligand field of a surface Ni ion changes when a molecule is coordinated to it, these d-d excitations are energetically further shifted upon adsorption. This observation may be used to identify adsorption sites. It has been demonstrated that differently coordinated Ni ions in the surface lead to different d-d excitations [49]. Coming back to the dissociation of water on NiO defects, addressed above, we can follow the process in the EL-spectra [8]. Since water only dissociates at the defect sites, the d-d excitations typical for the terrace sites are not influenced by the hydroxyl groups formed in the dissociation process.

The $Cr_2O_3(0001)$ surface is a polar surface as well, and, therefore - according to theoretical predictions -, we have to expect strong structural relaxations. In fact, a recent LEED I/V analysis [67] together with other spectroscopic evidences [62, 64] have clearly revealed that the surface is very far from being bulk terminated. Two possibilities for the bulk terminated surface are shown in Fig. 2a. Ion scattering spectrocopy (ISS) [62] indicates, however, the presence of only half the number of chromium ions on the surface. Electron energy loss spectroscopy of the d-d excitations [64] may be used to clearly determine the oxidation state of the chromium ions as Cr^{3+}. Therefore, in order to depolarize and thus stabilize the surface, the distribution of metal ions in the surface is likely to be as shown in Fig. 2b. This is actually the situation also corroborated by the LEED structure analysis [67]. In addition, however, the interlayer distances relax considerably. The surface chromium ions are pulled into the surface oxygen layer, and the buckling of the chromium ion layer just beneath the first oxygen plane is reduced. Even more than four layers beneath the surface interlayer relaxation can be observed. This relaxed $Cr_2O_3(0001)$ surface is still chemically very active. We have investigated adsorption of various molecules, such as O_2, CO, C_2H_4, H_2O, and also metallic adsorbates such as Na [62-67, 85]. We only want to mention that CO adsorption leads to a non-classic side-on coordination [66], similar to ethene which is classically [68, 69, 70] side-on coordinated. Under ultrahigh vacuum conditions adsorption is fully reversible in both cases. Upon raising the pressure to 1 bar, however, it is possible to polymerize ethene at slightly elevated temperature.

In contrast to the polar NiO(111) surface, the $Cr_2O_3(0001)$ surface does not readily dissociate water. The reason is clear: the surface is stabilized by removal of half of the metal ions and additional relaxations to such an extend that the surface energy is comparable to the non-polar surfaces. Thus, dissociative adsorption of water will not lead

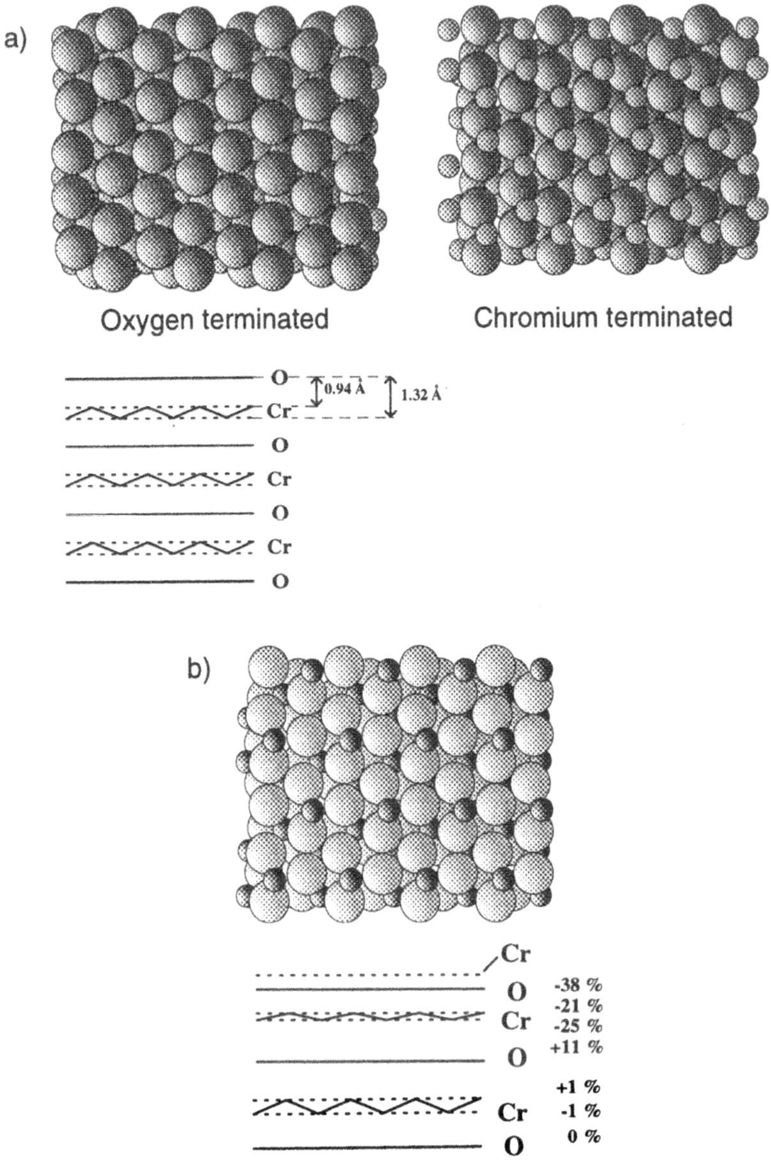

Figure 2. a) Possible terminations of the $Cr_2O_3(0001)$ surface. Termination by a complete layer of oxygen ions (left) and by a complete layer of chromium ions (right). Bottom: stacking of layers (schematically). b) Reconstructed and depolarized $Cr_2O_3(0001)$-surface (top) and inter-layer distances after reconstruction (bottom).

to further stabilization. It is possible to dissociate water via photoinduced processes [50]. At this point we would like to consider oxygen adsorption in somewhat more detail because it also allows us to understand how the oxide film grows on the surface. We shall mainly use results from infra-red reflection absorption spectroscopy [68, 70].

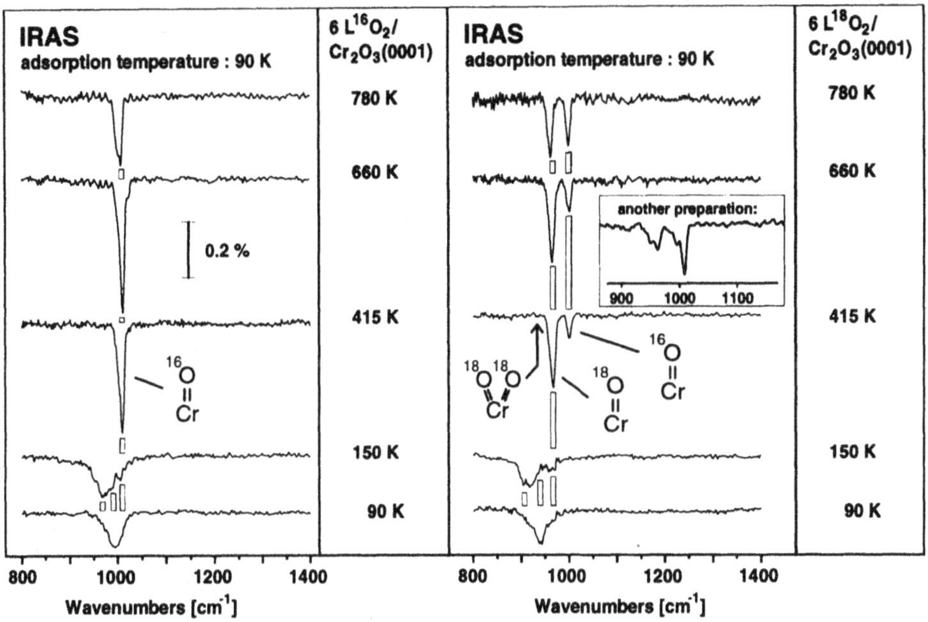

Figure 3. Infrared spectra of a $Cr_2{}^{16}O_3(0001)$-surface after dosing of 6 L $^{16}O_2$ (left) and 6 L $^{18}O_2$ (right) at 90 K and subsequent heating to given temperatures.

Figure 3 shows a set of infra-red spectra taken after dosing 6 L at 90 K and subsequent heating to temperatures between 90 K and 780 K. On the left hand side spectra obtained after dosing $^{16}O_2$, on the right hand side those obtained after dosing $^{18}O_2$ are depicted. At lowest temperature a broad peak centred around 990 cm^{-1} ($^{16}O_2$) and 930 cm^{-1} ($^{18}O_2$), respectively, shows up. While there is only little intensity at 965 cm^{-1} ($^{16}O_2$) and 905 cm^{-1} ($^{18}O_2$), this intensity increases to form another broad band as the surface temperature rises to 150 K. At this temperature or a little below, another spectral feature, finally developing into a rather sharp peak, starts to show up at 1005 cm^{-1} ($^{16}O_2$) and at 960 cm^{-1} ($^{18}O_2$). For the $^{16}O_2$ dosage, a single relatively sharp feature grows until the surface reaches a temperature of 660 K. Then it starts to decrease again. In the case of $^{18}O_2$ dosage, two sharp features at 960 cm^{-1} and 1005 cm^{-1} are observed. The peak at 1005 cm^{-1} gains intensity as the temperature increases. Depending somewhat on the preparation conditions of the Cr_2O_3 film, the sharp feature may be split by approximately 10 cm^{-1} as shown in the inset and published earlier [69]. Also, all measured frequencies are compatible with the previously published HREELS data [62]. We note at this point that the oxide film was grown using $^{16}O_2$, i.e. before the adsorption experiments were carried out. If the oxide film is grown using $^{18}O_2$, the infra-red spectra after adsorption of oxygen $^{16}O_2$ or $^{18}O_2$, respectively, did not show any difference in comparison to the spectra from the $Cr_2{}^{16}O_3$ film. On the basis of the TD-spectra not shown here it is quite obvious that the broad features observed below 150 K to 220 K are due to molecular adsorbed O_2. In fact, it is probably the feature with maximal intensity at 965 cm^{-1} ($^{16}O_2$) and 905 cm^{-1} ($^{18}O_2$), becoming more prominent in the IRA-spectra after heating to 150 K, that leads to a strong desorption feature at 300 K. The higher frequency peak at 990 cm^{-1} ($^{16}O_2$) and 930 cm^{-1} ($^{18}O_2$) could be associated with the O_2 molecules desorbing at lower temperature, indicative for a smaller interaction with the substrate. The shift can also be due to the change in intermolecular interaction. The adsorbed molecular species show a stretching frequency rather different from the gas phase value (1554.7 cm^{-1}) [85]. The frequency shifts with respect to the gas phase are similar to those observed for peroxo- or superoxo-compounds [100, 101], nominally containing an O_2^{2-} or an O_2^- moiety. While Cr 2p XPS data indicate a charge transfer from the surface [62] to the O_2 moiety, this charge transfer has to be rather small as revealed by electronic ELS data [69]. The data are compatible with an O_2 molecule moderately (covalently) interacting with the substrate and inducing only a small charge transfer. This would also be consistent with an oxygen molecule that leads to an O-O vibrational absorption in the region around 1000 cm^{-1}. It is important to note, of course, that upon heating above room temperature the transitions in the EL-spectrum (see [68, 69]) disappear but the Cr^{3+} surface excitations do not reappear indicating that there is still surface oxygen present. Effectively, the oxidation state of the surface chromium ions has increased after dissociation. This surface oxygen gives rise to the sharp feature in the IRA-spectra. The observed frequency is typical for Cr=O bonds as found in chromyl compounds [102]. The isotope shift for Cr=O is smaller as compared with the isotope shift for $^{18}O_2$ / $^{16}O_2$ adsorption. The fact that for certain conditions the Cr-O vibrational mode is split (see Fig. 3) is compatible with two different mono-oxo Cr=O species on the surface, whose relative concentrations would depend on preparation conditions. However, it would also be consistent with a CrO_2 species as in CrO_2Cl_2 with two Cr=O bonds at one chromium ion. In fact, the observed split is very close to the difference between the in- and out-of-phase coupled stretching normal modes of CrO_2Cl_2 as mentioned above. The present isotopic labelling data reveal instead the existence of single Cr=O bonds as discussed in the following: Consider again Fig. 3 and let us assume for the moment that the vibration at 1005 cm^{-1} is due to ^{16}O-Cr-^{16}O as in CrO_2Cl_2. If we expose a $Cr_2{}^{16}O_3(0001)$ surface with $^{18}O_2$, we would have to expect the formation of ^{18}O-Cr-^{18}O, ^{18}O-Cr-^{16}O and ^{16}O-Cr-^{16}O of the adsorbed dissociated oxygen, because it may exchange with the lattice oxygen. Indeed, as the temperature is increased, the right hand side of Fig. 3 indicates the formation of a second band, but a third band has never been observed under these conditions. The frequency of the first band (1005 cm^{-1}) occurring after adsorption of $^{18}O_2$ is identical to the one found for $^{16}O_2$ exposure. Since we assumed within the present argument that this band is due to ^{16}O-Cr-^{16}O with two Cr=^{16}O bonds, the observation of only two bands would indicate that there is no ^{18}O-Cr-^{18}O species which is unreasonable from a statistical point of view. The position of the expected ^{18}O-Cr-^{18}O mode is shown in Fig. 3. Therefore, we come to the conclusion that dissociation of O_2 leads to the formation of chromyl species

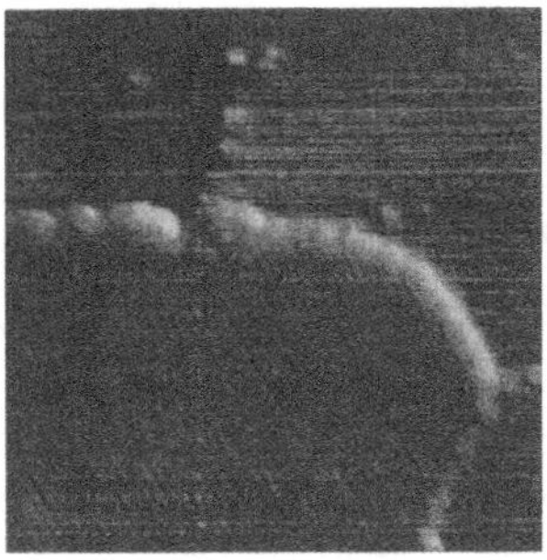

Figure 4. STM image (CCT, constant current topography, 500 Å× 500 Å, +4 V, 2 nA) of the growing Al_2O_3 film (patch with diagonal lines) near NiAl step edges [112].

with single Cr=O bonds on the surface similar to the situation found for polycrystalline samples [103-106].

While oxygen slightly withdraws charge from the surface, alkali is expected to enrich the electron density at the surface. This has, indeed, been observed [68]. Na grows on $Cr_2O_3(0001)$ in a Frank-van der Merwe mode very similar to its growth on a metal surface [63]. The full sodium layer is metallic and very effectively screens the strong Fuchs-Kliewer phonons of the dielectric substrate when excited with electrons. At sub-monolayer coverage there are two sodium species on the surface [68], namely an ionic as well as a neutral species, with the neutral species appearing beyond half a monolayer. Once the ionic species is formed, charge is injected into the surface, and the EL-spectra indicate the formation of a Cr^{2+} species on the surface [68]. We have shown [68, 69] that bonding of molecules such as ethene is different on the alkali free and the alkali predosed surface. It may be interesting to note that on the basis of bulk thermodynamic data we would predict a reduction of Cr_2O_3 to Cr metal by sodium. However, this did not happen under any circumstances tried. Earth alkali metals such as Mg, on the other hand, easily reduce Cr_2O_3 to Cr metal via formation of MgO [107]. With Na we may reduce NiO to Ni metal and simultaneously form sodiumoxide [89]. Other, more stable oxides, such as Al_2O_3 may not be reduced by Na. However, Na may be incorporated into the bulk of some oxides [108] as we have observed for Cr_2O_3 [68].

Al_2O_3 in its α–form is isostructural with Cr_2O_3. As mentioned above, a thin Al_2O_3 film may be formed via oxidation of a NiAl(110) single crystal surface. A STM picture of the growing film is shown in Fig. 4. The film is only 5 Å thick as revealed via XPS [109] and Auger data [110]. Eventually, it covers the surface completely growing even across steps of the metallic substrate like a carpet [111]. Such kinds of growth modes are well known from alkali halide epitaxy on elemental semiconductor surfaces [111]. ISS has been used to show that the surface is mainly oxygen terminated [131]. With photoelectron spectroscopy the two dimensional band structure of the film has been

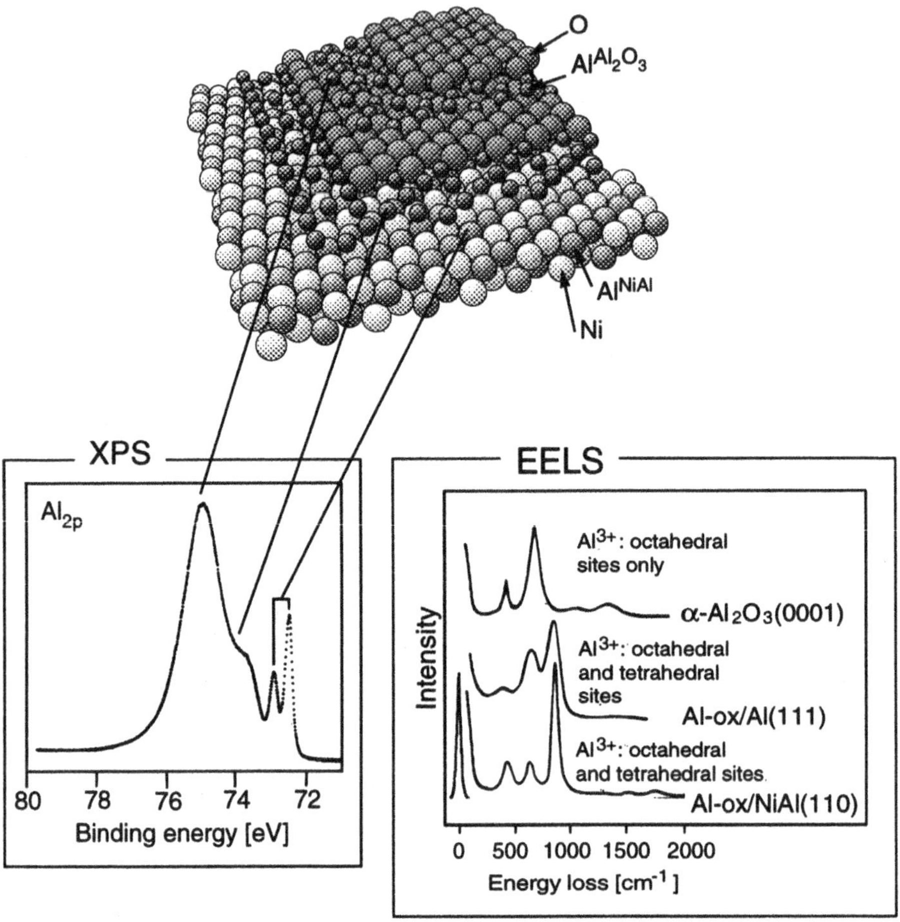

Figure 5. Left: Al 2p photoelectron spectum of Al_2O_3/NiAl(110) (hν = 150 eV, normal emission). Substrate-, interface- and oxide components are clearly resolved.
Right: High resolution electron energy loss spectra of various Al_2O_3 surfaces.

measured [78], and this is compatible with a hexagonal arrangement of oxygen atoms in the oxide film. XPS may be used to identify the coordination of the aluminium ions in the film [109]. Figure 5 shows an Al 2p spectrum of the film. The sharp double peak at lowest binding energy can clearly be assigned to ionization of Al in the alloy substrate. The shoulder at 74 eV turns out to be due to Al located at the interface between the substrate and the oxide film. The broad band at highest binding energy results from ionization of the Al ions sandwiched between the oxygen layers. However, from the spectra it is not possible to differentiate between Al ions tetrahedrally or octahedrally coordinated, because the inherent line widths are too large. Here the phonon spectrum, also shown in Fig. 5, helps in the assignment [78]. The fact that the spectrum exhibits

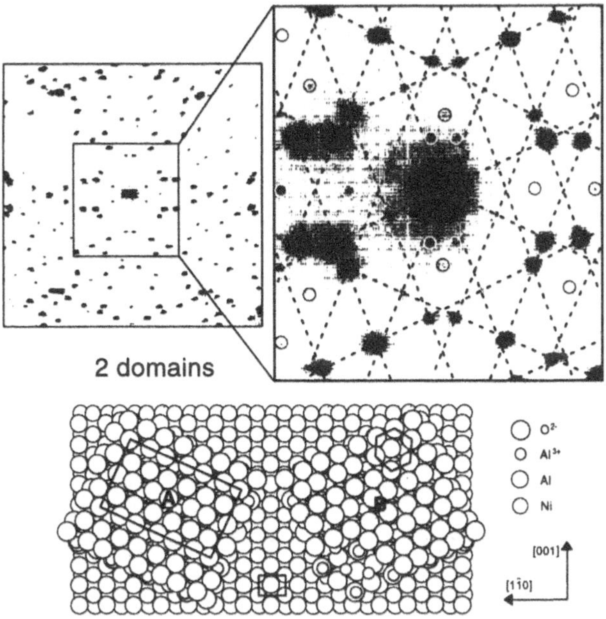

Figure 6. Top: LEED pattern (85 eV) of Al_2O_3/NiAl(110) and close-up of an area around the (00)-spot. In addition to the substrate and oxide spots (the oxide lattice is indicated by dashed lines) the incommensurability of the oxide superstructure gives rise to a number of double diffraction spots indicated in the close-up. Bottom: Structure model of Al_2O_3/NiAl(110), derived from a distorted γ-Al_2O_3 structure. The unit cell as derived from the LEED pattern is indicated.

three phonon bands is indicative of the presence of both tetrahedrally and octahedrally coordinated aluminium ions: α-Al_2O_3 which only contains octahedrally coordinated Al ions exhibits a phonon spectrum with two bands while γ-Al_2O_3 with ions in both coordinations exhibits a three band spectrum similar to the present film. The electron diffraction diagram indicates a very high degree of order in the film as shown in Fig. 6. The LEED pattern may be interpreted to arise from the superposition of two domains which are rotated by 42° with respect to each other [82]. The domains are commensurate with the substrate in the (110) direction and incommensurate in the direction perpendicular to this, i.e. (100). A typical defect on the Al_2O_3 film forms where the two domains intersect. Figure 7 shows a STM picture of such a domain boundary. Another defect observed on the film results from antiphase domain boundaries which form within one domain by a lateral shift of areas with identical rotational orientation on the surface. These antiphase domain boundaries may be detected via a detailed LEED spot profile analysis but can also be directly observed via STM as a white line pattern visible in Fig. 8 [83]. We shall see in the following that these defects may act as nucleation centres for metal particle growth on the surface [112-123]. Unfortunately, the STM topographs shown in Fig. 7 and Fig. 8 do not reveal atomic resolution of the

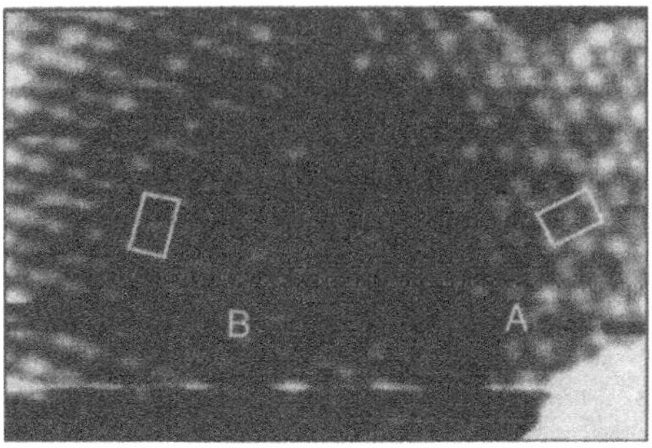

Figure 7. STM image (CCT, 210 Å× 130 Å, -2 V, 0.5 nA) of Al_2O_3/NiAl(110). Domains of orientation A and B can be observed, the unit cells are indicated.

Figure 8. STM image (CCT, 500 Å× 500 Å, 4 V, 0.5 nA) of Al_2O_3/NiAl(110) showing several anti-phase boundaries.

oxide film. The reason is the rather high tunnelling voltage which is neccessary because the smallest energy to tunnel from the oxide density of states into the STM tip is near 4 eV. At such a high voltage the tip has to be retracted from the surface so far that atomic resolution is lost. If we lower the tunneling voltage we may tunnel from the metal substrate underneath the oxide film into the tip [83]. Via this procedure, it is possible to image the interface between the alloy substrate and the oxide film as shown in Fig. 9. With the low tunnelling voltages applied, it is now even possible to achieve atomic resolution. The schematic diagram indicates how we interpret this topograph. It seems that the substrate accomodates the incommensurate oxide film by slightly shifting every fourth atom in the rows across the surface, oriented according to the angle of rotation of the domain. On the basis of these experimental results, we have a rather detailed picture of the morphology of the alumina film.

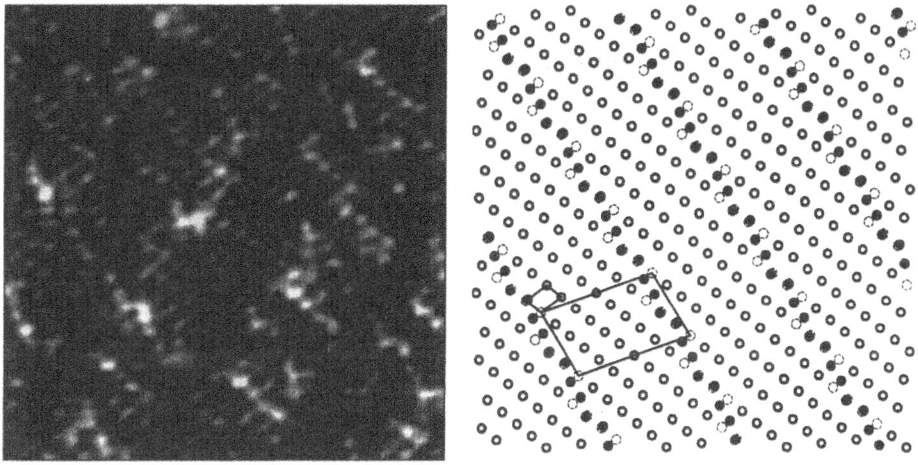

Figure 9. STM image (CCT, 90 Å× 90 Å, -1 V, 1.5 nA) of Al_2O_3/NiAl(110) showing the structure of the NiAl(110) interface in atomic resolution.

With respect to the chemistry of the alumina film, we find that it is rather inert with respect to molecules like CO and O_2 [80, 81]. Figure 10 contains EL spectra of CO adsorbed on the surface at low temperature [80]. The thermal desorption spectra also shown indicate that we need temperatures below 80 K to adsorb appreciable quantities of CO on the surface. The electronic EL spectra exhibit vibrational fine structure in the excited state and excitation energies very close to the gas phase values, indicating that the coupling between the molecule and the surface is rather weak. The coupling, however, is not weak for all molecules. While O_2 behaves similar to CO, other free radicals such as NO, NO_2 and, in particular, di-tert-butyl-nitroxide (DTBN) show pronounced interaction with the surface [124]. We have studied the interaction of the latter moelcule in detail via a newly designed UHV-ESR spectrometer and refer to the literature [124]. At low temperature, we observe a strong coupling of the spin on the free radical probably with Al ions in the surface. This seems at first sight to be in contradiction to the oxygen termination observed via ISS. However, if we consider the strong surface relaxations observed for the Cr_2O_3 surface, where the Cr and oxygen ions are almost in one plane at the surface, it is quite conceivable that a similar effect allows for the interaction between the lower layers and the molecules on top. Therefore, we may state that the question whether the Al_2O_3 surface will participate in the surface chemistry strongly depends on the molecules involved. In this respect, our alumina film behaves very similar to a typical alumina support material used in catalytic modelling. There are, however, differences. In particular, interaction of the film with water is rather weak. In fact, the film is - within the detection limit - free of hydroxyl groups. This, on the one hand, is very unusual because real alumina surfaces always carry hydroxyl groups [78-83]. On the other hand, this is a desirable property

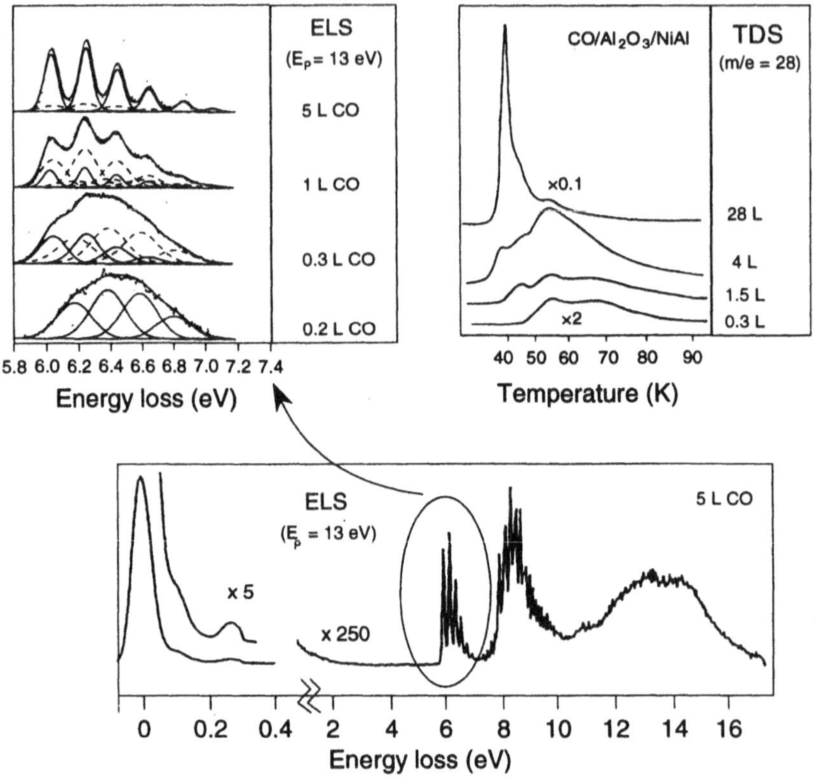

Figure 10. Electron energy loss spectrum for $CO/Al_2O_3/NiAl(110)$ (bottom). Left: Close-up of the region indicated. Right: Thermal desorption spectra for $CO/Al_2O_3/NiAl(110)$.

because it allows us to study directly the influence of adsorbed hydroxyl groups if we could manage to specifically hydroxylate the film. This, indeed, may be achieved as indicated in Fig. 11. At the top (A), the Al 2p spectra of the clean oxide film obtained with electron detection normal to the surface (left hand side) and grazing detection (right hand side) are shown [3, 112]. In the next set of spectra (B), the clean oxide film has been dosed with Al from an evaporation source and covered with less than a monolayer of the metal. The Al signal shows up particularly at grazing detection. The further step (C) involves the treatment of the evaporated aluminium with water in oder to form aluminium hydroxide at the surface leading to the disappearence of the metallic aluminium signal. At the same time there appears a characteristic shoulder in the O 1s spectrum, also shown, which is indicative of the formation of surface hydroxyl groups. The same procedure may also be used to grow a thicker aluminium oxide film, if, instead of water, oxygen is used for oxidation.

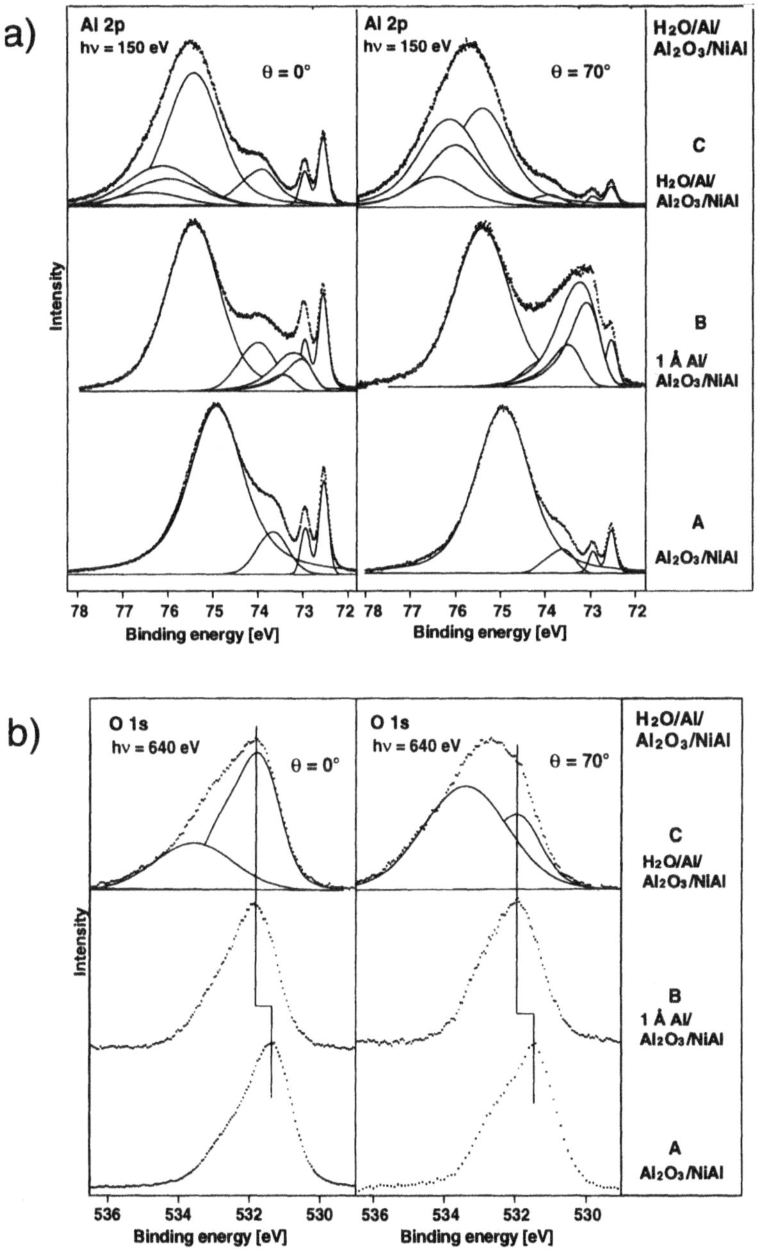

Figure 11. Al 2p (a) and O 1s (b) core level spectra for the chemically modified Al_2O_3/NiAl(110) surface. A: Clean Al_2O_3/NiAl(110). B: Deposition of 1 Å Al at 300 K. C: After H_2O exposure at 90 K and subsequent heating to 300 K.

4. Morphology and Electronic Structure of Metals on Oxide Surfaces

By deposition of additional chemical components, we may start to modify the above discussed oxide surfaces. We focus here on modification by evaporation of ultrathin metal films. As alluded to in the introduction we may consider such systems as model systems for disperse metal catalysts. Typical examples for supported metals are those where a transition metal is deposited onto a support such as SiO_2, Al_2O_3 or MgO. MgO [51, 125, 129] and Al_2O_3 [78-83, 130-132] have been prepared as well-ordered thin film supports. As far as the preparation of well-ordered thin SiO_2 films is concerned, there are still problems. We have concentrated on the preparation of alumina films of the type discussed in the previous chapter including the possibilities of chemical functualisation before metal deposition. In the following, we shall dwell on growth, morphology and electronic structure of such films.

Figure 12 shows a set of STM pictures of metal deposits onto the clean, chemically unmodified alumina surface. For the picture on the top, 1/3 of a monolayer of Ag has been deposited [114]. In the middle panel and on the bottom almost equal amounts of Rh [133] and Pt [112] have been evaporated onto this substrate, which was at room temperature in all three cases. Obviously, Ag has a very high mobility at room temperature and, consequently, rather large aggregates form, which nucleate at the very few steps of the alumina film. In the case of Rh, the mobility is still pronounced but less than in the case of Ag. The metal moves across the alumina terraces but only reaches the antiphase domain boundary defects. Pt exhibits the lowest mobility among the three examples, which is documented by its nucleation directly on the terraces. The relevant quantities that determine the adsorption energy (E_{ads}) of the metal on the substrate are connected via the classical Young equation [134-136]:

$$E_{ads} = E_{oxide/gas} + E_{metal/gas} - E_{oxide/metal} \tag{1}$$

where the first two terms referring to the oxide/gas and metal/gas interactions may be estimated, while the third term, referring to the metal/oxide interaction, is generally not known. Whether metal particles move at a given temperature, now strongly depends on E_{ads}. If $E_{ads} < 0$, the metal wets the surface, and we observe a layer-by-layer or Frank-van der Merwe growth mode. Provided that $E_{ads} > 0$, we find the formation of three-dimensional islands (Volmer-Weber-growth). Sometimes an intitial layer-by-layer growth mode is followed by three-dimensional growth at a given layer thickness. This is called a Stranski-Krastanov growth mode [136]. Of course, as is documented by Fig. 12, the detailed defect structure of the substrate is as important for the finally observed morphology as the strength of interaction with the ideal substrate. While in the case of Ag only the steps on the alumina surface exhibit sufficient interaction strength at room temperature to nucleate Ag particles, the interaction of Rh with the

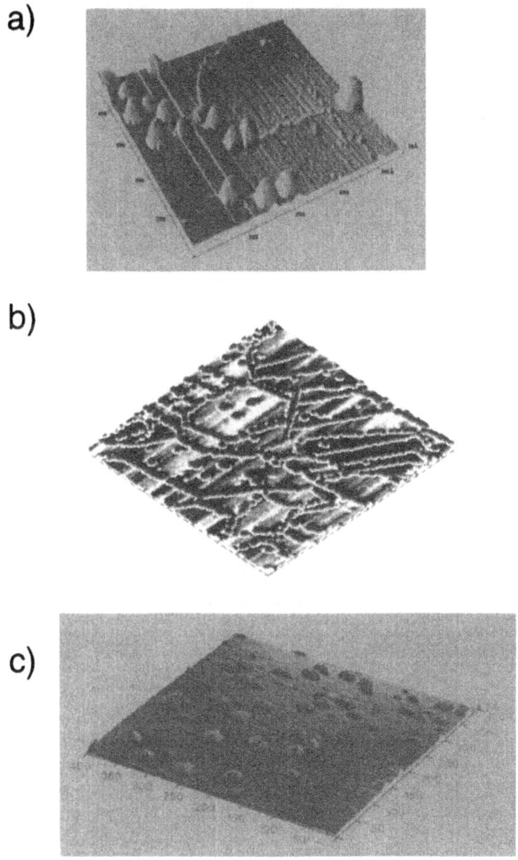

Figure 12. STM images of Al_2O_3/NiAl(110) after metal deposition:
a) Growth of Ag at 300 K (CCT, 900 Å× 900 Å, -0.6 V, 0.5 nA).
b) Growth of Rh at 300 K (CCT, 1500 Å× 1500 Å, +3 V, 1.6 nA).
c) Growth of Pt at 300 K (CCT, 400 Å× 400 Å, -4 V, 0.8 nA).

antiphase domain boundaries is strong enough to lead to nucleation at those defects [113, 126]. Another example, which we want to discuss in more detail in the following, shows nucleation and growth modes being intermediate between Pt and Rh. Figure 13 shows STM topographs of Pd deposited at 90 K (Fig. 13a) and at room temperature (Fig. 13b) [118]. Pd deposited at 90 K decorates the antiphase domain boundaries but also nucleates on the terraces, probably at point defects. In the case of deposition at room temperature, considerably larger aggregates are formed due to the higher mobility in the deposition process. There is no particular preference of formation at defects or on the terraces. The aggregates imaged in Fig. 13b partly exhibit the shape of small crystallites. The crystallographic orientation of the metal particle planes are revealed from electron diffraction diagrams as shown in Fig. 13c [118]. The oxide reflexes are

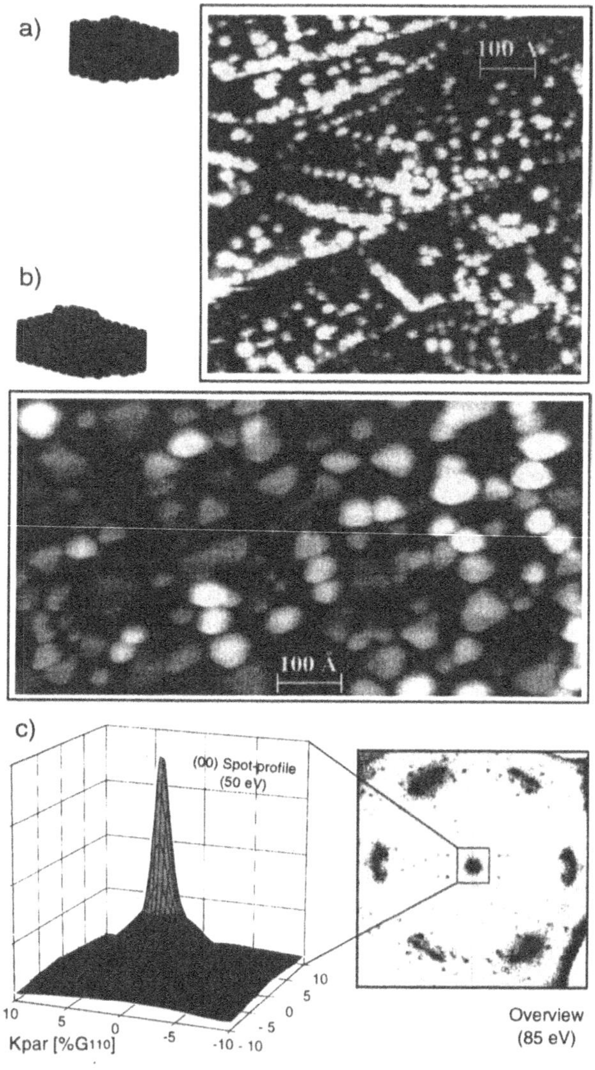

Figure 13. Pd growth on Al_2O_3/NiAl(110): a) STM image after Pd deposition at 90 K (CCT, 1.6 V, 2.2 nA). b) STM image after Pd deposition at 300 K (CCT, +0.4 V, 0.5 nA). c) LEED pattern after deposition of 20 Å Pd at 300 K (right) and intensity profile of an area close to the (00)-spot (left).

superimposed by two diffuse, but clearly visible, hexagonal (111) oriented LEED patterns which are rotated by 12° with respect to each other. The two rings result from growth on the two domains of the alumina substrate. An analysis of the LEED spot profiles yields detailed conclusions on the average metal island size, the average island separation, as well as of the island density and the number of atoms per island. Figure 14 shows as an example the (0,0) profile after Pd deposition at both temperatures.

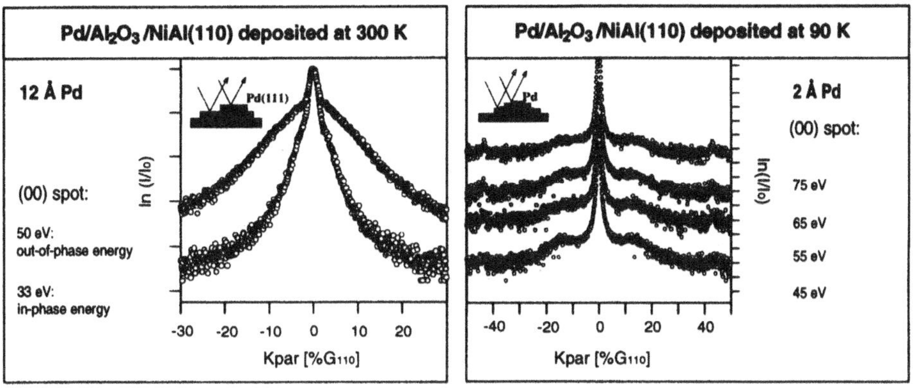

Figure 14. LEED spot intensity profile of the (00)-reflex at various electron energies after deposition of Pd on Al_2O_3/NiAl(110). Left: Pd growth at 300 K. Right: Growth at 90 K.

Shoulders are particularly emphasized if a logarithmic intensity scale is chosen. From the shape and the intensity variation of the shoulder as a function of the electron energy, a rather well documented analysis [137, 138] allows the desired information to be extracted. Figure 15 shows the island size as a function of the amount of deposited material as determined via a quartz microbalance. For 300 K, the island size reaches a plateau after a relatively steep ascent. At this point, equilibrium between nucleation, diffusion and growth is reached. Only if more than 12-15 Å material is deposited, the average island size increases via coalescence of smaller islands very rapidly. The islands then have a regular shape with well-ordered crystallographic planes. This shows up in the energy dependence in Fig. 14 (left hand side), where clear in-phase and out-of-phase relations among the scattered electron beams from the various terraces of the growing aggregate are detected. Such energy dependences are not found for the deposit prepared at 90 K (right hand side). Obviously, island growth is considerably different from the room temperature case. Probably due to the reduced mobility during evaporation, the typical merging into a plateau region cannot be detected. Figure 15 compares the island densities and numbers of atoms per island and shows the expected strong increase in island density in the range of 10 Å layer thickness for room temperature deposition. Apparently, many of the islands do not form at the beginning of the growth process by heterogeneous nucleation at preferential sites but rather at a later point in the process by homogeneous nucleation or nucleation at weakly preferential sites. The analysis of the number of atoms per island leads to similar conclusions. Note that even at low coverages the formation of aggregates with more than 1000 atoms is found. At 90 K deposition temperature, the situation changes significantly. We observe a high island density of (10^{13} cm^{-2}) already at very low coverages. Coalescence sets in at an early stage and leads to a steady decrease of island density. Consequently, the number of atoms per aggregate varies over a large range, indicating that under these circumstances very small as well as very large islands are accessible.

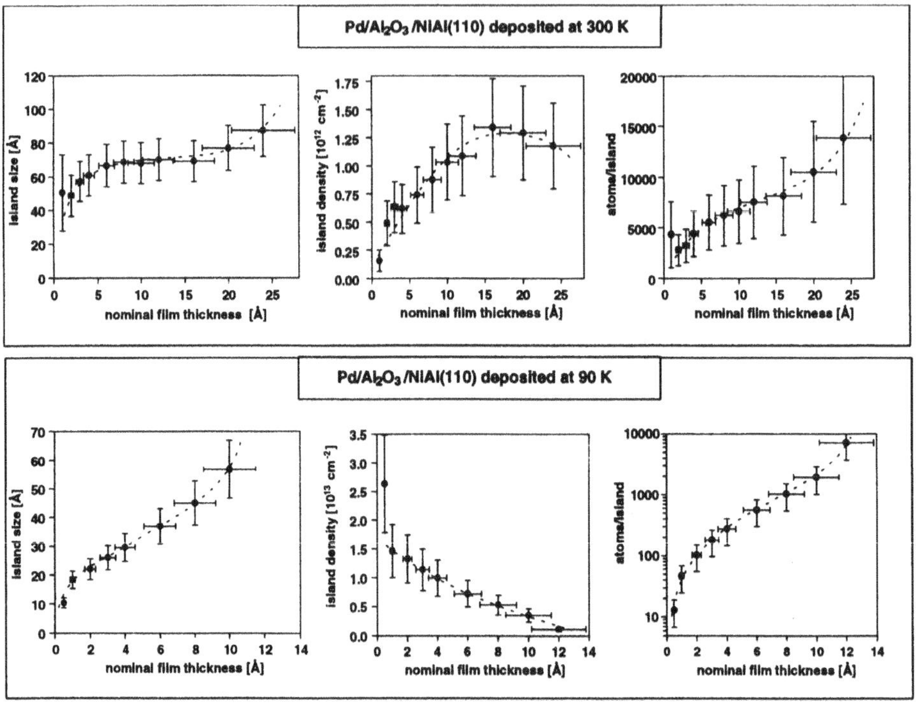

Figure 15. Dependence of several characteristic quantities for Pd growth on nominal metal coverage, as determined by LEED spot-profile analysis. Top: Average island size, island density and average number of Pd atoms per island for growth at 300 K. Bottom: Corresponding quantities for growth at 90 K.

We have investigated several different metal deposits along these lines. As pointed out above, Rh shows a similar growth mode as Pd except that it exhibits a strong tendency for nucleation at particular defects, namely at the antiphase domain boundaries of the oxide film (see Fig. 12) [133]. In Fig. 16 a LEED spot profile is plotted for a Rh deposit [113], showing the anisotropic distribution of the intensity in the shoulder as a function of direction within the Brillouin zone. The intensity maxima point in the two domain directions in line with the alignment of the antiphase domain boundaries which are decorated by the deposited metal.

A comparison of the reflex profiles found for the various metal deposits, as shown in Fig. 17, allows us to summarize the various growth modes [113]: Characteristic for Ag is the formation of rather large aggregates nucleating at the steps of the alumina film, leading to a profile where the central reflex and the shoulder are hardly discernible. Even at 90 K, the large diffusion length of Ag leads to large islands, and we note that this behaviour is very similar to the one observed for Ag on bulk single crystal surfaces [139]. The other extreme case is realized for Pt deposits: A very pronounced shoulder is observed at 90 K as well as at 300 K due to the strong interaction between metal and support leading to very small islands. Pd and Rh are intermediate cases with

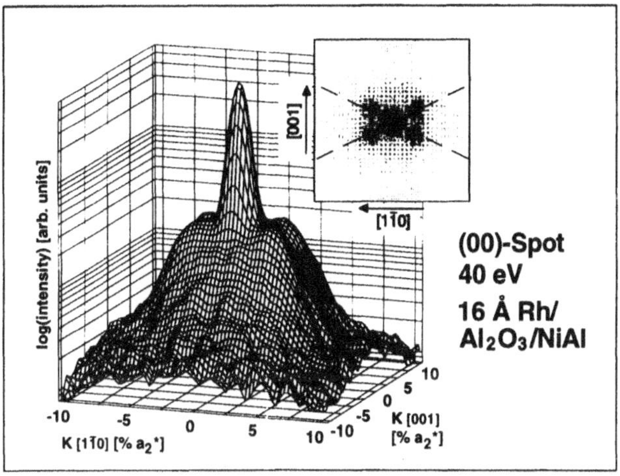

Figure 16. Two-dimensional intensity profile of the LEED (00)-spot for Al_2O_3/NiAl(110) after deposition of 16 Å Rh at 300 K. The orientation of the lattice vectors of both oxide domains is indicated in the inset.

Rh interacting slightly stronger with the substrate than Pd. In both cases, changes in the reflex profiles are observed as a function of temperature. With this comparison, we are in a position to order the metals with respect to their interaction with the substrate:

$$Pt > Rh \geq Pd > Ag.$$

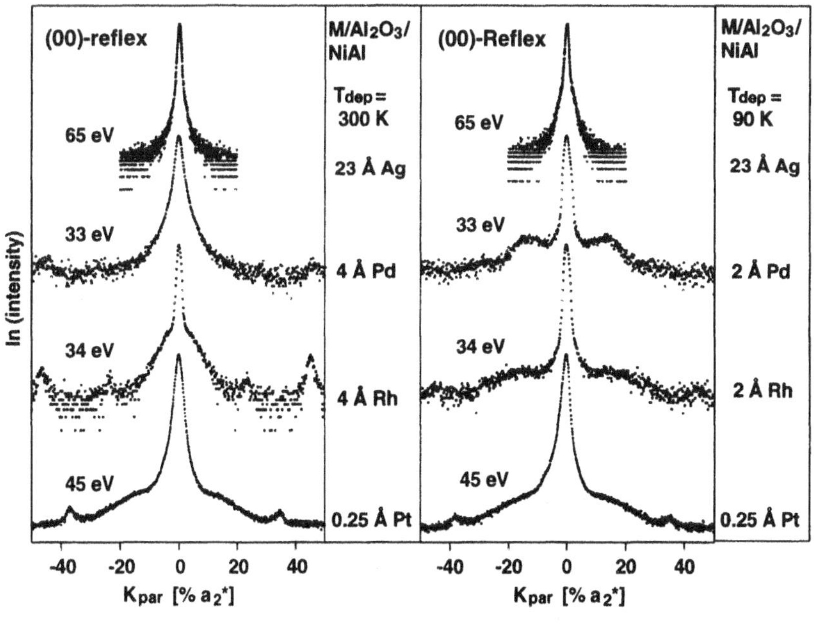

Figure 17. Comparison of characteristic intensity profiles of the (00)-LEED reflex for Ag, Pd, Rh and Pt growth on Al_2O_3/NiAl(110), each at 90 and 300 K.

TABLE 1.
Enthalpies of formation of oxides [140]

ΔH^0_f	Oxide
-12.1 kJ/mol	AgO
-85.4 kJ/mol	PdO
-90.1 kJ/mol	RhO
-172.0 kJ/mol	PtO_2

As alluded to above, it is very difficult to explain this observation on the basis of microscopic considerations. However, a hint for the observed order may be deduced from the comparison of enthalpies of formation of the corresponding transition metal oxides. They are given in Table 1, and it is clear by inspection that the enthalpies of formation increase in the same sequence as the apparent interaction strengths observed via the present morphological studies. In other words, it might be the affinity towards oxygen that dominates the interaction. Nevertheless, we would like to stress that this is pure speculation at this point.

It has been shown above how the morphology of the transition metal deposits varies as a function of interaction strength with the substrate. This was demonstrated by changing the metal deposited. Another possibility to vary the strength of interaction is to modify the surface chemistry of the substrate. We are only at the beginning of such studies but we would like to briefly mention some results. After hydroxylation of the film - applying the procedure discussed above -, the interaction with Rh for example, deposited at 300 K, changes considerably. In particular, photoelectron spectroscopy of the Al 2p and O 1s emissions indicates characteristic changes, which are interpreted via specific interactions between the deposited metal and the hydroxylated areas on the surface [113]. Through the relatively strong interaction between Rh and the surface hydroxyl groups, a highly dispersed Rh film is formed whose morphology is more similar to the film deposited on the clean alumina film at 90 K. Of course, the change in morphology is a consequence of the changes in the interaction between Rh and the modified substrate, and this in turn shows up in variations of the electronic structure. We come back to this in connection with the study of CO adsorption on such deposits.

At this point, we have an idea of the morphology of the metal deposits. Next, electronic structure as revealed via photoelectron spectroscopy should be discussed. Figure 18 collects some Pd 3d spectra of deposits with different average island size [119-121].

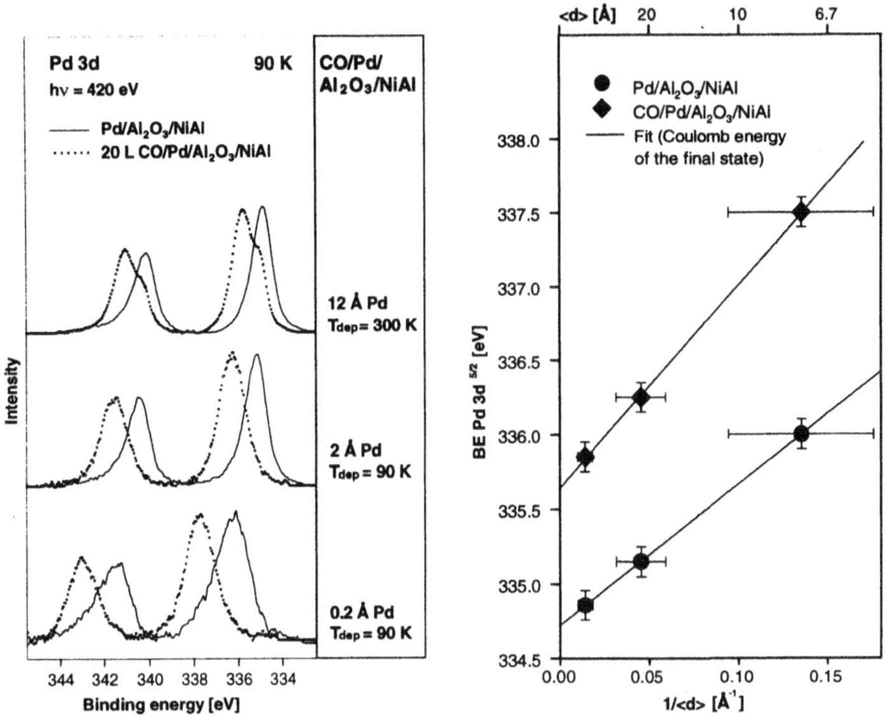

Figure 18. Left: Pd 3d core level photoelectron spectra for various Pd deposits on Al_2O_3/NiAl(110) before and after CO saturation. Right: Dependence of the binding energy of the Pd 3d surface component on reciprocal particle size for the clean and CO saturated metal films.

Two main effects are observed: Firstly, a shift to lower binding energy with increasing island size and, secondly and simultanously, a strong decrease in line widths is found. There may be several contributions to these effects, such as charge transfer, as an example for initial state effects, as well as final state effects [141-143]. An analysis of the line widths turns out to be particularly difficult [141, 142, 144-147]. At this time, we can only speculate that the inequivalence of different atoms within the aggregate combined with final state effects are the main contributions. In contrast to this, the interpretation of the binding energy shifts is more straightforward in the sense that it is possible to isolate final state contributions [143, 148]: The screening of the final state ionic charge by the valence electrons of the metal aggregate yields very characteristic binding energy shifts. While the positive charge is completely delocalized within a surface of a volume crystal, the delocalization is limited within a metal island insulated from the metallic substrate. In such a case the charge distribution depends on the mean radius of the island according to:

$$E_b \sim R^{-1} \qquad (2)$$

and thus shifts the binding energy to larger values for decreasing island radii. There are several experimental observations that are in line with such an interpretation [141-144]. One is shown in Fig. 18, where the observed binding energies are plotted versus the inverse radius with perfect scaling. Therefore, we conclude that final state effects dominate the shift. If charge transfer processes in the initial state would dominate, we would expect rather strong deviations from the 1/R dependence. In addition, we would expect chemical shifts of the support levels which, however, are rather small. The existing effects can be interpreted as weak band bending effects and are of the order of 10% of the binding energy shift. The conclusion reached above, as to the dominance of final state effects connected with electrical insulation of the metal islands on the support, can be substantiated through another interesting observation. If the final state shifts were not caused exclusively by the electrons of the metal island, but also by electrons tunnelling through the thin alumina film from the metallic substrate, the final screened state would be more stable compared to the situation where screening is provided by the metal island electrons alone due to the higher degree of delocalization in the former case. Therefore, the spectroscopy of the core hole decay allows us to investigate this question further [123]. However, we shall postpone this discussion to a later point in this paper, because it is advantageous to use the decay spectra of CO adsorbed onto the metal islands to study this question in more detail.

The spectroscopy of valence electrons has also been carried out for a variety of deposited metals [113]. Without going into the details at this point, we state that for low temperature deposition of metals a Fermi edge develops at a considerably higher metal coverage as compared to the room temperature preparations. The reason, of course, is the formation of larger aggregates at room temperature due to the larger diffusion coefficient. Moreover, for the larger aggregates the development of a band structure can be observed in angle resolved photoemission experiments. The formation of a band structure shows up as a dispersion of the metal levels. Quantum size effects expressing themselves as discrete level structures have so far not been observed [113].

5. Reactivity of Oxide-Supported Metals

The appropriate starting point to discuss reactivity of oxide supported metals is a detailed analysis of the adsorption characteristics of small molecules, as for example CO, for which detailed knowledge from investigations on compact metal surfaces exists [149-152]. In photoelectron spectroscopy, the binding energy shifts as well as the line shapes of the Pd 3d ionizations change considerably upon adsorption of CO as shown in Fig. 18. The spectrum after saturation exposure of CO onto the largest Pd islands (average diameter 70 Å) at 90 K is very similar to the spectrum encountered for saturation coverage of the Pd(111) metal surface [153-155]. The Pd 3d signal of the clean Pd(111) surface contains a surface and a volume component which are shifted with

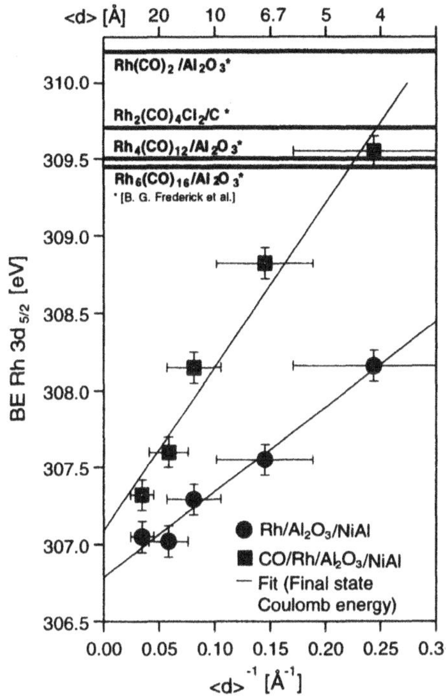

Figure 19. Particle size dependence of Rh 3d binding energies for clean and CO saturated Rh films.

respect to each other by the surface chemical shift of -0.3 eV [153]. This shift is below the present resolution. CO adsorption leads to a shift of the surface component by about 1.1 eV towards higher binding energy allowing us to separate volume and surface contributions after adsorption. The volume contribution amounts to 40% in this case, decreases to 13% for an average island size of 22 Å (2 Å, 90 K) and is below detection limit (by deconvolution of the spectra) for an average island size of 7.5 Å (0.2 Å, 90 K). Also, the CO induced binding energy shift of the Pd 3d levels increases with decreasing island size to 1.5 eV. We know from investigations of single crystal surfaces that the observed binding energy shifts depend on the number of coordinated CO molecules. A detailed analysis of the present data, together with thermal desorption data, led us to the supposition that for smaller aggregates the number of CO molecules coordinated towards a Pd metal atom increases from one to two [113]. Of course, the number of CO molecules coordinated towards a metal atom in carbonyl compounds is never reached. Nevertheless, in those cases where it is possible to compare metal ionizations in carbonyl compounds with those of deposited metal aggregates, the observed metal ionization energies in carbonyl compounds are consistent with those of the smallest deposited aggregates [156]. For Rh ionizations, Fig. 19 shows a comparison. The binding energies determined for the smallest aggregates,

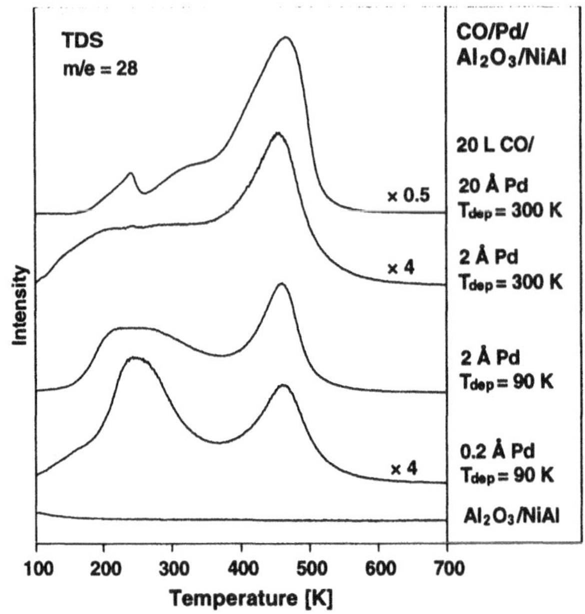

Figure 20. Thermal desorption spectra of CO adsorbed on $Pd/Al_2O_3/NiAl(110)$ (CO exposure at 90 K, heating rate ~1.5 Ks^{-1}). The desorption spectra shown in the plot are arranged in order of decreasing average particle sizes from top to bottom.

indeed, are compatible with those measured for polynuclear metal carbonyls [156]. Only the $Rh(CO)_2$ species adsorbed on alumina surfaces exhibits with 310.2 eV an even higher binding energy [156]. Due to the different substrates in those cases, the comparison has to be taken with care. In general, the observed shifts scale according to a 1/R dependence similar to the clean aggregates. The different shape results from the effectively reduced number of electrons within the metal particle. Those electrons are involved in the CO-metal bond and are not available for screening.

Further information about the electronic structure of the CO adsorbate systems is of course gained via C 1s ionizations of adsorbed CO as well as via the study of valence ionizations. In addition, thermal desorption spectroscopy is very useful. Before we further analyse the photoelectron spectra, we consider the TD spectra.

In Fig. 20 several CO TD spectra of various Pd deposits are compared [121]. As expected for large Pd deposits, we find a TD spectrum similar to the one known from the single crystal surface [157]. The small, relatively sharp feature at 240 K desorption temperature is due to hydrogen contamination of the material [113]. With decreasing size of the deposited islands, a desorption maximum between 250 K and 300 K appears. Rearrangement of the metal deposit during the heating cycles is less important as revealed by a corresponding experiment, where a CO saturated deposit has been warmed to 300 K, followed by another CO saturation cycle at 90 K and a second TD

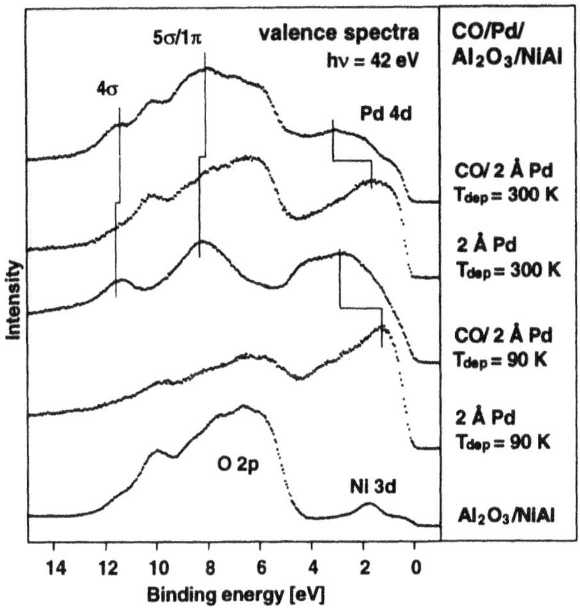

Figure 21. Comparison of valence photoemission spectra for clean and CO saturated Pd deposits on Al_2O_3/NiAl(110) grown at 300 and 90 K.

experiment. The second TD spectrum is almost identical to the first one, indicating that the desorption at lower temperature, i.e. between 250 K and 300 K, is an inherent property of small Pd deposits.

We now turn back to the electronic properties as revealed by photoemission. Figure 21 gives an impression of the changes induced by CO adsorption in the valence photoelectron spectra of the metal deposits [121]. We recognize at 11 eV and 8 eV binding energies the CO 4σ and CO 5σ/1π emissions, respectively. The considerably higher intensity of the adsorbate induced features for the deposit at 90 K as compared with the 300 K deposit may be interpreted - on the basis of the growth behaviour - as a larger surface contribution in the former case [113]. Also, considerable changes are induced in the region of the Pd 4d emissions. The size dependent effects are clearly apparent. Due to the small volume contribution for the low temperature deposition, CO adsorption leads to much larger changes near the Fermi energy for the 90 K deposit. A more detailed analysis has to await further experiments.

The C 1s spectra of the CO adsorbates are shown in Fig. 22 and are typical for molecular CO. Therefore, it is appropriate to compare the spectra of the metal deposits with those of carbonyl compounds [158-160] on one hand and with an adsorbate on a Pd(111) surface [161] on the other hand again. The weak fine structure of the surface adsorbate may be traced back to the co-existence of several adsorption sites. The C 1s spectrum taken for large Pd deposits is very similar to the adsorbate on the extended

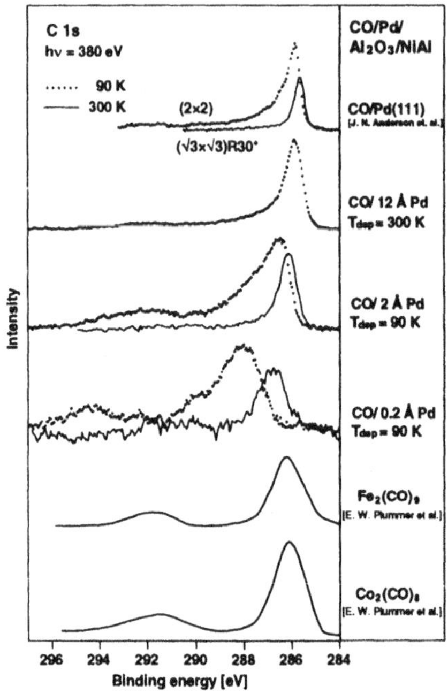

Figure 22. C 1s core level photoelectron spectra for $CO/Pd/Al_2O_3/NiAl(110)$ after CO saturation at 90 K and after heating to 300 K. Additionally, C 1s spectra of two ordered CO superstrutures on Pd(111) as well as spectra of two transition metal carbonyls are shown for comparison.

surface. As we turn to smaller metal islands, we find shifts to higher binding energies, increase of line width and the occurrence of an intense, so called shake-up satellite structure [159, 162-168]. After desorption of the weakly adsorbed CO by heating to 300 K, we find a shift to lower binding energies, decrease of line widths as well as asymmetry and shake-up satellite intensity. Shake-up satellites are a consequence of the complex many particle processes occurring during photoemission and have been discussed in the literature quite extensively [159, 162-168]. The satellites are due to the population of various ion states in the ionization process. The intensity in the spectrum strongly depends on the CO-metal bonding [158, 159, 168]. A detailed analysis shows that charge transfer satellites gain relatively large intensities if the CO-metal coupling is weak. This means that, in agreement with the TDS experiments, the change in the satellite intensities is indicative of a considerable change of the electronic structure as a function of aggregate size: While the strength of the chemisorption bond for small aggregates is sensitive to coverage, this is to a much lesser extent true for larger aggregates and single crystal surfaces. Due to the lower degree of ordering in the smaller aggregates, we have to expect a wider distribution of adsorption geometries and we believe that this fact contributes to the observed larger line widths. However, other contributions are expected [166, 169-171] but will not be discussed here. The observed

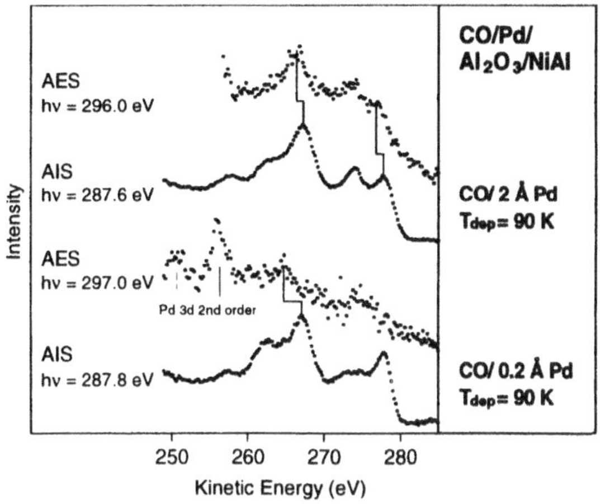

Figure 23. Comparison of autoionisation and Auger spectra for $CO/Pd/Al_2O_3/NiAl(110)$.

shifts of the binding energies are, as discussed for the Pd ionizations, mainly caused by final state effects, i.e. screening via charge transfer from the metal aggregates [142]. As announced above, we may use the CO core hole decay spectra to show that the screening charge only flows from the metal island and not from the metal substrate underneath the alumina film. Since a state with a core hole is highly unstable, it will rapidly decay for example by electron emission. By comparing the decay of the neutral core excited state, i.e. autoionization (AI) [172, 173] spectrum, with that of the core ionized state (the Auger electron (AE) spectrum) as shown in Fig. 23, the screening dynamics occurring within the core hole lifetime (typically 10^{-15} s) can be studied [123]. For a free CO molecule for example, the two hole (2h) states in the AE spectrum are shifted towards lower kinetic energies relative to the two hole one particle (2h-1p) states (spectator peaks) in the AI spectrum due to the absence of screening from the spectator electron in the $2\pi^*$ orbital. Thus, the AI and AE spectra are fundamentally different. For CO chemisorbed on a metal, on the other hand, the strong coupling of the $2\pi^*$ orbital with the metal states quenches this difference, leading to AI and AE spectra which are nearly identical. Comparing the AE and AI spectra of the small particles obtained at 90 K, we indeed find a "spectator shift" which increases as the particle size decreases. This shows that there is also no detectable charge transfer to the clusters occurring during the core hole lifetime.

In contrast to the photoemission experiments, which are dominated by final state effects, the x-ray absorption spectra are much less influenced by such effects as revealed by the spectra shown in Fig. 24 [113]. We consider the excitation of a C 1s or O 1s electron into the antibonding $2\pi^*$ orbital of CO [174-178]. Since the $2\pi^*$ electron remains on the molecule, the $2\pi^*$ electron itself screens the core hole. Therefore, while

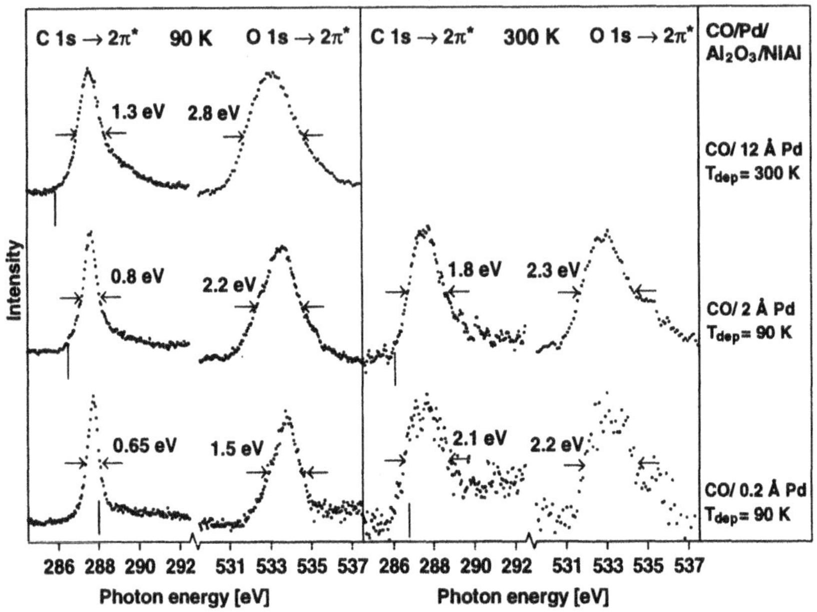

Figure 24. X-ray absorption spectra of the C 1s and O 1s → 2π* resonances for CO/Pd/Al_2O_3/NiAl(110).
Left: After CO saturation at 90 K. Right: After heating to 300 K.
The corresponding XPS binding energies relative to the fermi level of the NiAl substrate are marked with lines.

in the case of the core ionization the binding energy shift is strongly depending on screening, mainly provided by the electrons of the metal aggregate and thus on particle size, the x-ray absorption signal depends much less on these parameters. However, the signal is strongly influenced by the strength of the CO-metal bond as the following argument illustrates: The synergetic σ-donor-π-acceptor bonding mechanism leads to a strong interaction between CO-π-orbitals and occupied Pd-4d-orbitals. Since the 4d orbitals form a band, the interaction with the CO-2π orbitals lead to a CO-2π* density of states distribution whose width depends on the CO-metal interaction. Systematic investigations of adsorbates on single crystal surfaces have shown that life time of the excitation, vibrational excitation and intermolecular interaction play a less significant role for the x-ray absorption spectra [178]. Two effects dominate: Firstly, the width of the resonance increases with increasing chemisorption bond strength while the morphology does not change, and, secondly, it also increases with increasing coordination of the CO molecule, i.e. when the CO molecule changes site from a top to bridge to hollow site. These results may now be used to interpret the spectral changes observed in Fig. 24. While the ionization potential, indicated by the bar in the C 1s x-ray absorption spectra, shifts to larger values with decreasing aggregate size, the 2π* resonance stays constant in energy. Simultaneously, the resonance becomes narrower as expected for decreasing chemisorption energy. The same is true for the O 1s → 2π* resonance. After heating the sample from 90 K to 300 K the resonances become wider, in full agreement with the above discussion. Therefore, it is quite clear that also the x-ray absorption spectra indicate a stronger CO-metal bond for smaller CO coverage.

In fact, all applied experimental methods point for the Pd/Al_2O_3 system in the regime of low CO coverage towards a strong CO-metal bond for all island sizes. For small island sizes, however, the CO bond strength decreases considerably with increasing CO coverage, and it is for the smallest aggregates only situated in the range of a weak chemisorptive bond. It is now rather interesting to compare the results for the Pd/Al_2O_3 system with those for the Rh/Al_2O_3 system. Both, TDS as well as x-ray absorption indicate in this case that the CO adsorption properties are much less sensitive to the size of the aggregates. Only the smallest aggregates show a very weak dependence. A possible explanation involves different CO adsorption geometries on the smallest aggregates, because it is well known that Rh tends to form geminal dicarbonyl species [156]. This is also in agreement with investigations [179-182] according to which Rh restructures upon CO adsorption at 300 K which leads to higher adsorption capacities at 90 K. We have followed these processes by a LEED profile analysis coupled with TDS experiments [113].

For Rh deposits, we find a new reaction channel which is not observed in our experiments for the case of Pd deposits [29, 183-186]. Upon heating, the CO adsorbates on the Rh deposits, CO dissociation is experimentally observed. This reaction channel is activated between 300 K and 400 K and can be followed via the C 1s photoelectron spectra as shown in Fig. 25. The carbidic carbon signal due to the presence of atomic carbon on the Rh deposits appears at 284 eV binding energy upon heating. After completing the reaction, only atomic carbon remains, while the molecular signal has completely disappeared. X-ray absorption assures that the CO-$2\pi^*$ resonance has also disappeared [113]. The present Rh model systems, therefore, allow us to study CO dissociation as a function of the size of the aggregates [184-186]. A corresponding series of spectra where systems with different average aggregate sizes were subjected to the reaction are shown in Fig. 25b. The signals recorded at 90 K after CO saturation are compared with those after heating to 400 K and 600 K, respectively. If we consider the ratio of molecular CO intensity with respect to atomic carbon intensity for small aggregates versus larger aggregates, we realize that this ratio decreases as the aggregate size increases. A more detailed analysis [113] of the data also corroborates the conclusion that the smallest aggregates, indeed, exhibit the smallest dissociation activity [184-186]. Several factors contribute to this. One, of course, must be connected with the variation of the electronic structure of the aggregate as a function of the size. However, there are other effects, one of which may be a simple geometric one. The products of the dissociation process, atomic carbon and oxygen, prefer to adsorb in coordination sites of high symmetry. Calculations [187] predict that those sites have to be at a minimal separation. Therefore, the dissociation channel can only be operative if the aggregate has a particular minimal size. Such a process is schematically represented in Fig. 25c.

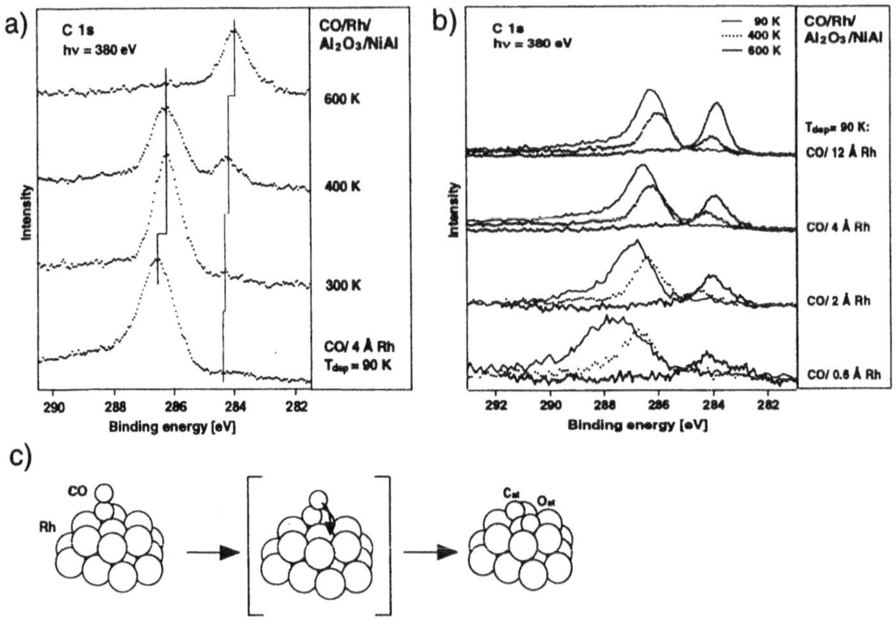

Figure 25. a) Series of C 1s core level photoelectron spectra for Rh/Al_2O_3/NiAl(110) after CO saturation at 90 K and after heating to the temperatures given in the plot.
b) Comparison of the normalized C 1s photoelectron spectra for various nominal Rh coverages after CO saturation at 90 K and after heating to 400 and 600 K. c) Schematic representation of the CO dissoziation process.

We have reached a point where we may correlate chemical reactivity with the size of the deposited aggregates. If we want to understand effects which occur in connection with catalyst preparation, we must try to influence the morphology of the metal deposit by chemical modification and study its consequences for the adsorption and reaction properties [179-182]. While we have already studied chemical modification of the alumina substrate on the morphology of the metal deposit, we now investigate the consequences on CO adsorption. As discussed above, there were indications that the Rh substrate interaction is strongly influenced if the alumina substrate was hydroxylated [188-193]. The main observation was that while on the clean substrate Rh deposits restructure under CO exposure, this is inhibited upon hydroxylation. We have followed the restructuring via CO adsorption at 300 K. In contrast to the non modified substrate where the restructuring shows up in a higher adsorption capacity upon reexposure at low temperature, this effect is not observed on the chemically modified substrate. A strong interaction between metal and support is compatible with these findings. If we finally consider the dissociation activities in the chemically modified system, we find a reduction by about a factor of two with respect to the non modified system. We can trace this back to the formation of smaller Rh aggregates on the hydroxylated substrate, which exhibit, as pointed out above, a reduced dissociation activity. Summarizing at this point, we may state that investigations on model systems

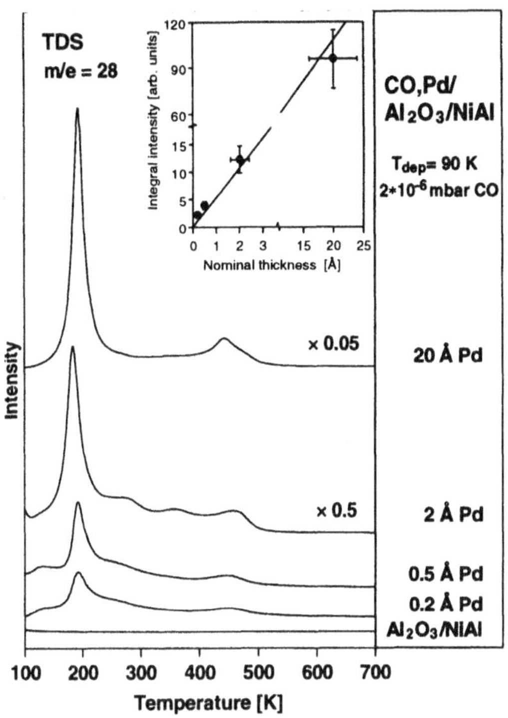

Figure 26. Thermal desorption spectra for various nominal Pd coverages on Al_2O_3/NiAl(110), deposited at 90 K in CO atmosphere (heating rate ~1.5 Ks^{-1}).
The dependence of the integral desorption intensity on nominal metal coverage is shown in the inset.

allow to a certain degree the simulation of typical situations encountered in connection with real catalyst preparation. Our results are in line with the well-documented CO induced Rh redispersion on Al_2O_3 [179-182] and other substrates [194-196]. In these investigations, it was shown that at high enough CO partial pressures a redispersion occurs, accompanied by the formation of geminal rhodium-dicarbonyl species. There are indications that in real systems the dicarbonyl formation also involves hydroxyls on the surface. Further investigations on the model systems, in particular infra-red investigations, have to be carried out.

Another option for chemical modification is to deposit the metal in the presence of a gas atmosphere. Under these circumstances, the adsorbate influences the growth process as a „surfactant" [197-200]. As an example, we have studied the deposition of Pd aggregates in a CO atmosphere [118]. Figure 26 shows TD spectra after deposition of various amounts of metal. The spectra are dominated by a relatively sharp signal at 200 K maximum desorption temperature typical for decomposition of a compound. The weak signals up to 500 K are due to desorption processes from fragments and agglomerates formed via decomposition. The intensity of the compound decomposition is

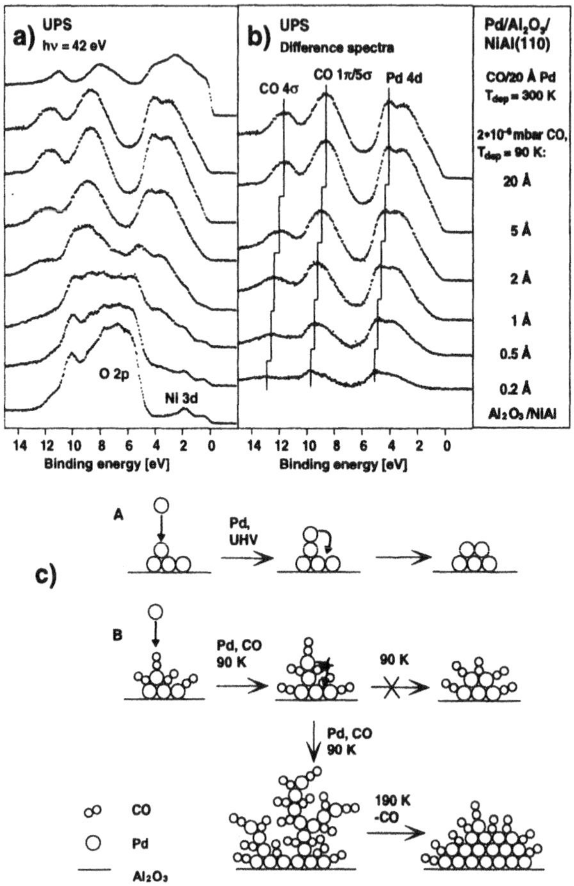

Figure 27. a) Valence photoelectron spectra for successive deposition of Pd in CO atmosphere. A valence spectrum for deposition of 20 Å Pd at 300 K in UHV after CO saturation is shown for comparison.
b) Difference spectra after subtraction of oxide- and substrate-induced emission.
c) Schematic representation of a possible growth model for Pd deposition in CO atmosphere.

directly proportional to the deposited amount of metal as indicated in the inset in Fig. 26. The valence photoelectron spectra as well as the core hole spectra support the claim that a transition metal carbonyl compound is formed [149, 158, 159]. Figure 27 shows spectra of the valence region, that were obtained after stepwise Pd deposition. By calculating the difference spectra, we can suppress the substrate emssion as also shown in Fig. 27. It becomes obvious that, except for a slight shift in binding energy, the spectra show little change over the whole coverage range. The spectra are dominated by the $5\sigma/1\pi$ emission at 8.8 eV and the 4σ emission at 11.9 eV [149, 201] as well as the ligand field splited Pd 4d signals [158, 202] at 4.1 eV and 3.3 eV for a Pd coverage equivalent to 20 Å. With decreasing coverage, the signals shift to lower binding energy by about 1 eV. An interesting observation may be made in the range of the Fermi energy. Even for the highest coverages the density of states in this energy range remains negligable. In agreement with the proposed formation of a compound, we can exclude the formation of large metallic aggregates. Schematically, the growth of the carbonyl compound may be represented as illustrated in Fig. 27c. This model of the formation of fractal, low dimensional structures takes account of the various observations by LEED, XPS and TDS. The formation of such species is quite unexpected because binary Pd carbonyl species in general exhibit rather low stabilities [203-209]. In contrast to a similar study for Pt deposition on TiO_2 [14], where the CO induced changes have been dicussed on a thermodynamic basis, kinetic mechanisms must be active in the present case.

6. Synopsis

It is possible to prepare oxide surfaces with different crystallographic orientations by applying techniques of epitaxial growth. The systems are well structured and allow a variety of spectroscopic investigations. Adsorption of molecules may be studied and some interesting conclusions with respect to catalytically relevant reactions on the oxide surfaces may be drawn. Most relevant, however, the epitaxial oxide films can be used as support materials for the deposition of metals. The morphology and growth modes of various transition metals have been studied on a model alumina surface. These studies also include chemical modification of the substrate and the investigation of the consequences for the growth of metals. Adsorption and reactivity of small molecules have been investigated with a variety of techniques. We have identified CO dissociation as a size dependent process in the case of small Rh aggregates. The influence of chemical modification of the substrate onto such processes has also been considered. Deposition of metals within an adsorbate atmosphere is an interesting option to prepare otherwise hardly accessible systems.

In summary, studies on model systems will continue to allow us to gain interesting insight into processes relevant in the preparation and function of catalysts.

7. Acknowledgements

We are grateful to a number of agencies who have supported our work: Deutsche Forschungsgemeinschaft, Ministerium für Wissenschaft und Forschung des Landes Nordrhein-Westfalen, Bundesministerium für Bildung und Wissenschaft, German -Israeli-Foundation, and Fonds der chemischen Industrie. Also, we would like to thank Anders Sandell, Sascha Stempel and Martin Frank for their contributions.

8. References

1. A.B. Stiles (ed., 1987) *Catalyst Supports and Supported Catalysts*, Butterworth, Boston.
2. B.C. Gates, L. Guczi and H. Knözinger (eds., 1986) *Metal Clusters in Catalysis, Studies in Surface Science and Catalysis*, Vol. 29, Elsevier, Amsterdam.
3. H.-J. Freund *Angew. Chemie*, submitted.
4. G.A. Somorjai and J. Carraza (1986) *Ind. Eng. Chem. Fundam.* **25**, 63.
5. H.-J. Freund and E. Umbach (eds., 1993) *Adsorption on Ordered Surfaces of Ionic Solids and Thin Films*, Springer Series in Surface Sciences, Vol. 33, Springer Verlag, Heidelberg.
6. C. Noguera (1996) *Physics and Chemistry at Oxide Surfaces*, Cambridge Univ. Press, Cambridge.
7. V.E. Henrich and P.A. Cox (1994) *The Surface Science of Metal Oxides*, Cambridge Univ. Press, Cambridge.
8. H.-J. Freund (1995) *Ber. Bunsenges. Phys. Chemie* **99**, 1261.
9. H.-J. Freund (1995) *Phys. Stat. Sol (b)* **192**, 407.
10. H.-J. Freund (1996) Nato Advanced Study Institute NATO ASI Series C, Vol. 232, Kluwer Acad. Press, p. 474.
11. S.K. Purnell, X. Xu, D.W. Goodman and B.C. Gates (1994) *Langmuir* **10**, 3057; (1994) *J. Phys. Chem.* **98**, 4076.
12. M.C. Wu, W.S. Oh and D.W. Goodman (1995) *Surf. Sci.* **330**, 61.
13. D.W. Goodman (1995) *Chem. Rev.* **95**, 523; (1995) *Surf. Rev. and Lett.* **2**, 9.
14. H.-P. Steinrück, F. Pesty, L. Zhang and T.E. Madey (1995) *Phys. Rev. B* **51**, 2427.
15. M.C. Wu and P.J. Møller (1989) *Surf. Sci.* **221**, 250.
16. P.J. Møller and M.C. Wu (1989) *Surf. Sci.* **224**, 265.
17. M.C. Wu and P.J. Møller (1990) *Surf. Sci.* **235**, 228.
18. M.C. Wu and P.J. Møller (1991) in S.Y. Tong, M.A. van Hove, K. Takayanaki and X.D. Xider (eds.), *The Structure of Surfaces III* , Springer Verlag, Berlin, p. 652.
19. M.C. Wu and P.J. Møller (1990) *Surf. Sci.* **250**, 179.
20. Q. Guo and P. J. Møller (1990) *Vacuum* **41**, 1114.
21. P.J. Møller and Q. Guo (1991) *Thin Sol. Films* **201**, 267.
22. P.J. Møller and J. Nerlov (1993) *Surf. Sci.* **307-309**, 591.
23. O. Ge and P.J. Møller (1994) *Appl. Surf. Sci.* **82-83**, 305.
24. Q. Guo and P.J. Møller, *Surf. Sci. Lett.*, in press.
25. A. Ludviksson, K.H. Ernst, R. Zhang and C.T. Campbell (1993) *J. Catal.* **141**, 380.

26. K.H. Ernst, A. Ludviksson, R. Zhang, I. Yoshihara and C.T. Campbell (1993) *Phys. Rev. B* **47**, 13782.
27. E.I. Altman and R.J. Gorte (1988) *Surf. Sci.* **195**, 392; (1988) *J. Catal.* **113**, 185.
28. H. Cordatos, T. Bunluesin and R.J. Gorte (1995) *Surf. Sci.* **323**, 219.
29. E.I. Altman and R.J. Gorte (1989) *Surf. Sci.* **216**, 386.
30. J.-M. Pan, B.L. Maschhoff, U. Diebold and T.E. Madey (1993) *Surf. Sci.* **291**, 381.
31. U. Diebold, J.-M. Pan and T.E. Madey (1993) *Phys. Rev. B* **47**, 3868.
32. J.-M. Pan and T.E. Madey (1993) *J. Vac. Sci. Technol. A* **11**, 1667.
33. C. Duriez, C.R. Henry and C. Chapon (1991) *Surf. Sci.* **253**, 190.
34. C.R. Henry, C. Chapon, C. Duriez and S. Giorgio (1991) *Surf. Sci.* **253**, 177.
35. C.R. Henry, C. Chapon, C. Goyhenex and R. Monot (1992) *Surf. Sci.* **272**, 283.
36. C. Goyhenex, M. Meunier and C.R. Henry (1996) *Surf. Sci.* **350**, 103.
37. C. Goyhenex, M. Croni, C. Claeys and C.R. Henry (1996) *Surf. Sci.* **352-354**, 475.
38. P. Légaré and B.R. Bilwes (1992) *Surf. Sci.* **279**, 159.
39. P. Légaré, F. Finck, R. Roche and G. Maire (1989) *Z. Phys. D* **12**, 19.
40. E. Gillet and V. Matolin (1991) *Z. Phys. D* **19**, 361.
41. B. Ealet and E. Gillet (1993) *Surf. Sci.* **281**, 91.
42. S. Ogawa and S. Ichikawa (1995) *Phys. Rev. B* **51**, 17231.
43. J.-W. He and P.J. Møller (1986) *Chem. Phys. Lett.* **129**, 13;
P.J. Møller (1994) in J. Nowotny (ed.) *Science of Ceramic Interfaces II*, p. 473.
44. T. E. Madey, U. Diebold and J.-M. Pan, (1993) in H.-J. Freund and E. Umbach (eds.) *Adsorption on Ordered Surfaces of Ionic Solids and Thin Films*, Springer Series in Surface Science, Vol. 33, Springer Verlag, Heidelberg, p. 147.
45. H. Kuhlenbeck (1994) *Appl. Phys. A* **59**, 469.
46. W. Weiss, A. Barbieri, M.A. van Hove and G.A. Somorjai (1993) *Phys. Rev. Lett.* **71**, 1848;
A. Barbieri, W. Weiss, M.A. van Hove and G.A. Somorjai (1994) *Surf. Sci.* **302**, 259.
47. H. Kuhlenbeck, G. Odörfer, R. Jäger, G. Illing, M. Menges, Th. Mull, H.-J. Freund, M. Pöhlchen, V. Staemmler, S. Witzel, C. Scharfschwerdt, K. Wennemann, T. Liedtke and M. Neumann (1991) *Phys. Rev. B* **43**, 1969.
48. M. Haßel and H.-J. Freund (1995) *Surf. Sci.* **325**, 163.
49. D. Cappus, M. Haßel, E. Neuhaus, M. Heber, F. Rohr and H.-J. Freund (1995) *Surf. Sci.* **337**, 268.
50. D. Cappus, X. Xu, D. Ehrlich, B. Dillmann, C.A. Ventrice Jr., K. Al-Shamery, H. Kuhlenbeck and H.-J. Freund (1993) *Chem. Phys.* **177**, 533.
51. a.) S.M. Vesecky, X. Xu and D.W. Goodman (1994) *J. Vac. Sci. Technol. A* **12**, 2114;
b) M.C. Wu, C.M. Truong and D.W. Goodman (1993) *J. Phys. Chem.* **97**, 4182;
c) C.M. Truong, M.C. Wu and D.W. Goodman (1993) *J. Am. Soc.* **115**, 3647;

d) J.W. He, J.S. Corneille, C.A. Estrada, M.C. Wu and D.W. Goodman (1992) *J. Vac. Sci. Technol. A* **10**, 2248;
e) M. C. Wu, C.A. Estrada, J. S. Corneille and D.W. Goodman (1992) *J. Chem. Phys.* **96**, 3892.
52. D. Cappus, J. Klinkmann, H. Kuhlenbeck and H.-J. Freund (1995) *Surf. Sci. Lett.* **325**, L 421.
53. M. Schönnenbeck, D. Cappus, J. Klinkmann, H.-J. Freund, L.G.M. Pettersson and P.S. Bagus (1996) *Surf. Sci.* **347**, 337.
54. N. Floquet and I.C. Dufour (1983) *Surf. Sci.* **126**, 543.
55. V.E. Henrich (1976) *Surf. Sci.* **57**, 385.
56. H. Oniski, C. Egawa, T. Aruga and Y. Iwasawa (1994) *Surf. Sci.* **310**, 135.
57. C.R. Henry and H. Poppa (1990) *Thin Sol. Films* **189**, 303.
58. J.S. Foord and R.M. Lambert (1986) *Surf. Sci.* **169**, 327.
59. C.A. Ventrice Jr., H. Hannemann, A. Brodde and H. Neddermeyer (1994) *Phys. Rev. B* **49**, 5773;
H. Hannemann, C.A. Ventrice, Th. Bertrams, A. Brodde and H. Neddermeyer (1994) *Phys. Stat. Sol. (a)* **146**, 289.
60. F. Rohr, K. Wirth, J. Libuda, D. Cappus, M. Bäumer and H.-J. Freund (1994) *Surf. Sci.* **315**, L977.
61. H. Kuhlenbeck, M. Frank, G. Lilienkamp, C. Koziol, T. Schmidt, E. Bauer and H.-J. Freund, private communications.
62. H. Kuhlenbeck, C. Xu, B. Dillmann, M. Haßel, B. Adam, D. Ehrlich, S. Wohlrab, H.-J. Freund, U.A. Dittinger, H. Neddermeyer, M. Neuber and M. Neumann (1992) *Ber. Bunsenges. Phys. Chem.* **96**, 15.
63. C.A. Ventrice, D. Ehrlich, E.L. Garfunkel, B. Dillmann, D. Heskett and H.-J. Freund (1992) *Phys. Rev. B* **46**, 12892.
64. M. Bender, D. Ehrlich, I.N. Yakovkin, F. Rohr, M. Bäumer, H. Kuhlenbeck, H.-J. Freund and V. Staemmler (1995) *J. Phys.: Condens. Matter* **7**, 5289.
65. C. Xu, M. Haßel, H. Kuhlenbeck and H.-J. Freund (1991) *Surf. Sci.* **258**, 23.
66. C. Xu, B. Dillmann, H. Kuhlenbeck and H.-J. Freund (1991) *Phys. Rev. Lett.* **67**, 3551.
67. F. Rohr, M. Bäumer, H.-J. Freund, S. Müller, L. Hammer and K. Heinz, in preparation.
68. B. Dillmann, O. Seiferth, G. Klivenyi, F. Rohr, I. Hemmerich, M. Bender, I. Yakovkin, D. Ehrlich and H.-J. Freund, *Farad. Disc. Chem. Soc.*, submitted.
69. H.-J. Freund, B. Dillmann, O. Seiferth, G. Klivenyi, M. Bender, D. Ehrlich, I. Hemmerich and D. Cappus, *Catal. Today*, in press.
70. B. Dillmann (1996) PhD-Thesis, Ruhr-Universität Bochum.
71. I. Hemmerich, F. Rohr, O. Seiferth, B. Dillmann and H.-J. Freund, *Z. Phys. Chem. NF*, submitted.
72. C. Xu (1991) PhD-Thesis, Ruhr-Universität Bochum.

73. F. Rohr, PhD-Thesis, Ruhr-Universität Bochum, in preparation.
74. H.-J. Freund, H. Kuhlenbeck and V. Staemmler (1996) *Rep. Progr. Phys.* **59**, 283.
75. U.J. Katter, H. Schlienz, M. Beckendorf and H.-J. Freund, *Ber. Bunsenges. Phys. Chem.* **97**, 340.
76. H. Schlienz, M. Beckendorf, U.J. Katter, T. Risse and H.-J. Freund (1995) *Phys. Rev. Lett.* **74**, 761.
77. Th. Risse, U.J. Katter, T. Hill, H. Hamann and H.-J. Freund, *Langmuir*, submitted.
78. R.M. Jaeger, H. Kuhlenbeck, H.-J. Freund, M. Wuttig, W. Hoffmann, R. Franchy, and H. Ibach (1991) *Surf. Sci.* **259**, 235.
79. M. Wuttig, W. Hoffmann, R. Jaeger, H. Kuhlenbeck and H.-J. Freund (1991) *Mat. Res. Soc. Symp. Proc.* **221**, 143.
80. R.M. Jaeger, K. Homann, H. Kuhlenbeck and H.-J. Freund (1993) *Chem. Phys. Lett.* **203**, 41.
81. R.M. Jaeger, J. Libuda, M. Bäumer, K. Homann, H. Kuhlenbeck and H.-J. Freund (1993) *J. Electron Spectrosc. Relat. Phenom.* **64/65**, 217.
82. J. Libuda, M. Bäumer and H.-J. Freund (1994) *J. Vac. Sci. Technol. A* **12**, 2259.
83. J. Libuda, F. Winkelmann, M. Bäumer, H.-J. Freund, Th. Bertrams, H. Neddermeyer and K. Müller (1994) *Surf. Sci.* **318**, 61.
84. M. Bender, K. Al-Shamery and H.-J. Freund (1994) *Langmuir* **10**, 3081.
85. K.P. Huber, G. Herzberg (1979) *Molecular Spectra and Molecular Structure, IV: Constants of Diatomic Molecules*, Van Nostrand, New York.
86. P.W. Tasker (1979) *J. Phys. C: Solid State Phys.* **12**, 4977.
87. J.W. Tasker (1979) *Phil. Mag. A* **39**, 119.
88. R. Lacmann (1965) *Colloq. Int. CNRS* **152**, 195.
89. D. Wolf (1992) *Phys. Rev. Lett.* **68**, 3315.
90. K. Refson, R.A. Wagelies, D.G. Fraser, M.C. Payne, M.H. Lee and V. Milan (1995) *Phys. Rev. B* **52**, 10833.
91. H. Papp and B. Egersdörfer, private communication;
B. Egersdörfer (1993) PhD-Thesis, Ruhr-Universität Bochum.
92. G. Pacchioni and P.S. Bagus (1993) in H.-J. Freund and E. Umbach (eds.) *Adsorption on Ordered Surfaces of Ionic Solids and Thin Films*, Springer Series in Surface Science Vol. 33, Springer Verlag, Heidelberg, p.180.
93. a.) K. M. Neyman and N. Rösch (1992) *J. Chem. Phys.* **168**, 267;
b.) K. M. Neyman and N. Rösch (1993) *Surf. Sci.* **297**, 223;
c.) K. M. Neyman and N. Rösch (1993) *Chem. Phys.* **177**, 561.
94. C. Pisani, R. Dovesi, R. Nada and S. Tamiro (1989) *Surf. Sci.* **216**, 267.
95. H.-J. Freund and M. Neumann (1988) *Appl. Phys. A* **47**, 3.
96. G. Pacchioni, G. Cogliandro and P.S. Bagus (1991) *Surf. Sci.* **255**, 344;
G. Pacchioni, P.S. Bagus (1992) in G. Pacchioni, P.S. Bagus and F. Parmigiani

(eds.) *Cluster Models for Surface and Bulk Phenomena*, NATO ASI Series, Series B: Physics, Vol. 283, Plenum, New York, p. 305.

97. M. Pöhlchen and V. Staemmler (1992) *J. Chem. Phys.* **97**, 2583.
98. G. Blyholder (1964) *J. Phys. Chem.* **68**, 2772.
99. A. Freitag, V. Staemmler, D. Cappus, C.A. Ventrice Jr., K. Al-Shamery, H. Kuhlenbeck and H.-J. Freund (1993) *Chem. Phys. Lett.* **210**, 10.
100. a) N.D.S. Canning and M.A. Chesters (1983) *J. Electron Spectrosc. Relat. Phenom.* **30**, 247;
b) J. Rolfe (1968) *J. Chem. Phys.* **49**, 963.
101. J.H. Lunsford, X. Yang, K. Haller, J. Laane, G. Mestl and H. Knözinger (1993) *J. Phys. Chem.* **97**, 13810.
102. F.A. Müller, G.L. Carlton and W.B. White (1959) *Spectrochim. Acta* **15**, 709; W.E. Hobbs (1958) *J. Chem. Phys.* **28**, 1220.
103. P.J.M. Carrott and N. Sheppard (1983) *J. Chem. Soc. Farad. Trans.* **79**, 2425.
104. A.A. Davydov, Y.M. Shchekochikhin and N.P. Keier (1972) *Kinet. Katal.* **13**, 1088.
105. I.E. Wachs, *Catal. Today*, in press.
106. A. Zecchina, G. Spotoy and G. Ghiotti (1992) *J. Mol. Catal.* **74**, 175.
107. M. Bender, I.N. Yakovkin and H.-J. Freund (1996) *Surf. Sci.*, in press.
108. H. Kadowaki, H. Takei and K. Motoya (1995) *J. Phys.: Condens. Matter* **7**, 6869 and references therein.
109. B. Adam (1991) *Diploma-Thesis*, Ruhr-Universität Bochum.
110. H. Isern and G. R. Castro (1989) *Surf. Sci.* **211/212**, 865.
111. H. Pfnür, C. Schwennicke and J. Schimmelpfennig (1993) in H.-J. Freund and E. Umbach (eds.) *Adsorption on Ordered Surfaces of Ionic Solids and Thin Film*, Springer Series in Surface Sciences, Vol. 33, Springer Verlag, Heidelberg, p. 124.
112. Th. Bertrams, F. Winkelmann, Th. Uttich, H.-J. Freund and H. Neddermeyer (1995) *Surf. Sci.* **331-333**, 1515.
113. J. Libuda (1996) PhD-Thesis, Ruhr-Universität Bochum.
114. H.-J. Freund, B. Dillmann, D. Ehrlich, M. Haßel, R.M. Jaeger, H. Kuhlenbeck, C.A. Ventrice, F. Winkelmann, S. Wohlrab, C. Xu, Th. Bertrams, A. Brodde and H. Neddermeyer (1993) *J. Mol. Catal.* **82**, 143.
115. F. Winkelmann, S. Wohlrab, J. Libuda, M. Bäumer, D. Cappus, M. Menges, K. Al-Shamery, H. Kuhlenbeck and H.-J. Freund (1994) *Surf. Sci.* **307-309**, 1148.
116. S. Wohlrab, F. Winkelmann, J. Libuda, M. Bäumer, H. Kuhlenbeck and H.-J. Freund, *Adsorption on Epitaxial Oxide Films as Model Systems for Heterogeneous Catalysis* in Springer Proceedings in Physics, in press.
117. J. Libuda, A. Sandell, M. Bäumer and H.-J. Freund (1995) *Chem. Phys. Lett.* **240**, 429.
118. M. Bäumer, J. Libuda, A. Sandell, F. Winkelmann, H.-J. Freund, G. Graw, Th. Bertrams and H. Neddermeyer (1995) *Ber. Bunsenges., Phys. Chem.* **99**, 1381.

119. A. Sandell, J. Libuda, P.A. Brühwiler, S. Andersson, M. Bäumer, A.J. Maxwell, N. Mårtensson and H.-J. Freund, *J. Vac. Sci. Technol. A*, in press.
120. A. Sandell, J. Libuda, P. Brühwiler, S. Andersson, A.J. Maxwell, M. Bäumer, N. Mårtensson and H.-J. Freund (1995) *J. Electr. Spectr. Rel. Phen.* **76**, 301.
121. A. Sandell, J. Libuda, M. Bäumer and H.-J. Freund (1996) *Surf. Sci.* **346**, 108.
122. J. Libuda, M. Frank, A. Sandell, S. Andersson, P.A. Brühwiler, M. Bäumer, N. Mårtensson and H.-J. Freund, *Springer Proceedings in Physics*, accepted.
123. A. Sandell, J. Libuda, P.A. Brühwiler, S. Andersson, M. Bäumer, A.J. Maxwell, N. Mårtensson and H.-J. Freund *Phys. Rev. B*, submitted.
124. U.J. Katter, T. Hill, T. Risse, H. Schlienz, M. Beckendorf, T. Klüner, H. Hamann and H.-J. Freund, *J. Phys. Chem.*, submitted.
125. J. Heidberg and D. Meine (1993) *Ber. Bunsenges. Phys. Chem.* **97**, 211.
126. A. Zecchina, D. Scarano, S. Bordiga, G. Ricchiardi, G. Spoto and F. Geobaldo (1996) *Catal. Today*, **27**, 403.
127. M. Henzler, A. Stock and M. Böl (1993) in H.-J. Freund and E. Umbach (eds.) *Adsorption on Ordered Surfaces of Ionic Solids and Thin Films*, Springer Series in Surface Science, Vol. 33, Springer Verlag, Heidelberg.
128. K.M. Schröder, F. Schäfer, J. Wollschläger and M. Henzler, *Surf. Sci.* submitted.
129. C. Schwennicke, J. Schimmelpfennig and H. Pfnür (1993) *Surf. Sci.* **293**, 57.
130. P.J. Chen and D.W. Goodman (1994) *Surf. Sci. Lett.* **312**, L767.
131. Y. Wu, H.-S. Tao, E. Garfunkel, T.E. Madey and N.D. Shinn (1995) *Surf. Sci.* **336**, 123.
132. Y. Wu, E. Garfunkel and T.E. Madey (1996) *J. Vac. Sci. Technol. A* **14**, 1662.
133. S. Stempel, PhD-Thesis, Ruhr-Universität Bochum, in preparation.
134. J.E. McDonald and J.G. Eberhart (1965) *Trans. Met. Soc. AIME* **233**, 512.
135. D. Chatain, I. Rivollet and N. Eustathopoulos (1986) *J. Chim. Phys.* **83**, 561; (1987) *J. Chim. Phys.* **84**, 201.
136. E. Bauer (1958) *Zeitschrift für Kristallographie* **110**, 372.
137. J. Wollschläger, J. Falta and M. Henzler (1990) *Appl. Phys. A* **50**, 57.
138. P.R. Pukite, C.S. Lent and P.I. Cohen (1985) *Surf. Sci.* **161**, 39; C.S. Lent and P.I. Cohen (1984) *Surf. Sci.* **139**, 121.
139. G. Beitel, K. Markert, J. Wiechers, J. Hrbek and R.J. Behm (1993) in H.-J. Freund, E. Umbach (eds.), *Adsorption on Ordered Surfaces of Ionic Solids and Thin Films* Springer Series in Surface Science, Vol. 33, Springer-Verlag, Heidelberg, p. 71.
140. A.F. Hollemann and E. Wiberg (1985) *Lehrbuch der Anorganischen Chemie*, de Gruyter, Berlin.
141. M.G. Mason (1983) *Phys. Rev. B* **27**, 748.
142. G.K. Wertheim (1987) *Z. Phys. B* **66**, 53.
143. G.K. Wertheim (1989) *Z. Phys. D* **12**, 319.

144. G.K. Wertheim, S.B. DiCenzo and D.N.E. Buchanan (1986) *Phys. Rev. B* **33**, 5384.
145. F.A. Marks, I. Lindau and R. Browning (1990) *J. Vac. Sci. Technol. A* **8**, 3437.
146. K. Schönhammer and O. Gunnarsson (1978) *Z. Phys. B* **30**, 297.
147. K. Schönhammer and O. Gunnarsson (1977) *Solid State Commun.* **23**, 691.
148. S.B. DiCenzo and G.K. Wertheim (1985) *Comm. Solid State Phys.* **11**, 203.
149. H.-J. Freund in G. Ertl, H. Knözinger and J. Weitkamp (eds.) *Handbook of Heterogenous Catalysis* Part A, Chapter 5.1, Section 5.1.1 Principles of Chemisorption, in press.
150. D.A. King and D.P. Woodruff (eds., 1990) *Chemisorption Systems* The Chemical Physics of Solid Surfaces and Heterogeneous Catalysis, Vol. 3, Part A, Elsevier.
151. N. Mårtensson and A. Nilsson (1995) in W. Eberhardt (ed.) *Applications of Synchrotron Radiation* Springer Series in Surf. Science Vol. 35, Springer.
152. H.-J. Freund and M. Neumann (1988) *Appl. Phys. A* **47**, 3.
153. J.N. Andersen, D. Hennig, E. Lundgren, M. Methfessel, R. Nyholm and M. Scheffler (1994) *Phys. Rev. B* **50**, 17525.
154. J.N. Andersen, M. Quarford, R. Nyholm, S.L. Sorensen and C. Wigren (1991) *Phys. Rev. Lett.* **67**, 2822.
155. O. Björneholm, A. Nilsson, H. Tillborg, P. Bennich, A. Sandell, B. Hernnäs, C. Puglia and N. Mårtensson (1994) *Surf. Sci.* **315**, L983.
156. B.G. Frederick, G. Apai and T.N. Rhodin (1987) *J. Am. Chem. Soc.* **109**, 4797.
157. X. Guo and J.T. Yates Jr. (1998) *J. Chem. Phys.* **90**, 6761.
158. E.W. Plummer, W.R. Salanek and J.S. Miller (1978) *Phys. Rev. B* **18**, 1673.
159. H.-J. Freund and E.W. Plummer (1981) *Phys. Rev. B* **23**, 4859.
160. H.-J. Freund, E.W. Plummer, W.R. Salaneck and R.W. Bigelow (1981) *J. Chem. Phys* **75**, 4275.
161. J.N. Andersen et al., unpublished results.
162. J.C. Fuggle, E. Umbach, D. Menzel and K. Brundle (1978) *Solid State Commun.* **27**, 65.
163. K. Schönhammer and O. Gunnarsson (1980) *Phys. Scr.* **21**, 575.
164. O. Gunnarsson and K. Schönhammer (1978) *Phys. Rev. Lett.* **41**, 1608.
165. E. Umbach (1982) *Surf. Sci.* **117**, 482.
166. A. Nilsson and N. Mårtensson (1989) *Phys. Rev. B* **40**, 10249.
167. H. Tillborg, A. Nilsson and N. Mårtensson (1993) *J. Electr. Spectr. Relat. Phen.* **62**, 73.
168. P.S. Bagus and M. Seel (1981) *Phys. Rev. B* **23**, 2065.
169. A. Nilsson, N. Mårtensson, S. Svensson, L. Karlsson, D. Nordfors, U. Gelius and H. Ågren (1992) *J. Chem. Phys.* **96**, 8770.
170. A. Sandell, J. Libuda, P.A. Brühwiler, S. Andersson, M. Bäumer, A.J. Maxwell, N. Mårtensson and H.-J. Freund, in preparation.

171. J. Bustad, C. Enkvist, S. Lunell and S. Svensson (1995) *J. Electron Spectrosc. Relat. Phen.* **70**, 233.
172. T. Porwol, G. Dömötör, I. Hemmerich, J. Klinkmann, H.-J. Freund and C.-M. Liegener (1994) *Phys. Rev. B* **49**, 10557.
173. J. Klinkmann, D. Cappus and H.-J. Freund (1995) *J. Electr. Spectrosc. Relat. Phenom.* **72**, 37.
174. J. Stöhr (1991) *NEXAFS Spectroscopy* Springer Series in Surf. Sci., Vol. 25 Springer-Verlag, Heidelberg.
175. U. von Barth and G. Grassmann (1982) *Phys. Rev. B* **25**, 5150.
176. A. Nilsson and N. Mårtensson (1995) *Physica B* **209**, 19.
177. A. Nilsson, O. Björneholm, E.O.F. Zdansky, H. Tillborg, N. Mårtensson, J.N. Andersen and R. Nyholm (1992) *Chem. Phys. Lett.* **197**, 12.
178. O. Björneholm, A. Nilsson, E.O.F. Zdansky, A. Sandell, B. Hernnäs, H. Tillborg, J.N. Andersen and N. Mårtensson (1992) *Phys. Rev. B* **46**, 10353.
179. H.F.J. van't Bilk, J.B.A.D. van Zon, T. Huizinga, J.C. Vis and D.C. Koningsberger (1983) *J. Phys. Chem.* **87**, 2264; (1985) *Am. Chem. Soc.* **107**, 3139.
180. F. Solymosi and M. Pásztor (1985) *J. Phys. Chem.* **89**, 4789.
181. P. Basu, D. Panayotov and J.T. Yates Jr. (1987) *J. Phys. Chem.* **91**, 3133; (1988) *J. Am. Chem. Soc.* **110**, 2074.
182. F. Solymosi and H. Knözinger (1990) *J. Chem. Soc. Faraday Transactions* **86**, 389.
183. D. N. Belton and S. J. Schmieg (1988) *Surf. Sci.* **202**, 238.
184. V. Nehasil, I. Stará and V. Matolín (1995) *Surf. Sci.* **331-333**, 105.
185. M. Rebholz, R. Prins and N. Kruse (1991) *Surf. Sci.* **259**, L797; (1992) *Surf. Sci.* **269/270**, 293.
186. V. Matolín, K. Mâsek, M.H. Elyakhloufi and E. Gillet (1993) *J. Catal.* **143**, 492.
187. A. de Koster and R.A. van Santen (1990) *Surf. Sci.* **233**, 366.
188. D. B. Almy, D. C. Foyt and J. M. White (1977) *J. Electron Spectrosc. Relat. Phen.* **11**, 129.
189. J. Paul and F. M. Hoffmann (1986) *J. Phys. Chem.* **21**, 5321.
190. J.G. Chen, J.E. Crowell and J.T. Yates Jr. (1986) *J. Phys. Chem.* **84**, 5906.
191. V. Coustet and J. Jupille (1994) *Surf. Sci.* **309**, 309; (1994) *Surf. Interface Anal.* **22**, 280.
192. B.G. Frederick, G. Apai and T.N. Rhodin (1991) *Surf. Sci.* **244**, 67.
193. M.A. Schildbach and A.V. Hamza (1993) *Surf. Sci.* **282**, 306.
194. A. Ludviksson, K.H. Ernst, R. Zhang and C.T. Campbell (1993) *J. Catal.* **141**, 380.
195. N. Kruse and A. Gaussmann (1993) *J. Catal.* **144**, 525.
196. M.H. El-yakhloufi and E. Gillet (1993) *Catal. Lett.* **17**, 11.

197. H.A. van der Vegt, H.M. van Pinxteren, M. Lohmeier, E. Vlieg and J.M.C. Thornton (1992) *Phys. Rev. Lett.* **68**, 3335.
198. M. Horn-von Hoegen, M. Pook, A. Al-Falou, B.H. Müller and M. Henzler (1993) *Surf. Sci.* **284**, 53.
199. H. Wolter, M. Schmidt and K. Wandelt (1993) *Surf. Sci.* **298**, 173.
200. S. Esch, M. Hohage, T. Michely and G. Comsa (1994) *Phys. Rev. Lett.* **72**, 518.
201. H.-J. Freund, H. Kuhlenbeck (1995) in W. Eberhardt (ed.) *Applications of Synchrotron Radiation,* Springer Series in Surface Science, Vol. 35, Springer, p. 9.
202. J.E. Reutt, L.S. Wang, Y.T. Lee and D.A. Shirley (1986) *Chem. Phys. Lett.* **126**, 399.
203. J.H. Darling and J.S. Ogden (1972) *Inorganic Chemistry* **11**, 666; (1972) *J. Chem. Soc. Dalton Trans.*, 2496.
204. P. Kuendig, M. Moskovits and G.A. Ozin (1972) *Can. J. Chem.* **50**, 3587; (1972) *J. Mol. Struct.* **14**, 137.
205. H. Huber, P. Kuendig, M. Moskovits and G.A. Ozin (1977) *Nature* **235**, 98.
206. L.-L. Sheu, H. Knötzinger and W.M.H. Sachtler (1989) *J. Am. Chem. Soc.* **111**, 8125.
207. Z. Zhang, H. Chen and W.M.H. Sachtler (1991) *J. Chem. Soc. Faraday Trans.* **87**, 1413.
208. Z. Zhang, F.A.P. Cavalcanti and W.M.H. Sachtler (1992) *Catal. Lett.* **12**, 157.
209. P.A. Hintz and K.M. Ervin (1994) *J. Chem. Phys.* **100**, 5715.

THE GROWTH AND STABILITY OF ULTRATHIN FILMS ON METAL AND OXIDE SURFACES

THEODORE E. MADEY
Rutgers, The State University of New Jersey,
Department of Physics and Astronomy
and Laboratory for Surface Modification,
PO Box 849, Piscataway, NJ 08855-0849, USA

Abstract

The interactions of ultrathin transition metal films deposited onto single crystal TiO_2 and Al_2O_3 surfaces have been studied using surface science methods. We characterize the morphology of the films during initial stages of growth, as well as the chemical and electronic properties of the metal/oxide interface. By comparing results for different metals (Ti, Cr, Fe, Cu, Pd, Hf, Pt, Au) general trends of overlayer growth on oxide surfaces are extracted. We find that the ability of the metal film to cover (wet) the oxide substrate correlates directly with the reactivity of overlayer metal toward oxygen.

Ultrathin films of metals on metals (model bimetallic catalysts) have also been studied; we have concentrated on the atomically rough bcc(111) substrates, W(111) and Mo(111). We find that these two surfaces coated with ~1 monolayer of certain metals (Pt, Pd, Au...) undergo extensive restructuring to form highly faceted surfaces with pyramidal microfacets after annealing to T > 750K. The dominant facet orientation is W{211}, and facet dimensions range from ~3 to 100 nm. Not all metal monolayer films cause faceting of W(111): Pt, Pd, Ir, Rh, Au induce faceting of W(111) upon annealing, whereas Ti, Co, Ni, Cu, Ag, Gd do not. Electronic effects appear to be the driving factors in the observed morphological instability of W(111) and Mo(111).

1. Introduction

The interfaces between metals and oxides are of crucial importance in heterogeneous catalysis, as well as fiber optics, chemical sensors, thin film technology and electronic device fabrication. Many of the properties of a metal/oxide system depend on the chemical and physical interactions that take place at the interface. An important scientific issue of studies of metals on oxide surfaces is the relationship between the chemistry of the interface

R.M. Lambert and G. Pacchioni (eds.),
Chemisorption and Reactivity on Supported Clusters and Thin Films, 105-116.

and the growth modes of the metal overlayers, i.e., the morphology of the films in the initial stages of growth [1, 2].

In the work summarized here, we have studied the initial growth of a series of 3d, 4d, and 5d transition metals on TiO_2 (110) surfaces [1-13] and on Al_2O_3(0001) oriented thin films [14, 15] using surface science methods, mainly Low Energy Ion Scattering (LEIS), X-ray photoelectron spectroscopy (XPS), low energy electron diffraction (LEED), medium energy backscattered electron diffraction (MEED), soft XPS using synchrotron radiation, high resolution scanning electron microscopy (HRSEM) and scanning tunneling microscopy (STM).

Our most extensive measurements of metals on oxides have been made on TiO_2(110) surfaces. TiO_2 has been chosen because of its interesting properties in catalysis, photocatalysis and other applications. The TiO_2(110) surface is the charge neutral surface of rutile TiO_2, and is believed to have a relaxed bulk truncation structure[16-18]. We compare the metal/TiO_2 behavior with recent data for Cu and Nb on thin films of Al_2O_3(0001).

We have also studied model bimetallic thin film catalysts composed of Pt-group and other metals evaporated onto single crystal surfaces, in particular, the atomically rough W(111) and Mo(111) substrates [19-22]. Both surfaces are morphologically unstable when covered with films of certain metals, i.e., each undergoes a transition from planar surface to a microscopically faceted surface upon annealing. *A major objective is to determine how the transition from planar to faceted affects the catalytic activity of these bimetallic systems.* The importance of bimetallic catalysts based on Pt group metals has been increasing in the past several decades [23] These catalysts exhibit important advantages over classical reforming catalysts, including better stability as well as better activity and selectivity. In particular, refractory metals (W, Mo, Re...) in combination with Pt - group metals are active catalysts for hydrogenation and hydrogenolysis reactions [24, 25].

The present paper is a brief survey of recent measurements mainly from the author's laboratory; it is *not* an exhaustive survey of the literature. More complete accounts of metals on oxides can be found in Refs. [26-30], as well as in the papers by Freund, Goodman, Hayden, Henry, Lambert, and Moller in the present volume; the structure and reactivity of metal films on metal surfaces are discussed more fully in Refs. [31-35].

The plan of this survey is as follows. A brief discussion of experimental procedures is given in Section 2. Section 3 begins with a discussion of ultrathin films of metals on oxides (TiO_2, Al_2O_3) and concludes with presentation of metals on metals. Conclusions are given in Section 4.

2. Experimental

The metal-on-oxide experiments have been carried out in several different ultrahigh vacuum (UHV) systems with base pressures ~1 to 2 x 10^{-10} Torr. The TiO_2 substrates are 5 x 10 x 1 mm^3 TiO_2 crystals with polished (110) surfaces, and are annealed in UHV up to 1000K for hours so that the bulk is reduced enough to become conductive; this insures that the oxide samples are free from charging problems. The stoichiometric surface region is restored by annealing the sample in 2 x 10^{-6} Torr of oxygen, followed by cooling in oxygen. To minimize sample charging, the Al_2O_3 surfaces used in this work are thin films (~20 Å) of Al_2O_3 deposited onto a Re(0001) substrate. The films are prepared by vapor deposition of Al in an O_2 atmosphere onto the heated (1173K) substrate. LEED demonstrates the films to be crystalline, with (0001) orientation [14, 15].

The metal overlayer films studied in this work are deposited from a liquid nitrogen cooled thermal evaporator source at a rate of 1-3Å/min. The thicknesses of metal films are monitored by a quartz crystal monitor (QCM). Rutherford backscattering (RBS) measurements are performed as a calibration of the true thickness. The thicknesses reported in this work are average values of thickness in Angstroms (Å) by assuming uniformly thick films with the appropriate bulk density.

The metal-on-metal experiments have also been carried out in several UHV systems equipped with surface science instrumentation (LEED, LEIS, Auger electron spectroscopy (AES), and temperature programmed desorption (TPD). Of particular note is a UHV STM that permits atomic resolution images of microscopic facets on the film-covered W(111) surfaces.

3. Results and Discussion

3.1 METALS ON OXIDES

3.1.1 *Metals on TiO_2(110)*

In all of our studies of metal films deposited onto TiO_2(110), we have focused on the *initial stages* of growth, i.e., for average metal coverages ranging from fractional monolayer coverages to tens of monolayers. The most extensive studies have involved the growth of a series of transition metals on TiO_2(110) surfaces using Low Energy Ion Scattering (LEIS) together with X-ray Photoelectron Spectroscopy (XPS) [1-12]. This is a powerful combination for studying ultrathin film growth [1, 2]: because of the high sensitivity of LEIS to the composition in the topmost layer, we can separate 3-D island (Volmer-Weber, or VW growth) from the layer + island (Stranski-Krastanov or SK growth) and layer-by-layer (Frank-van der Merwe or FM growth) modes by identifying the coverage necessary to complete the first monolayer.

Figure 1 illustrates the use of LEIS for characterizing the initial growth [2]. The attenuation of the substrate Ti (LEIS) signal is shown for increasing coverages of Cu, Fe, Cr or Hf deposited onto $TiO_2(110)$ at 300K. The intensities are normalized to the value for the clean TiO_2 surface. The attenuation of the substrate signal is an upper limit for the covering of the surface by overlayer atoms. For complete covering (wetting) of the substrate by the metal overlayer one would expect a linear decrease, with the signal approaching zero at monolayer coverage (~2-3 Å). Such behavior is observed for Hf on TiO_2 and in the initial stages of Cr deposition in Fig. 1, but is very different from the behavior for Fe and Cu shown in Fig. 1.

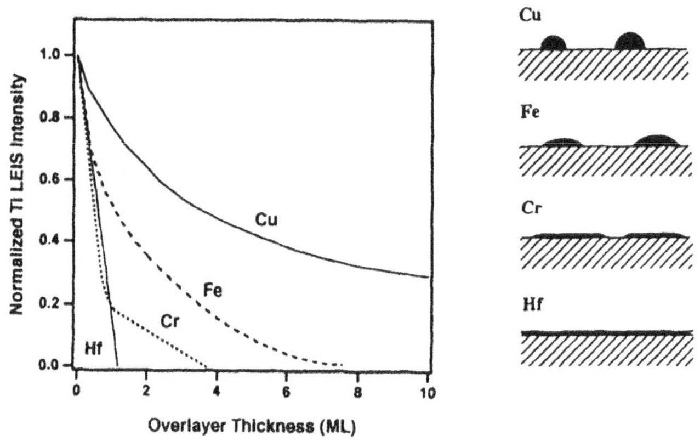

Figure 1. Trends for attenuation of Ti LEIS signals by Cu, Fe, Cr and Hf overlayers on $TiO_2(110)$, as a function of average overlayer thickness in monolayers (ML). Schematic drawings of initial stages of film growth are shown for Cu (3-dimensional clusters), Fe (flat 3-dimensional islands), Cr (two-dimensional islands, followed by 3-dimensional growth), and Hf (formation of continuous overlayer). From Ref. 2.

In the case of Cu, the fact that a LEIS substrate signal is visible up to high average coverages is clear evidence for Volmer-Weber (3-dimensional islands) growth [8] In the case of Cr, the LEIS signal initially is attenuated linearly up to a point where ~80% of the substrate is covered. It appears that Cr grows in a quasi-two-dimensional fashion, where the first monolayer covers ~80% of the substrate. For higher Cr coverages, the growth of the Cr film deviates from 2-D growth and becomes 3-dimensional [4]. The growth behavior of Cr/TiO_2 is similar to that reported recently for Cu/ZnO [36]. In the case of Fe at 300 K, the film grows clearly in 3-D islands initially, and its wetting ability is intermediate between that of Cr and Cu. Schematics of the initial growth modes are shown at the side of Fig. 1.

Figure 2 illustrates direct microscopic evidence for the formation of Au clusters with nanometer dimensions on TiO_2. The image is a HRSEM micrograph following deposition of 10 Å of Au at 300 K [37]. The "open

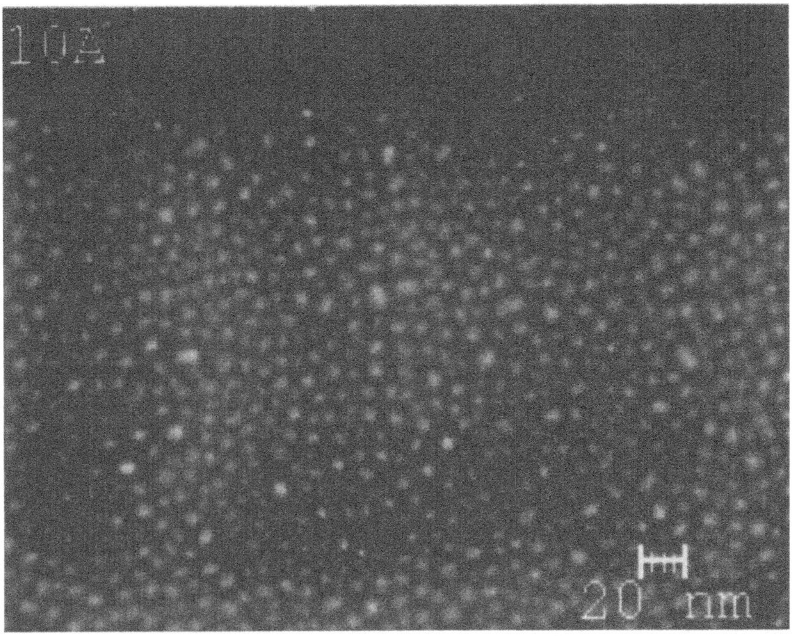

Figure 2. High resolution scanning electron micrograph (HRSEM) of Au on $TiO_2(110)$, showing formation of 3-dimensional clusters. A 10Å Au layer was deposited on $TiO_2(110)$ at 300K under UHV conditions before being transferred to the electron microscope. Note the 20nm length scale. Persaud, Cosandey, Zhang and Madey, in preparation.

area" fraction of the surface that is not covered by Au is the same as that determined in parallel LEIS studies [38].

We observe in this work a clear correlation between the ability of the metal to wet or cover the $TiO_2(110)$ substrate in the initial stages of growth, and the reactivity of the overlayer metal towards oxygen (the metal-oxygen bond strength) [1, 2, 26]. As discussed by Noguera [29], and Peden et al. [39], similar correlations have been reported for adhesion of metals to oxides. The chemical interaction at the interface between metal and oxide involves oxidation of the overlayer metal and reduction of the TiO_2 substrate. For the more reactive metals (Hf, Ti, Cr, Fe), the formation of a reduced substrate/oxidized metal overlayer is observed using XPS and SXPS [1, 2, 5, 6, 10]. At 300 K, the oxidation/reduction layer corresponds to ~0.5 ML for Fe, 1.0 ML for Cr, and ~4 ML for Hf on TiO_2 [2, 5, 9, 10]. Based on Fig. 1 and similar data, these metals are also observed to wet or cover the surface best. There is no oxidation/reduction behavior observed for fractional monolayers of Cu, Pd, Pt or Au on TiO_2, and these metals all form ordered three-dimensional clusters on the $TiO_2(110)$ surface even at 300 K [1, 2, 8, 12, 38].

The structures of the ultrathin films are determined using LEED, medium energy backscattered electron diffraction (MEED), and angle resolved XPS (ARXPS). Even in cases where LEED results indicate little or no long range order, clear evidence for short-range locally ordered structures can be seen using MEED and ARXPS [3, 12, 40, 41]. The fcc metals (Cu, Pd, Pt, Au) form clusters with the fcc (111) plane parallel to the $TiO_2(110)$ plane, while the bcc metals (V, Cr, Fe) form clusters with the bcc (100) orientation.

Further information about the electronic properties of the metal/TiO_2 interface can be obtained by means of soft x-ray photoemission (SXPS) and resonant photoemission using synchrotron radiation. As an example, we have used resonant photoemission to obtain insights into the partial density of states for Fe/TiO_2, and assign defect states in the bandgap as being Fe-derived and Ti-derived, respectively [10].

The thermal stability of metals on TiO_2 during annealing is rather complex, and depends on the properties of the metals [2, 4, 6, 13, 42]. Au and Cu form larger clusters upon annealing. For group VIII metals, encapsulation of metallic clusters by a reduced TiO_x layer is seen upon annealing to temperatures of several hundred °C [4, 13]. This encapsulation phenomenon is related to the "strong metal-support interaction" (SMSI) effect, a topic of long interest in catalysis [42]. Cr and Mn diffuse into the substrate and form ternary oxides whereas Ti and Hf films are oxidized, apparently by oxygen diffusion [2]. For Pt/TiO_2, a surfactant effect is also observed: when CO is adsorbed on clusters of Pt on $TiO_2(110)$ at 300K, the Pt island morphology changes. The Pt islands "spread out" and become more flat, due to the lowering of the surface energy by the adsorbate [12].

3.1.2. *Metals on Al_2O_3*

Recent measurements have been made of Cu and Nb Growth on Al_2O_3 films at both 80K and 300K, using XPS and LEIS. XPS results show decreases in the Cu $2p_{3/2}$ binding energy and increases in the Cu $L_3M_{45}M_{45}$ kinetic energy as a function of coverage. Analysis of these results indicates that Cu in *not* oxidized at the interface and that 3-d clusters form at low Cu coverages [14]. Coverage-dependent shifts in the Cu $2p_{3/2}$ peaks are interpreted in terms of the initial-state and final-state contributions that vary with Cu cluster size. Nb covers the Al_2O_3 surface effectively at Nb coverages slightly greater than 1 ML [43].

3.2 MORPHOLOGICAL INSTABILITIES IN GROWTH OF METALS ON METALS

3.2.1 *Experimental Results for Metals on W(111) and Mo(111)*

Monolayer films of certain metals (e.g., Rh, Pd, Pt, Au) deposited onto W(111) and Mo(111), followed by annealing to T>750 K, will cause the formation of three-sided pyramidal nanofacets with bcc {211} faces, which cover the entire surface [19-22]. Facet dimensions range from ~3 nm to as large as ~100 nm. The tool that we have used most often for identifying the formation of facets is low energy electron diffraction (LEED): the motion of LEED diffraction beams as incident electron beam energy is varied is a reliable diagnostic of faceting [19]. For a planar surface, the beams converge on the macroscopic surface normal as energy is varied. For a faceted surface, the diffraction beams appear to cross at various points, and beams converge on points or "poles" that correspond to specular reflections from microscopic facets. In this way, we [19-22] and others [44-47] have identified the formation of surface facets having {211} orientation for a number of metals (Rh, Pd, Ir, Pt, Au) and other elements (O, S, Cl) on Mo(111) and W(111) surfaces.

A more direct determination of faceting is by means of the ultrahigh vacuum STM [20, 22, 48], where we have examined the Pd/W(111) system *in situ* at coverages >1 ML, following annealing at various temperatures from 300 to 1400 K. Fig. 3 shows an atomic-resolution image (150Å x 150Å) of

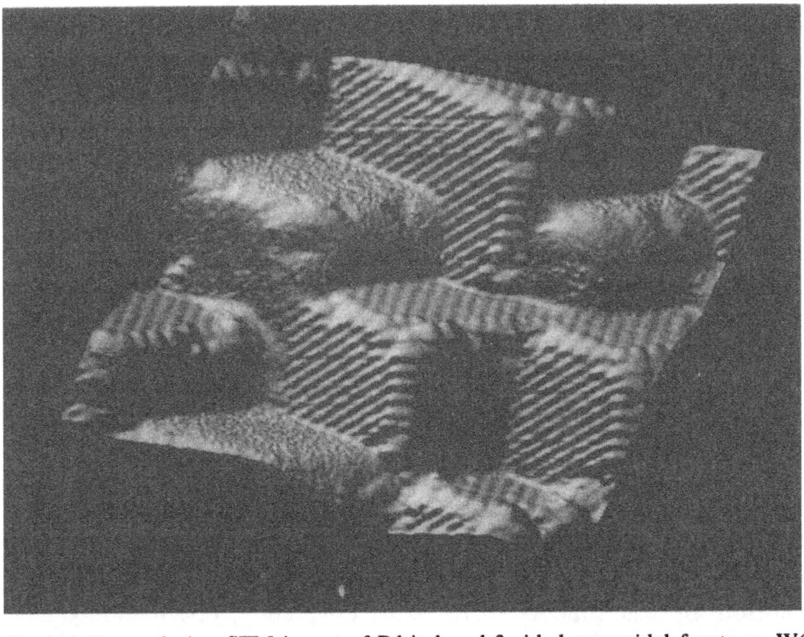

Figure 3. Atomic resolution STM image of Pd-induced 3-sided pyramidal facets on W(111). The rows of atoms characteristic of the bcc{112} surfaces are seen on the facets. From Ref. 46, Nien and Madey.

the Pd-induced facets, following annealing to 1100K. The real-space STM images of the faceted Pd/W(111) surfaces confirm the LEED conclusions about the formation of {211} facets [49]. Facets are observed following annealing from ~750 K to 1125 K.

After annealing Pd/W(111) at 1075 K the most probable facet size is ~50 Å while after annealing at 1125 K the most probable facet size is ~120 Å. The facet size is observed to grow with increasing annealing temperatures in the range of 700 K to 1125 K [20, 22, 48].

An important issue in the growth of the pyramidal facets concerns their microscopic structure. One could imagine the facets to be pyramidal clusters of the adlayer metal; alternatively, the pyramidal facets could arise from a reconstruction of the W(111) or Mo(111) substrate under the influence of a metal surfactant overlayer. Based on several different experiments, we believe that facet formation is better described by the latter picture: The metals that induce faceting are believed to act as surfactants and remain on the surface rather than forming 3-d clusters or a surface alloy [22].

3.2.2 *Discussion of Faceting*

For the metal overlayers that induce faceting (Pt, Pd, Au, Rh and Ir) a critical coverage of slightly greater than one physical monolayer (~1.7 x 10^{15} atoms/cm^2) is needed to induce faceting. The deposition of the admetals is believed to change the surface energetics dramatically and provide a thermodynamic driving force for the reconstruction of the surface [19, 22, 49]. For clean metal surfaces, the anisotropy in surface free energy for different crystal planes is generally too small for an atomically rough surface to form facets spontaneously (e.g., for clean W, the maximum anisotropy $\Delta\gamma/\gamma$ is ~3% and the planar (111) surface is stable against faceting). However, the presence of adsorbed layers on metals can lower the surface energy γ and can cause an increase in the anisotropy of γ. When facets are formed, the total energy $E = \int \gamma dA$ is lowered by the formation of low-energy facets even though the total surface area A increases. (When a bcc(111) surface reconstructs to form a completely faceted surface with [211] facets, the surface area increases by ~6%). Facet formation is limited by the kinetics of surface diffusion, i.e., the overlayer metal and the substrate W (Mo) atoms must be sufficiently mobile that atomic rearrangement can occur. The onset of facet formation at T ~750 K is related to diffusion on the atomically rough (111) surface. Thus, faceting is thermodynamically driven but kinetically controlled.

This view of faceting of W(111) is supported by recent embedded atom calculations for metal films on W surfaces performed by S. P. Chen [50], who was motivated by our measurements to do these calculations. He finds a lowering of the surface energy for all surfaces and a large increase of the anisotropy when the W surfaces are covered by one monolayer of Pd, Pt, or Au films (the surface energies of Pt, Pd and Au-covered W(112) are lower than Pt, Pd and Au-covered W(111) by 28%, 23% and 35%, respectively).

Although these results are encouraging, first principles calculations are sorely needed to compare with experimental results.

The faceting of W(111) and Mo(111) is remarkably specific to certain overlayer films. We have attempted to correlate the presence or absence of faceting with materials properties and electronic properties of the overlayer. Properties that have been examined include the atomic size, cohesive energy, surface free energy, melting point and elastic properties (Young's Modulus) for the metal overlayers. In each of these cases there is not a unique value or range of values associated with the elements that induce faceting, i.e., there is no correlation found between the overlayer property and faceting behavior [19, 22].

For both Mo(111) and W(111) there is one property with which a useful correlation is found: overlayer metals and other elements with Pauling electronegativities greater than 2.0 induce faceting of Mo(111) and W(111) while those with Pauling electronegativities less than 2.0 do not induce facets [19-22]. This observation suggests that charge transfer and surface electronic effects play a major role in the morphological stability of metal film-covered Mo and W(111) surfaces; this is motivation for a program of synchrotron radiation studies of surface electronic properties currently underway in our laboratory [51]. The role of overlayer electronic properties on faceting is a fascinating area for theoretical exploration.

A periodic table showing elements that do and do not induce faceting of W(111), as determined using LEED, is given in Fig. 4. In all cases where

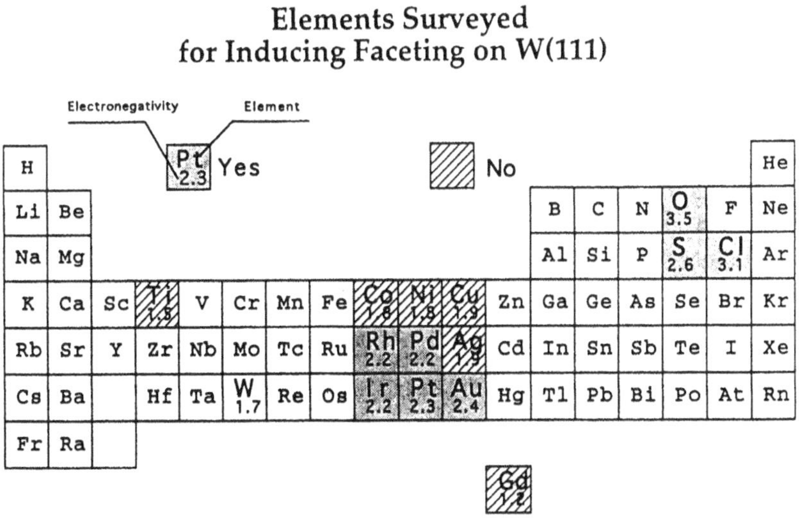

Figure 4. Periodic Table listing the Pauling Electronegativity, and whether or not faceting of W(111) occurs for overlayer films of the elements surveyed. From Ref. 19.

the same elements have been studied on W(111) and Mo(111), the results are the same for both surfaces [19-22, 44-47]. Note that S, O and Cl induce faceting of W(111), and all have Pauling electronegativity >2.0. For these elements only a fractional monolayer coverage is needed for faceting, as opposed to the 1 ML needed for the transition metals. Also S, O and Cl anions are larger in size than W or Mo in contrast to the metal overlayers that induce faceting, which have nearly the same atomic size as W or Mo. These differences suggest that the interactions between the oxygen, sulfur or chlorine and the (111) substrate are different from those of the transition metal films, and the formation of surface compounds may play a role in faceting.

Finally, we note that structure sensitivity is observed in a model catalytic reaction, n-butane hydrogenolysis, over planar and faceted Pt/W(111) [52]. Structure sensitivity is evidenced by reaction products in which the cleavage of the n-butane molecule, $n - C_4H_{10}$, will result in either the formation of methane and propane, or two ethane molecules, depending on the material and surface used, i.e.,

$$n - C_4H_{10} + H_2 \begin{array}{l} \nearrow CH_4 + C_3H_8 \\ \searrow 2C_2H_6 \end{array}$$

The hydrogenolysis kinetics, measured at ~100 Torr in a high pressure reactor attached to the surface analysis chamber, are quite different over the planar and the faceted Pt/W surfaces [53]. The planar surface has a higher selectivity towards ethane formation, whereas the faceted surface has a higher selectivity to methane. There appears to be a correlation between the concentration of fourfold coordination sites (C_4) sites on planar Pt/W(111) and the amount of ethane produced. The results are consistent with different intermediates on the planar and faceted surfaces, similar to a proposal by Engstrom, Weinberg and Goodman [54].

4. Summary and Conclusions

In conclusion, we find that the morphology of ultrathin metal films during the initial stages of overlayer formation on $TiO_2(110)$ substrates can be correlated with the metal overlayer reactivity, although the details vary for different systems. The formation of a reduced substrate/oxidized overlayer is observed for more reactive metals, with the reaction strongest for the most reactive film material. No such oxidation/reduction behavior is seen for Cu, Pd, Pt, Au overlayers.

The data presented for metals on W(111) and Mo(111) are among the first detailed microscopic studies of faceting of metals induced by ultrathin metal films. We suggest that these results have important implications for understanding the mechanisms of dynamic morphological restructuring of bimetallic catalyst surfaces under reaction conditions.

5. Acknowledgements

The author acknowledges the valuable contributions of many collaborators in this work, including U. Diebold, J. Guan, C. -H. Nien, J. -M. Pan, Raj Persaud, R. A. Campbell, F. Cosandey, C. -Z. Dong, E. Garfunkel, H. Orita, B. Maschhoff, J. Mayer, K. Pelhos, F. Pesty, K. -J. Song, H. -P. Steinrück, H. -S. Tao, Lei Zhang, Lizhong Zhang, and Yutong Wu. This work has been supported, in part, by the Division of Chemical Sciences, U. S. Department of Energy.

References

[1] T.E. Madey, U. Diebold and J.-M. Pan, Springer Series in Surface Sciences, Vol. 33, Eds.: E. Umbach, H. -J. Freund, *Adsorption on Ordered Surfaces of Ionic Solids and Thin Films* (Springer-Verlag Berlin Heidelberg, 1993) p. 147.
[2] U. Diebold, J.-M. Pan and T.E. Madey, Surface Sci. **331-333**, 845 (1995).
[3] J.-M. Pan, B.L. Maschhoff, U. Diebold and T.E. Madey, Surface Sci. **291**, 381 (1993).
[4] J.-M. Pan and T.E. Madey, Catal. Lett. **20**, 269 (1993).
[5] J.-M. Pan and T.E. Madey, J. Vacuum Sci. Technol. **A11**, 1667 (1993).
[6] J.-M. Pan, U. Diebold, L. Zhang and T.E. Madey, Surface Sci. **295**, 411 (1993).
[7] U. Diebold, J.-M. Pan and T.E. Madey, Surface Sci. **287/288**, 896 (1993).
[8] U. Diebold, J.-M. Pan and T.E. Madey, Phys. Rev. B **47**, 3868 (1993).
[9] J.-M. Pan, U. Diebold, B.L. Maschhoff and T.E. Madey, J. Vacuum Sci. Technol. **A10**, 2470 (1992).
[10] U. Diebold, H.-S. Tao, N.D. Shinn and T.E. Madey, Phys. Rev. B **50**, 14474 (1994).
[11] J. Mayer, E. Garfunkel, U. Diebold and T.E. Madey, J. Elec. Spec. Rel. Phen. **73**, 1 (1995).
[12] H.-P. Steinrück, F. Pesty, L. Zhang and T.E. Madey, Phys. Rev. B **51**, 2427 (1995).
[13] F. Pesty, H.-P. Steinrück and T.E. Madey, Surface Sci. **339**, 83 (1995).
[14] Y. Wu, E. Garfunkel and T.E. Madey, J. Vacuum Sci. Technol. **A14**, 1662 (1996).
[15] Y. Wu, H.-S. Tao, E. Garfunkel, T.E. Madey and N.D. Shinn, Surface Sci. **336**, 123 (1995).
[16] D. Novak, E. Garfunkel and T. Gustafsson, Phys. Rev. B **50**, 5000 (1994).
[17] A. Szabo and T. Engel, Surface Sci. **329**, 241 (1995).
[18] M. Ramamoorthy, R.D. King-Smith and D. Vanderbilt, Phys. Rev. B **49**, 16721 (1994).
[19] J. Guan, R.A. Campbell and T.E. Madey, Surface Sci. **341**, 311 (1995).
[20] T.E. Madey, J. Guan, C.-Z. Dong and S.M. Shivaprasad, Surface Sci. **287/288**, 826 (1993).
[21] J. Guan, R.A. Campbell and T.E. Madey, J. Vacuum Sci. Technol. **A13**, 1484 (1995).

[22] T.E. Madey, J. Guan, C.-H. Nien, C.-Z. Dong, H.-S. Tao and R.A. Campbell, Surf. Rev. & Lett. (1996). in press.
[23] J.H. Sinfelt, *Bimetallic Catalysts,* (J. Wiley, New York, 1983).
[24] B.N. Kuznetsov and e. al., J. Molec. Catalysis **4**, 49 (1978).
[25] A. Trunschke and e. al., J. Molec. Catalysis **56**, 95 (1989).
[26] V.E. Henrich and P.A. Cox, *The Surface Science of Metal Oxides* (Cambridge University Press, 1994).
[27] D.W. Goodman, Surf. Rev. & Lett. **2**, 9 (1995).
[28] R.J. Lad, Surf. Rev. & Lett. **2**, 109 (1995).
[29] C. Noguera, *Physics and Chemistry of Oxide Surfaces,* (Cambridge University Press, 1996).
[30] R. Persaud and T.E. Madey, in "The Chemical Physics of Solid Surfaces and Heterogeneous Catalysis", eds. D. A. King and D. P. Woodruff, in preparation.
[31] E. Bauer, in "The Chemical Physics of Solid Surfaces and Heterogeneous Catalysis", eds., D. A. King and D. P. Woodruff, vol. 3B (Elsevier, Amsterdam, 1994) p. 1.
[32] C.T. Campbell, Annu. Rev. Phys. Chem. **41**, 775 (1990).
[33] J.A. Rodriguez and D.W. Goodman, Science **260**, 1527 (1993).
[34] S. Chang and P.A. Thiel, Crit. Rev. Surf. Chem. **3**, 239 (1994).
[35] J.A. Rodriguez, Surf. Sci. Rpts. (1996). in press.
[36] K.H. Ernst, A. Ludviksson, R. Zhang, J. Yoshihara and C.T. Campbell, Phys. Rev. **B47**, 13782 (1993).
[37] R. Persaud, F. Cosandey, K. Bourdelle, L. Zhang and T.E. Madey, in preparation.
[38] L. Zhang, R. Persaud and T.E. Madey, in preparation.
[39] C.H. Peden, K.B. Kidd and N.D. Shinn, J. Vacuum Sci. Technol. **A9**, 1518 (1991).
[40] B.L. Maschhoff, J.-M. Pan and T.E. Madey, Surface Sci. **259**, 190 (1991).
[41] M. Sambi, E. Pin, G. Sangiovanni, L. Zaratin, G. Granozzi and F. Parmigiani, Surface Sci. **349**, L169 (1996).
[42] G.L. Haller and D.E. Resasco, Advances in Catalysis **36**, 173 (1989).
[43] Y. Wu, Ph. D. Thesis. Rutgers University (1995).
[44] J.C. Tracy and J.M. Blakely, Surface Sci. **13**, 313 (1968).
[45] F. Bonczek, T. Engel and E. Bauer, Surface Sci. **97**, 595 (1980).
[46] C. Zhang, M.A. vanHove and G.A. Somorjai, Surface Sci. **149**, 326 (1985).
[47] K.-J. Song, J.C. Lin, M.Y. Lai and Y.L. Wang, Surface Sci. **327**, 17 (1995).
[48] C.-H. Nien and T.E. Madey, in preparation.
[49] K.-J. Song, C.-Z. Dong and T.E. Madey, Langmuir **7**, 3019 (1991).
[50] S.P. Chen, Surface Sci. **274**, L619 (1992).
[51] H.-S. Tao, T.E. Madey, G.K. Wertheim and J.E. Rowe, Surface Sci. (1996).
[52] R.A. Campbell, J. Guan and T.E. Madey, Catal. Lett. **27**, 273 (1994).
[53] C.-Z. Dong, S.M. Shivaprasad, K.-J. Song and T.E. Madey, J. Chem. Phys. **99**, 9172 (1993).
[54] J.R. Engstrom, D.W. Goodman and W.H. Weinberg, J. Am. Chem. Soc. **110**, 8305 (1988).

SIZE EFFECTS IN HETEROGENEOUS CATALYSIS

A Surface Science approach

C.R. HENRY, C. CHAPON, S. GIORGIO AND C. GOYHENEX
CRMC2-CNRS, Campus de Luminy, case 913
13288 Marseille, cedex 09, France

Abstract. Often the activity and the selectivity of heterogeneous catalysts depend on the size of the metal particles. The origin of these size effects is not yet understood. A new way to investigate the origin of the size effects is to use supported model catalysts. They are prepared by vapour growth, under UHV, on oxide single crystals. By epitaxy at high temperature, homogeneous collections of clusters are obtained. The various techniques available to characterize the nucleation and growth, the structure and morphology and the electron properties of the metal clusters are described on practical examples. Two examples of size effects, for Pd clusters supported on oxide single crystals, are described: the CO chemisorption and the CO oxidation. In these two particular cases, the size effects are interpreted by a stronger binding energy of CO at low coordinated Pd atoms (i.e. edge atoms). More generally chemical and electronic properties of small metal clusters depend not only on their size but also on their morphology and surface structure.

1. Introduction to size effects in catalysis

1.1. INTRODUCTION

Industrial catalysts are constituted by metal clusters (1-10 nm in size) supported on an oxide powder (alumina, silica....). Metals with catalytic properties are, for the most part, late transition metals (VIII group) and coinage metals (IB group). In order to save precious metal, catalyst designers try to disperse finely the metal clusters since only surface atoms play a role in catalysis. The metal dispersion is the ratio of surface atoms to the total number of atoms. Ideally one seeks for a two-dimensional (2D) film that corresponds to a unity metal dispersion. However it has been observed, since a long time, that by increasing the metal dispersion (or decreasing the cluster size) the reaction rate per surface atom of the catalyst metal

R.M. Lambert and G. Pacchioni (eds.),
Chemisorption and Reactivity on Supported Clusters and Thin Films, 117-152.

(generally called turnover number, TON) can vary for some reactions. The TON can increase, decrease, go through a maximum or be constant (see fig.1 and 2). It is noticeable that the important variations of the reaction rate generally occur below 5 nm. In fact this size corresponds to the point where the proportion of surface atoms becomes non negligible (> 10 %), in other words the mean coordination number of the atoms in the clusters becomes different from the bulk one. We recognise here a general observation: the physical and chemical properties of metal clusters change relatively to the bulk when the ratio of number of surface atoms to number of bulk atoms becomes important. Catalytic reactions leading to variation of the TON with metal dispersion are called **structure sensitive,** those which do not depend of the dispersion are called **structure insensitive**. This classification, given by Boudart [1], resulted from the idea that by changing the particle size one changes the nature of the exposed facets (or the type of site geometry). Thus if a reaction does not depend on the macroscopic single crystal face it does not show a size effect, it is called structure insensitive.

1.2. EXAMPLES OF SIZE EFFECTS IN HETEROGENEOUS CATALYSIS

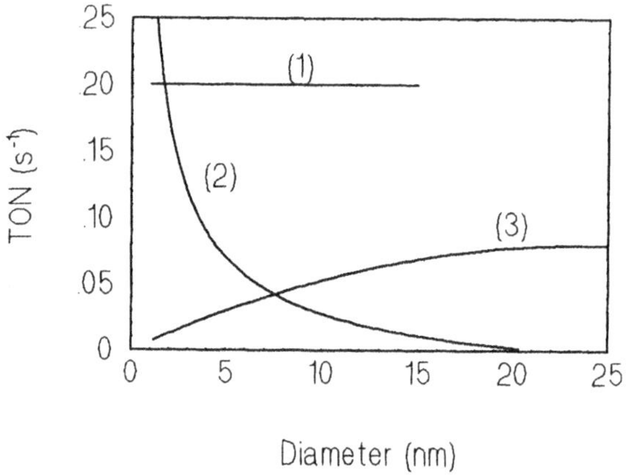

Figure 1. Turnover number for selected reactions versus particle size. (1) Benzene hydrogenation on Pt/SiO$_2$, (2) Ethane hydrogenolysis on Rh/SiO$_2$, (3) CO hydrogenation on Ru/Al$_2$O$_3$ (after Che and Bennett, fig. 7 [2]).

The size effects in heterogeneous catalysis have been extensively reviewed recently by Che and Bennett [2]. From that survey some characteristic features emerge. Hydrogenation reactions are generally considered as structure insensitive. It is the case of the hydrogenation of benzene on an Pt/SiO$_2$ catalyst, represented on curve (1) on fig.1. Curve (2) corresponds to the hydrogenolysis of ethane on a Rh/SiO_2 catalyst which is structure sensitive. The TON increases by nearly 2 orders of magnitude as the particle size decreases. In the last example, curve (3), corresponding to the CO hydrogenation on Ru/Al_2O_3, giving methane and water, the TON decreases as the cluster size drops. In the last example we see that in some cases corresponding to a decrease of the activity when clusters becomes smaller the size effect is observed from very large cluster sizes. Size effects are observed also in the selectivity (the preference for a given product among several ones produced by the reaction). This last point is very important for industrial reactions. Often both activity and selectivity vary with the metal dispersion.

It is often thought that hydrogenolysis reactions are structure sensitive reactions and hydrogenation structure insensitive reactions. In fact it is far from being a general law. For example, hydrogenation of butadiene or butyne exhibits a size effect while hydrogenation of butene is structure insensitive. However we have to note that: first, size effects for hydrogenation reactions correspond to a decrease of the activity when the clusters size drops, and second the variation of activity are relatively small (below a factor 10). Up to now we have considered only the size of the clusters without taking into account for the nature of the metal catalyst. Fig.2 shows the variation of the TON for the hydrogenolysis of ethane on Rh, Ir and Ni catalysts. It is clear that for the three metals, size effects are observed but they are qualitatively and quantitatively different. Also the reaction rate can depend on the nature of the support: compare curve (1) and (4), corresponding to Rh on SiO_2 and TiO_2 substrate, respectively.

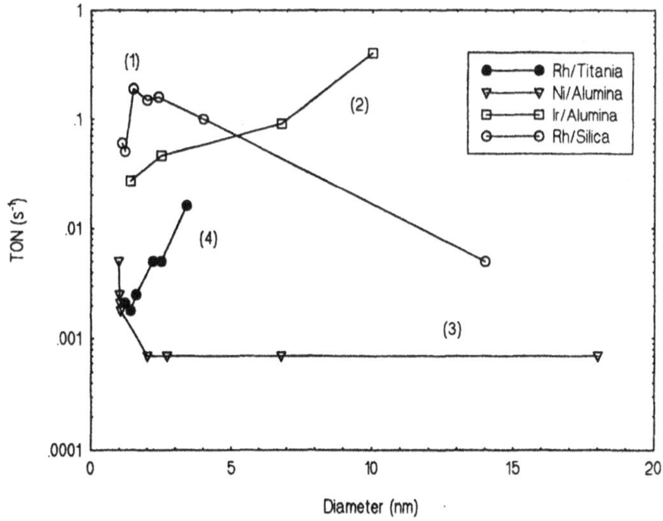

Figure 2. Hydrogenation of ethane on various catalysts (after Che and Bennettt, fig.14 [2]).

In summary size effects are observed in heterogeneous catalytic reactions in the activity as well as in the selectivity. It is very difficult to draw an unified picture of these size effects because they not only depend on the reaction but also on the nature of the metal particles and on the catalyst support.

1.3. POSSIBLE ORIGINS FOR SIZE EFFECTS

1.3.1. *Geometric Effect*

The first hypothesis put forward to explain size effects in heterogeneous catalysis is based on the variation of the proportion of active sites with the size of the metal clusters. This idea has been largely spread from the work of Van Hardeveld and Hartog [3]. They assumed that metal clusters have regular polyhedron shapes like octahedron, truncated octahedron, tetrahedron, cube or more exotic ones like icosahedron. Assuming a given shape it is possible to grow the clusters by successive shells. For example an octahedron gives the following series of numbers for the successive closed shells: 6, 19, 44, 85, 146, 231, 344, 489, 670, 891,......$1/3m(2m^2+1)$. m is the number of atoms in the equivalent edges (the number of shells is m-1). The simplest shapes to be considered are represented on fig.3 for the fcc structure.

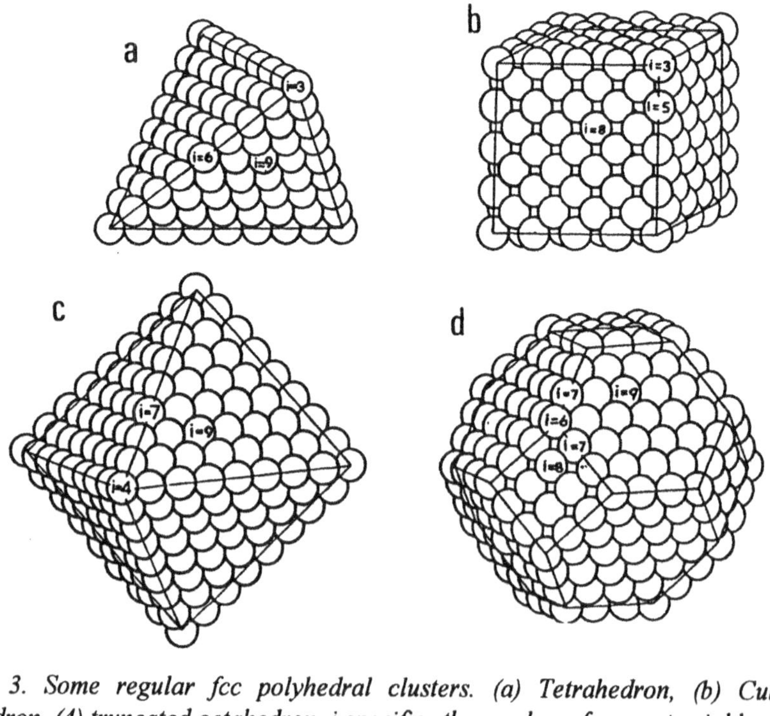

Figure 3. Some regular fcc polyhedral clusters. (a) Tetrahedron, (b) Cube, (c) octahedron, (4) truncated octahedron. i specifies the number of nearest neighbours.

They are the tetrahedron and the octahedron limited by (111) facets, the cube limited by (100) facets, the truncated octahedron limited by both (111) and (100) facets. These basic shapes are important because (111) and (100) facets have the lowest surface energies for fcc metals. It is possible to calculate the number of atoms on surface and on the edges following the formula given in the following table.

TABLE 1: Total number of atoms (N_T), number of surface atoms (N_S) and number of edge atoms (N_E) for closed shell clusters with various shapes.

Shape	tetrahedron	cube	octahedron	truncated octahedron
N_T	$1/6m(m+1)(m+2)$	$4m^3-6m^2+3m$	$1/3m(2m^2+1)$	$16m^3-33m^2+24m-6$
N_S	$2m^2-4m+4$	$12m^2-24m+14$	$4m^2-8m+6$	$30m^2-60m+32$
N_E	$6m-8$	$12m-16$	$12m-18$	$36m-48$

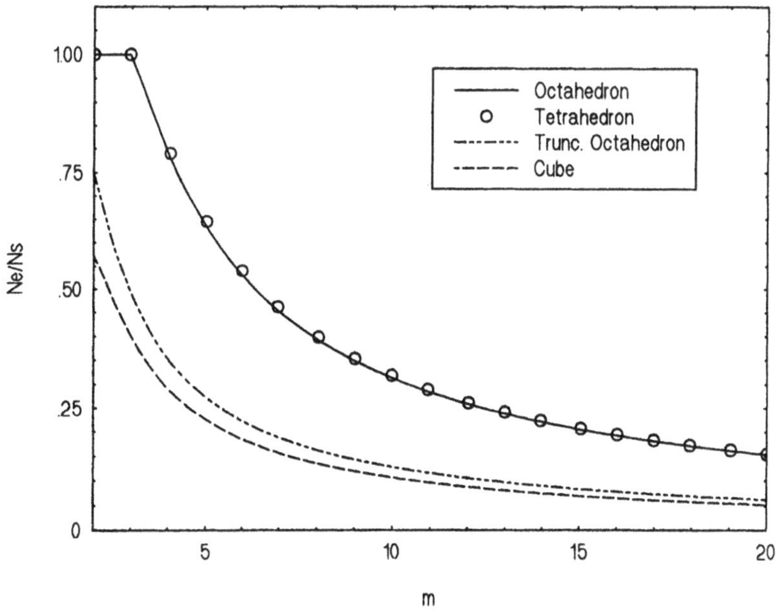

Figure 4. Number of edge sites to number of surface sites for some polyhedral clusters.

The basic idea under the « geometrical effect » is that the coordination number decreases going from a bulk atom to a surface atom, an edge atom and a apex atom. As the proportion of low coordinated atoms increases when the cluster size decreases one expects size effects if the reaction rate effectively depends on the coordination of the atoms involved in the reaction site. In fig.4 we have plotted the ratio N_E/N_S as a function of the cluster size for closed shell tetrahedra, octahedra, truncated octahedra and cubes as a function of m. It is clear that for m below about 20 (cluster size ~5 nm) the fraction of edge atoms increases strongly, it is precisely the range where size effects are the most important. A more rigorous test of the geometric hypothesis is hampered by serious difficulties. The first one is the lack of characterization of the actual shape of metal clusters in real catalysts. Second, the realism of the geometric models is questionable: the unicity of the shape, the smoothness of the cluster surface. Finally, it is not excluded that the cluster surface suffers restructuring during catalytic reactions.

1.3.2. *Electronic effect*

Small metal clusters are situated between molecule, having discrete energy levels, and bulk metal having continuous energy bands. For N electrons in a box the separation between electronic levels is:

$$\delta \sim \varepsilon_F / N \qquad (1)$$

where ε_F is the Fermi energy. Due to thermal energy the energy levels are broadened. If $\delta > 2$ kT (k is the Boltzman constant) the levels are separated. This phenomenon is called quantum size effect. For metals the Fermi energy is typically 5 eV thus at RT the level spacing is 50 meV. Then the critical size to observe quantum size effect will be ε_F / 2kT ~ 100 atoms. It corresponds to a small cluster size, about 1.5 nm. Therefore quantum size effects cannot explain all the size effects observed in catalysis. However clusters larger than the critical size, defined above, have not exactly the bulk electronic structure. When cluster size drops the valence band sharpens and the DOS at the Fermi level decreases (electrons are more localised). Also the maximum of the valence band and the core level show sizeable shifts. These features could be at the origin of the size effects in catalysis, more likely than a pure quantum size effect. Indeed strong correlations have been observed between reaction rate and binding energy shift or DOS at the Fermi level (measured by photoemission) for decreasing particle size. Fig. 5 shows an example for the hydrogenation of vinylacetylene on Pd catalysts [4]. However by themselves these correlations do not explain the size effect. Moreover the binding energy shifts are not always due solely to a particle size effect. Indeed for insulator substrate the charge left on the cluster during the photoemission process cannot be evacuated causing a shift of the measured binding energy. In addition it is clear that the electronic structure is related to the atomic structure of the clusters. Again, without a good structural characterisation of the catalyst, it is very difficult to investigate the origin of size effect in catalysis.

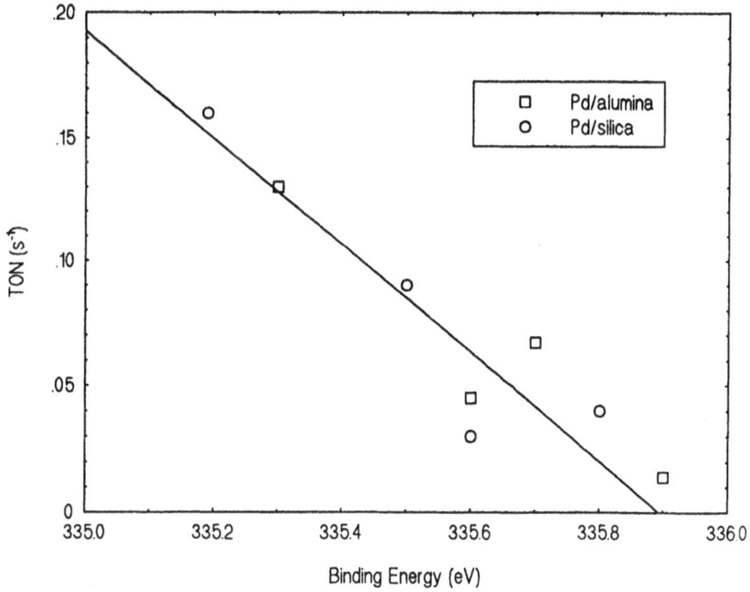

Figure 5. Correlation between TON and Pd $3d_{5/2}$ core level for the hydrogenation of vinyl acetylene on Pd catalysts (after Nosova et al. [4]).

2. Supported model catalysts: preparation and characterization

2.1. INTRODUCTION

From part 1 it becomes clear that our knowledge on the origin of size effects in catalysis is limited mainly because of the lack of characterization of catalysts. Indeed industrial catalysts are composed by small metal particles prepared (generally) by decomposition of metal salts on a high specific area oxide powder. Due to the preparation method and to their texture real catalyst cannot be (or with great difficulties) characterized by surface science techniques. Microscopy techniques can be in some case used on catalysts [5] but the observations are not generally made on a large collection of particles and the representativeness of the results is still questionable. Since about 20 years, basic investigations on heterogeneous catalysis are made on model catalysts. They are extended single crystal surfaces which are prepared under UHV, allowing the use of the huge set of surface science techniques. However single crystals of catalytic metals cannot take into account for two important characteristics of real catalysts which are the finite size of the metal particles and the presence of the support, precluding in particular investigations of size effects. To overcome this limitation surface scientists have developed a new type of model catalyst. It is obtained by growing, in UHV conditions, metal clusters on a clean oxide surface. Surface science tools can still be used to characterise the model catalysts. Meanwhile, most of surface science tools probe a sample area of a few mm² that contains about 10^{10} clusters, then the results can be interpreted at the atomic level (as for the case of a single crystal surface) only if the collection of clusters is homogeneous (sharp size distribution, uniform clusters density, single shape), at this scale, and well oriented on the substrate. This has been, for a long time, the bottleneck of the studies of supported model catalysts. However recently these difficulties have been overcome (at least partially) in the last years in two ways. The first one uses epitaxial growth of metal clusters on clean oxide single crystals or epitaxial thin films, leading (for particular experimental conditions) to uniform collections of well oriented clusters. The second way is to used the powerful new microscopy techniques: STM and AFM. Due to their (potential) atomic resolution it is not necessary to have a uniform collection of clusters but the investigation of cluster structure and cluster properties with these techniques, while very promising, is still in infancy.

A large number of oxide single crystals have been used like: Al_2O_3, MgO, TiO_2, ZnO, NiO. Due to their easy cleavage and cleaning procedures lamellar crystals have been widely used like mica, graphite, MoS_2. As an alternative to bulk oxide crystals several groups have used thin epitaxial oxide films prepared by evaporation of the oxide or by oxidation of the metal during or after deposition. Amorphous oxide thin films have been used but in this case it was not possible to get oriented well shaped clusters.

To insure a coherence, in the following we will concentrate mainly on the Pd /MgO(100) which have been extensively studied in our group. We will see how the clusters nucleate and grow on the substrate, how their shape can be controlled, what are their structure and epitaxial relationships. Finally electronic properties of Pd clusters will be shortly described.

2.2. SUBSTRATE PREPARATION

By cleavage in UHV, clean well ordered MgO(100) surfaces are easily obtained. Gold decoration reveals that the density of steps (mainly of atomic height) is low (mean length of the terraces 250 nm) [6]. The perfect order of the cleaved substrate can be tested by the diffraction of He atoms. Most of the incoming atoms (>80%) are elastically scattered in diffraction peaks [6]. Air cleavage has a dramatic effect on the nucleation of metal clusters, the step decoration desappears and the density of clusters increases by a factor 20 to 30 due to the presence of point defects, probably coming from the interaction with water. While on a UHV cleaved surface, clusters are perfectly oriented on the substrate (see fig.6a), on an air cleaved sample, after heating at 500°C, Pd clusters are randomly oriented. If the surface has been cleaned at 700°C, the orientation of the clusters has greatly improved but it is still not so perfect as for the case of UHV cleavage (see fig.6b).

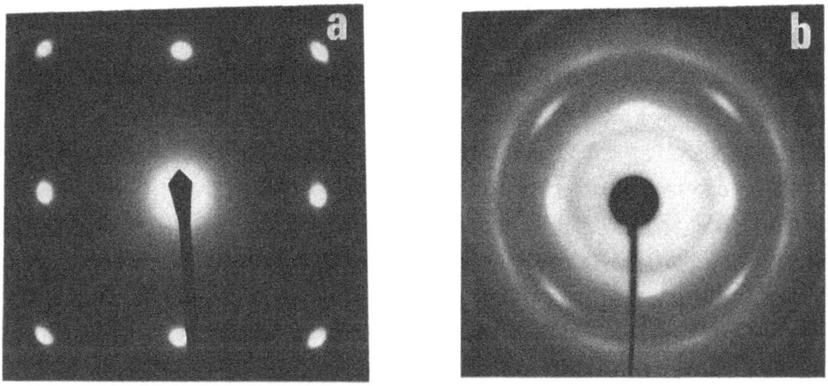

Figure 6. Electron diffraction patterns of Pd clusters grown on UHV (a) and air cleaved (b) MgO crystals.

2.3. NUCLEATION AND GROWTH

2.3.1. *Experimental conditions*

Metal clusters are grown by condensing a calibrated beam of atoms on the clean (UHV cleaved) surface. Three parameters control the nucleation and growth processes: the metal beam flux **R**, the deposition time **t** and the substrate temperature **T**. To have a constant and reproducible deposition rate, the best way is to use a Knudsen-cell evaporator, for which R is only related to the cell temperature. It is not recommended to use an electron gun evaporator oven nor an electron beam heating of the Knudsen-cell because charged particles creates defects on oxide substrates which become preferential nucleation centres.

2.3.2. *Experimental techniques*

TEM. Classically, nucleation and growth of metal clusters are studied by ex situ TEM. A deposit is obtained for some experimental conditions (R,t,T), then the sample is backed, in situ, by a thin carbon film (~20 nm). When the sample is removed from the vacuum chamber the carbon film, containing the metal clusters, is separated from the substrate by dissolution of the substrate at the interface in dilute HCl. The carbon film is observed in TEM and from the micrographies one obtains the size distribution of the clusters. If we are interested in the nucleation or growth kinetics it is necessary to do again the same procedure in varying the deposition time. These studies are very time consuming. Moreover TEM has some limitations for nucleation and growth works, especially in the early stages. It is difficult to measure accurately the size of clusters smaller that about 1.5 nm, moreover exposure to air can change the cluster shape by oxidation. For all these reasons in situ technique allowing to follow in real time the nucleation and growth processes are strongly needed.

AES-XPS. Electron spectroscopy techniques are from the early ones to be applied to in situ study of thin films growth. These methods are well established (especially AES) for the determination of growth modes. However they are more limited for quantitative studies of nucleation and growth of clusters on oxide crystals [7]. Below 1ML, it is very difficult to distinguish between different growth modes, especially at low temperature, when clusters are flat. Moreover, without additional informations from other techniques (like TEM or STM) it is generally not possible to derive quantitatively the nucleation and growth kinetics.

TEAS. We have developed recently a new method to study, in situ the nucleation and growth of metal clusters on ionic substrates [8]. This method is a particular application of the well known TEAS method widely used in metal/metal epitaxy. In the last method the attenuation of the specular peak of He scattered from the sample is measured during the growth at a constant rate. In the present case, taking advantage of the ionic substrate presenting a high corrugated He/surface potential, one measures the attenuation of a first order diffraction peak, the contribution of which providing only from the bare substrate. Thus the interpretation is relatively simple. The attenuation comes from the clusters and from a band of width ξ around them, where He atoms are diffusely scattered. Taking the case of Pd clusters on MgO, the shape of the clusters is, at not too low temperature, a truncated half-octahedron (see section 2.4). The attenuation of the diffracted beam of He is :

$$A(t) = n(t)\,[d(t) + 2\xi]^2 \qquad (2)$$

ξ is determined by the TEM measurement of the size distribution of the final deposit. Knowing the final attenuation and size distribution, ξ is obtained from eq.2. In the case of Pd on MgO(100) a value of 1.12 nm was determined [9]. By a log-log plot of the attenuation, versus deposition time, two regimes are clearly identified (see fig. 7a). The first one, corresponding to a fast attenuation, is due to simultaneous nucleation and growth, the second regime where the attenuation is slower corresponds to pure growth as the density of clusters has reached a saturation value. Knowing the saturation density by TEM observation of the final deposit, the growth kinetics is given directly applying eq. 2. in the second regime. If one assumes (in agreement with TEM measurements [10]) that the growth law is still valid before the saturation density has been reached, the nucleation kinetics is obtained again from eq.2 applied on the first regime. Fig 7 shows nucleation (b) and growth (c) kinetics obtained for Pd on MgO (100) [9]. Qualitatively the density of nuclei increases rapidly and reaches a plateau: the saturation density (as already observed by TEM [10]). The main advantage in this new method is that the transient part of the nucleation kinetic can be observed at temperature lower than 300 °C when the clusters are too small to be seen directly by TEM.

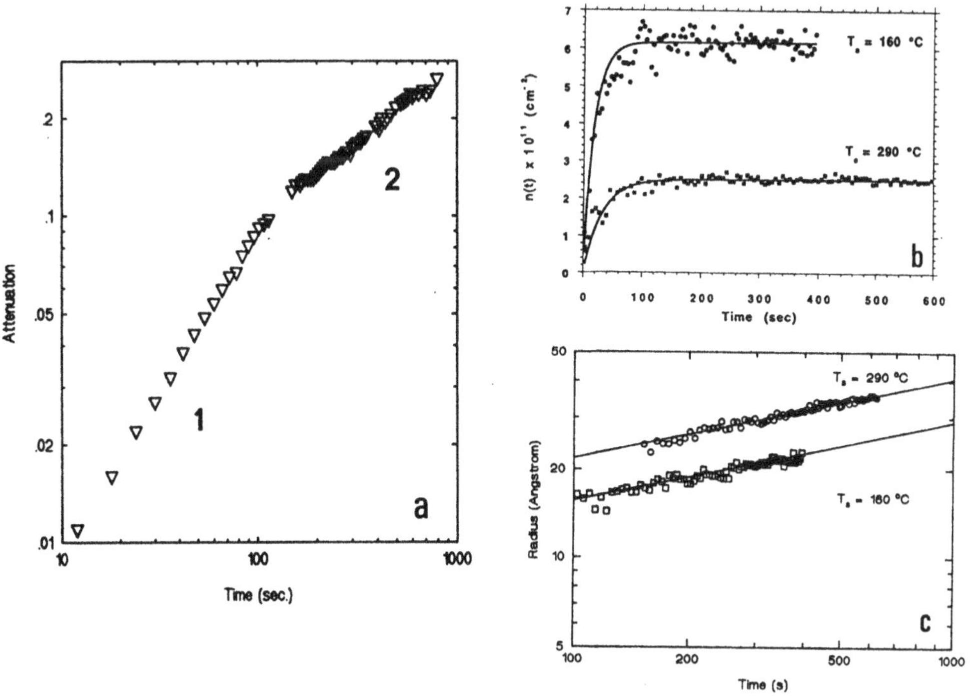

Figure 7. Application of the diffraction of He atoms to the nucleation and growth of Pd clusters on MgO(100). (a) Attenuation of the (11) diffracted peak versus deposition time, (b) nucleation kinetics, (c) growth kinetics (after Meunier and Henry [9]).

From a detailed analysis of the nucleation kinetics as a function of T and R it was concluded that the nucleation was mainly controlled by point defects. The growth kinetics follows a power law:

$$d = d_o\, t^p \qquad (3)$$

where p is around 0.3.

STM / AFM. New scanning microscopies: Scanning Tunneling Microscopy (STM) and Atomic Force Microscopy (AFM) are very promising for nucleation and growth studies due to their atomic resolution capability. STM needs conducting samples while AFM can be used with conducting and non-conducting samples. The atomic resolution is obtained currently in UHV by commercial STM. For the first time the size of the critical nucleus has been directly measured from STM images in the case of the 2D nucleation of Ag on a Pt (111) sample [11]. For substrates concerned by model catalysts no nucleation or growth kinetic studies have been yet published. However the capabilities of STM, in this field, has been demonstrated. Comparative studies by TEM and STM on the same samples have shown, in the case of Pd/graphite [12] that beyond the resolution limit of the TEM (d > 1 nm) the two techniques give the same number density of clusters. STM is definitely more efficient in the early stages of nucleation. As concerning the growth studies the cluster size was overestimated (by ~ 30%) relatively to TEM by a convolution of the cluster shape with the tip shape. For bulk insulating substrates STM cannot be used. An alternative solution is to use a thin oxide layer on a metal single crystal [13,14]. For bulk insulators AFM can be used but some difficulties arise. By using normal forces (~1 nN) small clusters are moved as seen for Pd/mica in air [15] or Ag/Al_2O_3 in UHV [16]. This problem is almost solved now by using low forces or non contact mode. The convolution with the type shape is more important than with STM because the tip is larger. Besides the actual limitations of the AFM and STM techniques it is clear that they will play a prominent role in the near future for nucleation and growth studies.

2.3.3. *Results and discussion*

Basic mechanisms. It is out of the scope of this paper to describe in detail the nucleation and growth theory of supported clusters, several review papers are available [17,18]. In this paper only the microscopic mechanisms and some theoretical results will be given. When a metal vapour impinges on a substrate (we take the example of a weak interacting substrate, like an oxide), atoms adsorb with a sticking probability α (it is generally assumed to be unity but it is not always true [18]). Adsorbed atoms diffuse on the substrate and finally desorb after a life time τ. The mean travel distance of an adatom before desorption is Xs which is given by the following formula:

$$Xs = (D\tau)^{1/2} = (a/2) \exp [(Ea-Ed)/2kT] \qquad (4)$$

where D is the diffusion coefficient, a the distance between two neighbouring adsorption sites, Ea and Ed the respective adsorption and desorption energies. On a uniform perfect substrate nucleation occurs when two diffusing atoms encounter. The smallest stable cluster is the dimer, because in practical conditions the supersaturation is very high (> 10^{10}). The nucleation frequency will be proportional to the square of the adatoms density. For a not too short time and a not too high coverage the density of adatoms is $\alpha\tau R$ (R is the atom flux reaching the substrate), then the nucleation rate is proportional to R^2. If point defects are present on the substrate (it is often the case on oxides) and if the adsorption energy of a metal atom on a defect is much larger than on a normal site a single atom adsorbed on a defect becomes the smallest stable cluster, then the nucleation frequency will show a linear dependence with the impinging rate R. In a pure nucleation on point defects the maximum cluster density is given by the density of defects while in the case of a perfect surface (random nucleation) the maximum density of clusters will depends on the substrate temperature and the deposition rate (R) [17].

Growth process occurs via two mechanisms: direct impingement on the clusters surface or capture of a diffusing adatom at the periphery of the cluster. The growth rate of a cluster has been theoretically solved in the case of a regular array of clusters [19]. It can be shown than, in some particular conditions, the clusters grow as a power law of the deposition time (eq. 3). In the case (called extreme incomplete condensation) where there is no competition between the clusters for the capture of adatom (i.e. the distance between two clusters is much larger than Xs) clusters are isolated and the exponent p grows continuously from 1/3 to 1 when the growth proceeds. The prefactor, d_o is proportional to Xs at the power 3/2, then it decreases when temperature increases. One other limiting case (called complete condensation) is obtained when all the adatoms are captured by growing clusters before they can desorbs (i.e. the distance between neighbouring clusters is much smaller than Xs). The exponent is equal to 1/3 and stays constant up to the coalescence regime. The prefactor is inversely proportional to the density of clusters.

Comparison with experiments. The most studied system for the nucleation of 3D clusters is certainly Au/NaCl [20]. For oxide surfaces results are scarce. In our, in situ, study of Pd/MgO by TEAS the initial nucleation rate showed an almost linear dependence of the deposition rate interpreted by a nucleation on defects mechanism [21]. However contrary to a pure point defect nucleation the saturation density of clusters depends on the temperature (see fig. 7b). On fig. 8 we have plotted the saturation density of Pd clusters as function of 1/T for various substrates. For the Pd/oxide systems studied so far, the saturation density depends on the temperature with a well defined activation energy (see following table):

$$n_s = n_o \exp(Eact/kT) \qquad (5)$$

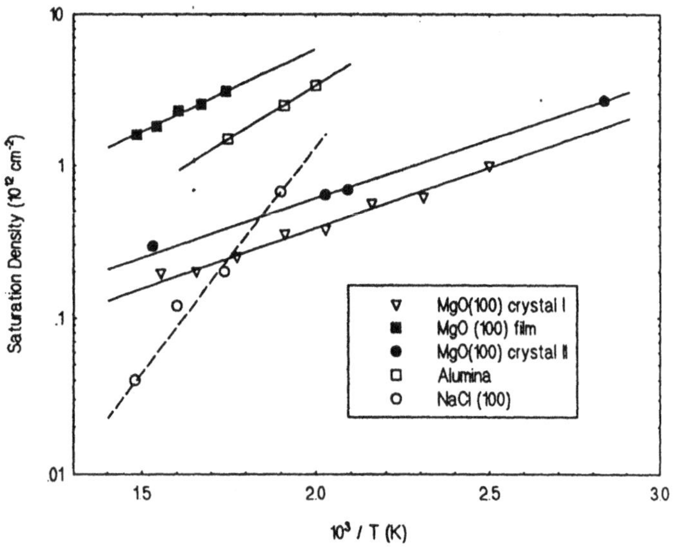

Figure 8. Saturation density versus temperature for Pd clusters on various substrates.

TABLE 2. Thermal activated variation of the saturation density of Pd clusters on various systems

Substrate	n_o (cm^{-2})	Eact(eV)	Reference
MgO(100) film	3.9×10^{10}	0.216	[10]
MgO(100) crystal I	1.7×10^{10}	0.154	[21]
MgO(100) crystal II	9.8×10^{9}	0.158	[21]
alumina	4.6×10^{9}	0.284	[18]
NaCl(100)	1.7×10^{6}	0.586	[22]

It is interesting to note that MgO single crystals coming from different batches give different curves having almost the same activation energy. This fact can be only explained by a variation of the crystal purity, resulting in different surface defects (or impurities) densities. In another study on epitaxial MgO films which are less perfects than single crystals the saturation density was much higher with a slightly higher activation energy. Polished alumina single crystals give saturation densities between MgO films and MgO cleaved single crystals. The behaviour of cleaved NaCl is different than the oxide case, the activation energy is very high. From the available data it seems that nucleation of metal clusters on oxide is strongly influenced by point defects (or impurities) but the temperature dependence of the saturation density cannot be explained by a pure nucleation on defects, it is possible that both random nucleation and defect nucleation occur simultaneously or that small clusters are only temporary trapped on defects as in the case of Au on NaCl [20].

Considering the growth kinetics, in situ experiments by TEAS for the Pd/MgO(100) system showed (see fig. 7c) a power law dependence of the diameter with the deposition time giving an exponent close to 0.3 on the whole deposition time range. This result suggests a complete condensation. In addition the prefactor d_o increases with temperature as expected for the complete condensation process because the number density decreases as the temperature increases. In fact exact calculations of the growth kinetics showed that the condensation was not complete but that the competition between clusters was still very important [21]. Fig. 9 shows log-log plots of the diameter of Pd clusters versus exposure for different insulating substrates. The data are well fitted with nearly parallel straight lines, indicating a power law dependence. Prefactor and exponents are reported in the following table.

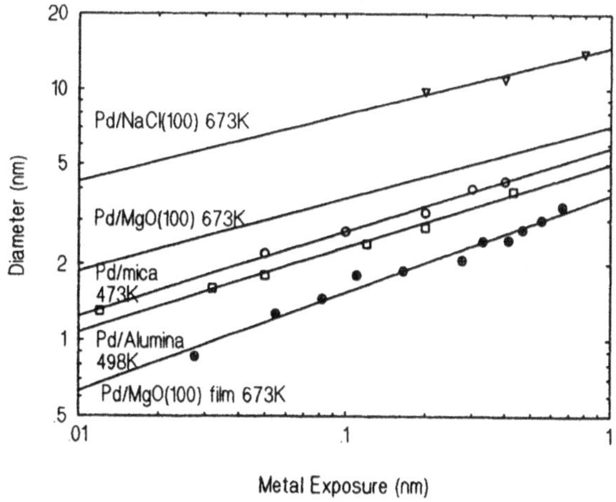

Figure 9. Growth kinetics of Pd clusters on various substrates.

TABLE 3. Growth kinetic parameters for Pd clusters on various substrates

Substrate	Temperature (K)	d_o (nm)	p
NaCl(100)	673	14.66	0.27
MgO(100)	673	7.11	0.29
MgO(100) film	673	3.77	0.39
Mica (0001)	473	5.83	0.33
Alumina	498	5.02	0.33

All the systems represented in table 3 show a power law dependence with nearly 1/3 exponent, indicating a complete or nearly complete condensation. It is due to the high density of clusters on these systems, the distance between clusters is always smaller than Xs.

From a practical point of view of preparing supported model catalysts it is possible to obtain homogeneous collections of metal clusters by vacuum condensation. By changing the temperature of the substrate one can adjust the density of clusters (beyond the short transient nucleation period). At a fixed temperature the size is adjusted by the deposition time. Knowing the size distribution of clusters (for example by TEM) for some temperatures and deposition times one can easily predict the size and the cluster density for any conditions (below the coalescence regime) with the simple laws given by eqs 3 and 5.

2.4. CLUSTER SHAPE

2.4.1. *Introduction*

In order to study size effects it is very important to know precisely the shape of the clusters. The shape of metal clusters is governed by two factors a kinetic one and a thermodynamic one. If clusters are grown far from equilibrium conditions the clusters shape will be dictated by the microscopic growth mechanisms, on the contrary, if growth occurs at very low supersaturation or if the clusters are annealed at sufficiently high temperature, their shape will result from the minimisation of the surface (and interface) energies which gives the equilibrium shape. The problem of the determination of the equilibrium shape of crystals was already addressed in 1901 by Wulff. He showed that the equilibrium shape of a macroscopic crystal is a polyhedron (at 0°K) which is limited by crystal faces of lowest energy.

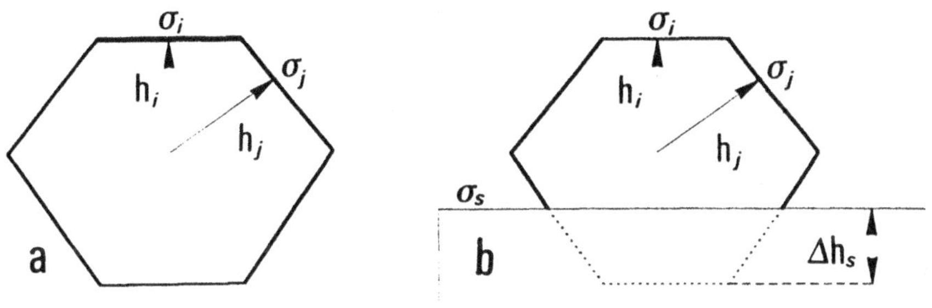

Figure 10. Schematic representation of the equilibrium shape of a crystal given by the Wulff's theorem. (a) free crystal, (b) supported crystal.

Practically the equilibrium shape can be constructed by applying the following relation known as the Wulff theorem:

$$\sigma_i / h_i = Cst \qquad (6)$$

where σ_i is the surface energy of a given face and h_i the central distance to this face. Fig. 10a illustrates this theorem. For an fcc metal the equilibrium shape is a truncated octahedron with hexagonal (111) facets and square (100) facets (see fig. 3d). At high temperature the extension of the facets decreases and the rounded parts appears, near melting point the equilibrium shape is almost a sphere [23]. If the crystal is supported the equilibrium shape is truncated at the interface. One can modify the Wulff theorem to take into account for the adhesion energy (see fig. 10 b). Then:

$$\Delta h_s / h_i = \beta / \sigma_i \quad (7)$$

where Δh_s is the amplitude of the truncation and β the adhesion energy. If β is zero, Δh_s is zero and the equilibrium shape is those of the free crystal, if $\beta > 2\,\sigma_i$, $\Delta h_s = 2\,h_i$ as if the crystal has sunk in the substrate, in reality it means that the crystal is a 2D layer on the substrate. Thus, when the cluster-substrate interaction increases the shape becomes flatter. Now we will consider the effect of the cluster size on the equilibrium shape. The major changes arising in decreasing the size is the relative increase of the proportion atoms on the edges (see fig. 4), then the edge (and corner) energy has to be taken into account to minimise the total surface energy. This problem has been recently approached by molecular dynamics. For very small clusters of fcc metals the more stable shape is an icosahedron but for the size relevant in catalysis (> 1 nm) the equilibrium shape is always a truncated octahedron [24]. However it is noteworthy that the energy difference between the various shapes is weak that may explain the TEM observation of shape fluctuations of small clusters. However the shape of minimum energy is strongly favoured in the case of epitaxial clusters.

2.4.2. *Characterisation of cluster shape*

TEM. By conventional TEM only the outline of the clusters can be observed. For clusters larger than 5 nm the Weak Beam Dark Field technique is used to determine the 3-dimensional shape of supported clusters. By this method thickness fringes appear in the image of the particles allowing the indexation of the different facets (if any). This technique has been applied in the field of model catalysts for Pd/MgO [25] and platinum group metals on NaCl [22].

For small clusters (<5nm) the WBDF technique can no longer be used but High Resolution TEM becomes very efficient if the clusters are deposited on microscopic substrates with a well defined shape. Indeed oxide microparticles like MgO cubes [26,5], ZnO prisms [27] or Al_2O_3 spheres[28] can be prepared by combustion of the metal in controlled atmosphere. The observation in high resolution (giving the position of atomic columns) of the clusters profile in different directions gives the actual shape of the clusters down to 1 nm.

STM/AFM. For large clusters STM and AFM techniques give distorded images due to the convolution of the actual shape by the tip shape. However if the actual shape of the tip or of the clusters is known it could be possible to deconvolute the tip effect. For small clusters in some cases (probably when a microtip is present at the apex of the tip) atomic resolution is obtained on the clusters. The ordering of surface atoms allows a direct identification of the facets. Tetrahedron Pd clusters with (111) facets [29] have been observed on graphite. The possibility to follow in situ the evolution of the cluster shape during adsorption or a chemical reaction is very promising.

2.4.3. *Effect of temperature on the cluster shape*

By changing the growth temperature of the clusters it is possible to switch from a kinetic growth shape to the equilibrium shape. Fig 11 shows the evolution of the shape of large Pd clusters on MgO (100) with the growth temperature. At room temperature (RT) the clusters are rough with no geometric shape although they are in a perfect epitaxial orientation on the substrate. At 433 K smaller particles become faceted, while the larger ones have still no regular shape. The well shaped particles are relatively flat but the top (100) facets are rough. At 673 K all the particles present well regular shapes, with smooth facets of orientations (111), (100) and at less extend (110). Smaller particles are square pyramids with (100) top truncation. Larger particles are elongated in the [110] or [100] directions of the MgO substrate, and they are also truncated on the edges.

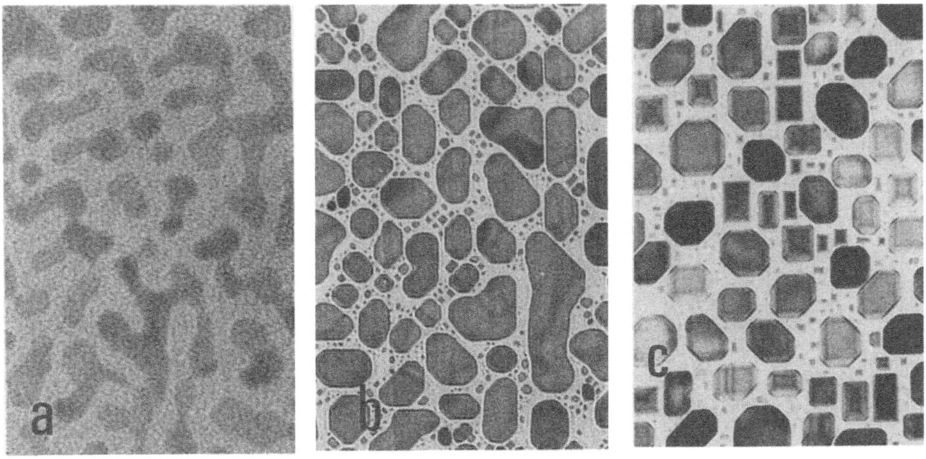

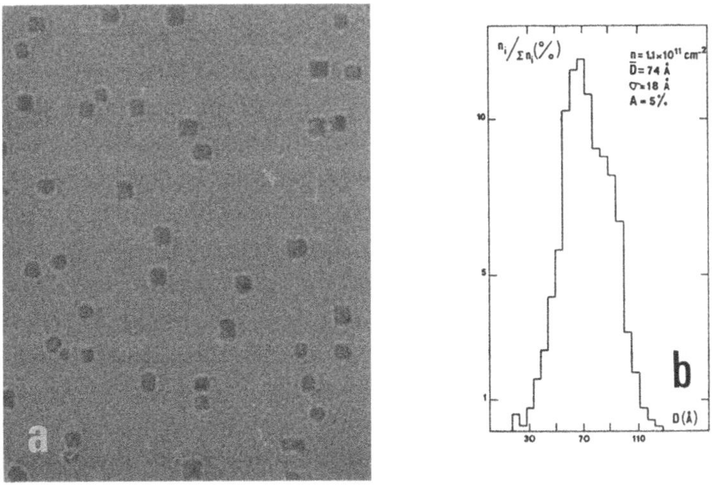

Figure 12. Homogeneous collection of Pd clusters deposited at 673K. (a) Micrography, (b) Size histogram.

For a thinner deposit (< 3 ML) a uniform collection of particles with the single shape of a truncated half octahedron and a narrow size distribution is obtained (see fig.12). The same shape with a height/base ratio of 0.4 was observed in the size range 1 to 10 nm [25]. Fig. 13 shows an HRTEM image of 2 nm Pd clusters grown at 673K on MgO microcubes (synthesised in the UHV chamber) [26].

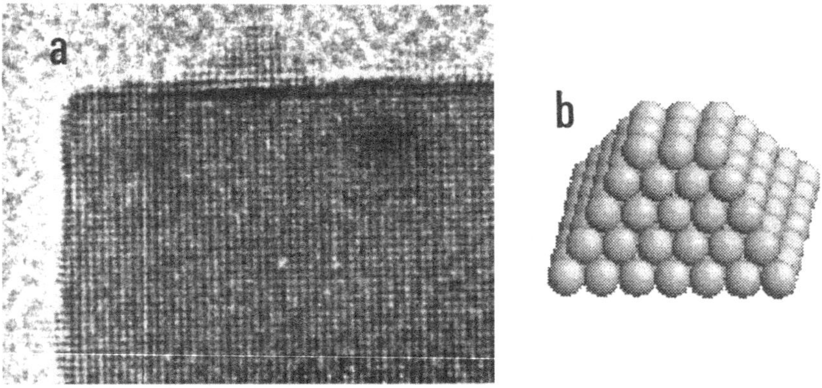

Figure 13. Pd clusters deposited on MgO(100)at 673K. (a) HRTEM profile, (b) ball model of a 2 nm Pd cluster.

These observations can be rationalised by considering the microscopic events during the growth of the clusters. We have seen that the clusters grow by atoms impinging directly on the clusters surface or by capture of adatoms diffusing on the substrate. If the temperature is low, the Pd atoms do not diffuse on the Pd surface and we are more or less in the case of ballistic deposition leading to rough clusters. At a given temperature (373-423 K) Pd atoms move on the cluster surface and regular facets appear to lower the total energy surface. At high temperature and for not too large particles, a single shape, independent of the size (1 to 10 nm) is observed that must be close to the equilibrium shape. At the temperature used during the growth experiments the vapour pressure of Pd is very small and the only way for the clusters to reach or keep the equilibrium shape is surface diffusion. To quantify this scheme one has to compare the diffusion rate of Pd atoms on the cluster surface with the arrival rate of Pd atoms. To keep the same shape during the growth we assume that between the arrival of two atoms on the same site, a Pd adatom has enough time to reach any site on the cluster surface. Practically we calculate the time necessary for an adsorbed Pd atom to travel through 5 nm (typical cluster size) which is given by the Einstein formula:

$$X = (Dt)^{1/2} \qquad (8)$$

Taking for the self diffusion coefficient $D_o = 0.001\ cm^2\ s^{-1}$ and Ed = 0.76 eV, the calculated time is reported in table 4. The time between the arrival, on the same site, of two Pd atoms from the vapour beam is typically 100 s. We see from table 4 that at room temperature surface diffusion is too slow to get a smooth regular shape, while from 573 K the diffusion is sufficiently fast as regarding the impinging rate to keep a well defined shape for the range of size considered here.

TABLE 4. Variation with temperature of the mean time for a Pd adatom to travel through 5 nm

T (°K)	300	423	573	623	673
t_{5nm} (s)	1430	0.3	.001	$3x10^{-4}$	$1x10^{-4}$

At 573 K one can observe regular shapes for small sizes (below 10 nm) but not for larger ones as observed on fig.11. However the previous conditions insure a smooth regular shape but not necessarily the equilibrium shape which is obtain at zero supersaturation. For a macroscopic crystal it is possible to calculate the time to reach the equilibrium shape from an arbitrary shape. Starting from a cube of size r and neglecting the nucleation of facets, the time to reach the equilibrium shape by surface diffusion is [23]:

$$2kT\ r^4 / v_o^{4/3}\ D\ \sigma \qquad (9)$$

with v_o and σ are the atomic volume and the mean surface energy (2000 erg/cm²) of Pd, respectively. This time depends on the cluster size with the power 4, this explains that the equilibrium shape can be obtained in some conditions for small clusters and not for large clusters. For a 5 nm cluster this time is 320 hours at RT, 2 min at 473 K and only 0.05 sec at 673 K, this time has to be compared to the typical deposition time 10 min and annealing time 30 min. For 70 nm clusters (large clusters in fig 11c) the time to reach the equilibrium shape at 673K is about 2000 sec. We understand that the equilibrium shape can be observed only at a critical temperature. These calculations are not necessarily correct for small clusters where the edge energy and edge diffusion are to be taken into account. However the agreement with observations is qualitatively good.

In summary, the shape of the clusters depends on the surface temperature and on the growth rate. At low temperature one obtains flat rough clusters. When the temperature increases the clusters become faceted, smaller ones have the equilibrium shape. Practically for Pd clusters and a deposition rate of 0.01 ML/s it is possible to obtain a uniform collection of clusters having a unique shape of a truncated half octahedron (assumed to be the equilibrium shape) by growth at 673 K. In the case of NaCl [22] the same shape has been obtained. On substrates with a three fold axis like mica [30], graphite or MoS_2 [12], the high temperature shape is a truncated tetrahedron which can be also the equilibrium shape as a part of the Wulff polyhedron.

2.5. STRUCTURE AND EPITAXY

The structure and the epitaxy of supported clusters is often studied ex situ by TEM. Electron diffraction inside an electron microscope gives directly the epitaxial relationship by comparing the diffraction spots of the substrate and of the clusters. For Pd clusters grown at all temperatures (300-800 K), on UHV cleaved single crystals one observes a single epitaxial orientation. On graphite, epitaxy is observed at temperature higher than 550 K. However two epitaxial orientations are observed [12]. Most of the metal clusters show a lattice contraction when their size decreases. This effect is explained by the surface stress which induces a pressure inside the clusters. The amount of contraction is given by the following formula:

$$\Delta a/a = -4/3\ K\ F/d \qquad (10)$$

where K is the compressibility of the metal, F the surface stress and d the size of the clusters (assumed to be a sphere). The 1/d law is well verified for most of clusters supported on amorphous substrates. However in the case of Pd several studies showed an expansion for decreasing particle sizes [26]. However recently contraction has been also observed for Pd clusters embedded in a polymer matrix [31]. In fact the dilatation observed on Pd was due to pseudomorphism in the case of epitaxial clusters or diffusion of impurities (carbon, hydrogen or oxygen) in the other cases when the Pd clusters were exposed to air or supported on carbon. In order to avoid contamination by air exposure, in situ techniques are needed. Several techniques for surface structure determination like, SEELFS, SEXAFS, RHEED and SPA-LEED have been already applied to metal clusters supported on oxide single crystals. Epitaxial relationships can be, in principle, determined by STM (or AFM) in the direct space if both clusters and substrate lattices can be observed simultaneously. This can be achieved for 2D islands on metal or semiconductors. For 3D clusters this has been recently performed for Pd clusters on MoS_2 (0001) [32]. However care must be taken with STM observations because surface atoms can be observed as bumps or hollows depending on the imaging conditions (voltage, tip/sample distance).

In summary the structure and the epitaxy of metal clusters supported on oxide substrates can be studied by various ex situ or in situ techniques. Integral techniques using diffraction methods (RHEED, LEED) or local interferences in electron waves (EXAFS, SEELFS) need a uniform distribution of the clusters to give meaningful data, while techniques with high spatial resolution (TEM, STM), which are generally more difficult to use because they need special care in the sample preparation, can be used on isolated clusters.

2.6. ELECTRONIC PROPERTIES

As we have seen in introduction very small clusters should present discrete electronic levels. At what size this separation becomes effective? The answer is not so short cut, photoemission on free mass selected copper clusters showed a continuous evolution of the electronic structure, already for about 20 atoms a well defined band structure was observed. For supported Pd clusters, photoemission shows a continuous evolution of the band structure, with three characteristic features in the valence band when the clusters size decreases: (i) a narrowing of the band, (ii) a shift of the maximum of the band towards higher binding energy, (iii) a decrease of the density of states at the Fermi level. However in the preceding works the size distribution of the clusters was not known and their actual shape even less.

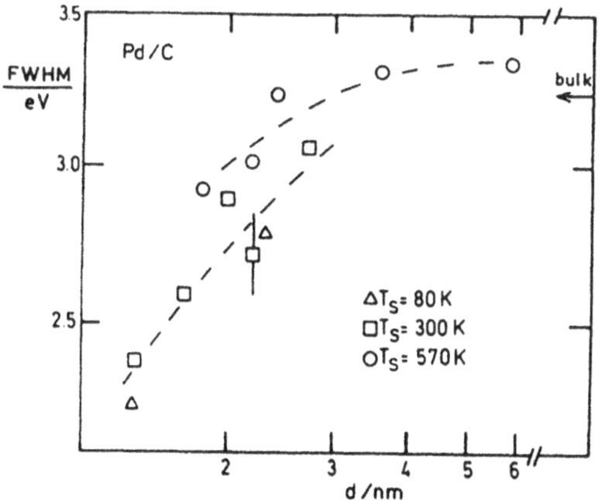

Figure 14. Valence band width versus diameter of Pd clusters grown on amorphous carbon at various temperatures (from Kuhrt and Harsdorff [33]).

A very interesting study was done in Hamburg in studying the valence band (and core levels) by photoemission on Pd clusters grown, at various temperatures, on amorphous carbon [33]. The size distribution and the cluster shape of the studied samples was investigated by TEM. In fig.14 the width of the valence band is plotted again the size of Pd clusters deposited at various temperatures. It is striking that the valence band depends on the cluster size but also on the deposition temperature. The valence band is larger on clusters deposited at high temperature. The band width depends roughly on the square root of the mean coordination number of the Pd atoms in the clusters. Indeed when the cluster size drops the coordination decreases (below around 5 nm). But, as we have seen previously, clusters grown at low temperature are flatter than those grown at high temperature. Thus, for the same size, the mean coordination number of a cluster grown at high temperature will be larger and the valence band will be wider than for clusters grown at low temperature. It is clear from this experiment that to analyse correctly the electronic properties of supported clusters we have to know not only their size but also their shape. Measuring by photoemission the valence band as a function of the amount of deposited metal is by no means sufficient. It is clear from these experiments that the electronic structure of metal clusters is strongly dependent of the mean coordination number which vary both with the number of atoms and the cluster shape. However when we are concerned by the chemical reactivity of metal clusters the local electronic structure near a reaction site will be the pertinent parameter and not the averaged band structure. One expects that changing the coordination of a surface atom the local Density of States will change.

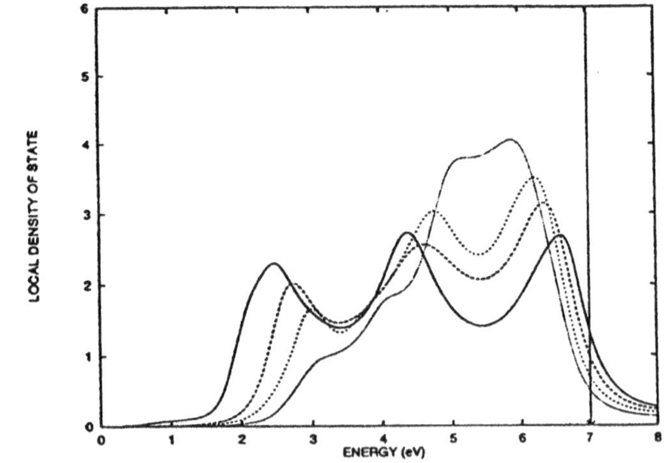

Figure 15 LDOS at Pd atoms of a 3871 Pd coboctahedron cluster at various sites: (____) centre, (——) (111) facet, (------) edge, (........) vertex (from Mottet at al. [34]).

Fig. 15 shows the local DOS in the valence band calculated by the Tight Binding Method on several sites of a Pd cluster having 3871 atoms (d=5.7 nm) and a cubo-octahedron shape [34]. Calculations have been done with spd hybridisation. From the figure it is clear that the valence band strongly depends of the site coordination. One finds again that the band becomes narrower when coordination decreases. Contrary to observations on supported clusters the middle of the band shifts towards lower binding energy when the coordination is reduced. The negative shift is due to initial state effects while the overall positive shift, observed in photoemission, should result from a dominant final state effect. Another characteristic point is that the DOS at Fermi level decreases with the coordination number. At the moment there is no measurements of the local DOS on supported clusters but this should be possible in the near future by STM.

3. Reactivity

3.1. CO ADSORPTION

3.1.1. *Introduction*

Chemisorption is the simplest surface reaction. CO adsorption has been studied on most of metal surfaces in various orientations. Moreover CO is a reactant in several important catalytic reactions. These are the reasons for the great interest in CO chemisorption on model catalysts. The classical technique used in chemisorption studies on metal surfaces is the Thermal Programmed Desorption. Its main advantage is that it is easily set up. Moreover it gives in one experiment the number of molecules adsorbed in the different adsorbed states and an estimate of the adsorption energies. The main difficulties to determine accurately the adsorption energies by TPD is its nonequilibrium character and the necessity to assume a value for the frequency factor. In the case of supported model catalyst one has to face to another problem. During the temperature ramping the clusters can suffer restructuring (if they were in a metastable state, after low temperature growth, for example). This phenomenon is characterized by important changes of the TPD curves during the first adsorption-desorption cycles. In the case of supported clusters isothermal methods avoid these difficulties.

3.1.2. *Molecular Beam Relaxation Spectrometry (MBRS) measurements*

The difficulty in using an isothermal method to study chemisorption is that we need to have simultaneously adsorption and desorption of gases. One possibility is to monitor the quantity of adsorbed molecules by integral measurements by Auger or Photoemission spectroscopies but this techniques are difficult to use quantitatively for coverages smaller than 0.1 ML. Infrared spectroscopy can be used below 0.01 ML on extended surfaces but only at relatively low temperature (to have a good Signal/Noise ratio) and despite the great progress in the speed of the FTIR spectrometers few time resolved experiments have been undertaken. The other way is to monitor the desorbed flux of molecules by a mass spectrometer. But we need to separate the desorbed flux from the adsorbing flux. This can be achieved in using a molecular beam source which gives an intense directed beam of molecules. Then, if the mass spectrometer is movable one can measure the spatial and temporal distributions of desorbed molecules. In the MBRS technique the incident beam of molecules is modulated periodically (generally by a square wave) and the pulses of desorbed molecules are detected synchronously. Knowing the flight time of the molecules one can directly measure the life time of the adsorbed species. This method has been used to study the adsorption-desorption kinetic of CO on Pd/MgO(100) model catalysts [35].

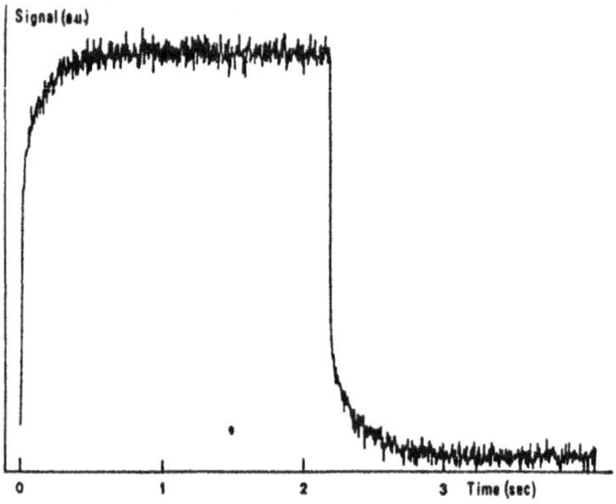

Figure 16. Pulse of CO molecules desorbed at 453K from a collection of 4 nm Pd clusters supported on MgO(100), (after Henry et al. [36]).

Fig. 16 shows a typical pulse of CO desorbed from a sample with 4 nm Pd clusters. As the clusters covered less than 10 % of the MgO surface the outcoming CO results from the interaction on the bare substrate and on the clusters. Angular distributions showed two components: a quasispecular lobe and a cosine contribution [35]. The lobe arises from CO inelastically reflected from the bare MgO and the cosine part from CO desorbed either from MgO or from Pd clusters. However the adsorption energies are very different: ~ 10 kcal/mol for MgO and ~ 30 kcal/mol for Pd. Therefore the life times of CO molecules are very different. On fig. 16 the step increase/decrease of the signal when the beam is on/off corresponds to CO reflected or desorbed from the bare MgO, while the slow variations of the signal correspond to CO desorbing from the Pd clusters. The ratio of the amplitude of the slow variation to the total amplitude gives the initial sticking coefficient. Surprisingly this parameter showed an increase by decreasing the temperature, although the initial sticking coefficient of CO is close to one. Moreover this behaviour depends also of the size and the density of clusters. This phenomenon is due to the capture by the Pd clusters of physisorbed CO molecules diffusing on the MgO surface [35]. As the temperature decreases the mean diffusion length of physisorbed CO increases (see eq.4) and the capture zones around the clusters increase. We will see later that this precursor state effect of physisorbed CO on the support also plays an important role in reaction kinetics.

The slow variations (fig. 16) corresponding to CO chemisorbed on Pd clusters can be fitted by exponentials if the clusters are not too small (> 5 nm). It is easy to show in the case of a first order desorption and a single adsorption state that the time constant of the exponential is exactly the life time of the CO molecules adsorbed on the clusters. Representing the temperature variation of the life time in an Arrhenius plot gives, at low coverage, a straight line. From the slope and the ordinate at the origin one obtains the desorption energy (E_{des}) and the frequency factor (ν_{des}) for desorbing CO molecules. For large particles one obtains E_{des} = 30 kcal/mol and $\nu_{des} = 10^{15}$ s^{-1} in agreement with measurements on a Pd film. For small Pd clusters the signal is no longer exponential. However, assuming still a first order desorption rate, the ratio of the coverage (obtained by integration of the desorbed rate) to the desorbed rate gives the instantaneous life time corresponding to a given coverage. Thus for life time measurements at very low coverages (<0.01 ML) one obtains the desorption energy from an Arrhenius plot [36].

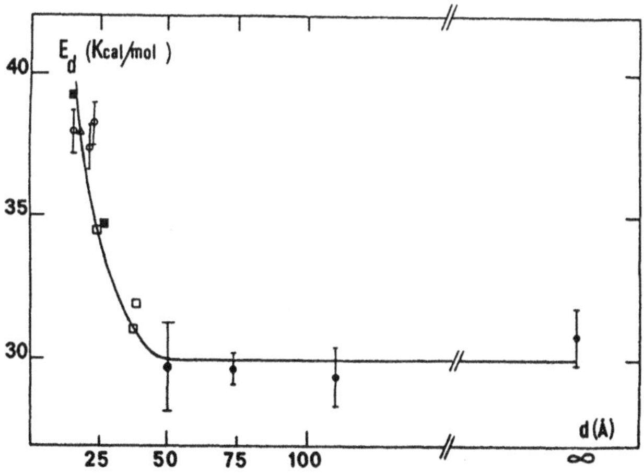

Figure 17. Desorption energy of CO, at low coverage, as a function of the size of Pd clusters supported on MgO (100), (after Henry at al. [36]).

Fig. 17 shows the variations of the desorption energy, at low coverage, as a function of the cluster size. It is clear that below 5 nm the desorption energy increases strongly up to 40 kcal/mol at 2 nm. This size effect is called intrinsic as it does not seem to depend on the substrate. For Pd on mica we observed the same effect. Moreover calorimetric measurements on powder catalysts showed the same size effects for various supports like silica, alumina, titania [37].

3.1.3. *Discussion*

CO adsorption on supported Pd clusters has been extensively studied on various substrates: mica [38, 39], alumina [40,41]. At first sight the results seem contradictory. Some show size effects in the desorption energy other not. In addition some papers reports on the dissociation of CO. In fact a careful examination of the preparation conditions of the model catalysts clarifies the situation. If Pd clusters are perfectly epitaxially oriented and exhibit a well defined regular shape, they do not show any dissociation of CO for different substrates: mica (0001) [42], MgO (100) [36] and α-Al_2O_3 [41]. Probably CO dissociation occurs on some low coordinated sites present on rough clusters. Indeed dissociation of CO is not observed on low index extended Pd surfaces while some dissociation was evidenced on ion bombarded Pd surfaces [43]. These observations illustrate that it is very important to have well defined particle shapes for size effect studies. In the case corresponding to curve on fig. 17, the Pd clusters had the unique shape of a truncated half octahedron, therefore when decreasing the size the only difference is the proportion of edge sites. The observed size effect was tentatively interpreted in assuming that CO was more strongly bound on the edges than on the facets. In fact, lower is the coordination, higher is the adsorption energy. This hypothesis is supported by recent theoretical calculations (see contributions of J. Norskov and R. Van Santen in this volume). Now when the cluster size drops the fraction of strongly bound CO increases and, in the mean, the adsorption energy increases.

To test this hypothesis, in situ IR measurements have been performed on these Pd/MgO(100) model catalysts [44]. For 4 nm Pd particles on MgO(100) two absorption bands are observed, an intense one at 1957 cm^{-1} and a weak one at 2055 cm^{-1}. The former band corresponds very well with that observed in the same conditions on Pd(111) (at 1950 cm^{-1}) and attributed to bridged CO [45]. The small one is not observed in these conditions on extended surfaces. Linear CO, corresponding to a loosely bound state, is observed on Pd(111) (2110 cm^{-1}), at low temperature and high coverage, simultaneously with ternary CO [45]. However the weak band observed on the small Pd clusters is stable at high temperature well after the bridged CO is desorbed, thus it must correspond to strongly bound CO.

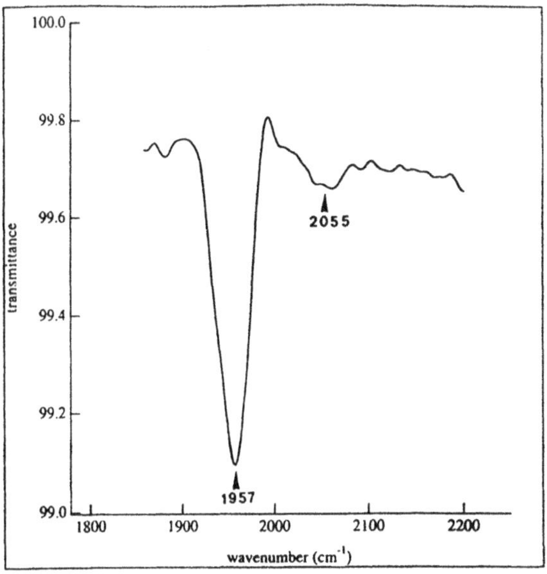

Figure 18. FTIR spectrum of CO adsorbed at 310 K on 4 nm Pd clusters supported on MgO (100). (after Goyhenex et al. [44]).

The adsorption of CO on supported Pd clusters depends on their shape. Keeping the shape constant the adsorption energy increases as the cluster size drops below 5 nm. This effect is interpreted by a stronger adsorption on low coordinated sites.

3.2. CO OXIDATION

3.2.1. *Introduction*

The CO oxidation is a structure insensitive reaction and thus is not expected to present size effect. The first study of this reaction on a supported model catalyst was performed in the Boudart group's on Pd clusters on (1012) α-Al_2O_3 [46]. At 518K the TON was observed to increase below 5 nm. Originally this effect was interpreted as due to a greater accessibility for edge and corner atoms to gas phase molecules increasing the CO coverage and the reaction rate. Later it was shown that the TON depends also on the cluster number density, at constant cluster size [47,48]. This effect was in fact due to an increase of the adsorption rate on the Pd clusters by the capture of CO molecules physisorbed on the substrate. A kinetic model explained very well the measurements with temperature but some disagreements was still present for clusters smaller than 4 nm [49]. Recently molecular beam experiments have been undertaken on a Pd/MgO (100) model catalyst in order to see if the detailed mechanism of the CO oxidation reaction is the same as on extended Pd surfaces .

3.2.2. *MBRS experiments*

In these experiments the CO impinged on the substrate as a pulsed modulated molecular beam and oxygen was introduced as an isotropic pressure through a leak valve. Desorbed CO and CO_2 product molecules are detected by a mass spectrometer. The advantage of MBRS experiments is to measure the reaction rate both at steady state and in transient conditions. Fig. 19 shows the steady state turnover number measured as a function of the substrate temperature for 2.8, 6.8 and 13 nm clusters [50]. From fig. 19 it is clear that the TON depends on the cluster size. Before to conclude to an intrinsic size effect it was necessary to correct from the precursor state effect of CO physisorbed on the substrate. From the diffusion model used before [49], a very good fit of the experimental data was obtained in assuming that large clusters (13 nm) behaved like extended Pd surfaces (see continuous curve on fig. 19). But on small clusters, even after correction of the capture of physisorbed CO, the observed reaction rate is very different than those extrapolated from extended surfaces. The different behaviour of the small Pd particles regarding the CO oxidation was more clearly evidenced in transient measurements. During these experiments the oxygen pressure was kept constant and the CO molecular beam was modulated.

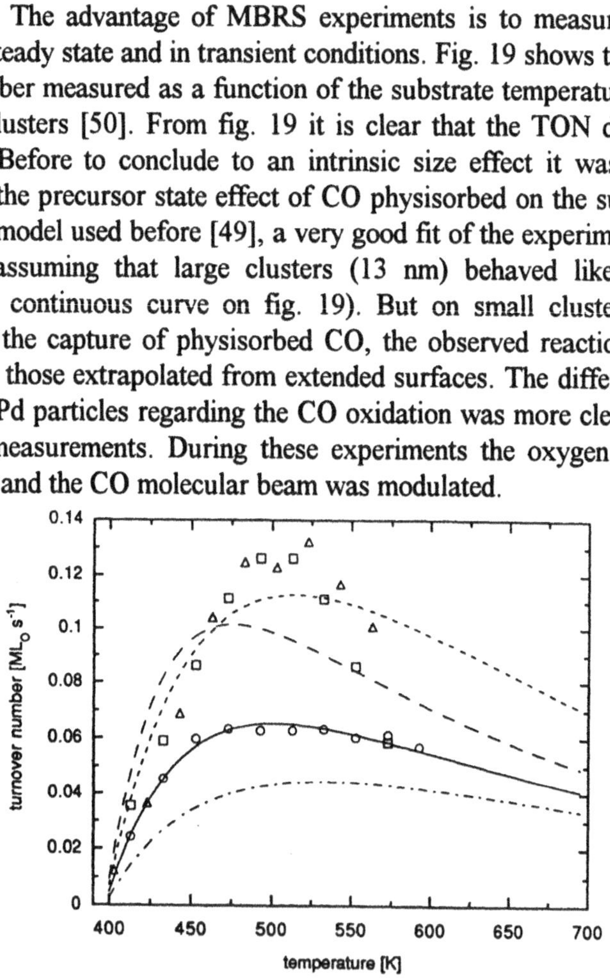

Figure 19. Steady state TON of CO_2 versus temperature during oxidation of CO on Pd clusters, of various sizes, supported on MgO(100): (O) 13 nm, (Δ) 6.8 nm, (□) 2.8 nm. The dotted-dashed line corresponds to the case of a Pd(111) surface. The solid, dashed, and dotted lines are fits calculated with the data from Pd(111) and correction of the effect of the capture of CO adsorbed on the substrate, for 13 nm, 6.8 nm and 2.8 nm clusters, respectively. $P_{CO} = 2.5 \times P_{O_2} = 1.3 \times 10^{-5}$ Pa. (after Becker and Henry [50]).

At high temperature, after turning the CO beam off, the CO_2 production decreased rapidly, because the unreacted CO desorbs in a fraction of second at this temperature. On the contrary at low temperature (see fig. 20), after shutting down the CO beam the CO_2 production first decreases and increases again, to reach a maximum and finally drops to zero.

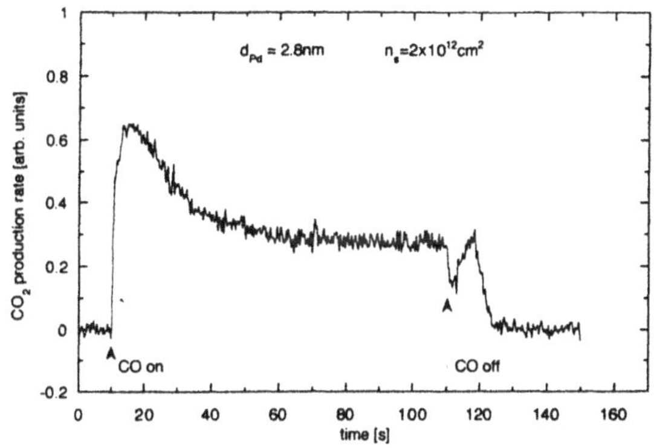

Figure 20. CO_2 transient for CO oxidation, at 433 K on 2.8 nm Pd clusters on MgO(100). (After Becker and Henry [50]).

When the temperature decreases further the second peak of CO_2 becomes larger and its maximum shifted towards longer times. The area under the second peak increases slowly up to a maximum value which increased by decreasing the cluster size. In another type of experiments the CO beam was sent on a saturated layer of oxygen (0.25 ML) previously adsorbed (irreversibly at the used temperature). On large particles the CO_2 production showed a rapid increase folllowed by a slow decrease in good agreement with calculations based on the Langmuir-Hinshelwood mechanism with parameters obtained from measurements on Pd (111). However on small particles the kinetics was different. After a sharp increase the CO_2 production continues to grow, but at a slower rate, and finally decreases slowly as for large particles.

3.2.3. *Discussion*

The CO oxidation reaction on Pd, although structure insensitive, presents important differences with extended surfaces. First, CO physisorbed on the substrate is a second source for chemisorption of CO on Pd clusters. This effect depends on the clusters size and of the cluster number density. It increases when the temperature decreases up to a saturation. This effect is not very different on Al_2O_3 and on MgO because the adsorption and diffusion energies of CO on these two supports are very close. When this effect has been quantitatively taken into account there are still differences between extended Pd surfaces and small Pd clusters. Theses differences appear on steady state measurements of the reaction rate, but the more striking difference is observed in transient conditions which reveal two different kinetics. The more intriguing feature of this size effect is that CO_2 production appears when CO is no longer supplied on the sample. This observation can be explained only in assuming that some CO is still present on the sample several seconds after closing the CO beam. We can eliminate CO species adsorbed on the MgO substrate because the study of the interaction of CO with a clean MgO surface, has clearly shown that between RT and 700 K no CO could stay adsorbed more than 1 millisecond [35]. The only possibility is that some CO is trapped for a long time on the small Pd clusters. Probably it is the strongly bound CO observed in the CO adsorption-desorption experiments. Indeed the increase of the adsorption energy by few kcal/mol is enough to increase the life time to several seconds. However if we assume that this strongly bound CO is located on the edges, the most important point for the considered reaction is the diffusion barrier from a step site to a normal face site. In fact this barrier must be considerably increased to explain the observed effects. Another point in favour of the edge sites as the source of the strongly bound CO, is that the low temperature limit of the area under the second CO_2 peak is roughly equal to the number of edge sites.

3.3. PERSPECTIVES

Due to the limited space available in these proceedings we have only discussed on two examples of reaction leading to size effects on Pd particles. A few studies have been made to investigate size effects on supported model catalysts for some other reactions. However several groups have recently started research programmes on the reactivity of metal clusters supported on thin epitaxial oxide layers see in particular the contributions of H.J. Freund, W. Goodman and T. Madey and R. lambert in this volume.

There is no doubt that reactivity studies on supported model catalysts have grown in interest during the recent years, mainly because the number of techniques for the preparation and the characterization is increasing fast. However the necessity to put together, in the same laboratory, all the techniques and the knowledge useful in these studies is a strong limitation. The minimum required to obtain significant results is the combination of reactivity tests with a microscopic characterization of the clusters structure. Among the new techniques available STM (and AFM) seems the more promising because it can potentially give informations, at the atomic level, on the structure and the electronic properties of supported model catalysts. Moreover with the possibility of working at high temperature and in various atmospheres some catalytic reactions could be studied directly at the atomic scale on these model catalysts.

4. CONCLUSION

Size effects play an important role in heterogeneous catalysis. Their study was precluded for a long time because of the lack of characterization of real world catalysts. In the recent years the situation changed dramatically by the possibility to elaborate, under UHV, model catalysts by growth of metal clusters on oxide single crystals or thin films, that can be further characterized by all the available surface science techniques. The main difficulty of this approach is the preparation of clusters with a well defined stable shape. This can be achieved by epitaxial growth at high temperature. In these conditions, and if the growth rate is low, the clusters can achieve and keep the equilibrium shape. For fcc metals this shape is deduced from the Wulff polyhedron by a truncation at the interface. The nucleation of the clusters, and then their number density, is strongly affected by surface defects, the preparation of the substrate is of particular importance. The structure of the model catalysts is generally characterised, ex situ by electron microscopy, however the use of in situ STM and AFM will certainly increase in the next years. From the results obtained recently on supported model catalysts the classical dual explanation of the size effects by an electronic or a geometrical origin must be reconsidered. Electronic quantum size effects can be neglected for clusters larger than 1 nm. However the local electronic structure which is closely related to the local geometric structure seems to be more important to understand reactivity of metal clusters. Rather than the exact number of atoms in a cluster (larger than ~ 1 nm) the cluster shape (and the arrangement of surface atoms) determines its reactivity. It is in fact the old idea of reaction sites tested on extended surfaces. Part of the size effects could result from the variation of the proportion of some specific sites (like edges and corners) with cluster size. In the two examples chosen in this paper, the CO adsorption and CO oxidation on Pd clusters, the observed variation of the reactivity when the cluster size decreases below 5 nm has been explained by the increase of the relative number of edge sites. To confirm this interpretation calculations on comparable cluster size are needed. For the experimental side, studies of size effects need to develop local characterization of the atomic structure and the electronic properties of supported clusters. New scanning probes (STM, AFM) will certainly play a major role in these investigations. For relatively small clusters (<3 nm) the effect of the metal support interaction can be another source of size effects. Systematic studies of metal clusters deposited on various well characterized support are of great interest. Another particular behaviour of small clusters could be the influence of the cluster size in the segregation phenomenon of alloy clusters. Up to now we have considered that the clusters keep their original shape and surface ordering during a reaction. It is not necessary true because a number of surfaces reconstruct reversibly during chemical reactions. The dynamical character of the surface structure of a supported catalyst during a reaction should be also studied. The combination of FTIR with scanning probes or even Near Field Optical microscopy in infrared are promising.

References

1. Boudart, M. (1969), *Adv. Catal.* **20**, 153
2. Che, M. and Bennett, C.O. (1989) The influence of Particle Size on the Catalytic Properties of Supported Metals, *Adv. Catal.* **36**, 55-172
3. Van Hardeveld, R. and Hartog, F. (1969) The statistics of surface atoms and surface sites on metal crystals, *Surf. Sci.* **15**, 189-230
4. Nosova, L.V., Stenin, M.V., Nogin, Yu.N. and Ryndin, Yu.A. (1992) EXAFS and XPS studies of the influence of metal particle size, nature of support and H_2 and CO adsorption on the structure and electronic properties of palladium, *Appl. Surf. Sci.* **55**, 43-48
5. Datye, A.K. (1992), The Study of Heterogeneous Catalysts by High-Resolution Transmission Electron Microscopy, *Catal. Rev. Sci. Eng.* **34**, 129-178
6. Duriez, C., Chapon, C., Henry, C.R. and Rickard, J.M. (1990) Structural Characterisation of MgO(100) Surfaces, *Surf. Sci.* **230**, 123-136
7. Goyhenex, C., Meunier, M. and Henry, C.R. (1996) Limitation of Auger electron spectroscopy in the determination of the metal-on-oxide growth mode: Pd on MgO(100), *Surf. Sci.* **350**,103-112
8. Henry, C.R., Meunier, M. and Morel, S. (1993) He diffraction study of the nucleation and growth of Pd clusters on MgO(100), *J. Cryst.Growth* **129**, 416-420
9. Meunier, M. and Henry, C.R. (1994) Nucleation and Growth of metallic Clusters on MgO(100) by helium Diffraction, *Surf. Sci.* **307-309**, 514-518
10. Chapon, C., Henry, C.R. and Chemam, A. (1985) Formation and characterisation of small Pd particles deposited on MgO(100) as Model Catalyst, *Surf. Sci.* **162**, 747-754
11. Brune, H., Röder, H., Boragno, C. and Kern, K. (1994) Microscopic View of Nucleation on Surfaces, *Phys. Rev. Lett.* **73**, 1955-1958
12. Granjeaud, S., Yckache, K., Dayez, M., Humbert, A., Chapon, C. and Henry, C.R. (1993) STM/TEM comparative study of Pd clusters epitaxially grown on Highly Oriented Pyrolitic Graphite, *Microsc. Microanal. Microstruct.* **4**, 409-418
13. Freund, H.J., Dillmann, B., Ehrlich, D., Hassel, M., Jaeger, R.M., Kuhlenbeck, H., Ventrice, C.A., Winkelmann, F., Wohlrab, S., Xu, C., Bertrams, Th., Brodde, A. and Neddermeyer, H. (1993) Adsorption and reaction of molecules on surfaces of metal-metal oxide systems, *J. Mol. Catal.* **82**, 143-169
14. Xu, X., Vesecky, S.M., He, J.W. and Goodman, D.W. (1993) Surface spectroscopic studies of a model silica-supported copper catalyst: adsorption and reaction of CO,

H_2O and NO, *J. Vac. Sci.Ttechnol.* **A11**, 1930-1935

15. Colchero, J., Marti, O., Mlynek, J., Humbert, A., Henry, C.R. and Chapon, C. (1991)
Pd clusters on Mica: a study by atomic force microscopy, *J. Vac. Sci. Technol.* **B9**, 794-797
16. Beitel, G., Markert, K., Wiechers, J., Hrbek, J. and Behm, R.J. (1993) Characterization of Single-Crystals α-Al_2O_3 Surfaces and Ag/Al_2O_3 Model Catalysts by Atomic Force Microscopy, in H.J Freund and E. Umbach (eds;), *Adsorption on ordered Surfaces of Ionic Solid and Thin Films*, Springer, Berlin, pp. 71-92
17. Stoyanov, S. and Kashchiev, D. (1981) Thin film nucleation and growth theories: a confrontation with experiments, in E. Kaldis (ed.) *Current Topic in Material science Vol. 7*, North-Holland, pp69-141
18. Poppa, H. (1993) Nucleation Growth and TEM Analysis of Metal Particles and Clusters deposited in UHV, *Catal. Rev.-Sci.Eng.* **35**, 359-398
19. Stowell, M.J. (1972) Capture numbers in Thin Film Nucleation theories, *Phil. Mag.* **26**, 349-360
20. Robins, J.L. (1988) Thin Film Nucleation and Growth theories, *Appl. Surf. Sci.* **33/34**, 379-394
21. Meunier, M. (1995) Etude par Diffraction d'atomes d'hélium des premiers stades de la croissance d'agrégats de Pd sur MgO(100), *Thesis (Marseille)*
22. Rupprechter, G., Hayek, K., Rendon, L. and José-Yacaman, M. (1995) Epitaxially grown model catalyst particles of platinum, rhodium, iridium, palladium and rhenium studied by electron microscopy, *Thin Solid Films* **260**, 148-155
23. Kern, R. (1987) The equilibrium form of a crystal, in I. Sunagawa (ed.) *Morphology of Crystals*, Terra, Tokyo, pp. 77-206
24. Pinto, A., Pennisi, A.R., Faraci,G., D'Agostino, G., Mobilio, S. and Boscherini, F. (1995) Evidence for truncated octahedral structures in supported gold clusters, *Phys. Rev.* **B 51**, 5315-5321
25. Henry, C.R., Chapon, C., Duriez, C. and Giorgio, S. (1991) Growth and morphology
of Pd particles epitaxially deposited on MgO(100), *Surf. Sci.*, **253**, 177-189
26. Giorgio, S., Henry, C.R., Chapon, C. and Penisson, J.M. (1990) Structure and morphology of small Pd particles supported on MgO microcubes, *J. Cryst. Growth,* **100**, 254-260
27. Giorgio, S., Henry, C.R. and Chapon, C. (1995) HRTEM study of the epitaxial growth of Pd particles on ZnO microprisms, *Microsc. Microanal. Microstruct.* **6**, 237-248
28. Fuchs, G., Neiman, D. and Poppa, H. (1991) Three Dimensional observation of supported Pd particles, *Langmuir* **7**, 2853
29. Humbert, A. Dayez, M., Granjeaud, S., Ricci, P., Chapon, C. and Henry, C.R. (1991) UHV and air observations of Pd clusters on clean graphite, *J. Vac. Sci. Technol.* **B9**, 804-805
30. Gillet, M.F. and Channakhone (1986) Crystallographic structure and chemisorption activity of Pd/mica model catalysts, *J. Catal.* **97**, 427-436

31. Lamber, R., Wetjen, S. and Jaeger, N.I. (1995) Size dependence of the lattice parameter of small Pd particles, *Phys. Rev.* **B51**, 10968-10971
32. Perrot, E. (1996) Etude par microscopies à sonde locale de petits agrégats de Pd supportés sur MoS_2 (0001) and MgO(100), *Thesis (Marseille)*
33. Kuhrt, Ch. and Harsdorff, M. (1991) Photoemission and electron microscopy of small supported Pd clusters, *surf. Sci.* **245**, 173-179
34. Mottet, C., Treglia, G. and Legrand, B. (1996) Electronic structure of Pd clusters in the tight binding approximation: influence of spd-hybridization, *Surf. Sci.* **352-354**, 675-679
35. Henry, C.R., Chapon, C. and Duriez, C. (1991) Precursor state in the chemisorption of CO on supported Pd clusters, *J. Chem. Phys.* **95**, 700-705
36. Henry, C.R., Chapon, C., Goyhenex, C. and Monot, R. (1992) Size effect in the CO chemisorption on Pd clusters supported on MgO, *Surf. Sci.* **272**, 283-288
37. Chou, P. and Vannice, M.A. (1987) Calorimetric heat of adsorption measurements on Palladium: II influence of crystallite size and support on CO adsorption, *J. Catal.* **104**, 17-30
38. Doering, D.L., Poppa, H. and Dickinson, J.T. (1982) UHV studies of the interaction of CO with small supported metal particles, Pd/mica, *J. Catal.* **73**, 104
39. Gillet, E., Channakhone, S. and Matolin, V. (1986) Crystallographic and chemisorption activity of Pd/mica model catalysts, *J. Catal.* **97**, 437-447
40. Stara, I. and Matolin, V. (1994) The influence of particle size on CO adsorpton on Pd/alumina model catalysts, *Surf. Sci.* **313**, 99-106
41. Cordatos, H., Brunlesin, and Gorte, R.J. (1995) Study of CO, NO and H_2 on model Pd/α-Al_2O_3 (0001) catalysts, *Surf. Sci.* **323**, 219-227
42. Koch, R. and Poppa, H. (1987) The influence of the mica surface composition on the growth and morphology of discontinuous epitaxial Pd vapor deposits, *J. Vac. Sci. Technol.* **A5**, 1845-1848
43. Matolin, V., Rebholz, M. and Kruse, N. (1991) Defect-induced dissociation of CO on palladium, *Surf. Sci.* **245**, 233-243
44. Goyhenex, C., Croci, M., Claeys, C. and Henry, C.R. (1996) FTIR studies of the adsorption of CO on supported metallic clusters Pd/MgO(100), *Surf. Sci.* **352-354**, 475-479
45. Szanyi, J., Kuhn, W.K. and D.W. Goodman (1993) CO adsorption on Pd(111) and Pd(100): low and high pressure correlations, *J. Vac. Sci. Technol.* **A11**, 1969-1974
46. Ladas, S., Poppa, H. and Boudart, M. (1981) The adsorption and catalytic oxidation of CO on evaporated Pd particles, *Surf. Sci.* **102**, 151-171
47. Matolin, V. and Gillet, E. (1986) The surface diffusion in CO oxidation on small supported Pd particles: experimental evidence, *Surf. Sci;* **166**, L115-L118
48. Rumpf, F., Poppa, H. and Boudart, M. (1987) Oxidation of carbon monoxide on palladium: role of the alumina support, *Langmuir* **4**, 722-728
49. Henry, C.R. (1989) On the effect of the diffusion of CO on the substrate during CO oxidation on supported Pd clusters, *Surf. Sci.* **223**, 519-526
50. Becker, C. and Henry, C.R. (1996) Cluster size dependent kinetics for the oxidation of CO on a Pd/MgO model catalyst, *Surf. Sci.* **352-354**, 457-462

SUPPORTED CLUSTERS, STRUCTURE, REACTIVITY AND MICROSCOPIC PROCESSES IN CATALYSIS

MASARU ICHIKAWA

Catalysis Research Center, Hokkaido University,
Sapporo 060, Japan.

1. Introduction

Organometallic cluster compounds with metal frameworks containing more than three metal atoms are akin to metal particles and stepped metal surfaces having neibouring multi-metallic centers. These metal centers are chemically accomodated by appropriate ligand molecules such as carbonyls, phosphines, aryls and olefins. The similarity between very small metal particles and molecular clusters was firsly proposed by Muetterties(1) and Chini(2) on the basis of the chemical behavior of lignads such as CO coordinated to a cluster framework and chemisorption onto the metal surface. The adjacent metal sites in polynuclear metal clusters make available multi-metallic coordination environments(e.g., face, edge and kink) that can not be realized at a single metal atom or ionic site typical of

Fig.1. Molecular cluster compounds as precursors for tailor-made metal catalysts (a)$Os_3(CO)_{12}$; (b)$Rh_4(CO)_{12}$; (c)$Fe_2Rh_4(CO)_{16}^{2-}$; (d)$Pt_{12}(CO)_{24}^{2-}$; (e)$[Pt_{38}(CO)_{44}H_x]^{2-}$; (f)$[Pt_{19}(CO)_{18}]^{2-}$; (g)$[Ni_{38}Pt_6(CO)_{48}H_{6-n}]^{n-}$.

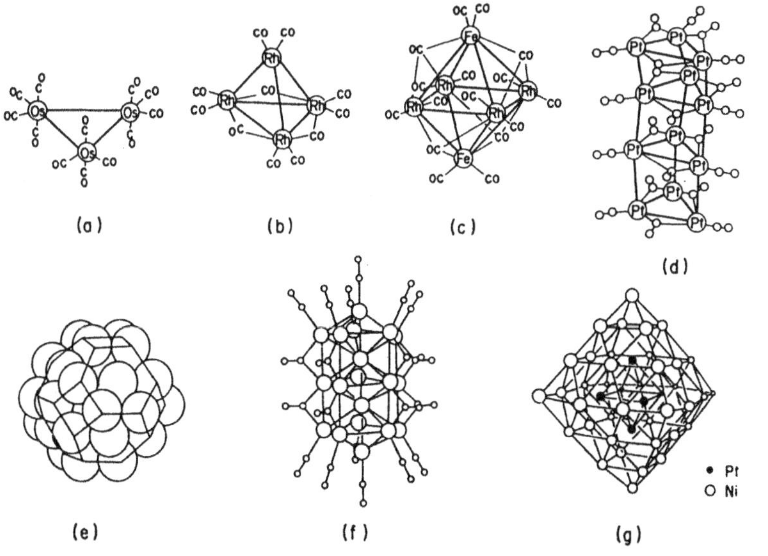

R.M. Lambert and G. Pacchioni (eds.),
Chemisorption and Reactivity on Supported Clusters and Thin Films, 153-192.

most homogeneous metal complex catalysts. Transition metal cluster complexes are ideal objects for the study of collective behavior in stoichiometric and catalytic reactions. In this sense, metal clusters occupy an intermediate position between molecular metal complexes(homogeneous catalysts) and bulky metals such as films and crystals (classic heterogeneous catalysts) in catalysis. Heterogeneous catalysts which involve colloidal, crystalline, and supported metal particles have the advantage of being readily separable from the products, which gives a practical advantage over homogeneous metal complex catalysts even with a high selectivity and uniformity. Catalytically active metal/alloy particles of several nanometer(nm) diameters are usually dispersed on supporting materials such as metal oxides, sulfides or carbon having large surface area to prevent further metal agglomeration ("sintering"). Several examples of supported multi-metal catalysts used in industrial proccesses are Fe/K-Al_2O_3 for ammonia synthesis, Cu/ZnO for methanol synthesis, Ag/Re-Al_2O_3 for selective epoxidation of ethylene, and Pt/Re(or Ru/Cu)-Al_2O_3 which is used for naphtha reforming. The activity and selectivity of supported metal catalysts generally depend on the state of metal dispersion(ensemble sizes), structure(shape and morphology), metal composition(geometric distribution of multi metals), and metal-support(metal-oxide/sulfide) interactions.

Metal dispersion depends on selection of the metal precursors and supporting materials and also on the method of preparation. For the conventional catalyst preparation, a solution of inorganic salts such as a metal nitride and chloride is adsorbed ("impregnation") in a supporting material such as SiO_2 gel, Al_2O_3 , zeolite or carbon, followed with oxidation in air("calcination"). This results in dispersed metal/bimetallic oxide crystalines, which can then be reduced by hydrogen or hydride reagents to metal/alloy particles(1-10 nm sizes) dispersed on supporting materials. Although these methods are conventionally used for production of commercial catalysts, they do not provide good control of particle size distribution and metal composition. Additionally, residual contaminants derived from inorganic salts, such as Cl^-, SO_4^{2-} and SCN^- may act as a catalyst poison. Some early transition metals such as Mo, Cr, Fe and Co as catalytic elements after calcination in catalyst preparation, require extremely higher temperatures

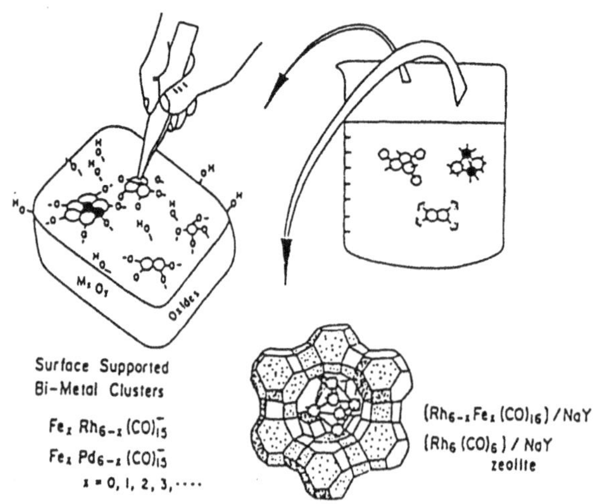

Fig.2. Pictrorial representation of heterogenizing metal/bimetal clusters on amorphous metal oxides(sulfindes or carbons) and encapsulated in NaY zeolite micropores

(>500°C) for H_2-reduction to metallic state, which results in considerable metal agglomeration and sintering. Because of these problems and drowbacks, since 1970 chemists have tried to use some organometallic precursors for the advanced catalyst preparation and chemical design of solid surface. The alternative of using transition metal carbonyls for better control for supported metal catalysts was first proposed by Parkyns(3), Burwell and Brenner(4) And Howe(5). They showed that $Ni(CO)_4$, $Fe(CO)_5$ and $Mo(CO)_6$ could be decomposed on fully hydrated SiO_2 gel, Al_2O_3 , to give metallic Ni, Fe and Mo particles in high dispersion, which exhibited marked catalytic activities. Subsequently, low oxidation state carbonyl cluster compounds(see Fig.1) have been explored by Anderson(6), Ichikawa(7), Basset(8) and Gates(9) in 1975-1980 as postential precursors for tailor-made metal catalysts having uniform dispersion(less than 1 nm size) and well-defined metal composition of metal/bimetal ensembles in keeping their cluster frameworks. More recently, microporous cavities of zeolite and layered clays are used as a "ultimate reaction vessel" for selective synthesis of metal cluster compounds by means of "ship-in-bottle" technique(10), which provide naked metal/alloy clusters in uniform distribution, which are encapsulated in micro [illegible] and CO having multi-bonds. The strength of the binding interaction with the metal surface provides the thermodynamic drive for these cleavage reactions. Reactions such as N_2 and CO appear to require more than one surface metal atom, and this phenomenon is referred to as a "metal ensemble effect". Thus, the bridging CO chemisorption modes and conversion of CO+H_2 to methane are substantially suppressed by reducing the size of the surface ensembles of Ru or Ni atoms with inactive Cu or Ag in alloy formation(11). Similar ensemble effects are also known in metal cluster chemistry(12). The proton-induced reduction of CO to CH_4 strongly depends on the size of the carbonyl clusters, thus trinuclear clusters such as $Ru_3(CO)_{12}$ do not yield CH_4 whereas tetranuclear and higher nuclearity clusters, for example, $Fe_4(CO)_{13}^{2-}$ yield significant quantities of methane. Similarly, the bonding of some hydrocarbon fragments is shown in Table 1, where molecular analogs of organometallic transformation on metal, metal oxides, and metal sulfides are represented in conjunction with possible intermediates in the heterogeneous catalytic reactions. thought to require metal ensembles to catalyze the skeletal isomerization of hydrocarbons (13). Analogous to this is a wide variety of metal cluster compounds invloving multi-metal attachment to hydrocarbon fragments(Table 1).

Table 1. Molecular Analogs of Organometallic Transformation on Metal, Metal Oxides and Metal sulfides Incorporated in Typical Catalytic Reactions.

Catalytic reaction	Proposed intermediates on surfaces	Model molecular clusters
1. Hydrogenation $RC{\equiv}CR + H_2 \xrightarrow[Ni,\ Pt,\ Rh]{} RHC{=}CHR$	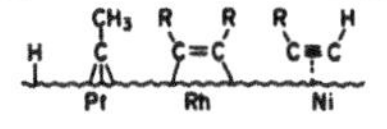	$Cp_3M_3(CO)_3(RC{\equiv}CR)$ M = Co, Rh, Ir $Fe_3(CO)_{12}(C{\equiv}CCH_3)$ $Co_3(CO)_9(\mu_3\text{-}C{-}CH_3)$ $Fe_3(CO)_{12}(\mu_3\text{-}C{-}CH_3)$
2. Hydrogenation of benzene $C_6H_6 + H_2 \xrightarrow[Pt,\ Rh]{} C_6H_{12}$		$Os_{10}(CO)_{24}(C_6H_6)$ $Ru_6(CO)_{14}(C)(C_6H_3Me_3)$
3. Hydroformylation $C_2H_4 + CO + H_2 \xrightarrow[Rh/MgO,\ SiO_2]{} C_2H_5CHO$	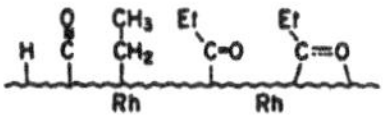	$[Rh_6(CO)_{15}(COEt)]^-$ $Ru_3(\mu\text{-}H, \mu\text{-}O{=}C)(Et)(CO)_{10}$ $Fe_3(CO)_9[\mu\text{-}O{=}C)(Me)]$
4. Hydrodesulfurization 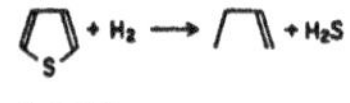		$HOs_3(CO)_{10}(SR)$
CoS, MoS_3	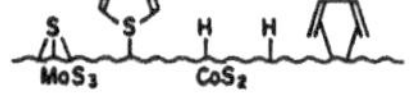	$HOs_3(CO)_{10}(S)$ $Co_6(CO)_{12}S_2$
5. Fischer–Tropsch reaction $CO + H_2 \xrightarrow[Ni,\ Fe,\ Ru]{} CH_4, C_nH_{2n+2}$ + oxygenates	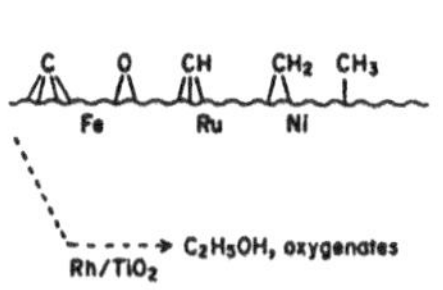	$[Os_3(\mu_2\text{-}H)_2(C)(CO)_9]$ $Ru_3(CO)_5(C)][Ru_3(\mu_3\text{-}O)(CO)_{10}]$ $[Fe_4(CO)_{14}(C)]$ $[Rh_{15}(CO)_{28}(C_2)]^-$ $Os_3(CO)_{10}(\mu\text{-}CH_2)$ $HOs_3(\mu\text{-}CH_3)(CO)_{10}$ $[Co_{13}(CO)_{12}(\mu CO)_{12}(C_2H)]^{n-}$
	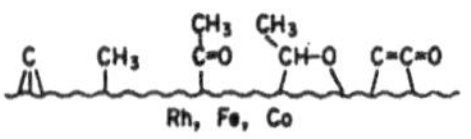	$Os_3(\mu\text{-}H)(\mu\text{-}C{=}C{=}O)(CO)_9$ $[Fe_4(CO)_9(\mu_4\text{-}C{-}C)({=}O)OMe]$ $[Ru_3(CO)_9(\mu_3\text{-}C{=}C{=}O)]^{2-}$
6. NO reduction with CO and H_2 $NO + CO \xrightarrow[Rh,\ Pt,\ Ru]{} N_2(N_2O) + CO_2$ $NO + H_2 + H_2 \xrightarrow[Pt,\ Rh,\ Pd]{} N_2 + NH_3 + H_2O$	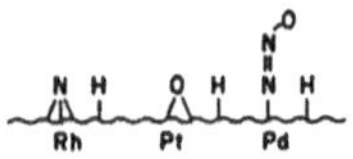	$FeRu_3(CO)_{12}(NO)^-$ $FeRu_3(CO)_{12}(\mu_4\text{-}N)$ $Ru_3(CO)_{10}(\mu_3\text{-}O)$

Fig.3. Butterfly Cluster Compounds Useful for Molecular Modeling of Intermediates at Edge and Kink Sites on Ir(100) Surface.

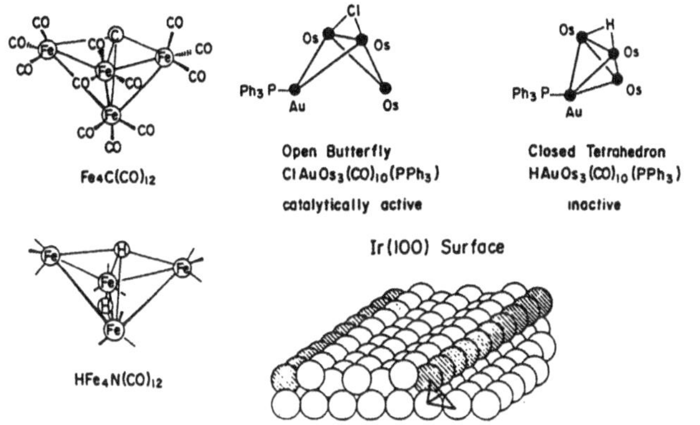

In addition to planar(two-dimensional) faces, the surfaces of supported metal particles consist of edges between sufaces as well as steps and kinks on the otherwise planar surfaces. These sites are thought to be highly active for the dissociative adsorption and bond cleavage of reactant molecules. Analogies have been drawn between the binding of small molecules(CO, N_2, and CH_4) to coordinatively unsaturated sites(edge and kink) of metal surfaces and the binding molecules and atoms(C, N and O) in clefts of metal clusters. The simplest examples are four-metal butterfly cluster compounds(Fig.3), where the cluster geometry can be considered as a model for the dissociative chmisorption of CO, N_2, and CH_4 on the steps (or kinks) of metal films. The M_4 butterfly metal skeleton appears to be a structurally versatile unit with dihedral angles and metal-metal bond strengths varying over the wide range even for the same metal core. There is growing evidence that small ligands such as CO, C, and N coordinated within the butterfly framework exhibit unusual patterns of chemical reactivity(14).

2.2. Surface organometallic chemistry of surface-grafted metal clusters

Structures and reactivities of supported metal clusters have been extensively investigated(15) mostly by spectroscopic and chemical methods and the available data are generally consistent with the statement of the chemistry of the supported cluster species, which is often nearby comparable to that of precursor clusters and their molecular analogs in solution. These supported "molecular" clusters even offer the prospect of direct characterization of the metal-metal and metal-suport bonds by means of infrared(IR), ^{1}H and ^{13}C nuclear magnetic resonance(NMR), ultraviolet-visible(UV-vis) and Raman spectroscopy, combined with surface or solid spectroscopies such as extended X-ray absorption fine structure(EXAFS), X-ray photoelectron(XPS) and Mossbauer spectroscopies. In addition, scanning tunneling(STM) and high resolution electron microscopy are powerful techniques to identify the location, nuclearity, morphology and dynamic behaviors of metal/aloy clusters bound to surface or entrapped in micropores.

A. Type of Cluster-Support Intractions

The intrinsic reactivities of supports such as phosphin-functionalized polymer and metal oxide/sulfides, which have chemically active centers for the organometallics to offer the specific supported species by impregnation with the different metal ligand clusters:

Since 1975, the interaction between organometallic cluster complexes and the different surface sites of metal oxides has been much collected(15,16). Various oxide-supported metal cluster complexes have been examined from the standpoint of their metal nuclearity, structures and reactivities, in changing their ligands, the concentration and natures of surface chemical groups such as O^{2-}, OH^- and metal cations. The subsequent chemical and thermal activation of these supported clusters may induce a new class of the organometallic surface species having coordinative unsaturation. It is generally assumed that thc transition metal ensembles are directly linked to the oxide and sulfide supports by surface ligands such as surface oxide(and sulfide)ions or by ligands from the cluster cordination compounds that interact with atoms and ions of the supports. Thus, a metal oxide surface may simply cordinate to metal or ligand centers or it may undergo more elaborate electron and proton transfer to accomodate the reactive metal cluster species at the solid surfaces. The cluster ligands such as carbonyl and allyl groups are potentially susceptible to surface oxides, hydroxyl and protons of metal oxides and zeolites. The hydrides in the carbonyl cluster complexes can display both protonic and hydridic character. It has been often assumed and occasionally demonstrated that surface reactions of metal cluster complexes proceed in the manner similar to the reactions in solution. The following examples are typical of those posturated for the reaction of clusters with surfaces having the specific chemical groups.

(1) Coulombic attraction between anionic clusters and surface positive sites M^{n+} may occur.

$$(Na^+)_2[Pt_{15}(CO)_{30}{}^{2-}] + \underset{Cl^-}{NR_4^+} \sim P \xrightarrow[-NaCl]{} [Pt_{15}(CO)_{30}][NR_4^+{-}P]$$

$$[HOs_3Ru(CO)_{11}]^-[NR_3^+{-}CH_2CH_2{-}O{-}Si{\equiv}] \quad \text{on } SiO_2$$

$$2\,Na_2[Pt_{15}(CO)_{30}] + [O{-}\underset{\underset{O}{|}}{Al^{3+}}{-}O] \rightarrow [Pt_{15}(CO)_{30}][{-}O{-}\underset{\underset{O}{|}}{Al^{3+}}{-}O] + 2\,Na^+$$

$$[Mg^{2+}][HOs_{10}C(CO)_{24}]^{2-}, \qquad [Rh_6(CO)_{15}][Al^{3+}]$$

$$[Mg^{2+}][Re_4C(CO)_{12}]^-/Al_2O_3,\ MgO,\ TiO_2,\ \text{etc.}$$

$$[HRu_3(CO)_{11}]^-[PyN^+H_2(CH_2)_3OSi{\equiv}]$$

$$[HFe_3(CO)_{11}]^-[NH_2^+CH_2H_5{-}(CH_2)_3OSi{\equiv}] \quad \text{on } SiO_2$$

(2) Ligand exchange may occur with bases such as O^{2-}, S^{2-} and P-anchored on the oxide surfaces.

$$Rh_6(CO)_{16} + [PPh_2{-}Pol]_3 \xrightarrow{-3\,CO} Rh_6(CO)_{13}[PPh_2{-}Pol]_3$$

$$Ir_4(CO)_{12} + [PPh_2{-}CH_2CH_2{-}Si{\equiv}] \xrightarrow{-CO} Ir_4(CO)_{11}[PPh_2{-}CH_2CH_2SO]$$

$$Co_4(CO)_{12} + [OAl{=}]_2 \xrightarrow{-2\,CO} Co_4(CO)_{10}[OAl{=}]_2$$

(3) Adduct formation may occur between Lewis base sites on the cluster and cations on the surface(Lewis cations or protonic M-OH)

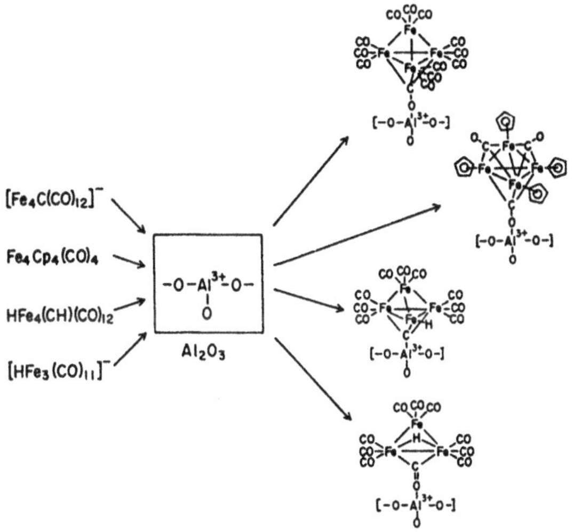

(4) The oxidative addition of the OH(SH) group may occur with zero-valent metal atoms/ensembles having M-M bonds.

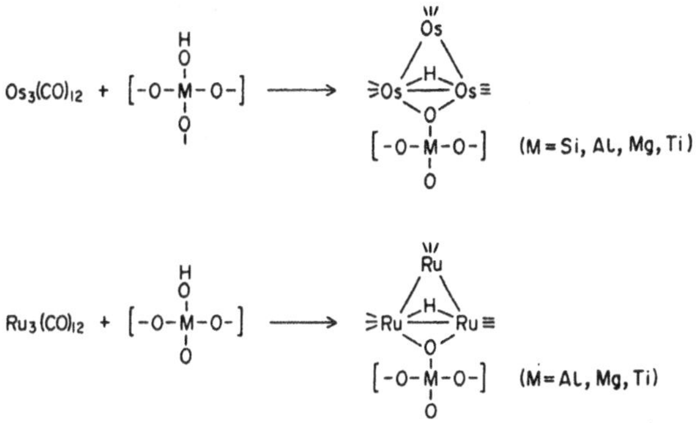

(5) This oxidative addition may be coupled with reductive elimination of H_2

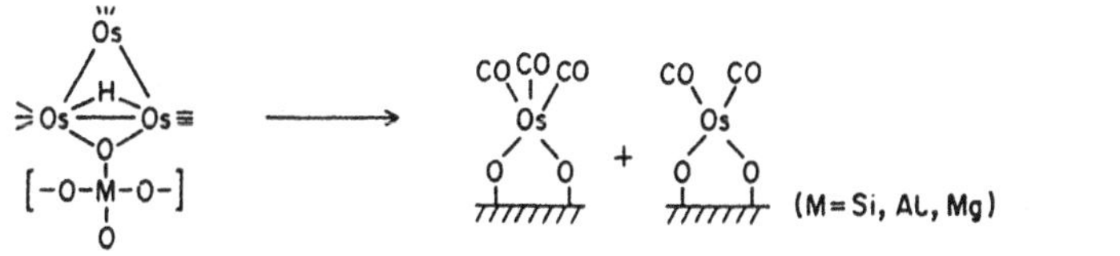

$$Os_3(CO)_{12} + 6[HOM] \longrightarrow 3Os(CO)_3(OM)_2 + Os(CO)_3(OM)_2\ 3H_2 + 3CO$$

$$Rh_6(CO)_{16} + 6[HOM] \longrightarrow 3[Rh(CO)_2(OM)]_2 + 3H_2 + 4CO$$

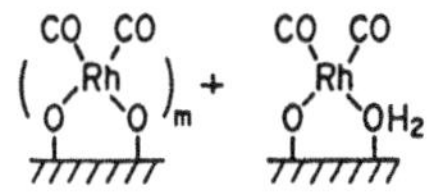

(6) The nucleophilic attack on coordinated CO followed by a β-hydrogen elimination occurs on neutral clusters to produce anionic clusters.

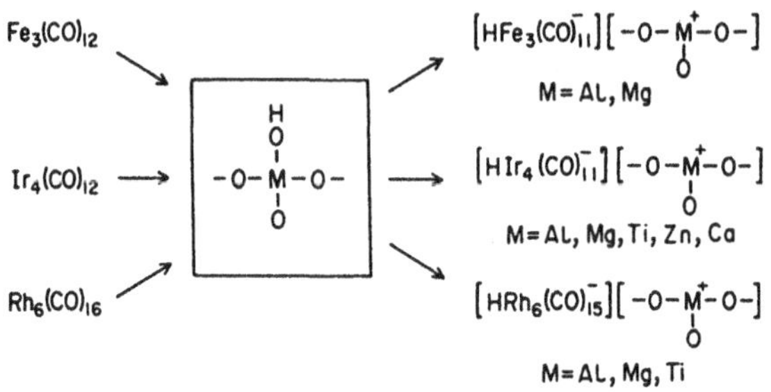

The experimental(17-20) data are available to confirm a variety of metal clusters bound to surface by the different types of bondings and structures of the surface-bound cluster species as the versatile precursors for heterogeneous catalysts. The strength and surface concentration of acidic and basic centers on oxide supports depend strongly on the nature of the M-O bond of $MO_m(OH)_n$ oxides and the local environment of the metal ion sites, which are affected by the thermal activation and pretreatment of the oxide supports. This is reflected in the control of surface species derived from the cluster precursors which are interacted with the different surface oxide ions and metal centers.

B. Surface-mediated cluster formation, fragmentation and recontruction

Judging from IR, EXAFS and UV-visible spectra(21) larger nuclearity Os clusters such as $Os_6(CO)_{18}$ and $H_2Os_{10}C(CO)_{24}$ are preferably retained on alumina and MgO even on thermal activation at 523K in $CO+H_2$ atmosphare. The stable Os_6 and Os_{10} clusters are bound to one/two oxygen atoms sharing the support silica and alumina in keeping their metal frameworks even on hydroxyl surface and at elevated temperatures.

Recently, Gates et. al.(22) found that by heating $Os_3(CO)_{12}$ and $Os_3(CO)_{12}$ bound to MgO under a $CO+H_2$ atmosphere at 200-280°C, a thermally stable $Os_{10}(CO)_{24}(C)^-$, irrespective of their starting precursors, is formed in high yield and it may be extracted as the PPN+ salt. This may offer a very prospective way for the surface-promoted synthesis of the specific and contrained polynuclear cluster complexes from the smaller carbonyl precursors on the oxide surfaces and zeolite matrices, as exemplied by the following :

a) $Co_2(CO)_8 \xrightarrow[\text{50-150°C >80\% yield}]{Al_2O_3, SiO_2} Co_4(CO)_{12}$

b) $Fe(CO)_5 \xrightarrow[\text{CO/H}_2\text{O, 50-70°C >35\% yield}]{Al_2O_3, MgO, NaY} HFe_3(CO)_{11}^-$

c) $Rh_4(CO)_{12} \xrightarrow[\text{20-50°C MgO, NaY}]{Al_2O_3, MgO, NaY} Rh_6(CO)_{15}^{2-}$

d) $Os_3(CO)_{12}$ ---------------> $HOs_{10}(CO)_{24}(C)^-$
H_2+CO, 200-280°C 23% yield

In fact, amorphous oxide surfaces and zeolite matrices having the suitable acidity and basicity promote the syntheses of structurally adapted polynuclear carbonyl metal clusters with higher yields and higher selectivity, comparing those by the analogous inorganic synthesis in solution. The oxidized monomeric Os and Ru species on SiO_2, Al_2O_3 and MgO described above can be reduced by H_2 at 400°C to yield highly dispersed metal aggregate which are in less than 10Å in diameter, which was meaured by EXAFS indicating that metal aggregates consist of 6-8 Os(Ru) atoms. It is probable that they exist in the raft-structure of aggregates located at the oxide interface.

The surface chemistry of both $Rh_6(CO)_{16}$ and $Rh_4(CO)_{12}$ has been extensively studied because of their high reactivity with surfaces and of their unique catalytic performances in CO conversion to some useful oxygenated compounds(23). $Rh_6(CO)_{16}$ on SiO_2 is stable under CO or Ar, but decomposes slowly under evacuation even at room temperature due to decarbonylation and reaction between cluster carbonyls and acidic surface OH on the different oxides, resulting in breaking of Rh-Rh bonds to form Rh(I) dicarbonyl species having two IR bands at ca 2030 and 2090 cm^{-1}, analogues to those of $[Rh(CO)_2Cl]_2$ and $[Rh(CO)_2(OSiPh_3]_2$.

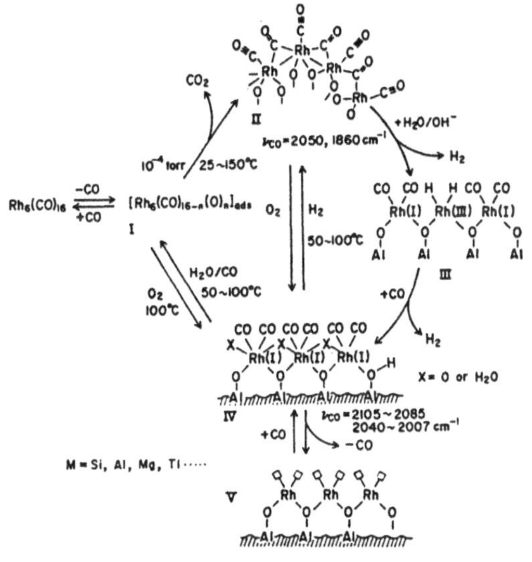

Fig.4. Transformation and successive decomposition of $Rh_6(CO)_{16}$ on dehydrated alumina surface as deduced by FTIR, EXAFS and TPR studies.

C. Morphology, Mobility and Tranformation of Supported Metal Clusters on Various Oxides

Transmission electron spectroscopy has yielded some striking results in the characterization of metal clusters on supports. Iijima and Ichikawa[24] observed the Rh aggregates derived from the $Rh_6(CO)_{16}$ which has been impregnated on a spinel-type alumina crystal particles(200-500 A diameter). Figure 5 reproduces an high resolution electron micrograph of the spherical particles of this Al_2O_3 consisting of facing and surface steps occuring on the atomic scale. $Rh_6(CO)_{16}$ was impregnated from CH_2Cl_2 solution, on this alumina particles, followed by solvent evaporation and heating at 150°C at 10^{-9} torr to remove CO. Small speckles on (111) alumina plane are measured to be less than 6 A in diameter and are attributed to individual Rh_6 clusters. The clusters are revealed clearly on the periphery of the particles, whereas they are not seen in the region showing the lattice fringes. Two sizes of clusters, 6 and 10 Å in a diameter, labelled "a" and "b" in Fig 5 can be recognized. The former is close to the ideal size of six atom clusters derived from hexanuclear Rh carbonyl cluster. All the clusters appear to be hemispherical, possibly due to the strong metal-oxide support interaction. Individual metal atoms are discernible in some examples, but nearby the best available technology for high resolution microscopy is required. Schmidt et. al. reported(25) recently a high resolution electromicrograph of $[Au_{55}(PPh_3)]_{12}Cl_6$ on carbon in proving the Au_{55} cluster having 7.1 Å diameter. By contrast, when $Rh_6(CO)_{16}$ and $Pt_{15}(CO)_{30}{}^{2-}$ was impregnated on a thin silicon oxide film which was prepared by oxygen exposure to Si particles at 1000°C, they showed by the HRTEM observation that intact spherical Rh and Pt particles less than 10A diameter readily move around on the silicon oxide memblane and collapse in real time to make larger clusters as shown in Fig.6. This migration and agglomerization may be due to a weaker interaction of Pt(or Rh) aggregates with silica surface oxygen atoms compared with Al_2O_3 surfaces. The Rh-Rh bonds in supported Rh clusters on alumina are successively ruptured on CO chemisorption to give $Rh(CO)_2$, which is confirmed by EXAFS and IR. Yates et. al.(26) have proposed that the CO ligands on reduced Rh aggregates less than 10 Å in diameter interact with the peripheral acidic hydroxyl groups(3500-3650 cm^{-1}) on alumina surface, which leads to an oxidative breaking of Rh-Rh bonds apart the twin carbonyl species. Such a carbonyl species has facile migration on oxide surfaces to be resulted in agglomeration of clusters in higher nucleality.

Moreover, Scanning Tunneling Microscopy(STM) has emerged as a powerful in-situ spectroscopy giving the atomic scale observation of solid surface(metal crystal and films, silicon) and surface species (supported metal catalysts), irespective of the reaction atmosphere. Recently several interesting STM observations have been conducted for some metal carbonyl clusters such as $[NEt_4]_3[Re_7C(CO)_{21}]$(27), $[NEt_4]_2$ $[Pt_{12}(CO)_{24}]$

Fig. 5. High-resolution electron micrograph showing "Rh_6" clusters scattered on terace, edge and kink sites of the (111) surfaces of spinel-type Al_2O_3 crystalines.

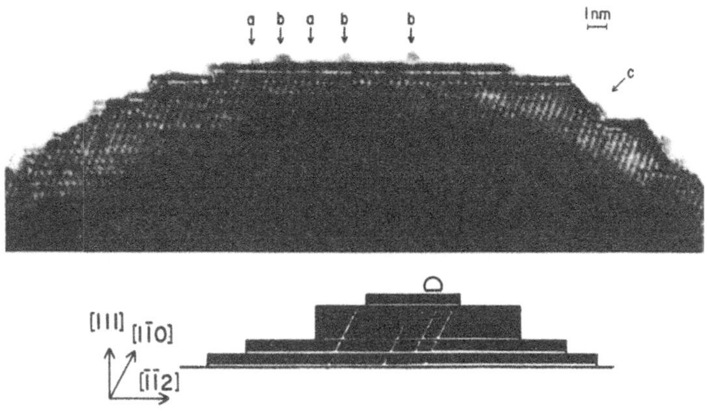

Fig.6. High-resolution electron micrograph showing spherical particles of approximately 8-10Å diameter derived from $[Pt_{15}(CO)_{30}]^{2-}$ deposited on SiO_2/Si(111) crystalline (500Å).

and $Rh_4(CO)_{12}$[20] bound to highly orientated highly oriented Pyrolytic graphite(HOPG) under the inert N_2, Ar and CO atmosphere. As presented in Fig 7, the scattered humps in the STM images were obtained, which are very much akin to the original Rh_4 carbonyl clusters in terms of their molecular shapes and sizes determined by the x-ray crystalographic data. This STM technique may be prospective in applying to a variety of surface-bound metal clusters for atomic-scale observation of their morophological transformation of clusters under the reaction atmospheres depending on the nature of supports such as metal and semi-conductive oxide films. Additionally, some snap-shots of STM images were monitored(29) for an isolated aggregate of Pt carbonyl clusters $[NEt_4]_2[Pt_{12}(CO)_{24}]$, which were adsorbed onto the surface of a freshly cleaved HOPG under the nitrogen atmosphere. In situ STM images demonstrated the morphological

Fig.7. STM images of $Rh_4(CO)_{12}$ deposited on a HOPG surface(140X140Å dimension;under N_2)

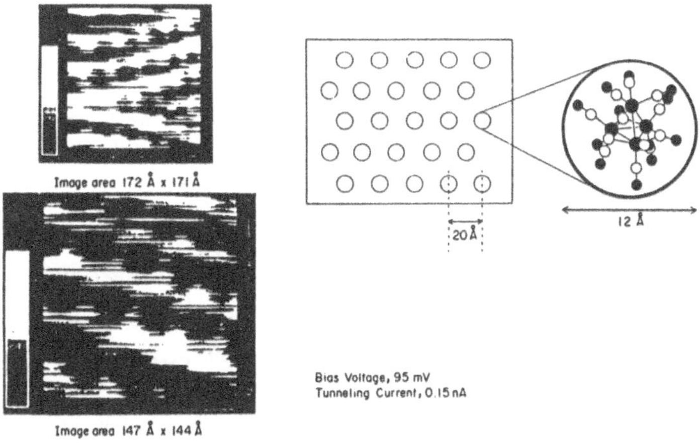

changes of the surface-bound Pt_{12} carbonyl clusters under the pulse laser irradiation.The Pt_{12} carbonyl clusters were converted by the successive pulse activation into a highly dispersed Pt crystaline(18X50Å), which was imaged by STM as shown in Fig.8.

In addition, it is worthy to note that a great advantages of many of these techniques is that they can be used for catalyst samples in a reactive atmonsphere, even at high temepratures and at high pressures. Infrared spectroscopy is the method used most successfully with samples in the presence of reactive atmosphere even high pressure. Raman and EXAFS, Mossbauer, STM spectroscopies can be also used for monitoring the microscopic transformation of supported metal clusters in catalysis under the prevailing reaction conditions such as CO+NO and $CO+H_2O$ atmosphere(30).

Fig.8. Changes of the STM images of $[NEt_4]_2[Pt_{12}(CO)_{24}]$,on graphite under laser irradiation. The uper figures are topographic images and the lower ones are line scan STM images. The image area of all figure; x=160Å, y=150Å. Bias voltage 91 mV tunneling current 0.14nA. (a)Before irradiation,(b) 10 laser pulses,(c)20 pulses, (d)30 pulses, and (e)40 pulses.

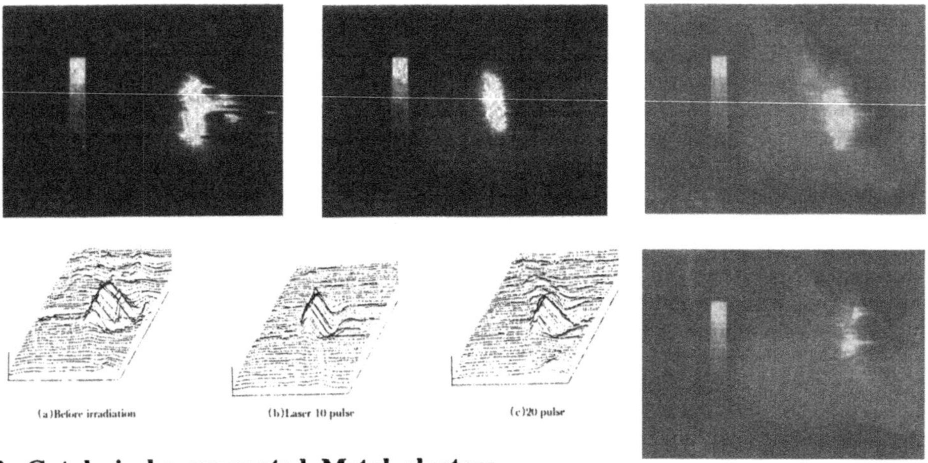

3. Catalysis by supported Metal clusters

3.1. Cluster-size effect.

Ichikawa(31) has studied cluster-size effects for the catalytic activities in the vapor-phase hydroformylation of ethylene and propene on some cluster-derived catalysts by varying the precursor clusters such as Rh_4-Rh_{13} carbonyls impregnated on ZnO:

$$H_2C{=}CH_2 + CO + H_2 \text{ -------->} \quad C_2H_5CHO\ (+C_3H_7OH)$$

$$CH_3CH{=}CH_2{+}CO{+}H_2 \text{ -------->} \quad i/n\text{-}C_3H_7CHO\ (+i/n\text{-}C_4H_9OH)$$

$Rh_6(CO)_{16}$ on ZnO was competely inactive for both hydroformylation and hydrogenation reactions, but it turned out to exhibit high activities for the hydro-formylation after a parial removal of CO by evacuation at 50°C or activation under the atmosphere of C_2H_4+H_2(or+CO) upto 90°C. The IR spectral change of $Rh_6(CO)_{16}$ on ZnO in the hydroformylation atosphere are indicative of intensity decrease of linear band at

2070 cm^{-1} and triply bridging CO of $Rh_6(CO)_{16}$ at 1795 cm^{-1}, accompanied with a development of a new band at 1680 cm^{-1} in the prolonged reaction. At this stage in the procedure propionaldehyde as the hydroformylation product appeared in the gas phase. The resulting IR carbonyl spectra resembled those of the coordiantively unsaturated $[Rh_6(CO)_{15}(RCO)^{2-}]$ (R=Et, Pr). The acyl species has been reported to be synthesized in the reaction of $Rh_4(CO)_{12}$ with $C_2H_4+H_2$ at 50-70°C

$$3Rh_4(CO)_{12}+4C_2H_4+4H_2 \text{ -----> } [Rh_6(CO)_{15}(EtCO)][NEt_4] \text{(80\% yield)}$$

By contrast, $Rh_6(CO)_{16}$ on the partially dehydrated Al_2O_3 at the elevated temepratures such as 120°C it was almost completely converted into a oxidized mononuclear $Rh(I)(CO)_2$ species and Rh metal particle, which was found to only give an inmeasurable activity for the hydroformylation under the same conditions. Regardless of the different treatments for partially decarbonation to activate the impregnated $Rh_6(CO)_{16}$ and $Rh_4(CO)_{12}$. It was found that the rates of formation in terms of TOF and selectivities toward linear-butyl aldehyde in the propene hydroformylation at 150°C are increased by the increase of cluster-nuclearity as summerized in Fig.9 depend on the nuclearity of Rh carbonyl precursors supported on ZnO, showing a maximum yield of linear-chained product obtained for Rh_4-Rh_6 clusters as follows;

$$Rh_2(CO)_3CP < Rh_4(CO)_{12} > Rh_6(CO)_{16} >> Rh_7(CO)_{16}3NEt_4$$

$$>>[Rh_{13}(CO)_{23}H_2]2NBu_4 >>> \text{"Rh"}(RhCl_3,\ H_2 \text{ red at } 400^{\circ}C)$$

This cluster-size dependency of catalytic olefin hydroformylation may be interpreted in conjunction with the partial electron deficiency of ZnO-supported Rh clusters which effectively promote migratory CO insertion. Among the metal oxide suports, high activity

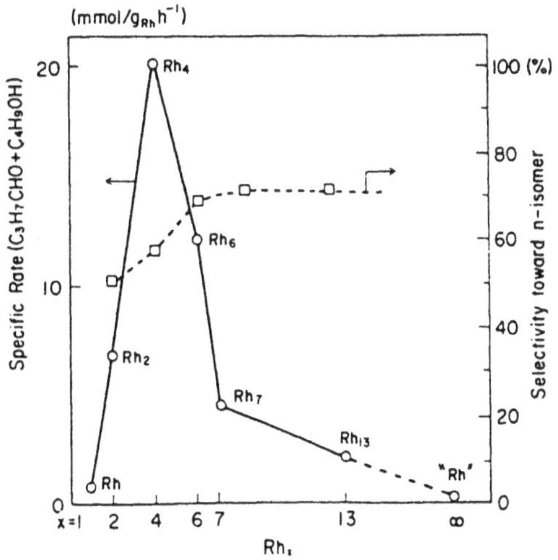

Fig.9. Effect of the size of Rh carbonyl clusters as precursors supported on ZnO on the propene hydroformylation activities at 120°C. The following precursors were used: $RhCp(CO)_2$, $Rh_2Cp_2(CO)_3$, $Rh_4(CO)_{12}$, $Rh_6(CO)_{16}$, $Rh_7(CO)_{16}^{3-}$ and $Rh_{13}(CO)_{24}H_3$. The conventional Rh catalyst"Rh" was prepared from $RhCl_3/ZnO$ by H_2 reduction at 200°C.

in the olefin hydroformylation is observed on the Rh_4 and Rh_6 carbonyl clusters which are supported on amphoteric base oxides such as ZnO, MgO, La_2O_3 and ZrO_2, compared with Al_2O_3 and SiO_2(31) It was found that suitable supports anchoring Rh carbonyl clusters exhibit higher activities for the olefin hydroformylation. It was proposed that the basic sites e.g., O^{2-} and/or OH^- groups on these oxides are favorable for the formation of hydride Rh carbonyl cluster species which are catalytically active for the olefin hydroformylation reaction, and additionally Lewis acid metal cations in congestion with Rh clusters promote the CO insertion to give higher oxygenates(32,33).

Table 2. Propene Hydroformylation over Various Metal Carbonyl clusters Impregnated on Metal Oxides Compared with That over Conventional Rhodium Supported Catalysts.

Catalyst (0.5 wt% loading)		Hydroformylation characteristics	
Metal carbonyl	Metal oxide	Rate, V^b	n-Isomer selectivity (%)^c
$Rh_4(CO)_{12}$	ZnO	21	59
$Rh_6(CO)_{16}$	ZnO	11	71
$Rh_6(CO)_{16}$	MgO	5	38
$Rh_6(CO)_{16}$	TiO_2	2	62
$Rh_6(CO)_{16}$	ZrO_2	3.8	72
$Rh_6(CO)_{16}$	La_2O_3	3.5	75
$Rh_6(CO)_{16}$	SiO_2	0.4	63
$Rh_6(CO)_{16}$	Al_2O_3	0.01	50
$Co_4(CO)_{12}$	ZnO	1.2	91
$[Pt_{15}(CO)_{30}]\cdot 2Et_4N$	ZnO	—	—
$Ir_4(CO)_{12}$	ZnO	Trace	—
Rh^d	ZnO	0.01	—

Recently, the structure of attached $Rh_4(CO)_{12}$ on tris(hydroxymethyl)-phosphine ($P(CH_2OH)_3$, THP) silica was invenstigated by EXAFS and IR spectroscopy(34). The catalyttic behavior of the Rh_4 carbonyl cluster coordinated with THP ligands anchored on $SiO_2(P(CH_2OH)_{3-x}(CH_2O)_x(x=1,2)$ in gas phase hydroformylation reactions was investigated. The number of THP ligands coordinating to the Rh_4 loaded with 6.2wt%THP on which the average distance of P atoms on the neiboring THP is 8Å, were coordinated by two ligands, and the Rh_4 cluster was distorted. The Rh_4 clusters on SiO_2 laoded 1.6wt% THP, on the other hand, were coodinated by one THP ligand and the clusters were not distorted. The attached cluster was active for olefn hydroformylation. The reaction rate of hydroformylation on this catalyst was comparable to that on Wilkinson catalyst in solution. The distorted Rh_4 framework, which is coordinated by two THP, was much more active in hydroformylation reactions, than that of monosubstituted clusters of THP/SiO_2 or $Rh_4(CO)_{10}(THP)_2$ in solution. In situ IR spectroscopy and EXAFS revealed that a reversible coordiation of CO ligands attached to the Rh_4 cluster was associated with the selective hydroformylation reaction. In hydroformylation atmosphere, the Rh_4 cluster was almost saturated by CO ligands and effectively suppresed the simple hydrogenation of olefins as a side reaction. In contrast, under the reaction condition of hydrogenation, CO ligands were removed to increase the reaction rate of hydrogenation. The reversible CO bonding in the surface-grafted Rh_4 site was related to the dynamic control of selective hydroformylation of olefins, as shown in the following scheme.

Fig.10 Proposed structures of distorted Rh_4/THP(6.2)/SiO_2 catalytically active for olefin hydroformylation, as deduced by EXAFS, FTIR and TPR studies.

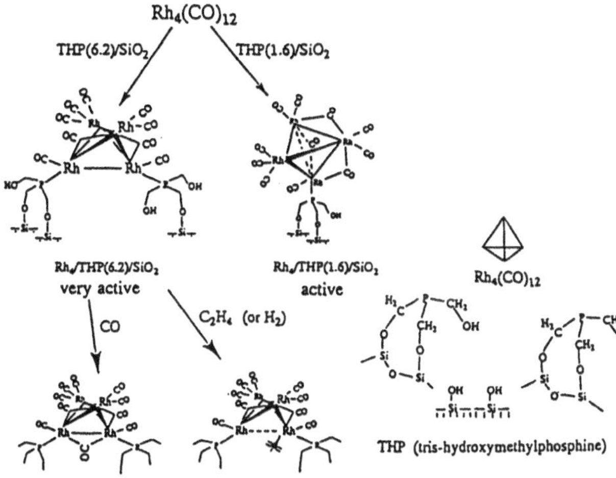

3.2. Cluster-derived heterogeneous catalysts for CO hydrogenation

Olefin hydrogenation, ethane hydrogenolysis, and methanation have been tested for evaluating the catalytic activities of cluster-derived metal catalysts(31-33). All the test reactions, the catalysts derived from noble metal carbonyl precursors such as $Rh_6(CO)_{16}$, $Ru_3(CO)_{12}$ and $Os_3(CO)_{12}$ were at most ten time more active than their traditional counterparts. In particular, the catalysts of metals which are difficult or incapable of reduction with H_2 even at 500°C, e.g., Mo, W and Mn carbonyl-derived ones exhibited quite extraordinary activities(upto 10^4 times more than those of salt-derived conventional catalysts. The zerovalent metal cluster complexes provide advantages as the precursors in giving easily the reactive metal ensembles in lower oxidation states impregnated on the dehydrated/dehydroxylated support surfaces.

Methane and other C^+ hydrocarbons were evolved with H_2 and CO_2 by heating metal carbonyl cluster complexes, such as $Rh_6(CO)_{16}$, $Fe_3(CO)_{12}$, $Co_4(CO)_{12}$, $Ru_3(CO)_{12}$, $Os_3(CO)_{12}$ and $Ir_4(CO)_{12}$ impregnated on the strong acid metal oxides such as SiO_2/Al_2O_3, HY zeolite and MgO at the temperatures below 200°C in vacuo or in a helium flow. Such a hydrocarbon production derived from the metal carbonyl cluster complexes impregnated on the hydrated oxides could be interpreted due to a proton-induced CO reduction of the oxide-supported carbonyl clusters, as suggested by Whitmire and Shriver(23). They found that $Fe_4(CO)_{13}^{2-}$ reacts slowly with a strong Brønstead acid such as $H_2SO_3CF_3$ at room temeperature, resulting in the formation of ca 1 mole of CH_4 per cluster.

$$Fe_4(CO)_{13}^{2-} + H^+(H_2SO_3CF_3) \text{ -----> } CH_4 + 3Fe^{2+} + CO + H_2 + \text{Fe clusters}$$

In their ^{13}C labelling experiments, they demonstrated that the formed CH_4 was originated from a ligand CO(likely a triply CO) of the starting Fe_4 carbonyl cluster, which was directly reduced with a proton source, not with H_2 evolved in the reaction. In this process the necessary electrons for the reduction of ligand CO may be provided from the metal cluster framework, with the consequent conversion of which to Fe^{2+},leading to the formation of CH4.The prerequisite of metal nuclearity more than M_4 in the proton-induced CO reduction is proposed to be associated with the necessary numbers of electron provided from a cluster unit and multiple-coordination of CO leading to CO cleavage through the transformation to methane and other higher hydrocarbons.

Ichikawa previously demonstratcd(35,36) that the product selectivity in CO hydrogen-ation catalyzed on the Rh catalysts derived from a series of Rh carbonly clusters markedly depended upon not only nuclearity of precursor Rh carbonyl clusters, but also strongly on the nature of the oxide supports. This is illustrated in Table 3, where highly dispersed Rh crystallites prepared by decomposing $Rh_4(CO)_{12}$ impregnated on suitable oxides such as La_2O_3, Nd_2O_3, ZrO_2, TiO_2, Nb_2O_5 and MnO_2 give C_2-oxygenates such as ethanol in higher effieciencies. On ZnO, MgO and CaO Rh cluster-derived catalysts provided almost exclusively methanol with a minor of hydrocarbons, while with SiO_2 and Al_2O_3 hydrocarbons such as methane are the preferential reaction products with a poor selectivity for the oxygenates. On the other hand, the product selectivities toward oxygenates offered a moderate change for a variety of cluster nuclearity from Rh_4 to Rh_{13} carbonyl clusters on La_2O_3 and ZrO_2(37). In situ XPS srudies on the $Rh_4(CO)_{12}$ impregnated on the different metal oxides has been conducted by Kawai et al.(38) the observed B.E.(binding energy) of the Rh 3d3/2,5/2 lines are shifted to relatively higher energy values, i.e., 307.0, 307.3, 307.8, and 308.4 ev on Rh derived from $Rh_4(CO)_{12}$ impregnated on SiO_2, TiO_2/SiO_2, ZrO_2/SiO_2, ZrO_2 and ZnO, respectively, where TiO2/SiO2 and ZrO2/SiO2 were prepared by pyrolysis and calcination of $Ti(iso\text{-}C_3H_7O)_4$ and $Zr(n\text{-}C_3H_7O)_4$ deposited on SiO_2 gel. In comparing B.E. value of reference Rh samples, the XPS data suggested that Rh aggregates derived from $Rh_4(CO)_{12}$ on ZnO and MgO which catalyze methanol formation exist in the oxidation state close to Rh^{+1}, whereas the catalysts on SiO_2 and Al_2O_3 active for hydrocarbons in a Rh^0 state. In this context, TiO_2/SiO_2, ZrO_2/SiO_2 and ZrO_2 are reasonably favorable oxide supports to manage CO dissociation and promote CO insertion in forming C_2 oxygenates such as ethanol.

Table 3. Product Distribution for $CO\text{-}H_2$ Conversion at 1 atm Pressure over $Rh_4(CO)_{12}$ -Derived Catalysts Impregnated on Various Metl Oxides

Catalyst[b]	Temperature (°C)	CO conversion (% h^{-1})	Carbon basis selectivity ($iC_i/\Sigma iC_i \times 100$) (%)					
			CH_3OH	C_2H_5OH	$CH_3CHO + CH_3COOR$	CH_4	C_2–C_4 hydrocarbons	CO_2 + others
$Rh_4(CO)_{12}$-ZnO	220	1.6	94	+[c]	—	4	—	2
$Rh_4(CO)_{12}$-MgO	220	2.6	88	2	+	7	+	3
$Rh_4(CO)_{12}$-CaO	230	0.8	92	1	—	2	+	5
$Rh_4(CO)_{12}$-La_2O_3	205	3.0	38	42	1	8	2	9
$Rh_4(CO)_{12}$-Nd_2O_3	210	3.8	24	47	1	17	4	7
$Rh_4(CO)_{12}$-ZrO_2	215	4.4	13	45	2	31	5	4
$Rh_4(CO)_{12}$-TiO_2	210	6.0	6	32	8	30	21	3
$Rh_4(CO)_{12}$-Nb_2O_5	195	5.8	7	30	9	32	19	3
$Rh_4(CO)_{12}$-Ta_2O_5	190	4.2	5	22	8	43	20	2
$Rh_4(CO)_{12}$-MnO_2[d]	205	1.2	4	10	15	35	28	8
$Rh_4(CO)_{12}$-SiO_2	235	1.7	1	3	3	66	26	1
$Rh_4(CO)_{12}$-γ-Al_2O_3	250	8.6	+	+	+	67	32	1

[a] Reaction conditions: CO/H_2 = 20:45 cmHg in closed circulating reactor of 420 cm^3 capacity.

[b] $Rh_4(CO)_{12}$ was deposited from hexane solution onto each oxide powder (20 g) at 0.5 wt% Rh loading, followed by heat treatment at 120–200°C *in vacuo* or under an H_2 atmosphere.

[c] + indicates formation of trace.

[d] MnO_2 was partially reduced to MnO in the $CO\text{-}H_2$ reaction.

4. Cluster-derived bimetal catalysts

4.1. Structures and metal composition

The recent trials by using the molecular bimetallic clusters as the precursors have been carried out for the tailored heterogeneous catalysts, because of the advantages maily accounted as follows:

1) Preexistence of heterometallic bonds in the discrete precursor clusters may favor the formation of bimetallic(and multimetallic) particles of well-controlled metal compositions and higher dispersion. Systematic variation in particle stoichiometry may be possible by employing precursor complexes of different known metal compositions in the homologue cluster frameworks, e.g., $Rh_{4-x}Co_x(CO)_{12}$(x=0-4), $Ru_{3-x}Fe_x(CO)_{12}$(x=0-3), $Pd_6Fe_6(CO)_{24}^{2-}$ and $[Rh_{6-x}Fe_x(CO)_{16}]^{2-}$(x=0-4)

2) Fairly homogeneous mixed phase systems such as $[Ru_6Cu(CO)_{17}]^{2-}$ and $Rh_4Fe_2(CO)_{16}^{2-}$ may be achieved by using mixed metal cluster complexes as the precursors whose metals are immisible in the bulk, i.e., Cu+Ru, Os+Cu, Rh+Fe etc.

3) The heterometallic combination of the precursor complexes having the different susceptible to oxygen, e.g., oxide supports, one of which acts as an anchor in keeping the other metal ensembles to prevent a sintering. On the other hand, the heteronuclear adjacent metallic sites are coperative for multi-centered activation of reactant molecules such as CO, involving acid-promoted CO dissociation and a migratory CO insertion.

The recent advances in using EXAFS, coupled with Mossbauer spectroscopies have made better insight into their locations and coordination circumstances including their oxidation states of the resulting multimetallic ensembles attached with the different surface goups of the supporting oxides. Although relatively little is known at present about the actural shapes and structures of the mixed metal species resulting from the thermal decomposition of the precursor clusters, particularly to what extent they keep their original cluster frameworks, metal compositions and ligand coordinations.

Using molecular bimetal clusters of a miscible combination of elements such as $Rh_2Co_2(CO)_{12}$ and $RhCo_3(CO)_{12}$ as the precursors impregnated on Al_2O_3 and SiO_2. Infrared studies indicated that such precursor bimetal clusters are strongly chemisorbed on dehydrated alumina through an ionic-covalent bonding between their bridging CO and Lewis acid site e.g., Al^{3+}($1650\ cm^{-1}$). The EXAFS data(39) suggested that the resulting impregnated species are still keeping their original cluster frameworks in terms of coodination numbers and atomic distances, as shown in Table 4. EXAFS spectroscopic studies also demonstrated that after the mild oxidation of the impregnated Rh_2Co_2 and $RhCo_3$ carbonyl complexes to eliminate CO, followed with H_2-reduction at 350-400^oC produced the catalysts consisting of highly dispersed bimetallic particles of less than 10A in size having the Rh/Co compositions similar to those for the starting presursor clusters(38). When a physical mixture of $Rh_4(CO)_{12}$ and $Co_4(CO)_{12}$ was similarly impregnated from the THF solution on the dehydrated alumina, followed with mild oxidation plus H_2-reduction at 400^oC, the reduced catalysts referred as $Rh_4(CO)_{12}$+$Co_4(CO)_{12}$(1:1) showed the EXAFS data having only Rh-Rh and Co-Co bonds but a negligible contribution of Rh-Co bond. The results reveal no scrambling of Rh and Co atoms/ions (with no appreciable segregation of metal phase) in the RhCo

Table 4. Curve Fitting Results for RhCo Carbonyl cluster-Derived Catalysts Supported on γ -Al_2O_3

Sample (metal loading)	Rh—O n	Rh—O r (Å)	Rh—Co n	Rh—Co r (Å)	Rh—Rh n	Rh—Rh r (Å)	Co—O n	Co—O r (Å)	Co—Co n	Co—Co r (Å)	Co—Rh n	Co—Rh r (Å)
$Rh_2Co_2(CO)_{12}$ crystal[a]			~2	2.63	~1	2.73			~1	2.55	~2	2.63
Impregnation (4 wt%)[a]	—	—	~2	2.66	~1	2.73	—	—	~1	2.57	~2	2.61
Pyrolysis (2 wt%)	1-2	2.14	—	—	~1	2.64	2-3	1.96	—	—	—	—
H_2 reduction (2 wt%)	1-2	2.20	1-2	2.49	2-3	2.64	~2	1.97	1-2	2.49	1-2	2.53
$RhCo_3(CO)_{12}$ crystal[a]			~3	2.60					~2	2.53	~1	2.61
Impregnation (4 wt%)[a]	—	—	1-2	2.64	—	—	—	—	1-2	2.64	~1	2.61
Pyrolysis (4 wt%)	~1	2.22	1-2	2.56	—	—	1-2	1.99	1-2	2.43	—	—
H_2 reduction (4 wt%)	~2	2.18	~3	2.58	—	—	1-2	1.98	1-2	2.43	—	—
$Rh_4(CO)_{12}$ + $Co_4(CO)_{12}$ H_2 reduction (2 wt%)	~2	2.18	—	—	4-6	2.66	~2	2.02	~1	2.45	—	—
Rh-Co 1:1 salt (4 wt%)[b]	—	—	4-6	2.58	7-8	2.67	1-2	2.05	3-4	2.47	1-2	2.62
Rh-Co 1:3 salt (4 wt%)[b]	—	—	4-6	2.59	7-8	2.66	—	—	~10	2.48	—	—

[a] In the case of the crystal and impregnated specimens, other contributions attributed to Rh—O and Co—O where the oxygen belongs to carbonyl ligands are found.

[b] $RhCl_3$ and $CoCl_2$ were impregnated on γ-Al_2O_3 from methanol solution, prior to H_2 reduction at 400°C.

Fig.11. Structures and metal compositions of catalysts derived from Al_2O_3-impregnated $Rh_2Co_2(CO)_{12}$, $RhCo_3(CO)_{12}$, $Rh_4(CO)_{12}$+$Co_4(CO)_{12}$, and conventional $RhCl_3$+ $CoCl_2$ as deduced by EXAFS and IR data.

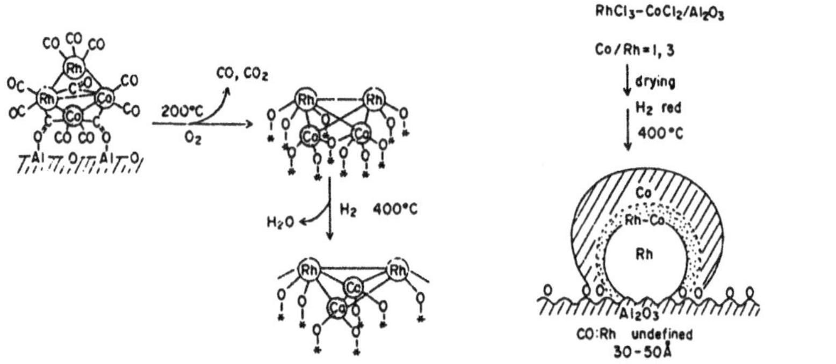

"Raft structure" of a Rh_2Co_2 unit derived from $Rh_2Co_2(CO)_{12}$

Surface composition of conventional $RhCl_3$-$CoCl_2$ supported catalyst

cluster-derived catalysts. By contrast, the conventional RhCo catalysts using $RhCl_3$ and $CoCl_2$ on alumina, after the H_2 reduction at 673K gave the EXAFS parameters to be interpreted by a metal segregation of of metal particles in 50 Å diameter in which Co is enriched in the surface layer regardless of their relative metal compositions of starting metal salts, as the pictoral presentation in Fig 11. On the other hand, the high resolution EDAX studies recently revealed that the terahedral clusters $H_2FeRu_3(CO)_{13}$ and $HFeCo_3(CO)_{12}$ were chemisorbed on dehydrated MgO and their thermal decomposition led to the formation of bimetallic particles having the metal compositions similar to the precursor complexes. For instance, coimpregnation of two different homometallic clusters, $Fe_3(CO)_{12}$ +$Os_3(CO)_{12}$ or $Fe_3(CO)_{12}$+$Ru_3(CO)_{12}$ has been reported not to result in the formation of bimetallic particles, whereas others have reported that $Fe_3(CO)_{12}$+$Ru_3(CO)_{12}$ leads to the formation of a undetermined bimetallic catalysts.

Using a variety of bimetallic clusters to prepare the heterogeneous bimetal catalysts, which have been tested to some typical catalytic reactions and exhibited activities and selectivities different from those afforded by conventionally prepared catalysts.

Anderson and Mainwaring(6) first used $Rh_2Co_2(CO)_{12}$ impregnated on Al_2O_3 to prepare a dispersed bimetallic catalysts. They demonstrated that the catalysts gave a higher metal dispersion(12-28Å in size) having a ratrher uniform Co/Rh composition (Co/Rh atomic ratio=0.51), estimated from their magnetic susceptibility $\chi s(Co)$. The carbonyl derived RhCo bimetallic catalyst exhibited higher selectivities toward the skeletal rearrangement of methylcyclopenetane to a mixture of 2- and 3-methylpenetanes, whereas on the conventional counterpart the preferential hydrocraking proceeded to give lower molecular weight products. Such a promotion on the RhCo bimetallic catalysts for the skeletal rearrengement is expected to be related with decreasing Rh ensemble sizes by breaking Co atoms in the Rh particles.

4.2. Catalytic Activities of Bimetal cluster-Derived Catalysts

Ichikawa (35) has extended a series of works using homologous tetranuclear clusters, $Rh_4(CO)_{12}$, $Rh_2Co_2(CO)_{12}$, $RhCo_3(CO)_{12}$ and $Co_4(CO)_{12}$ as precursors to prepare $Rh_{4-x}Co_x$(x=0-4 bimetal clusters)bound to ZnO and active carbon as the suitable supports. The hydroformylation rates of ethene and propene increased comparatively on the Rh_2Co_2 and $RhCo_3$ catalysts, compared with Rh_4 catalyst as shown in Fig.12, whereas Co_4 showed lower activity. In the propene hydroformylation on the carbon-supported RhCo bimetal cluster-derived catalysts, the precursors with larger cobalt content exhibited the enhancement of activities and selctivities toward a linear aldehyde. This may reflect the advantage resulting from a more intimate bimetal interaction favorable for cogesting the reactive molecules at the adjacent heterometallic sites which could be derived from the carbonyl mixed metal clusters(10,40).

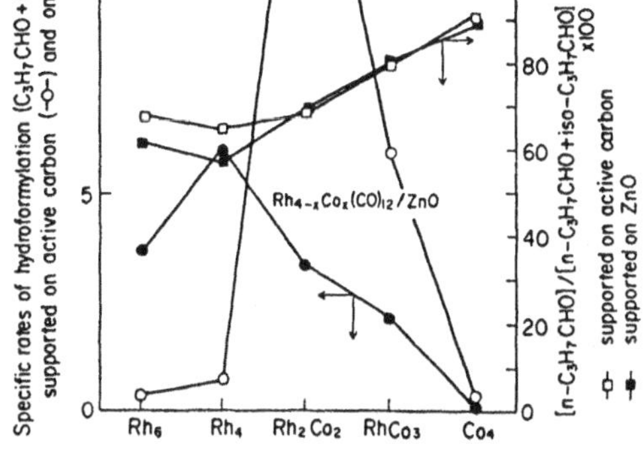

Fig.12. Catalytic performances of Rh, RhCo bimetallic and Co carbonyl clusters grafted on ZnO for propene hydroformylation reaction. [propene/H_2/CO=1/1/1, total pressure0.8 atm at 425K].

Esteban Puges et al(41) also have used $Pd_2W_2Cp_2(CO)_6(PPh_3)_2$ impregnated on Al_2O_3 to prepared the Pd-W bimetallic catalysts after H_2 reduction at 623K, on which they found that methyl-cyclopenane(MCP) was converted exclusively to benzene. XPS and EXAFS studies on their structures and oxidation states suggested that the central Pd was surrounded with Pd and Cr atoms in the resulting PdCr bimetallic catalysts. The isomerization of 2-methylpentane gave a mixture of 3-methylpentane and n-hexane in the molar ratio of ca 30. On the other hand, the H_2-reduced catalysts derived from $Pd_2Cr_2Cp_2(CO)_6PMe_3$ adsorbcd on Al_2O_3 exhibited the higher selectivties of MCP formation from 2-methylpentane through the C_5 cyclic mechanism proposed on the well-dispersed Pd catalysts. In these cases of the bimetallic catalysts, the Pd ensemble sizes may be controlled with the contegeous W and Cr atoms decorated on Pd particles, which is reflected in promoting the skeletal rearangement of hydrocarbons.

It is of interest to note that supported Pt-Co catalystrs (42) prepared from linear and non-linear Pt-Co carbonyl cluster complexes, e.g., $Pt\{Co(CO)_4\}_2(CNC_6H_{11})_2$, $Co_2Pt_2(CO)_8(PPh_3)_2$ and $Co_2Pt_3(CO)_9(PEt_3)_3$ impregnated on Al_2O_3(42). The Co_2Pt_2 and Co_2Pt_3 catalysts showed a higher selectivity for demethylation of MCP ($C_6 > C_5+C_1$) than the Co_2Pt catalyst and conventional Pt and Pt-Co catalysts. Pt and Co alone on alumina did not exhibited this selectivity for the demethylation reaction. At presence, however, the unusually higher selectivities for demethylation on the Co_2Pt_2 and Co_2Pt_3 cluster-derived catalysts are discussed to be associated not with their Pt-Co heteronuclear interaction, but with the phosphine ligands of the precursor complexes. In fact, in stead of the PtCo precursor complexes, the catalysts from $Pt_3(CO)_3(PPh_3)_4$ on alumina, which was readily converted to Pt5(CO)6(PPh3)4 in the activation, followed with H_2-reduction gave also higher selectivities for the demethylation of MCP. It appears probable that P atoms derived from phosphine ligands under the catlyst preparation are partially covered on Pt particles in blocking the higher coodinatively unsaturated sites, which prefor the cyclic mechanism of isomerization and hydrogenolysis on Pt crystallites.

Yermakov et al(43) have first tried to prepare the bimetallic Pt-Sn catalysts derived from $H_4[Pt_3Sn_8Cl_{20}]$, $(COD)_3Pt_3(SnCl_3)_2$ impregnated on Al_2O_3. These Pt-Sn catalysts are characterized by a lower actvity for hydrocraking of MCP and n-hexane to lower hydrocarbons (C_1-C_5), compared with the conventional Pt and Pt+Sn salt-derived ones, and exhibited high selectivties toward aromatics. As the result, it may suggest that the C5 cyclic mechanism for conversion of n-hexane was strongly suppressed on the Pt-Sn catalysts, and at the same time coke formation was decreased. Two type of active sitres were assumed, M1, which was active for ethane hydrogenolysis, and M_2 which is active for C-C bond isomerization, but not for hydrogenolysis. As the Sn/Pt ratios were increased, the number of ensemble M_1 sites decreased, but increased an isolated Pt atom M_1 sites, where aromatic and hydrocraking products decreased whereas skeletal rearrangement was enhanced. The modifyer Sn plays a role of breaker of the Pt ensembles in the Pt-Sn bimetallic catalysts.

Shapley et al(44) have prepared the Ir-W bimetallic catalysts from pseudo-terahedral clusters $CpWIr_3(CO)_{11}$ and $CpW_2Ir_2(CO)_{10}$. The resulting Ir_3W bimetallic particles of less than 10 Å exhibited a high activity for the central C-C scission of butane to over 70% ethane, similarly on the Ir_4 and $[Ir_4]+[Cp_2W_2(CO)_6]$ catalysts, but $[Ir_2W_2]$ catalyst gave less than 50% ethane in the product. This different cracking pattern for the Ir_2W_2 catalysts was accounted as strong evidence for the Ir-W heteronuclear interaction. According to their larger decrease of activation enrgies for the butane hydrogenolysis on

Table 5. Preparation of Mixed Metal Cluster--Derived Catalysts and applications to Catalytic Reactions.

Bimetallic composition	Precursor cluster*	Support	Applied reaction
Pd-Fe	$Fe_2Pd_2(CO)_5(NO)_2(DPPM)_2$	SiO_2	$ArNO_2$ carbonylation
	$[Pd_6Fe_6(CO)_{24}H][TMBA]_2$	SiO_2	CO + H_2 reaction
	$[Fe_3Pd(CO)_{16}][TMBA]_2$	SiO_2	CO + H_2 reaction
Pd-Cr	$Cr_2Pd_2Cp_2(CO)_6(PMe_3)_2$	γ-Al_2O_3	Hydrocarbon rearrangement
Pd-Mo	$Mo_2Pd_2Cp_2(CO)_6(PPh_3)_2$	γ-Al_2O_3	$ArNO_2$ carbonylation
Pd-W	$W_2Pd_2Cp_2(CO)_6(PPh_3)_2$	γ-Al_2O_3	Hydrocarbon rearrangement
Rh-Fe	$[FeRh_5(CO)_{16}][TMBA]_2$	SiO_2	CO + H_2 reaction
	$[FeRh_4(CO)_{15}][TMBA]_2$	SiO_2	Olefin hydroformylation
	$[Fe_2Rh_4(CO)_{16}][TMBA]_2$	SiO_2	Olefin hydroformylation
	$Fe_3Rh_2(CO)_{14}C$	SiO_2	Olefin hydroformylation
	$Fe_2Rh_4(CO)_{16}^{2-}$	NaY	CO + H_2 reaction
Rh-Co	$Co_2Rh_2(CO)_{12}$	γ-Al_2O_3, SiO_2	Hydrocarbon rearrangement
		ZnO	Olefin hydroformylation
		ZrO_2	CO + H_2 reaction
		Carbon	Olefin hydroformylation
	$RhCo_3(CO)_{12}$	ZnO	Olefin hydroformylation
		Carbon	Olefin hydroformylation
Rh-Os	$H_2Os_3Rh(CO)_{10}(acac)$	γ-Al_2O_3	CO + H_2 and hydrogenation
Pt-Sn	$H_4[Pt_3Sn_8Cl_{20}]$	γ-Al_2O_3	Hydrocarbon rearrangement
	$(COD)_3Pt_3(SnCl_3)_2$	γ-Al_2O_3	Hydrocarbon rearrangement
Pt-Fe	$[Fe_3Pt_3(CO)_{15}][TMBA]_2$	SiO_2	CO + H_2 reaction
	$[Fe_4Pt(CO)_{16}][TMBA]_2$	SiO_2	CO + H_2 reaction
	$Fe_2Pt(CO)_6(NO)_2(t\text{-}BuNC)_2$	γ-Al_2O_3	Hydrocarbon rearrangement
Pt-Ru	$Ru_3Pt(CO)_{12}Py_3$	Carbon, oxides	CO (CO_2) + H_2 reaction
Pt-Re	$Re_2Pt(CO)_{12}$	γ-Al_2O_3	Hydrocarbon arrrangement
Pt-Co	$Co_2Pt(CO)_8(c\text{-}C_6H_{11}NC)_2$	γ-Al_2O_3	Hydrocarbon
	$Co_2Pt_2(CO)_9(PPh_3)_2$	γ-Al_2O_3	rearrangement
	$Co_2Pt_3(CO)_9(PEt_3)_3$	γ-Al_2O_3	Hydrocarbon rearrangement
Ir-Fe	$[FeIr_4(CO)_{15}][TMBA]$	SiO_2	CO + H_2 reaction
	$[FeIr_5(CO)_{15}][TMBA]$	SiO_2	CO + H_2 reaction
	$[Fe_2Ir_4(CO)_{16}][TMBA]_2$	SiO_2	CO + H_2 reaction Olefin hydroformylation
Ir-W	$WIr_3Cp(CO)_{10}$	γ-Al_2O_3	Butane hydrogenolysis
	$W_2Ir_2Cp_2(CO)_{10}$	γ-Al_2O_3	Butane hydrogenolysis
Ir-Pt	$Ir_2Pt(CO)_7Py_2$	Carbon	Hydrocarbon
	$IrPt(CO)_{15}Py_2$	Carbon	rearrangement

the Ir_2W_2 catalysts, rather than Ir_4 or Ir_4+W_2 catalysts, the Ir-W heteronuclear interaction induced a promotion of C-C bond cleavage of hydrocarbons.

Supported bimetallic catalysts derived from the sulfido cluster complexes $Mo_2Fe_2Cp_2(CO)_6$ and $Mo_2Co_2S_3Cp_2(CO)_4$ impregnated on Al_2O_3 have been found to be active for CO+H_2 exclusively into methane. In contrast, the catalysts prepared from the same complexes adsorbed on MgO gave highly selective formation of C_2H_4+C_2H_6, the result of which are completely different from those of Mo/Al_2O_3, MoS_3+Fe/Al_2O_3 or Co/Al_2O_3. Infrared, EXAFS and Mossbauer studies suggested that no structural change occurs upon impregnation on Al_2O_3 and MgO, and furthermore without any fragmentation and reaggregation to make larger crystallites on MgO. The higher selectivity toward C_2 hydrocarbon formation could be based on the MoFe ns MoCo heteronuclear interactions of the bimetallic sites. S atoms may play a role of retaining the bimetallic frameworks as the interstial ligands of the bimetal clusters.

5. Metal/alloy clusters in nanometer-size pores, -synthesis, structures and catalysis-.

5.1. Ship-in-bottle synthesis of metal clusters in micro/mesoporous space

Zeolites are aluminosilicate crystallines consisting of microporous cages of molecular dimensions(8-12 Å) which are interconnected with smaller windows and channels(6-8Å). Such micropores can supply "templating" circumstances for the selective synthesis of some bulky metal carbonyl clusters which fit the interior cages as ultimate "nanometer reaction vessels". Because of the size limitation of the zeolite pore windows(ca 7.4 A) it is usually difficult to directly introduce the bulky carbonyl clusters such as $Rh_6(CO)_{16}$(ca 10 A in molecular size) and $Ir_4(CO)_{12}$(8 Å) into them. To overcome this difficulty, the metal cluster complexes may be subsequently synthesized and rebuilt from the smaller size precursors e.g., metal cations by ion-exchang technique, while subcarbonyls and mononuclear complexes are preadsorbed or occuled inside zeolite cavities by monolar dispersion technique, respectively. Some carbonyl species which are deposited on the external surface of zeolite migrate and enter inside the microcages. The intrazeolitic mononuclear and subcarbonyl metal complexes are converted by the reductive carbonylation reaction with $CO+H_2$ or CO/H_2 towards the metal carbonyl cluster complexes, similarly in solution. In this context, in situ internal prepration of the metal complexes is inferred to "ship-in-bottle synthesis", as pictorally presented in Fig 13.
This "ship-in-bottle" synthesis may open new oppotunities for the rational design of tailor-made catalysts of descrete metal/alloy clusters having uniform sizes and metal compositions and with sufficient stability against a metal sintering and leaching under the prevailing reaction conditions.

Previously, uni- and bimetal carbonyl clusters such as $Rh_{6-x}Ir_x(CO)_{16}$/ NaY (0-6)(45), $[Pt_3(CO)_6]_n^{2-}$(n=3,4)/NaY(46) and $[Ru_6(CO)_{18}]^{2-}$/NaX(47), $[Fe_2Rh_4(CO)_{15}]^{2-}$/NaY(10,48) and $[HRuCo_3(CO)_{12}]$/NaY(38), which were characterized by EXAFS /XANES, FTIR, ^{129}Xe NMR, HRTEM and Raman spectroscopies. They are useful for preparing discrete metal/alloy clusters(less than 10Å size) which catalyze the alkane hydrogenolysis(45), CO hydrogenation towards C_1-C_5 alcohols(45,49) and olefin hydroformylation reaction(38). There are current interests in folded sheet-mesoporous materials such as MCM-41(50) and FSM-16(51) having their honeycomb structures with ordered enormous channels of 20-100 Å diameters, which are larger than microporous cavities of conventional zeolites such as NaY, ALPO-5 and ZSM-5. They are potential hosts for inclusion of bulky organic and inorganic species for new application to design of tailored metal catalysts accessible for larger substrates. There are the current interests in their mobility/flexibility and reactivity of the intrazeollitic clusters, it was recently extended to proceed a "Ship-in-Bottle" synthesis of novel carbonyl clusters such as

Fig.13. Pictorial representation of ship-in-bottle synthesis of metal carbonyl clusters in NaY microcavities via reductive carbonylation of Rh^{3+}/NaY

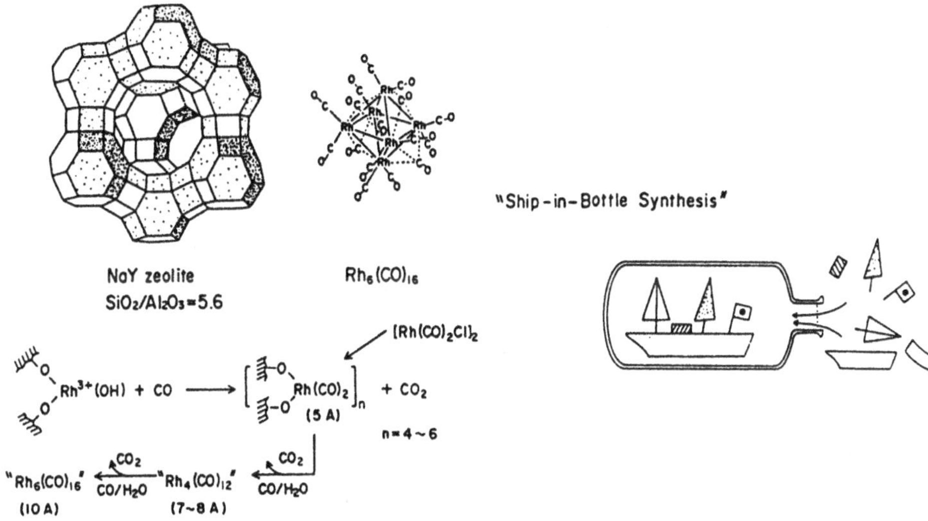

$Ru_3(CO)_{12}$, $H_4Ru_4(CO)_{12}$, $[HRu_6(CO)_{18}]^-$ and $[Pt_{12}(CO)_{24}]^{2-}$ in NaY micropores and a robust $[Pt_{15}(CO)_{30}]^{2-}$ encapsulated in the ordered mesoporous channels of FSM-16(52). Their structural flexibility and thermostabilities of metal clusters are related with their structural accomodation in micro and meso porous constraint.

The ion-exchanged Rh^{3+} in NaY zeolites was reacted with CO and H_2O, undergoing the transformation same as in aqueous alkaline solution at the temperatures of 50-120°C by the successive formation of various carbonyl species, $[Rh(CO)_2]$/NaY and $[Rh_4(CO)_{12}]$/NaY(2085,1830cm^{-1}), which were eventually converted into $[Rh_6(CO)_{16}]$ /NaY. $Rh_6(CO)_{16}$ was selectively(>85%) produced inside NaY cavity by the reductive carbonylation reaction of Rh^{3+} in NaY with CO+H_2 or CO+H_2O at 393-473K. The structures of Rh_6 clusters inside NaY was characterized by means of FTIR and Rh-K edge EXAFS. The data as shown in Table 6 provides a direct evidence for the stoichiometric formation of hexanuclear Rh carbonyl clusters in a good agreement of free molecule in terms of coordination numbers and atomic distance around Rh-Rh(2.74Å) and Rh-CO(1.88Å) bondings, as pictorial representation in Fig. 14. Gallezot et al. previously also reported by the RED measurement that the formed Rh6(CO)16 inside the NaY cages showed the characteristic Rh-Rh bonds of 2.77A.

Similarly, using ion-exchanged Ir^{4+}/NaY with a mixture of CO and H_2O at 440K $Ir_6(CO)_{16}$ was obtained via the intermediate formation of $Ir_4(CO)_{12}$. Gallezot et al (52) reported the predominant feature of the RED pattern is a strong peak at 2.77A which is typical first neighbor Ir-Ir distance of hexanuclear carbonyl complex in crystal.

Table 6. Results of Curve-fitting Analysis of Rh K-edge EXAFS Data Observed at 300K for NaY-Encapsulated Rh cluster sample.

Sample	Rh—Rh n	Rh—Rh r (Å)	Rh—CO_t n	Rh—CO_t r (Å)	Rh—$CO_{t\ or\ b}$ n	Rh—$CO_{t\ or\ b}$ r (Å)	Rh—O n	Rh—O r (Å)
$Rh_6(CO)_{16}$-NaY	3.1	2.74	1.5	1.88	1.6	2.15	1.8	2.06
$[Rh_6]_{ox}$-NaY							6.8	2.06
$[Rh_6]_{red}$-NaY(473 K H_2)	4.6	2.70					0.7	2.10
$[Rh_6]_{red}$-NaY(673 K H_2)	4.6	2.70					0.7	2.09
$CO_{(ads)}Rh_6$-NaY(473 K H_2)	3.2	2.72	1.4	1.85	1.4	2.15	0.8	2.03
Reference samples								
Rh foil (*fcc*)	12.0	2.69						
Rh_2O_3 (bulk)							6.0	2.05
$Rh_6(CO)_{16}$	4.0	2.76	2.1	1.87	2.0	2.17		
$Rh_6(CO)_{16}$[b]	4.0	2.776	2.0	1.864	2.0	2.168		

[a] Estimated experimental errors are ±0.02 Å for atomic distance r and ±0.2 for coordination number n in the EXAFS data evaluation.
[b] Results based on X-ray diffraction analysis.

Moreover, by the analogy of $[Rh_6(CO)_{16}]$/NaY that $Ir_6(CO)_{16}$/NaY $Co_6(CO)_{16}$ was also formed from the preadsorbed $Co_2(CO)_8$ inside the NaY zeolite cage. In this sense, the micropores of NaY zeolite act as a "templating microreactor" to synthesize hexanuclear homologeous carbonyl clusters.

After the mild oxidation of $[Rh_6(CO)_{16}]$/NaY with dry O_2 by heating from 293 to 473K to eliminate its cluster carbonyls, followed with hydrogen reduction of the oxidized sample at 473 and 673K. The EXAFS analysis for reduced $[Rh_6]$/NaY demonstrated the retention of Rh cluster unit in the zerovalent state in terms of the lower coordination number similar to that of the original $[Rh_6(CO)_{16}]$/NaY and the atomic distance closed to metallic Rh(Rh-Rh: C.N. =4.6, R=2.70 Å). The reduced $[Rh_6]$/NaY chemisorbed CO in a stoichiometric amount(CO/Rhtotal=2.6) and showed the characteristic carbonyl IR bands associated with shap bands of linear and bridged CO at 2087(vs), 2042(w) and 1835(s), 1760(w) cm-1) of a molecular carbonyl cluster. The EXAFS data of the sample in CO chemisorption, as also presented in Table 6, suggested the regeneration of $Rh_6(CO)_{16}$ in keeping a Rh-Rh bond (C.N.=3.2, R=2.72Å) and being partially linked to the zeolite oxide wall(Rh-O: C.N.=0.8, R=2.03Å), as shown in Table 6. The results afford the preparative advantages of zeolite-entrapped clusters as precursors for the well-defined metal catalysts, whatever the treatment and temperatures through several sequence of oxidation and reduction, without any possible formation of large crystallites on the external zeolite surface(45,53).

In the presence of zeolite water, $Fe_2(CO)_9$ adsorbed on the external zeolite NaY and NaX was readily converted into $HFe_3(CO)_{12}$ at 297-333K, similarly as the reaction of $Fe_2(CO)_9$ or $Fe(CO)_5$ in an aqueous alkaline solution. It was proposed that the active Fe(CO)4 radical species generated by decomposing $Fe_2(CO)_9$ or $Fe_3(CO)_{12}$ enter inside zeolite frameworks to rebuild more stable carbonyl clusters such as $HFe_3(CO)_{11}^-$.

The bimetallic RhFe carbonyl cluster inside NaY by the reaction of [HFe(CO)11]/ NaY with Rh4(CO)12 at 343K in vacuo as the analogous stoichiometric reaction between $Fe_3(CO)_{11}^{2-}$ and $Rh_4(CO)_{12}$(or $[Rh(CO)_2Cl]_2$) in THF solution to synthesize $[Rh_4Fe_2(CO)_{16}^{2-}]$. The resulting sample gave the IR carbonyl spectra, quite identical to those of $[Rh_4Fe_2(CO)_{16}^{2-}][TMBA]_2$ in CH_3CN solution or its deposition on NaY zeolite, showing the IR carbonyl spectrum(2093(s), 2044(w),1985(w), 1741(m), 1711(s), 1698(w) cm^{-1}. Rh-K and Fe-K edge EXAFS studies are under way to unravel their structures and Fe/Rh metal stoichiometries(48).

The metal catalysts derived from the zeolite-enthrapped metal cluster complexes have been studied because of the interests in a uniform distribution and a high degree of metal dispersion through the zeolite frameworks. Nevertheless, little information is available on the structural and chemical behaviors of the entrapped metal cluster complexes, particularly on the retention of the cluster character under the working reaction conditions, e.g., $CO+H_2$ and alkane reforming reactions.

The stabilities of the entrapped cluster species are found to be much higher than that of the clusters bound to the amorphous oxide supports such as alumina and silica, in keeping the cluster frameworks in the presence of CO and H_2.

5.2. Ship-in-Bottle synthesis of $[Pt_3(CO)_6]_n^{2-}$/NaY and FSM-16

Recently, the new series of trigonal prismatic platinum carbonyl cluster dianions suceeded to be synthesized uniformaly in NaY micro(12Å) and FSM-16 meso(28Å) porous channels by ship-in-bottle technique. They have been characterized by EXAFS, UV-vis and Fourier tranform IR spectroscopy. They exhibit high catalytic activity in NO reduction by CO to give N_2 and N_2O at 300-473K. A calcined Pt^{2+}/NaY was heated from 298 to 373K under CO with a trace of H_2O, resulting in the successive formation ofdifferent carbonyl species which show their characteristic carbonyl IR bands as shown in Fig.8(a)-8(c). It is suggested from the analogy with the IR bands of previously reported platinum carbonyl complexes e.g., $Pt(CO)Cl_3$ and $Pt_3(\mu_2\text{-}CO)_3(PPh_3)_3$ that Pt^{2+}/NaY reacts with CO formying Pt(CO)/NaY(2100 cm^{-1}) and a proposed trigonal species $[Pt_3(CO)_3(\mu_2\text{-}CO)_3]$/NaY(2112s,1896m and 1824s cm^{-1}) . They are eventually converted into the dark-gree $Pt_{12}(CO)_{24}^{2-}$(2080vs and 1824s cm^{-1}) . On the other hand, the $[Pt(NH_3)_4]^{2+}$/NaY provided stoichiometrically the organge-brown cluster species showing the intense IR bands at 2056vs and 1798s cm^{-1}), which has been fully characterized by EXAFS(46).

The reductive carbonylation of Pt^{2+}/NaY proceeds in the presence of water. The water. The water gas shift reaction with $CO+H_2O$ promote the formation of Pt hydride which is inforporated into the successive formation of the Pt carbonyl clusters such as $[Pt_{12}(CO)_{24}]^{2-}$/NaY. When the sample of Pt^{2+}/NaY was evacuated at 573K for 2 hrs to sufficiently remove water, the dehyrated Pt^{2+}/NaY gave dark-gey complex by reaction with $CO+H_2$(1:1 v/v mixture) of 600 Torr at 300-373K, showing the intense IR carbonyl bands at 2110(s), 2088(sh)cm^{-1} for linear CO and 1902(m), 1870(sh), 1850(s), 1800(sh) for bridging CO. Concerning the carbonyl IR bands this Pt carbonyl species differs from $[Pt_{12}(CO)_{24}]^{2-}$/NaY, but rather resembles those of $Pd_{13}(CO)_x$/NaY which was reported by Sachtler et. al.(54). The similar carbonyl species was formed in CO chemisorption on the reduced Pt_9/NaY derived from $[Pt_9(CO)_{18}]^{2-}$/NaY, which showed the IR bands at 2104(s), 2090(s) cm^{-1} for linear CO and 1924(m), 1886(sh), 1860(s), 1810(sh) cm^{-1} for bridging CO. The results suggested that Pt^{2+} located in the hexagonal sites of NaY is readily redueced with H_2 to Pt^0 which migrated in the supercage, resulting in the neutral carbonyl cluster like $Pt_{13}(CO)_x$ by aggregation.

These Pt_9 and Pt_{12} carbonyl cluster dianions in NaY microcavities exhibit high catalytic activities in NO reduction with CO towards N_2 and N_2O at lower temperatures of 300-473K. The conversion of NO+CO and selectivity for formation of N_2 on these catalysts samples. It is interesting to find that upon exposure of $[Pt_{12}(CO)_{24}]^{2-}$/NaY to NO(150 Torr), in situ FTIR studies demonstrated that NO breaks the intra- and inter-trigonal Pt-Pt bonds on the Pt_{12} carbonyl cluster even at 298K to give a trigonal intermediate "$Pt_3(CO)_6$"(2112,1896 and 1841 cm^{-1}) and Pt(CO)(2110 cm^{-1}) informing $N_2(N_2O)$ and CO_2 in the gas phase. Furthermore, it was found that $[Pt_{12}(CO)_{24}]^{2-}$/NaY was reversibly regenerated from the Pt carbonyl fragments such as "$Pt_3(CO)_6$" and Pt(CO) inside NaY by the reaction with CO and water at 300-353K. In contrast, in situ FTIR and GC studies suggested that NO reduction by CO proceeds catalytically to give N_2 and N_2O with the original Pt_{12} cluster framework being retained inside the zeolite micropores. under the stationary state of the reaction.

Fig.14. IR spectra of carbonyl species formed in the reaction of Pt^{2+}/NaY with CO:(a)after 3 min at 293K;(b) after 5 min at 353K:(c) after 6-10 hr at 353K:(d) in the reaction of $[Pt(NH_3)_4]^{2+}$/NaY with CO after 20 hr at 373K.

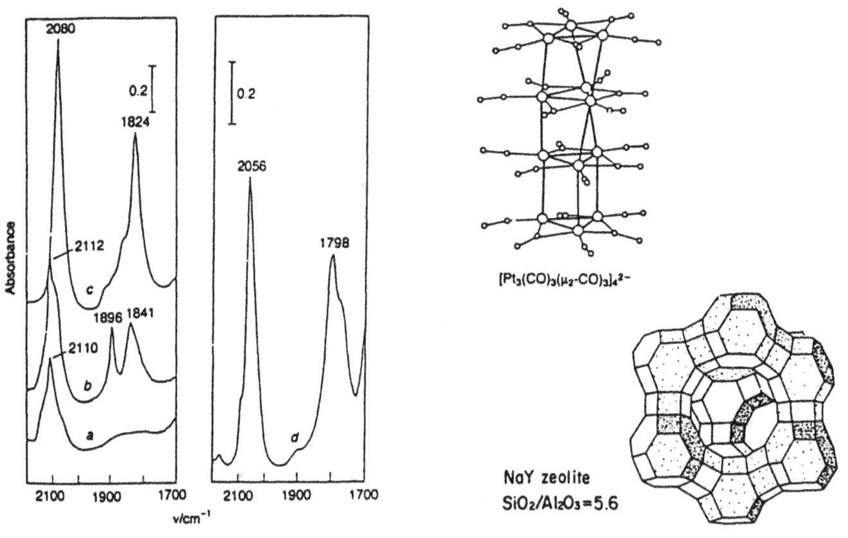

5.3. Catalytic Activity of Bimetal clusters in Zeolites

An atomospheric pressure hydroformylation of ethylene and propene was conducted at 373-453K on the reduced $[Rh_6]$/NaY and RhFe /NaY. The results were shown in Table 24, where acetaldehyde was catalyticaly obtained as the hydroformylation products on the $[Rh_6]$/NaY, but in contrast, it is of interest to find that the bimetallic RhFe/NaY catalyst gave much higher activties and selectivities for the normal-rich alcohols, compared with those on [Rh_6]/NaY. In particular, RhFe/NaX (prepared from the reaction with [$HFe_3(CO)_{11}$]/NaX+$Rh_4(CO)_{12}$) exhibited the highest linear alcohol selectivities close to 83% mole in the propene hydroformylation. A simple hydrogenation of olefins was relatively suppressed on the RhFe/NaY, rather than on [Rh_6]/NaY, resulting in the marked improvement of the oxygenate selectivities. [$HFe_3(CO)_{11}$]/NaY derived catalyst was completely inactive for both hydroformylation and hydrogenation of ethylene and propene under the similar reaction conditions. On the other hand, a physical mixture catalyst derived from NaY-entrapped [$Rh_6(CO)_{16}$]+[$HFe_3(CO)_{11}$] which was prepared by the reaction between the presynthesized [$Rh_6(CO)_{16}$]/NaY and $Fe_2(CO)_9$ in the presence of water vapor at 343K. The resulting sample showed the IR carbonyl spectra apparently superimposed with those of $Rh_6(CO)_{16}$ and $HFe_3(CO)_{11}^-$ inside NaY zeolite. The reduced sample of [Rh_6]+[Fe_3]/NaY was active for the hydroformylation of ethyelen and propene with the specific rates and selectivities almost same as those obtained on [Rh_6]/NaY. Additionally, it is of interest to find that the bimetallic RhFe/NaY(and RhFe/NaX) provided a good yield of oxygenates, mainly consisnting of ethanol and methanol at the expense of decreasing the hydrocarbon contents. While, [Rh_6]/NaY and [Rh_6]+[Fe_3]/NaY gave the preferntial formation of methane and higher hydrocarbons with a small selectivites toward acetaldehyde as only the oxygenate product in the CO+H_2 reaction. The results also rationalize that the adjacent RhFe bimetal ensembles as the active sites for the promotion of hydroformylation of olefins and oxygenates from CO+H_2 due to the enhancement of migratory insertion of CO with M-alkyl and M-H bonds, similarly as we have previously proposed on the catalysts derived from the oxide-support RhFe, RhCo and RuCo bimetal carbonyl clusters.

Fig.15. Preparation of hexanuclear RhIr alloy clusters inside NaY cavities from [$Rh_{6-x}Ir_x(CO)_{16}$/NaY (x=0-6) by the ship-in-bottle technique.

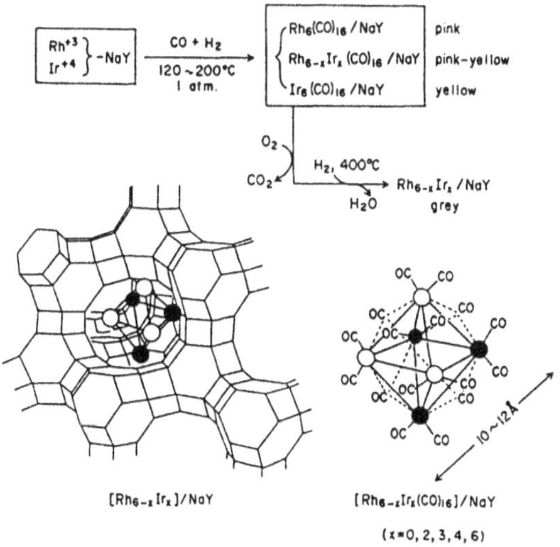

Recently, a series of RhIr bimetal clusters infered to [$Rh_{6-x}Ir_x(CO)_{16}$](x=1,2,3,4) was prepared inside NaY by the reaction of double-exchanged Rh^{3+}+Ir^{4+} NaY zeolite in flowing CO+H_2 at 423-500K. They showed the linear and three-centred CO bands characteristic of the hexanuclear metal clusters. As the testing probing reactions, hydrogenolysis of n-butane and ethene and hydrogenation of benzene were conducted on the cluster-derived catalysts. It is of interest to observe the dramatic decrease of hydrogenolysis activity by four order of magnitude across the series of clusters as the catalyst precursors by increasing the number of a modest Ir atom in the active Rh ensembles, while a slight enhancement of activity for the benezene hydrogenation. The remarkable activity-suppression of hydrogenolysis has been interpreted in terms not only simply of the geometric Rh ensemble-size effect by breaking a Ir atom, but of decreasing the electron difficiency of RhIr bimetal particles inside NaY. For the n-butane hydrogenolysis, the smaller Rh ensembles in the RhIr bimetal clusters gave the maximum selectivities towards the central C-C bond scission to give C_2H_6, higher than the Rh_6 and Ir_6 inside NaY. The turnover frequency for ethane hydrogenolysis at 473K and n-butane hydrogenolysis at 453K for the catalyst series $[Rh_{6-x}Ir_x(CO)_{16}]$(x=0-6) is plotted in Fig.16 as a function of percentage Rh for the cluster-derived catalysts. Ir has a very large effect on the turnover frequency of hydrogenolysis for both ethane($R=RoX^n$; n=5) and butane(n=6) on the RhIr heterometallic clusters. The activity decrease in this series of $[Rh_{6-x}Ir_x]$(x=0-6) for the hydrogenolysis activities can be interpleted in terms of either the breaking of active Rh ensemble sites by incooporation of inactive Ir atoms or drastic changes in the electronic states of the clusters. The hydrogenolysis of alkanes is classed as a "structure sensitive" reaction in which the sites invlove relatively large ensembles of metal atoms. By contrast, the rates and activation energies for benzene hydrogenation on the catalyst series $[Rh_{6-x}Ir_x]$(x=0-6) show that both Rh and Ir cluster catalyst exhibit hydrogenation activities of similar magnitudes. Benzene hydrogenation displays a quite different trend(basically insensitive) to the RhIr metal composition across the $Rh_{6-x}Ir_x$(x=0-6)/NaY catalyst series compared with those of ethane and butane hydrogenolysis.

Fig.16. Rates n-butane(-o-) and ethane hydrogenolysis(-Δ-)(mol/mol/metal/10^3 min) at 453K and 473K, respectiviely, compared to benzene hydrogenation(-□-) (mol/mol/metal 10^3min) on catalysts derived from theseries $[Rh_{6-x}Ir_x(CO)_{16}$ /NaY(x=0,1,2,3 and 6) as a function of atom% Rh[Rh/Rh+Ir]

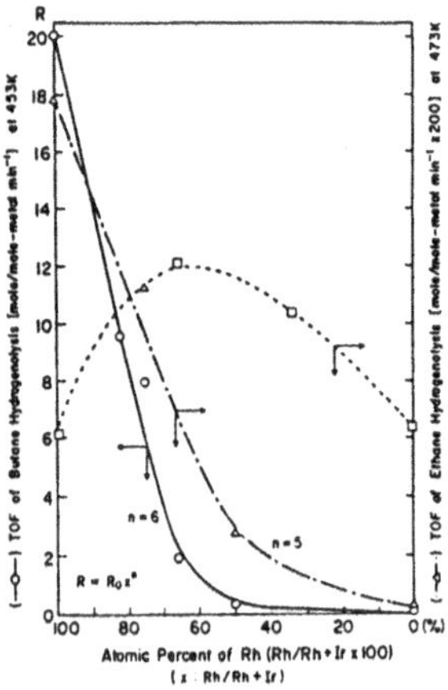

5.4. Ship-in-bottle synthesis of robust clusters such as $Pt_{15}(CO)_{30}{}^{2-}$ Encapsulated in mesoporous Channel of FSM-16

A sample contaning 5.0 mass% Pt was prepared by impregnation of FSM-16 with an aqueous solution of H_2PtCl_6. This sample was exposed to CO(200 torr) in a closed circulating system by ramping the temperature from 300 to 323K, resulting in the IR

bands at 2188, 2149 and 2119cm^{-1}, as presented in Fig.17(a). From the analogy of the previously reported Pt carbonyl species, the IR bands at 2188, 2149 and 2119 cm^{-1} can be ascribed to cis-$Pt(CO)_2Cl_2$(2188 and 2148 cm^{-1}) and $Pt(CO)Cl_3$(2121 cm^{-1}), respectively. The Pt carbonyls were converted by subsequent admission of H_2O vapor(15 torr) onto the CO atmosphere to make an olive-green product(sample A) exhibiting a steady-state spectrum(Fig.17(b)) of carbonyl bands(2086s and 1882 m cm^{-1}) and UV-vis reflectance (λ_{max};452 and 805 nm). The final spectrum closely resembles those of $[Et_4N]_2[Pt_{15}(CO)_{30}]$ in MeOH(2056s(terminal) and 1872m (bridged) cm^{-1}, λ_{max}; 408 and 697 nm) and crystal, but with shifting of both carbonyl bands to higher wavenumber(Δv=29-10 cm^{-1}) and the visible absorption peaks to higher wave length, respectively. These shifts are consistent with the formation of contact ion pairs between the Pt carbonyl anions and proton or Lewis acid sites in the solvating FSM-16 channels. Attempts to extract the platium carbonyl species from sample A with THF(tetrahydrofuran) and MeOH were unsuccesful, but with $[(Ph_3P)_2N]Cl$ in THF proceeded, which gave specifically an appreciable amount of $[(Ph_3P)_2N]_2[Pt_{15}(CO)_{30}]$ (2056s and 1872m cm^{-1};l_{max}; 405 and 702 nm in THF). This is consistent with an uniform entrappment of $[Pt_{15}(CO)_{30}]^{2-}$ in mesoporous channels of FSM-16, which was prepared by a reductive carbonylation of H_2PtCl_6/FSM-16, analogous to its synthesis in solution. It was found that $[Pt_{15}(CO)_{30}]^{2-}$ in FSM-16 is relatively unstable, and evacuation of the sample A at 10^{-4} torr and 300-323K led to the irreversible transformation owing to partial removal of CO to give the brownish product (2063s and 1820w cm^{-1}), which resembles those of higher nuclearity Pt carbonyl clusters reported as $[Pt_{55}(CO)_x]^{n-}$ and $[Pt_{38}(CO)_{44}]^{2-}$ (2060-2043s and 1832-1820w cm^{-1}) in THF solution. As presented in Fig.19, the electron micrograph of the evacuated sample A showed that platinum aggregates having 10-15 Å diameter, probably composed of 35-55 Pt atoms were uniformly distributed along the ordered mesoporous channels of FSM-16 crystals with a negligible formation of the external particles. This was in good agreement of the observed EXAFS parameters (C.N.=7.6;R(Pt-Pt) =2.74Å) for the sample A after evacuation at 323K. On the other hand, the thermostable $[Pt_{15}(CO)_{30}]^{2-}$ in FSM-16 was successfully prepared by using the FSM-16 which was coimpregnated with H_2PtCl_6 and quarternary alkyl ammonium salts (R_4NX;R= Methyl, Ethyl, Butyl and Hexyl; X = Cl, Br and OH) and methyl viologen dichloride ($[CH_3N$ - $NCH_3]2Cl^-$) from each aqueous solution. The reductive carbonylation of each coimpregnated sample resulted in an olive-green product (UV-vis reflect-ance 450 and 805 nm), showing the intense CO bands which appeared at 2075-2079s and 1875-1884m cm^{-1}(R_4NCl;R=Methyl, Ethyl, Butyl and Hexyl), relatively shifted to higher frequencies by using the larger quaternary alkyammonium cations. It is worthy to note that those organic cations play a role to stabilize the robust Pt_{15} carbonyl cluster dianion
and their thermostabilities in the FSM-16 channels at 323-393K decreased by varying the used quaternary ammonium cations(R_4N^+ and MV^{2+}) in the following order;

Ethyl>Butyl >Methyl >MV >Hexyl >> non.

Fig.17. In-situ FTIR spectral changes in reductive carbonylation of H_2PtCl_6 /FSM-16 with CO and CO+H_2O at 323K

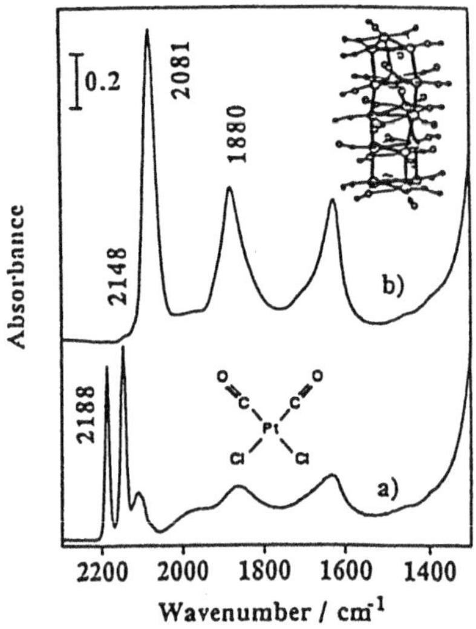

Fig.18. FT of $\chi(k)k^3$ of $[Pt_9(CO)_{18}]^{2-}$ /NaY and $[Pt_{15}(CO)_{30}]^{2-}$/NEt4/FSM-16

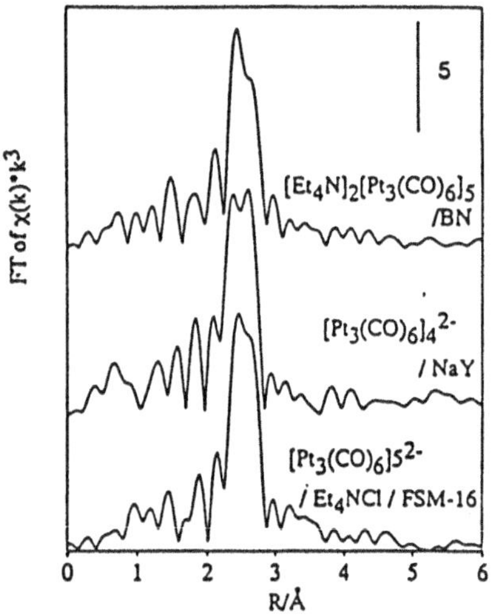

Table 7. Structural Parameters of Pt Carbonyl clusters Encapsulated in NaY and FSM-16

Compound / shell	C.N.	R / Å	ΔE_0/ eV	$\Delta\sigma^2$ / Å	R factor (%)
$[NEt_4]_2[Pt_{15}(CO)_{30}]^a$					
Pt-Pt	2.1	2.68	6.5	-1.1×10^{-3}	
Pt-Pt	1.5	3.07	-8.5	5.8×10^{-3}	7.5
Pt-C	1.5	2.10	0.4	2.0×10^{-3}	
Pt-C	1.0	1.89	-5.9	1.3×10^{-3}	
Pt-O	2.3	2.99	-4.6	4.5×10^{-3}	
$[Pt_{12}(CO)_{24}]^{2-}$/NaY					
Pt-Pt	2.2	2.67	6.5	-1.6×10^{-3}	
Pt-Pt	1.4	2.99	-8.5	7.4×10^{-3}	12.2
Pt-C	1.6	2.14	0.4	2.0×10^{-3}	
Pt-C	0.9	1.99	-5.8	0.9×10^{-3}	
Pt-O	2.2	2.99	-4.6	1.3×10^{-3}	
$[Pt_{15}(CO)_{30}]^{2-}$/Et_4NCl/FSM-16					
Pt-Pt	2.3	2.68	3.7	-0.7×10^{-3}	
Pt-Pt	1.7	3.08	-6.4	7.0×10^{-3}	9.6
Pt-C	2.2	2.08	-7.6	5.8×10^{-3}	
Pt-C	1.3	1.90	-10.0	2.6×10^{-3}	
Pt-O	2.3	3.01	-2.3	3.0×10^{-3}	

R/Å:interatomic distance, CN:coordination number, ΔEo:inner potential correction, $\Delta\sigma^2$:the dfference Debye correction from that of reference compounds.

Fig.19. High resolution TEM of $[Pt_{15}(CO)_{30}]^{2-}$/FSM-16 after evacuation(343K)

EXAFS data show the following; The observed Pt L-edge FT shapes for $[Pt_{15}(CO)_{30}]^{2-}$/NEt_4/FSM-16(samples B) and $[Pt_{12}(CO)_{24}]^{2-}$/NaY (sample C) were similat to that of $[NEt_4]_2[Pt_{15}(CO)_{30}]$/BN(sample D) as shown in Fig.18, and EXAFS analysis in Table 7 of the neighbouring Pt-Pt frameworks in terms of co-ordination number(N_c) and atomic distances(R) provided direct evidence for the stoichiometric formation of trigonal prismatic $[Pt_3(CO)_6]_n^{2-}$(n=4,5) cluster complexes within FSM-16 channels and NaY cavities. It is of interest to find that for $[Pt_{15}(CO)_{30}]^{2-}$ encapsulated in FSM-16(sample B) the avaraged interdistance of Pt-Pt between adjoining triplatinum planes (R_2=3.08Å; N_c^2=1.6) are slighly enlonged(ΔR=0.02Å), compared with that of $[NEt_4]_2[Pt_{15}(CO)_{30}]$/BN(sample D) ($R_2$=3.07 Å; N_c^2=1.5) and the X-ray analysis of $[PPN]_2[Pt_{15}(CO)_{30}]$ in crystal (R_2=3.05 Å; N_c^2 =1.5). By contrast, the inter-triplatinum distances of Pt_{12} cluster frameworks in NaY micropores (sample C; N_c^2=1.5, R_2=2.99Å) are significantly shorter(ΔR=0.07Å) than that of $[PPN]_2[Pt_{12}(CO)_{30}]$($R_2$=3.06Å; N_c^2=1.5) within the experimental errors (ΔR=0.02Å). ^{13}C NMR study on the structure of $[Pt_{15}(CO)_{30}]^{2-}$ in solution showed that the Pt_3 triangle was rotated fluxionally in the Pt_3 plane, which cause the wide distribution of the Pt-Pt distances of inter-Pt_3 triangle planes. Hence, the slight enlongment of Pt-Pt distance of inter Pt_3 planes for sample B may be caused by the fluxional rotation of Pt_3 triangle of the Pt_{15} carbonyl clusters (8X12.3Å van del Waals diameter) in the enormous channels of FSM-16(27.5Å). By contrast, the Pt_{12} carbonyl clusters(8X10 Å rod) encapsulated in NaY micropores may be frozen owing to the intrazeolitic constraint(10-12Å effective diameter).

5.5. WGSR on Pt Carbonyl Clusters in NaY and Mesoporous FSM-16.

The WGS reaction($CO+H_2O \rightarrow CO_2+H_2$) was performed at reduced pressure(P_{CO}=50 Torr,P_{H2O}=20 Torr) by using a closed circulating Pyrex glass reactor charged with the powdered samples of $[Pt_{15}(CO)_{30}][R_4N^+]$/ FSM-16(R=Me, Et, Bu), $[Pt_{15}(CO)_{30}][MV^{2+}]$/FSM-16, $[Pt_{12}(CO)_{24}]^{2-}$/NaY and $[Pt_9(CO)_{18}]^{2-}$/NaY at 300-373K. As shown in Table 8, $[Pt_{15}(CO)_{30}]^{2-}$ clusters with organic cations in FSM-16 exhibited remarkably high activities in the WGS reaction to form an equimolar mixture of CO_2 and H_2, compared with the Pt_{12}-Pt_9 cluster in NaY and conventional Pt/Al_2O_3 catalyst(4 mass %Pt). It was worthy to note that the Pt_{15} carbonyl anions coupled with methyl viologen cation as a redox charge-carrier in FSM-16 showed the highest activities for the reaction. The rates of WGS reaction at 300K based on TOF(turn-over-frequency) on $[Pt_{15}(CO)_{30}]^{2-}$ in FSM-16 are higher(20-100 times) than those on $[Pt_{12}(CO)_{24}]^{2-}$ and $[Pt_9 (CO)_{18}]^{2-}$(4 mass%Pt) entraped in NaY cavities[14]. Moreover, it was demonstrated that the carbonyls of $[Pt_{15}(CO)_{30}]^{2-}$ in FSM-16 underwent facile isotopic exchange of ^{13}CO at 300K. On exposure of $[Pt_{15}(CO)_{30}]$ $[Et_4N^+]$/FSM-16 to ^{13}CO(100 torr) for 2 h, the bands of linear CO at 2078 cm^{-1} was completely replaced by new band at 2013 cm^{-1}. As shownin Table 8, the exchange rates in TOF varied upon the sorts of organic cations in FSM-16, whereas the carbonyl exchange of Pt_9 and Pt_{12} clusters in NaY micropore proceeded very slowly. From these evidences, it is suggested that the robust carbonyl cluster anions such as $[Pt_{15}(CO)_{30}]^{2-}$(8X13Å rod) are accomodated and stabilized with organic cations entrapped in the ordered mesoporous channels of FSM-16(27.5Å diameter). They exhibit higher activities in the WGSR and ^{13}CO exchange reactions probably owing to their flexible cluster fameworks and sufficient diffusibility of reactant gases, compared with the Pt_9 and Pt_{12} carbonyl clusters which are restrected in NaY micropore constrain.

5.6. Trasformation of $Pt_{15}(CO)_{30}^{2-}$ in FSM-16 by evacuation to Pt aggregates and catalytic behaviors in hydrogenation of ethene and butadiene.

It was demonstrated by the in-situ FTIR and EXAFS observzation that the thermostable $[Pt_{15}(CO)_{30}]^{2-}$/Et_4N/FSM-16 (2075s and 1875m cm^{-1}) was gradually transformed by evauation at 10^{-4} torr by lamping temperatures from 300-343K, resulting in a partially decarbonylated Pt_{15} characteristic of CO IR bands (2060s and 1880w cm^{-1}) . EXAFS data show the following; the resulting Pt cluster species kept the prismatic triangular Pt framework and linear CO ligands unchanged(Nc^1=2.0;R_1= 2.68Å; Nc^2=1.5;R_1=3.10Å) but their bridging carbonyls are almost completely disappeared by the thermal evacuation upto 343K. A further evacuation of the Pt_{15} clusters in FSM-16 at 363K led to the substantial transformation to spherical aggregates of ca 15Å diameter coordinated with CO ligands(2065s and 1820w cm^{-1}) which were characterized by EXAFS parameters of CN(Pt-Pt)=7.6;R=2.74Å), likely as shown in TEM images of Fig.19. The exceeding thermal evacuation of the sample at 393-473K resulted in naked Pt aggregates of 15Å size owing to the complete removal of CO.

According to the analysis of the EXAFS and FTIR data, the transformations of Pt carbonyl clusters were proposed in the ordered mesoporous channel hosts FSM-16 as

Table 8. Catalytic performances in ^{13}CO Exchange and Water-Gas-shift ReactionExchange and Water-Gas-shift Reaction $[Pt_{15}(CO)_{30}]^{2-}/NR_4/FSM-16$(R=Me, Et, Bu, Hex and MV), $[Pt_9(CO)_{18}]^{2-}/NaY$, $[Pt_{12}(CO)_{24}]^{2-}/NaY$ and Pt/g-Al_2O_3(4 wt%Pt).

Pt carbonyl clusters/ FSM-16 or NaY	^{13}CO exchange reaction k/min(300K)[a]	WGSR k'/min(300K)[b]
$[Pt_{15}(CO)_{30}]^{2-}$-$[NEt_4]^+$/FSM-16	123	12
$[Pt_{15}(CO)_{30}]^{2-}$-$[NBu_4]^+$/FSM-16	89	4.8
$[Pt_{15}(CO)_{30}]^{2-}$-$[MV]^{2+}$/FSM-16		22
$[Pt_{12}(CO)_{24}]^{2-}$/NaY	7	0.42
$[Pt_9(CO)_{18}]^{2-}$/NaY	9	0.75
Pt/γ-Al_2O_3[c]	-	0.02

a) ^{13}CO(100 torr);TOF(mmol/Pt atom/min)
b) CO(200 torr)+H_2O(15torr);TOF(CO_2)(mmol/Pt atom/min)X10^{-2}
c) The catalyst was prepared by H_2-reduction at 673K for 2 h after H_2PtCl_6 impregnated on γ-Al_2O_3(4 mass%Pt).

Fig.20. Rates(TOF) of ethene hydrogenation on $[Pt_{15}(CO)_{30}]^{2-}$/FSM-16 after evacuated at various temperatures (300-473K) ($P(C_2H_4)=P(H_2)=13.3$ kPa.

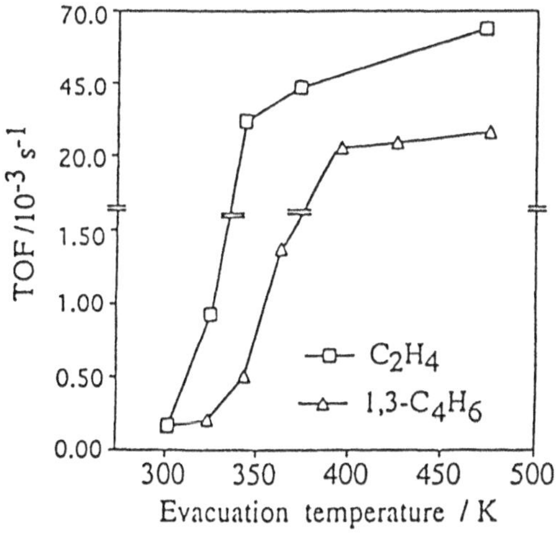

dipicted below, owing to the controlled removal of CO by evacuationat the temperatures of 300-473K.

As catalyst probing reactions, hydrogenation of ethene and butadiene has been conducted at 300K on the $[Pt_{15}(CO)_{30}]^{2-}$/FSM-16 after the controlled removal of CO by evacuation at various temperatures. As shown in Fig.20, it was interesting to find that the original $[Pt_{15}(CO)_{30}]^{2-}$ encapsulated in the channel host FSM-16 was inactive for the catalytic hydrogenation of ethene and butadiene when an mixture of olefin and H_2(1:1 molar ratio) was exposed at 300K. Ethene hydrogenation was initiat-ed but negligibly for butadiene on the clustersin FSM-16 after partial removal of CO by evacuation at less than 343K, where the Pt_{15} cluster framework was kept remained. By contrast, butadiene was firstly hydrogenated on the Pt carbonyl aggregates of 15Å size after the evacuat-ion at 364K, which proceeded selectively to give butenes(1-butene(75%)+2-butenes(24%)) with a negligible formation of butane. The naked Pt aggregates obtained by complete removal of CO at 400-475K exhibited non selective hydrogenation of butadi-ene towards a mixture of butanes and butane under the atmosphere of butadiene and hydrogen(1:1v/v) at 300K, as indicated in Fig.21.

Fig.21. Proposed transformation of $[Pt_{15}(CO)_{30}]^{2-}$ in FSM-16 by themal evacuation as deduced by IR and EXAFS.

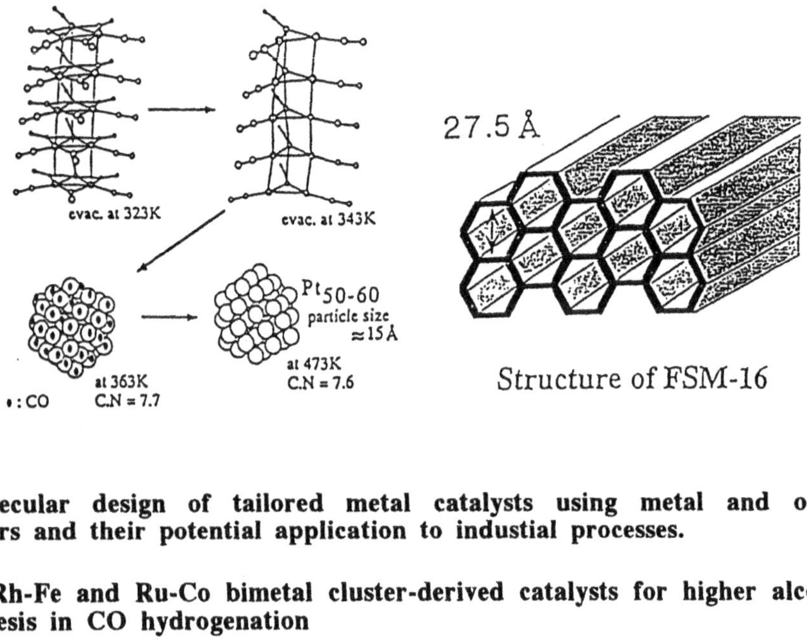

6.Molecular design of tailored metal catalysts using metal and oxide clusters and their potential application to industial processes.

6.1. Rh-Fe and Ru-Co bimetal cluster-derived catalysts for higher alcohol synthesis in CO hydrogenation

Recently, Ichikawa et. al.(55) demonstrated the substatial promotion of Fe for hydroformylation of etheyelen and propene on the SiO_2-supported bimetallic catalysts derived from the carbonyl cluster complexes having the different Fe/Rh atomic ratios such as $[TMBA]_2[FeRh_5(CO)_{16}]$, $[NMe_4]_2[FeRh_4(CO)_{15}]$, $[TMBA]_2[Fe_2Rh_4(CO)_{16}]$ and $Fe_3Rh_2(CO)_{14}C$. The results show that the marked promotion of Fe in enhancing rates which were dramatically increased by the magnification of 100-300 times larger on the catalysts derived from $FeRh_5$, $FeRh_4$, Fe_2Rh_4, and Fe_3Rh_2 carbonyl clusters, respect-ively, compared with that on Rh_4 derived one. Propanol as a further hydrogenation product was obtained on Fe-rich Rh bimetal cluster-derived catalysts, whereas $Rh_4(CO)_{12}$-derived one gave only C_2H_5CHO as the hydroformylation product.

It is difficult to simply explain such a remarkable enhancement of hydroformylation activity and substantial increase of alcohol selectivity by the addition of products catalysed on the individual Rh and Fe atoms in the catalysts.

To rationalize the promotion of Fe, a physical mixture $[Rh+Fe]/SiO_2$ catalyst was prepared from a THF solution of $Rh_4(CO)_{12}$ and $[TMBA]_2][Fe_3(CO)_{11}]$(Fe/Rh atomic ratio=0.26) impregnated on SiO_2. The resulting H_2-reduced catalyst gave much lower activity for the ethylene(or propene) hydroformylation and with a negligible alcohol conversion than that for $[TMBA]_2]_2[FeRh_4(CO)_{15}]$(Fe/Rh=0.25). For comparison, a mechanical mixted catalyst (Rh/SiO_2+Fe/SiO_2) gave negligible effect to enhance the yields of alcohol products, which are similar as those on the catalyst derived from $Rh_4(CO)_{12}$ alone. Accordingly, the results suggest that RhFe bimetal carbonyl clusters provided the discrete intact concerning site congestion constisting of RhFe atoms substantially active for the olefin hydroformylation, and for the successive hydrogenation to give higher alcohols. The higher order of the enhancement for the hydroformylation over the simple hydrogenation of olefins on the Fe-contaning Rh cluster-derived catalysts, resulted in the appreciable improvement of oxygenate selectivities.

The structure and metal compositions of the RhFe and PtFe bimetal carbonyl cluster-derived catalysts have been studied by means sof EXAFS, Mossbauer and FTIR in CO chemisorption.

The Fourier transform of Rh-K and Fe-K EXAFS on the H_2-reduced samples prepared from SiO_2-supported $[Rh_4Fe_2(CO)_{15}][TMBA]_2$(4 wt%Rh) suggest taht the RhFe bimetal ensembles are formed in a high dispersion(around 10A in size) having a direct Rh-Fe-O bondings(Rh-Rh*C.N.=7.2, R=2.65Å and Rh-Fe C.N.=1.8, R=2.54Å). Relatively larger contribution of Fe-O bond(Rh-O C.N.=3.0, R=1.99Å) was obtained, compared with that of Rh-O bond(Rh-O C.N.=less than 1.0, R=2.03Å). In the Mossbauer measurement of the sample gave the absorption band which could be resolved into a singlet and pair of quadrupole doublet. They were reasonably assinged to Fe^{o} and Fe^{3+}, respectively. No band for Fe^{2+} was observed. The results imply that Fe atoms in the RhFe bimetallic ensembles derived from Rh_4Fe_4 and Pt_3Fe_3 carbonyl complexes are mostly in the state of Fe^{3+} even after the H_2 reduction at 400°C, which is located at the metal-support oxide interface. It seems likely that the electropositive Fe play a role in anchoring the small ensembles of Rh(or Pt) to prevent sintering under the reduction and CO based reactions. On the other hand, the FTIR spectrum was observed in CO chemisorption on the reduced Rh_4Fe_2/SiO_2. Besides the strong linear CO bands at 2058 cm^{-1} and a brdging CO band at 1806 cm^{-1} comparatiuvely weaker than that on the Rh4(CO12-derived one, it is intersting to find a lower frequency band appearing at 1628 cm-1. This may arise from C- and O-bonded interaction of CO chemisorbved on the adjacent Rh-Fe^{3+} sites, as the analogy of the adduct formation between metal carbonyl complexes and Lewis reagents including Lewis acid site on alumina surface. In this context, it has been suggested by Shriver and Burwell that small metal cations such as Li^+ and Na^+, or Lewis acids such as AlX_3, greatly accelerate CO insertion(10^2-10^6 times larger than those without promoter ions) with alkyl group of $Mn(CO)_5(CH_3)$ converted to acyl complex through C- and O-bonded interaction between Mn^{4+} and Al^{3+}. In a similar manner, the analogous acetyl surface species were instanteneously formed, showing the characteristic IR bands at 1570-1780 cm^{-1}, when $Mn(CO)_5(CH_3)$ and $Fe(CO)_2Cp(CH_3)$ were in contact with dehdrated Al_2O_3.Accordingly, it is conceivable that an electropositive

Fe^{3+} in the RhFe ensembles shows a bifunctional role as Lewis acid-promotion to enhance CO migratory insertion to stabilize acetyl intermediate in the olefin hydroformylation, as proposed scheme(56). The PtFe and PdFe bimetallic catalysts were also prepared using $[Fe_3Pt_3(CO)_{15}][TMBA]_2$, $[Fe_4Pt(CO)_{15}][TMBA]_2$, $[Fe_6Pd_6(CO)_{12}][TMBA]_3$ and $[Fe_4Pd(CO)_{16}][TMBA]_2$ impregnated on SiO_2. The results in the CO hydrogenation on the H2 reduced catalysts derived from the PtFe and PdFe bimetal carbonyl clusters, including the SiO2-supported Rh_4Fe_2 and Rh_5Fe catalysts. It was of interest to find that the rates of oxygenates mainly consisting of methanol and ethanol were substantially enhanced on the Fe-contaning Rh, Pt and Pd cluster-derived catalysts, rather than those from the homometallic clusters e.g., $Rh_4(CO)_{12}$, $[Pt_9(CO)_{12}][NEt_4]_2$ and even higher than the conventional salt-derived catalysts using $RhCl_3$, $H2PtCl_6$ and $PdCl_2$. In contrast, the hydrocarbon formation was considerably suppressed on the RhFe cluster-derived catalysts, resulting in the appreciable improvement of oxygenate selectivities in the $CO+H_2$ reaction. Fe promotion toward methanol formation(in the selectivity close to 100% CO base) is quite remarkable on the Pt_3Fe_3 and Pd_6Fe_6 catalysts. For the catalysts prepared from Fe-rich Pt and Pd mixed metal clusters such as $Fe_4Pt(CO)_{15}^{2-}$ and $PdFe_4(CO)_{15}^{2-}$ an appreciable amount of methane and higher hydrocarbons was obtained, accompanied with methanol as the oxygenate product. The result suggested that the metal segregation copuld proceed through the catalyst preparation, in forming the isolated Fe particles which are active for the typical Fischer-Tropsch synthesis of hydrocarbons. Such a promotion of Fe for the oxygenate formation in the CO hydrogenation catalysed on the bimetallic catalysts is similarly as for the olefin hydroofrmylation interpreted in terms of the acid promoted CO insertion catalyzed on the hetrometallic active sites consisting of the contegious Rh-Fe^{3+}, Pt-Fe^{3+} and Pd-Fe^{3+} centers located at the metal-oxide interfaces(10,48).

Synthesis gas conversion was also studied(470K, CO/H_2=0.5 v/v mole)with the catalysts prepared from Rh_2Co_2 and $RhCo_3$ impregnated on ZnO, ZrO_2 and La_2O_3 at lower metal loadings(0.2-0.5 wt%). The total selectivities of oxygenates which are mainly consisting of methanol and ethanol decreased relatively on the catalysts from the precursors of increasing Co contents, although the propotion of ethanol in the oxygenate products appreciably increased. The heteronuclear Rh-Co sites promote not only CO dissociation to enhance the concentratiuon of surface carbon, but also migratory CO insertion similarly as described in the olefin hydroformylation promoted by the same RhCo bimetallic cluster-derived catalysts, rather than $Rh_4(CO)_{12}$, resulted in the enhancement of C_2-oxygenates such as ethanol in CO hydrogenation(35).

Braunstein et al(57) have recently reported the preparation of Pd-W and Pd-Fe bimetal catalysts derived from $Pd_2W_2Cp_2(CO)_6(PPh_3)_2$ on Al_2O_3 and $Fe_2Pd_2(CO)_5(NO)_2(PPh_2CH_2PPh_2)_2$ on SiO2. They gave rise to higher selectivity due to the Pd-W and Pd-Fe bimetallic interactions in phenylisocyanite than the conventional catalysts prepared by mixing the individual components.

6.2. Molecular Modeling of Supported Metal Catalysts;SiO_2-grafted $[(\eta^3\text{-}C_4H_7)_2Rh]_2V_4O_{12}$ and $[RhCp^*]_4V_6O_{19}$ Catalytic-ally Active in Selective Oxidation of Propene towards Acetone.

The IR spectra of $[n\text{-}Bu_4N]_2[\{(\eta 3\text{-}C_4H_7)_2Rh\}_2(V_4O_{12})]$ as a molecular modeling of Rh/V_2O_5-supported catalysts gave bands at 3078 and 3056 cm^{-1} characteristic of the stretching ν(C-H) of ($\eta 3$-C_4H_7)$_2$Rh groups capping the oxide cluster. TPD/Mass studies were performed using [n-Bu_4N]$_2$[{($\eta 3$-C_4H_7)$_2$Rh}$_2$(V_4O_{12})] /SiO_2 and [Rh (acac)($\eta 3$-C_4H_7)$_2$]/SiO_2 (acac= acetylacetonate) which were charged in a microreactor and heated by flowing He from 300 K-600 K with temperature increasing rate is 2K/min. In the TPD spectra, it was found that 2-metylpropene derived from about a half of h3-C_4H_7 ligands bound to $Rh_2V_4O_9$ and Rh(acac) moieties on silica was evolved at 350-413K and all of them completely desorbed below 450 K. Other species were formed on [n-Bu_4N]$_2$[{($\mu 3$-C_4H_7)$_2$Rh}$_2$ (V_4O_{12})]/SiO_2 at higher temperatures, resuled from Hoffman degradation of [n-Bu_4N]$^+$.

The Fourier Transform(FT) peaks of Rh-C and Rh-O contribut-ion were obtained from the raw Rh edge EXAFS data for [n-Bu_4N]$_2$ [{($\eta 3$-C_4H_7)$_2$Rh}$_2$(V_4O_{12})]/SiO_2 with a k^3-weighting over the range 3.5<k<15 $Å^{-1}$. When the sample was evacuated at 413 K, intensity of the Rh-C peak decreased (R=2.17Å, C.N.=3.2) as a half of that in the precursor sample(R=2.19Å, C.N.=5.6), whereas the C.N. of Rh-O(R=2.07Å) bonding in the precursor sample relatively increased from 1.8 to 2.1 possibly due to surface-attachment of capped $\eta 3$-C_4H_7)$_2$Rh with silanol groups. Data shows structural parameters for k^3 weighted V-K edge EXAFS of $Rh_2V_4O_{12}/SiO_2$, **1** and $Rh_4V_6O_{19}/SiO_2$, **2** before and after the thermal activation in vacuo or O_2 atmosphere of 13.3 kPa. When the sample **1** was evacuated at 300 K-413 K, the intensity and shape of the FT peaks of V=O and V-O remained unchanged, which agree well with those of the precursor cluster e.g. [n-Bu_4N]$_2$[{($\eta 3C_4H_7$)$_2$Rh}$_2$(V_4O_{12})] in the crystalline state. When the sample **1** was evacuated at 513 K, the resulting FT spectra substantially changed, resembling that of the conventional V_2O_5/SiO_2 which was prepared by calcination of ammonium vanadate on silica. Accordingly, the results of FTIR, EXAFS and TPD studies suggest that chair-form vanadate tetramer and hexamer clusters capped with ($\eta 3$-C_4H_7)$_2$Rh and RhCp* groups were grafted with Rh end on silica surface, as depicted in the following scheme; in keeping the vanadium oxide(V_4O_{12} and V_6O_{19}) cluster frameworks basically unchanged, at least by thermal activation in vaccuo or under the oxygen atmosphere below 413K.

Proposed Structural Transformation of SiO_2-Grafted [n-Bu_4N]$_2$[{η^3-C_4H_7)$_2$Rh}$_2$(V_4O_{12})] and Active Species for Selective Propene Oxidation to Acetone

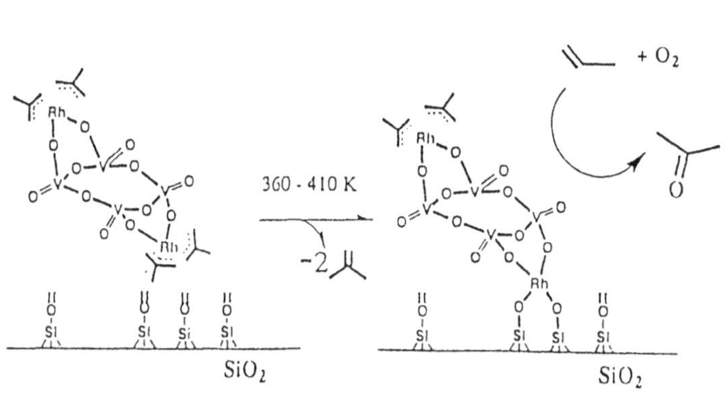

A mixture gas of propene and oxygen was admitted onto the samples of $Rh_2V_4O_{12}$ /SiO_2, **1** and Rh-V_6O_{19}/SiO_2, **2**. It was of interest to find that although propene oxidation negligibly occurred on both Rh-capped vanadium oxide cluster compounds in crystal, acetone was produced catalytically in keeping higher selectivities.(43-85%mole) on $Rh_2V_4O_{12}$/SiO_2, **1** and Rh-V_6O_{19} /SiO_2, **2** after grafted on silica surface. As shown in Table 2, by exceeding the activation temperature at 513K, the catalytic activity and selectivity for acetone formation substantially decreased, possibly due to the degradation of Rh capped vanadate clusters. For the Arrhenius plots in propene oxidation on $Rh_2V_4O_{12}$/SiO_2, the Activation energies were obtained as 46.8 kJ/mol for acetone production, 35.5 kJ/mol for CO_2 production., respectively. As shown in Table 2, the $Rh_4V_6O_{19}$/SiO_2, **2** exhibited higher selectivities towards acetone rather than $Rh_2V_4O_{12}$/SiO_2 **1** . By contrast, the oxidation of propene to acetone gave poor selectivities and conversion with higher activation energy of 73.7 kJ/mol over Rh(acac)(η^3-C_4H_7)$_2$/SiO_2. By contrast, a negligible amount of acetone was obtained in the reaction on [$V_{10}O_{28}$]/SiO_2 derived from non-Rh capped vanadate dodecamer cluster [(n-Bu)$_4$]$_2$($V_{10}O_{28}$) and the conventional V_2O_5/SiO_2 under the similar reaction conditions. The results suggested that the coordinately unsaturated Rh site capped with vanadium oxide cluster frameworks are associated with the selective catalysis for propene oxidation to give acetone, probably through a peroxy intermediate similarly on the surface-grafted Rh derived from Rh(π-allyl)$_3$ impregnated on Al_2O_3(59).

7. Summary

To promote our understanding towards a microscopic process in catalysis on metal/oxide interfaces, it is important and useful to study on their chemistry and dynamic behaviors in atomic/molecular-scale. Nevertheless,. due to the limitations and shortcomings of studying on catalysis by surface-bound metal clusters and metals derived from molecular metal clusters, up to the present, we have simply overviewed in the literatures on their structural characerization and catalytic perfomances. Extension of the present approaches to include the use of hetero-multimetallic clusters as precursors may lead to tailored metal catalysts having enhanced stabilities and relevant to achieve the prospective design of the selective catalysts. Few studies have been carried out with the clusters having ligands other than CO. Future directions in this area are therefore most likely to involve new synthetic approaches and collect the information on their surface chemistry with the reactive groups of well-characterized supporting materials. In this contex, some organometallic giant oxo compounds such as polymorbudates, e.g., [(CH_3O)$_4$ Mo_8O_{24}]$^{4-}$, $PMo_{12}O_{40}R_2^-$(R=CH_3 and C_2H_5) including a series of heteropolyacids such as $Nb_2Mo_4O_{19}^{4-}$ are prospective molecualar supports as modeling oxides available for .fixing metal clusters. We expect a new field of supported metal clusters in near years come for understanding heterogeneous catalysis in terms of molecular chemistry. The organometallic cluster precursors offer exciting and prospectively explored opportunities for creation of new surface structures and new catalysts, but it is still reemphasized that the lack of stability and lack of systematic investigation on their characterization in this class pose a major challenge and subject of lively debate. A variety of multinuclear metal clusters, homo/heterometallic, have been used to prepare the family of surface-bound clusters that would help identity the structural patterns and provide the basis of for determing how to understand the surface chemistry of metal clusters and to design supported metal/bimetallics. Spectroscopic approaches and organometallic chemistry could be reccomended to collect the data for molecular understanding of surface-bound clusters and their application to heterogeneous catalysis.

References:

1. Muetterties.E.L.,Bull.soc.Chim.Belg. 84,959(1975);science 196,839(1977);Band, E, and Mutterties, E., Chem. Rev. 78, 639(1978);Mutterties, E., Pure Appl., Chem., 54, 83(1982).
2. Chini, P., Gazz. Chim., Ital., 109, 225(1975).
3. Parkyns, N.D., Proc.Int. congr. Catal.,III, 1964 vol.2, p914(1965).
4. Burwell, R.L.,and Brenner A.,J. Mol. Catal., 1, 77(1975-1976).
5. Anderson, J.K., Elms,P.S.,Howe, R.F.,and Mainwaring, D.E., J. Catal., 50, 508(1977)
6. Anderson, J.K.,and Mainwaring, D.M., J. Catal., 35, 162(1974).
7. Ichikawa, M., J. Chem. Commun., 11, 26(1976).
8. Smith, A.K., and Basset, J.M., J. Mol. Catal., 2, 229(1977).
9. Gates, B.C., and Lieto, J., CHEMTECH p195(1980).
10. Ichikawa, Polyhedron, 7, 2351(1988); Adv. Catal., 38, 283(1992)
11. Araki,M., and Ponec, V., J. Catal., 57, 397(1981); Bilonen P., and W.M.H. Sachtler, Adv. Catal., 30, 165(1980).
12. Whitmer, K., Shriver, D.F., J. Am. Chem.Soc., 102, 1456(1980).
13. Anderson, J.R., Adva. Catal., 23, 1(1973).
14. Sappa,e.,Tripicchio,A.,Carty, A.J., and Toogood, G.E., Prog. Inorg. Chem.,35, 437(1987).
15. Gates, B.C.,Guczi,L., and Knozinger, H., "Metal Clusters in Catalysis"Elsevier, Amsterdam.,(1986).
16. Wyman, R., Basset, J.M.,(ed)"Surface Organometallic Chemistry" NATO Ser.C., 231, pp75-126(1988).
17. Hughes, F.,delmon, J.A., Busseier, P., Smith, A.K., Basset, J.M., and Oliver, D.J., J. Phys. Chem., 86, 5136(1982).
18. Theolier, A., Choplin, A., D'Ornelas, L., Basse, J.M., Sourisseau, S., Zhanderighi, G.M., Ugo, R., and Psaro, R., Polyhedron, 2, 119(1983).
19. Ichikawa, M., Sekizawa, K., Shikakura, K., and Kawai, M., J. Mol. Catal., 11, 167(1981).
20. Pierantozzi,R., Valagene,E.G.,Vardequist,A.F., and Dyer, P.N., J. Mol. Catal., 21, 189(1983).
21. Collier,A.,Hunt,D.J.,Jackson,S.D.,Mayers,R.B.,Pickering,I.A.,Wells, P.B., Shimpson, A.F., and Wynman, R.J., J. Catal., 80, 154(1983)
22. Lamb, H.H., and Gates, B.C., J. Am. Chem. Soc., 108, 81(1986).
23. Bein, T., Jacobs, P.A., and Schmidt,F., Stud. Surf. Sci. Catal., 12, 111(1982)
24. Iijima, S., and Ichikawa, M., J. Catal., 94, 313(1985).
25. Schmidt, G., Struct.Bonding(Berlin)62, 51(1985);Mater..Chem. Phys., 29, 133(1991).
26. Yates, J.T. , J. Phys. Chem., 91, 3133(1987).
27. Lyding, J.W., Hubacek, J.S., Grammie, G., Skala, S., Brokengrough, R., Shapley, J.R., and Kayer, M.P., J. Vac., Sci., Technol., A[2]6, 363(1988).
28. Fujimoto, T., Fukuoka, A., Nakamura, J. and Ichikawa, M., J. Chem. Soc., Chem.Commun., 845(1989).
29. Fujimoto, T. , Fukuoka, A., and Ichikawa, M., Chem.Mater., 4, 104(1992).

30.Fujimoto, T., Thesis,HokkAido University, pp37-79(1994).
31.Ichikawa, M., J. Catal., 59, 67(1979); ibid., 56, 127(1979).
32.Ichikawa, M., "Tailored Metal Catalysts"(ed.Y. Iwasawa) pp183-263, Reidel Pub., Dordrecht, The Netherlans, 1984.
35.Ichikawa, M., CHEMTECH, 12, 674(1982).
36.Ichikawa, M., Bull. Chem. Soc.,Japan, 51, 2268(1978).
37.Ichikawa, M., Bull. chem. Soc.,Japan, 51, 2273(1978).
38.Kawai, M., Uda, M., and Ichikawa, M., J. Phys. Chem., 89, 1654(1985).
39.Yokaoyama,M., Yamazaki, K., Kosugi, N.,Kuroda, H., Ichikawa, M., and Fukushima, T., J. Chem. soc., Chem. commun., 962(1986).
40.Fukuoka, A. Ichikawa, M., Hriljac, J.A., and Shriver, D.F., Inorg. Chem., 26, 3643(1987).
41.Esteban Puges, P., Garin, F., Girad, P., Bernhardt, P., and Maire, G., Proc., Iberoam., Congr., Catal., vol2, 1111(1984).
42.Galin, G]f., Zahrra, O., Crouzet, L., Schmitt, J.L., and Maire, Stud. Surf. Catal., 106, 466(1981).
43.Yermakov, Yu.I., and Kuznetsov, B.N., J. Mol., Catal., 9, 13(1980).
44.Shapley, J.R., Hardwuck, S.J., Foose, D.S., Stucky, G.D., Churchill, M.R., M.R. Bueno, C., and Muchison,P.J., J. Am. Chem. Soc., 103, 738(1985).
45.Ichikawa, M., L.F. Rao, N. Kosugi, J., Fukuoka, A., J.C.S. Faraday Diss., 87, 232(1989);Ichikawa, M., L.F. Rao, Kimura, T., and Fukuoka, A., J. Mol. Catal., 62,15(1990).
46.Li, G.J., Fujimoto, T., Fukuoka, A., and Ichikawa, M., J.C.S. Chem. Commun., 1331(1991);Wang, R.J., Fujimoto, T., Fukuoka, A. and Ichikawa, M., Catal., Lett., 12, 171(1992).
47.Liu, M., Shido, T., and Ichikawa, M., J.Chem.Soc., Chem. Commun., 507(1995).
48.Fukuoka, A., Rao, L.F., Kosugi, N., Kuroda, H., and Ichikawa, M., Appl., Catal., 50, 295(1989).
49.Ichikawa, M., Liu, A.M., G.S. Shen,and Shido, T., Tops. Catal., 2, 141(1995).
50.Beck, J.S.,Vartuli, J.C., Roth, W.J., Leonowicz, C.T., Kreag, S.K.,J. Am. Chem. Soc., 114, 10843(1992).
51.Yanagisawa, T., Shimizu, T., Kuroda, K., and Kato, C., Bull. Chem. Soc., Japan., 62, 763(1990); Inagaki,S., Fukushima, Y, and Kuroda, K., J.C.S., Chem. Comm 680(1993).
52.Yamamoto, T., Shido, T.,.Inagaki, S., Fukushima, Y., and , Ichikawa, M., J. Am. Chem. Soc., 118, 5810(1996).
53.Rao, L.F., Fukuoka, A., Kosugi, N., Kuroda, H., and Ichikawa, M., J. Phys. Chem., 94, 5317(1990).
54.Shen, L.L., Knozinger, H., and Sachtler, W.M.H., Catal. Lett., 2, 35(1989) ; J.Am.Chem.Soc., 111, 81259(1989).
55.Fukuoka, A., Kimura, T., Kosugi, Kuroda, H. Minai, Y., Sakai, Y., Tominaga, T., and Ichikawa, M., J. Catal., 126, 434(1990).
56.Ichikawa, M., "Dayanamic aspects in Heterogneous Catalysis"(ed.K. Tamaru), Plenum Pub., pp149-216(1994).
57.Braunstein, R., Bender, R., and Kervennual, L., J. Organometallics, 1, 1236(1982).
58.Yamaguchi, M., Shido, T., Ohtani, H., Isobe, K., and Ichikawa, M., J.C.S., Chem. Commun., 717(1995).
59.McMillian, J.W., Fischer, H.E., and Schwarz, J., J. Am. Chem. Soc., 113, 4041 (1991).

QUANTITATIVE DETERMINATION OF MOLECULAR ADSORBATE STRUCTURES

D.P.WOODRUFF
Physics Department,
University of Warwick,
Coventry, CV4 7AL, UK

1. Introduction

Understanding the local geometry of reactant species at a catalyst surface lies at the heart of any complete understanding of heterogeneous catalysis. Catalyst literature makes frequent reference to the idea of an *active site*, yet there is rarely any microscopic understanding of what is meant by such a phrase. Local bonding geometries are probably of particular importance in so-called *structure sensitive* reactions, because although this phrase is often used in relation to reactions whose rates or selectivity is sensitive to the morphology of the catalyst particles (size and/or shape), it is generally believed that these microscopic structural changes are accompanied by changes in the local atomic geometries which are the crucial factor determining the catalytic properties. Despite the importance of this underlying structural information, there is remarkably little truly quantitative data of this kind. In the case of dispersed powder catalysts, this is a consequence of the fact that very few techniques are available which are capable of providing this information. On extended single crystal surfaces, the situation is somewhat better; there is now an extensive literature of atomic adsorbate geometries. The problem of obtaining quantitative structural information on molecular adsorbates is more difficult, but here too the literature of solved structures is growing. In this chapter a short and somewhat selective review of this literature will be presented.

The organisation of the chapter is as follows. First, a very brief overview is given of the different true quantitative structural probes is given, with an emphasis on the proven or likely utility of the methods for molecular adsorbate systems, and particularly those containing mainly C, N and O (and H!). This group of techniques is then contrasted with *spectroscopic fingerprinting* methods which have been widely used in the past to infer structure, in order to highlight the limitations and dangers of this approach. In the second main section, two particular techniques, low energy electron diffraction (LEED) and photoelectron diffraction (PhD) are described in more detail and contrasted to some related methods. Finally, a brief survey is given of some of the novel or characteristic findings of the results obtained so far.

One final introductory remark is appropriate. Implicit in this chapter is the study

R.M. Lambert and G. Pacchioni (eds.),
Chemisorption and Reactivity on Supported Clusters and Thin Films, 193-214.

of equilibrium structures. As such, the geometries determined may not always be the ones which are involved in the key steps of a surface reaction. For example, one can only study reaction intermediates if these can be 'frozen' on the surface (i.e. be given a long - of order hours - lifetime) typically by cooling the surface to an appropriate temperature. As yet there are no quantitative structural probes capable of investigating very short-lived intermediate species or of following the geometrical aspects of surface dynamical processes directly.

2. Structural probes - a brief survey

2.1 TRUE QUANTITATIVE PROBES

In essence almost all surface techniques [1] can be categorised according to whether the incident probe and the detected signal are ions, electrons or photons. In the case of quantitative structural probes, this same classification is also useful if we categorise according to the probe which provides the structural information. This leaves us with three general methods; ion scattering, in which *shadowing* provides the structural information, X-ray diffraction in which electromagnetic wave interference phenomena are used, and electron scattering (produced either by incident electrons or by incident photons which in turn produce photoelectrons) which defines a group of techniques exploiting electron wave interference phenomena.

Although there are range of ion scattering methods covering an energy range in the incident particles from a few hundred eV to 1-2 MeV, the basic structural information in all of these techniques arises from the fact that a region exists behind each atom (relative to the incidence direction) into which no ions can enter due to the scattering properties of this atom. If a second atom lies within this *shadow cone* it cannot contribute to the scattered ion signal; by varying the incident ion geometry it is thus possible to move subsurface atoms in and out of the shadow cones of the surface atoms, providing simple geometrical information on their juxtapositioning. The most refined versions of this experiment (the so-called double alignment experiment used particularly at the higher energies) exploit not only incident ion shadowing but also the phenomenon of *blocking* in which the detector may fall in the shadow cone of a surface atom for the ion signal scattered from a subsurface atom. In this case both the incident and detector directions can be varied to give additional structural information. So far ion scattering does not appear to have been used for any molecular adsorbate structure determinations. One reason is that the methods are generally most effective when the adsorbate atomic species has a higher mass number than the substrate; under these circumstances the adsorbate scattering signal is intense and does not compete with a background signal from the substrate which occurs only at lower scattered particle energies. Low atomic number species (typically the main ingredient of molecular adsorbates) have a low backscattering cross-section for ions, although they do still cause some shadowing of underlying substrate atoms. There are a few cases in which this shadowing effect has been used at low energies to infer some aspects of molecular adsorbate geometries, but no full quantitative studies.

Until relatively recently, X-ray diffraction was always regarded as a means of determining bulk solid structures, but not surface structures, because of the low scattering cross-sections for all atoms by X-rays. By using the intense and intrinsically collimated X-ray source offered by synchrotron radiation, however, and sampling the scattered signal in directions well away from those of bulk substrate diffracted beams, it is now possible to determine surface structures using surface X-ray diffraction (SXRD). The technique is especially well-suited to the determination of complex surface reconstructions because the weak interaction of the X-rays means that the theoretical description is simple (single scattering compared with the multiple scattering needed in LEED - see below) and is thus able to cope with large surface net structures. In SXRD the relative atomic positions within the overlayer are particularly easy to determine, but adsorbate-substrate registry can also be found by sampling scattering directions in which both the substrate and the overlayer contribute to the scattering. Because X-ray scattering cross-sections scale with the square of the atomic number of the scatterer atom, the technique has not been thought of as very appropriate for low atomic-number-containing molecular adsorbates, but a recent experiment using the exceptionally bright European Synchrotron Radiation Facility has shown that the geometry of adsorbed CO on Ni(110) can be obtained from this technique [2]; so far this appears to be the only example of an SXRD study of a molecular adsorbate. An important limitation of SXRD for the study of molecular adsorbates is that it is only applicable to systems showing very good long-range order.

One X-ray diffraction-based method which relies only on diffraction within the substrate and thus is potentially much simpler than SXRD is the X-ray Standing Wave (XSW) method. The basic physics of this method is that when one excites an X-ray Bragg reflection in a crystal the incident and scattered beams interfere in their region of overlap inside the crystal to produce a standing wave with a periodicity in intensity equal to that of the scatterer planes with which it is associated. Because of the finite penetration of the X-rays at a Bragg condition, there is a finite range of wavelengths or angle (the 'rocking curve width') over which the diffracted beam is excited, and within this range there is a well-defined variation in the phase of the standing wave relative to the scatterer planes of atoms. If one therefore measures the absorption of the standing wave on a particular atom, the variation of absorption with wavelength or angle is characteristic of the position of this atom relative to the scatterer planes. This absorption profile thus provides a measure of the location of the absorbing atom relative to the bulk crystal structure. Because adsorbate atoms on a surface are immersed in this substrate-generated standing wavefield, the method can thus be applied to locate surface atoms relative to the substrate. By detecting an element-specific monitor of X-ray photoabsorption - a core level photoelectron, Auger electron, or fluorescent X-ray - one can thus obtain element specific site information for adsorbed species. So far this method has been applied mainly to atomic adsorbates, but a few molecular systems have been studied in this way. In one such experiment, PF_3 adsorbed on Ni(111) was investigated [3], using absorption in both the P and F atoms to obtain the local adsorption site and molecular orientation. A somewhat different example involved the methoxy species, CH_3O-, on Al(111) [4]. In this case only the O site was determined. This particular example defines one of the potential limitations of the method for molecular adsorbates. In particular, the X-ray energies involved in Bragg reflections are typically rather high relative to the core level binding energies of the deepest core levels of the first row elements of the Periodic Table. In the specific Al(111)/CH_3O- case the normal incidence (111) reflection occurs at a particularly low energy of 2660 eV, but even this is much higher than the C and O 1s binding energies (284 eV and 530 eV). The fact that the Bragg energy is high relative to the binding energy means the photoabsorption cross-section is rather low. This may produce a limitation of the application of XSW to C, N and O-containing molecules in general.

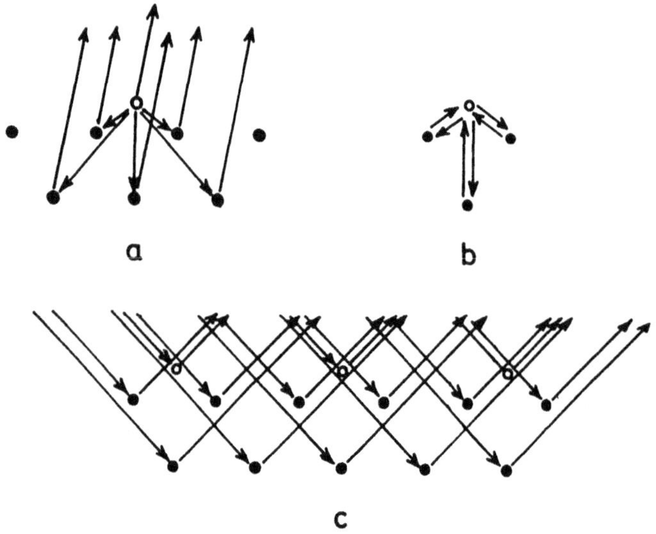

Figure 1. Schematic diagram showing the principle single scattering electron trajectories contributing to (a) photoelectron diffraction, (b) (S)EXAFS and (c) LEED

Electron scattering methods have so far provided the basis for the great majority of quantitative molecular adsorbate structure determinations. There are three basic techniques: LEED, PhD and X-ray absorption fine structure (XAFS) - notably surface extended XAFS (SEXAFS). The basic underlying electron scattering processes involved in these three methods are shown schematically in fig. 1. LEED is the direct electron diffraction analogue of X-ray diffraction and an incident electron plane-wave is diffracted by the two-dimensionally periodic surface structure giving rise to diffracted beams whose intensities are determined by the relative atomic positions of the surface atoms. Because low energy (e.g. 30-300 eV) electrons interact very strongly with atoms, proper inclusion of the effects of multiple scattering is an important ingredient of the interpretation of LEED data, as will be discussed in section 3. By contrast, both PhD and SEXAFS involve the scattering of photoelectrons emitted from a core level of a near-surface atom. In PhD these photoelectrons are detected outside the crystal in a particular emission direction, and as the schematic diagram shows this emitted signal must involve interference of the components of the photoelectron wavefield elastically scattered by surrounding surface atoms with the directly emitted component. The scattering pathlengths depend on the relative positions of the emitter and the scatterers and on the emission direction, so by varying either the emission direction or (as we shall consider in detail in the next section) the photoelectron energy, interference information is provided from which the local emitter geometry can be deduced. In general PhD is recorded using core level photoemission from the atoms which comprise the (molecular) adsorbate, thus providing local structure information on these constituent atoms. SEXAFS also exploits photoelectron backscattering interference with the directly emitted wavefield, but in this case the interference is detected not outside the crystal, but at the emitter site itself. The backscattering-induced modification of the wavefield amplitude at the emitter site enters directly into the matrix element defining the photoionisation cross-section. The backscattering interference thus leads to a modulation in the X-ray absorption cross-section (the so-called 'fine structure') as the photoelectron energy, and thus the photoelectron wavelength changes, causing successively constructive and destructive interference. Measurement of this modulation therefore provides a measure of the near-neighbour distances of scatterer atoms surrounding the photoabsorbing atom. Here we should remark that (S)EXAFS is the only technique of all those discussed in this section which can, at least in principle, provide local structural information on a real (i.e. polycrystalline or powder) catalyst. All the other methods require a well-defined substrate crystallography with known single real-space directions, a set of conditions only achievable with single crystal substrates. On the other hand, most SEXAFS experiments also make use of variations in the intensity of the EXAFS modulations which occur if the direction of the incident X-ray polarisation vector is varied relative to the absorber-scatterer real-space directions, an effect which also requires a single crystal substrate. Of course, an underlying problem in the interpretation of even the very highest quality EXAFS data from a dispersed catalyst is that even if the adsorbate coordination can be obtained, understanding the local geometry ultimately requires a knowledge of the substrate structure which in this situation is not available. A fuller discussion of this core group of electron scattering methods will be given in section 3.

2.2 SPECTROSCOPIC FINGERPRINTING PROBES

By contrast to the rather small number of detailed quantitative structural studies of molecular adsorbates on surfaces, there is a much longer tradition of surface investigation of such systems using electronic and vibrational spectroscopic fingerprinting, and it has been common to draw structural conclusions in much of this work. There is no doubt that vibrational spectroscopy, in particular, is a rich source of characterisation of surface molecular species. For the purposes of identifying the molecular species present on a surface (in many cases involving several co-adsorbed molecular fragments) the very high spectral resolution and strong dipole selection rules of reflection-absorption infra-red spectroscopy (RAIRS) is unrivalled, while complementary information on addition non-dipolar transitions from high-resolution electron energy loss spectroscopy (HREELS) can add further information. These data can also be valuable to clarify the internal conformation of the adsorbate species. A major strength of this approach is the existence of a large data base of vibrational spectroscopic data on simple and complex molecules including metal coordination compounds, the detailed structures of which are well-established by other methods (notably X-ray diffraction). On the other hand, the use of similar comparisons to 'determine' the adsorption site on a surface is now known to be a less generally reliable procedure.

The use of RAIRS and HREELS data to fingerprint molecular adsorption sites arises mainly from studies of CO adsorption and the very extensive data base of vibrational data on metal carbonyls of known structure. Most of the early surface vibrational data of this kind are from dispersed catalysts for which a simple transmission infra-red spectrum can be obtained, although very similar data now exist for many single-crystal adsorption experiments. A characteristic of such studies is a significant variation in the C-O stretching frequency which typically correlates with the coordination of the CO ligand. An extensive survey of this literature led to the proposal [5] that stretching frequencies above about 1950 cm^{-1} are associated with singly coordinated CO (atop adsorption), frequencies in the range of about 1800-1900 cm^{-1} are associated with two-fold or bridging coordination, whereas even lower frequencies are correlated to higher coordination. This initial characterisation has been used with increasing emphasis over the last decade or so to produce a large body of conventional wisdom on CO adsorption sites on metal surfaces. In several cases subsequent true quantitative structural studies (notably by LEED) have confirmed these assignments, but recent work has also identified examples of major failures of the way in which this methodology has been applied.

A particularly clear example of this kind concerns the adsorption of CO on Ni(111); CO adsorption on Ni surfaces has often been seen as the 'fruitfly of surface science' and it is somewhat startling to find that conventional wisdom for such a basic parameter as the adsorption site should have been in error. Extensive studies by HREELS and RAIRS of the Ni(111)/CO adsorption system, which show a C-O stretching frequency in the range 1830 to 1910 cm^{-1} (increasing with increasing coverage) have been widely interpreted as indicating bridge site adsorption, and a rather attractive model of the well-ordered 0.5 ML c(4x2) phase had been proposed (and widely accepted) on this basis. Furthermore, if CO is adsorbed on an oxygen-precovered Ni(111) surface the C-O stretching frequency is found to be significantly higher at around 2098 cm^{-1}, an observation interpreted as being due to atop adsorption of CO in the presence of the oxygen coadsorbate. We now know from true quantitative structure determinations by PhD [6,7] and (for the c(4x2)-CO phase) by SEXAFS [8] and LEED [9], that the CO adsorption site over a substantial coverage range and coadsorbed with oxygen [10] is in the three-fold coordinated hollow sites. With hindsight, one can understand how the vibrational data were misinterpreted. For example, in the case of coadsorbed oxygen, one might anticipate that the presence of this highly electronegative coadsorbate may cause some net charge transfer from the CO resulting in a depletion of charge from the anti-bonding 2π state (which is generally believed to be partially occupied on adsorption), leading to a stiffening of the C-O bond and an increase in the stretching frequency. On the otherwise clean surface, on the other hand, the C-O stretching frequency at the lowest coverages is in the range which could perhaps be assigned to a coordination of greater than two; the increase in frequency with coverage is partly due to dynamic dipole coupling, but is probably also partly due to a 'chemical' effect. Thus in both cases, the frequency change is probably assignable to 'chemical' (i.e. electronic) effects in the CO-metal and CO-metal-coadsorbate bonding. What this discussion highlights, however, is that vibrational spectroscopy is *not* primarily a structural probe, and structural deductions based on this technique are therefore not generally reliable (despite that fact that many such assignment have turned out to be correct).

Electronic spectroscopies have also been used with some success to identify, and provide partial structural information, on molecular adsorbates. The clearest electronic molecular fingerprints are provided by valence level spectroscopies - particularly (angle-resolved) ultra-violet photoelectron spectroscopy ((AR)UPS) as a probe of occupied valence levels, but also near-edge XAFS (NEXAFS) as a probe of unoccupied bound or quasi-bound states. Although these techniques have a much lower intrinsic bandwidth than vibrational spectroscopies they can provide some identification of the molecular species present on the surface. Moreover, through the use of selection rules, particularly when used with a polarised source of ultra-violet or X-ray radiation, they can provide information on the orientation of the adsorbed species. By the use of full multiple scattering simulations there are a few cases of site determinations from NEXAFS data, although the data range is so small that such determinations must suffer from real problems of uniqueness.

In addition to these valence spectroscopic methods, X-ray photoelectron spectroscopy (XPS) of adsorbate core levels has also been widely used to investigate surface chemical processes. The key information in this case is the *chemical shift* of the measured photoelectron binding energies. There is a history of successful spectral fingerprinting based on an extensive database of information on model compounds - the whole basis of what was originally dubbed ESCA - electron spectroscopy for chemical analysis. Here too, however, there are very real dangers in extending such fingerprinting from model compounds to chemisorption systems. The observed chemical shift is actually the sum of two effects; a change in the initial state one-electron binding energy due to the change in the valence electron environment, and a change in the extra-atomic final state relaxation due also to the modified environment. In free molecules, in particular, the changes in the chemical shift can correlate well with predicted initial state effects alone ('charge transfer'), but in chemisorption, particularly on metals, final state effects can be extremely important and far from trivial to predict. Until recently such data have not generally been used to infer much about molecular adsorbate bonding geometries, but high resolution data do now exist for CO adsorption on Pt(111), for example, which show that chemical shifts in the C 1s and O 1s photoelectron binding energies do appear to be related to local site changes on the surface [11]. High resolution XPS fingerprinting of this kind could become invaluable in the future for studying surface chemical processes and have the significant advantage over the vibrational spectroscopies that the peak intensities can be directly assignable to the relative occupation of different states. The quantitative interpretation of these different chemical shifts in terms of the actual associated geometries, however, is certainly not yet possible and is unlikely to become a primary source of quantitative structural information.

3. Electron scattering methods for structure determination

3.1 LEED: THE BENCHMARK FOR SURFACE STRUCTURE

LEED was the first technique to be developed for the quantitative determination of surface adsorption structures [12]. As such its practitioners have had to establish a methodology which provides a natural reference for many of the newer methods. A very healthy conflict between the practitioners of this technique and those of other less mature but complementary methods has led to significant improvements in the practice of all these different techniques (including LEED), but its maturity means that it is now seen as the benchmark against which other methods are judged. Radical new results from newer methods have commonly been checked by LEED before initial acceptance, although ultimately the complementary aspects of the different methods means that LEED will not always be capable of such tests; even in this preeminent position, LEED does have its intrinsic weaknesses. Although this brief review will concentrate ultimately on the technique of photoelectron diffraction, a description of the basic methodology and achievements of LEED is essential to put PhD in context.

In principle, at least, the LEED experiment is rather straight-forward. An incident mono-energetic beam of electrons, typically in the energy range 30-300 eV, is directed at the sample surface, and the two-dimensionally periodic system leads to a set of backscattered diffracted beams. The directions of these diffracted beams provides the initial information of the unit mesh of the surface, while an analysis of the diffracted beam intensities provides information on the location of the surface atoms within this unit mesh. Because the surface is only two-dimensionally periodic, the third Laue condition of a normal X-ray diffraction experiment (defining momentum conservation perpendicular to the surface) is lost, and the full set of diffracted beams are visible at all (sufficient) energies (wavelengths) of the electrons. For this reason it is usual to collect diffracted beam intensities in the form of intensity-energy spectra for each of a series of diffracted beams. A set of spectra of this kind are the primary data set on which the quantitative structure analysis is based.

The structure determination then proceeds via a trial-and-error iteration process involving comparison of the experimental spectra with the results of model calculations based on a series of 'guessed' trial structures. Because of the very strong interaction of the low energy electrons with atoms, these calculations include the role of multiple scattering within the surface; this multiple scattering is one of the key reasons why simple Fourier inversion of the data (as may be used in X-ray diffraction) is not helpful in LEED. Recent attempts to find more appropriate algorithms to invert LEED data directly to obtain at least an approximate structure as a starting point for the trial-and-error simulations do seem to hold some promise, but no established methodology is yet in place. A key aspect of the trial-and-error process to find the best model structure is the use of an objective measure of the quality of agreement between the experimental spectra and the results of the simulation for a given model structure. A number of these R-factors (reliability factors) have been developed, typically designed to be sensitive to correlations in the energies of peaks in the intensity-energy spectra, and tests indicate that the final structural conclusions are rather insensitive to which of these is used. In recent years there has been considerable effort devoted to improving the efficiency of the trial-and-error iteration procedure, mainly through the use of approximation schemes (e.g. 'tensor LEED') which allow limited changes in structural parameters without the need to run a full calculation for every combination of the parameter values, and through the use of automated searches of the structural parameter space [13,14]. A key problem in this process is that there is no absolute measure which determines the success of the structure determination, and that local minima can occur in the R-factor in regions of parameter space well-removed from the values corresponding to the true best-fit structure. Such a problem is not unique to LEED, but rather is true of all techniques which rely on such trial-and-error search methods. A related general problem is that of uniqueness; can one be sure that there is not another structural model (which may not have been tested) which gives agreement with the experimental data of at least as good a quality as the structure found so far? Apart from ensuring that a very wide range of structures are searched, the other key factor to minimise this problem is to ensure that the data set to be fitted is large. In LEED rather large data sets (typically involving several diffracted beam intensity spectra covering a combined energy range of 1000-2000 eV) are now normally used to minimise this problem.

As an illustration of the kind of results which emerge from recent LEED studies of this type [15], fig. 2 summarises the results of a study of ethylidyne on Pt(111). Sectional and plan views of the (2x2) ordered overlayer structure are shown together with a summary of the key parameter values found. Notice that the H atoms are shown schematically only; H is a weak scatterer of electrons and is typically very difficult to locate by electron scattering methods. Studies of ordered overlayers of atomic hydrogen have determined the adsorption site, but in the present case no significant improvements to the fit could be obtained by including possible H sites; of course, it is also possible that the CH_3 group is rotating freely about the C-C axis. The primary information emerging from this study is that the ethylidyne occupies the so-called fcc hollow sites (the hollow sites directly above a third layer Pt atom and thus the same site that an additional Pt atom would adopt in the fcc structure) with the C-C axis essentially perpendicular to the surface (a small tilt of 6°, which could be dynamic in origin, is actually favoured). The analysis also gives values for the C-Pt and C-C distances. In addition, however, the analysis involved a search of a range of substrate distortion parameters, the values of which are summarised in the figure. Specifically there appears to be a relaxation of the outermost Pt layer (bulk layer spacing 2.26 Å) and a slight 'rumpling' and lateral distortion of this outermost Pt layer. Similar distortions on the second Pt layer are less than the estimated errors. The ability to search this multi-parameter space associated with these substrate distortions is a consequence of the automated search procedure using the 'tensor LEED' approximation scheme. It is also an important feature of LEED that the diffracted beam intensities are sensitive to all such structural parameters in a complex and coupled fashion through the influence of the strong multiple scattering. For this reason it is generally regarded as important to search all such parameters in order to ensure proper convergence on the correct values of the primary structural values of interest (e.g. the C-C and C-Pt distances). On the other hand, one often finds that the optimum values of some of the distortion parameters (e.g. the second Pt layer rumpling b_2) are not statistically significant relative to the estimated errors. We will return to this point in section 4.

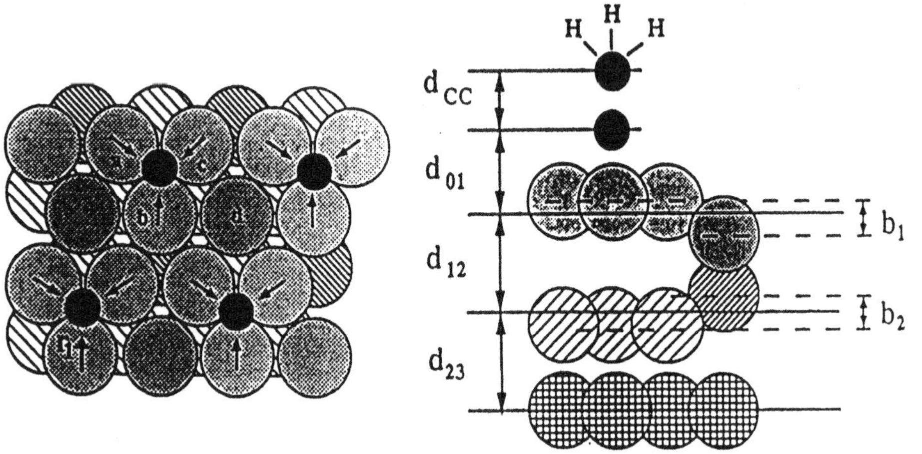

Figure 2. Plan and sectional views of the structural model arising from a LEED study of the structure of ethylidyne on Pt(111) [15], defining the parameters investigated. The values obtained were: $d_{CC}=1.49\pm0.05$ Å; $d_{01}=1.21\pm0.03$ Å; $d_{12}=2.30\pm0.03$ Å; $d_{23}=2.26\pm0.04$ Å; $b_1=0.11\pm0.05$ Å; $b_2=0.08\pm0.09$ Å; $r_1=0.10\pm0.08$ Å; $r_2=0.00\pm0.09$ Å.

3.2 LOCAL SCATTERING METHODS: GENERAL REMARKS

An important distinction between LEED on the one hand, and PhD and SEXAFS on the other, is that while in LEED the source of electrons lies outside the surface, in the other two techniques the source lies inside the surface and is localised on the adsorbate atoms. This distinction is important in two ways. Firstly, because the source is local and comprises an outgoing spherical wave (rather than the incident plane wave of LEED), the structural information content is intrinsically local. The spherical outgoing wave attenuates as it propagates away from the emitter atom, so the most important scattering contributions are those which occur from the atoms closest to the emitter. These local techniques neither require nor exploit the long-range order which is a formal requirement of a conventional diffraction technique such as LEED. This localisation means that the associated techniques are selectively sensitive to certain structural parameters (notably emitter-scatterer near-neighbour distances) and far less sensitive to other parameters such as those involving displacements of atoms well-removed from the emitter. The second key distinction deriving from the use of core level photoemission as the electron source is that the structural information is also element specific. If one has an atomic adsorbate, PhD or SEXAFS based on photoemission from this adsorbate species provides structural information on the scattering environment of this atom. Moreover, if one has an adsorbed molecule containing atoms of different elements (e.g. CO), one can obtain separate structural information on the local structural environment of each constituent atom by tuning to the appropriate element core level.

Before considering the photoelectron diffraction technique in more detail, we should comment briefly on a few more comparative remarks concerning SEXAFS and a further local scattering technique, diffuse LEED (DLEED). As may be seen in fig. 1(b), SEXAFS involves not only a local source of electrons but also a local detector - the emitter atom itself, and this causes the technique to be even more dominated by the very nearest neighbour scatterers. This especially high degree of localisation helps to minimise the role of multiple scattering in SEXAFS, and because all the single scattering events involve the same 180° backscattering, it also produces a marked simplification in the data interpretation. In particular, simple Fourier transform methods can provide useful direct structural information in SEXAFS. A correction to the apparent pathlength difference obtained in this way to deduce true nearest neighbour distances is necessary, because of the phase shifts incurred by the electrons in scattering off the neighbouring atom and through the emitter potential is necessary, but this is characteristic of the emitter-scatterer pair of atoms and can even be obtained experimentally from model compounds of known structure. At its simplest level SEXAFS is therefore in marked contrast to LEED. The data set is much smaller (just one energy spectrum for this basic information) but uniqueness is not a major problem because the first-order fit only involves a single parameter - the nearest neighbour distance. By contrast, LEED requires a fitting to many parameters, but if successful gives significantly more information. In fact SEXAFS is capable of providing somewhat more information than this simple description suggests. If the signal-to-noise quality is sufficient, more distance neighbour contributions can also be analysed. Moreover, limited directional information is also available through control of the direction of the X-ray polarisation vector relative to the surface normal. For example, if one studies K-shell adsorption (i.e. adsorption involving 1s state photoemission) the outgoing photoelectron wave is of p-character and is directed along the polarisation vector. By varying this polarisation direction some information on bond directions relative to the surface normal, as well as bond lengths, can therefore be obtained.

One further local electron scattering structural technique which has not yet been discussed is so-called diffuse LEED. The basic idea behind this technique is that if one directs an incident electron beam at a surface, the scattered wavefield emerging from an adsorbate atom will undergo exactly the same scattering events as those depicted in fig. 1(a) for photoelectron diffraction. For a surface which lacks long-range order in the adsorbed layer, the scattered (diffuse) electron signal thus contains information similar to that in PhD. There are, of course, other scattering signals contributing to this diffuse signal; most notably these include primary scattering from the substrate onto the adsorbate. All of these contributions are modelled in a conventional LEED calculation and it has been found that if a calculation is performed for a hypothetical ordered overlayer with an adequate adsorbate-adsorbate separation, the calculated diffracted beam intensity spectra provide a good description of the experimental diffuse intensities measured in the same directions. Evidently, this offers the possibility of obtaining local structural information from disordered overlayers using most of the standard LEED methodology. A small number of investigations have exploited this potential, at least one of which (benzene on Pt(111)) has been for a molecular adsorbate. The full potential of this method has yet to be assessed. Of course, although this technique is based on very similar scattering processes to those which characterise photoelectron diffraction, DLEED lacks the elemental specificity of PhD.

3.3 SCANNED-ENERGY MODE PHOTOELECTRON DIFFRACTION (PhD)

Although there is a group of techniques which all exploit the basic phenomenon of photoelectron diffraction, not all of these are comparably well-suited to fully quantitative molecular adsorbate structure determination. For example at high (typically >500 eV) kinetic energies, electron scattering is dominated by near-forward scattering and most of the applications of so-called X-ray photoelectron diffraction (XPD) are in the evaluation of epitaxial growth phenomena or in the determination of adsorbed molecular orientations only. At lower kinetic energies substrate backscattering provides the key adsorbate-substrate registry information required for full structure determination. In principle such experiments can be conducted at variable emission angle and fixed energy, but in practice it is difficult to build up a sufficient data set for a full and unique structure determination without measuring these angular scans at many energies, a procedure yet to be exploited in practice. Scanned-energy mode photoelectron diffraction (PhD) is closest in methodology to that of LEED and is currently the most fully developed and certainly the version of the technique which has yielded the greatest number of molecular structure determinations. The methodology described here is that adopted by a Warwick/Berlin collaboration exploiting facilities on the BESSY synchrotron radiation source in Berlin [16].

In the PhD technique the basic data set comprises a set of modulation spectra for the intensity of an adsorbate core level photoemission signal as a function of kinetic energy in each of a series of emission directions. The structure determination is ultimately achieved through a similar trial-and error modelling process to that used in LEED, and is also based on optimisation of an appropriate R-factor and an extensive (partially automated) search of a multi-parameter space aided by approximate methods as well as full, essentially exact, multiple scattering simulations. There are, however, a number of important differences in the methodology as well as the intrinsic strengths and weaknesses of the technique. A key difference is that because of the local character of the structural information content associated with the spherical wave electron source, multiple scattering is significantly less important than in LEED and some first-order structural information can be obtained through a direct data inversion procedure which is based only on the single scattering component.

Two somewhat different direct inversion methods have been developed by us, and both rely on the same underlying observation, namely that when a PhD spectrum is recorded in a direction corresponding to that in which a near-neighbour substrate atoms lies directly behind the emitter (i.e. in a geometry which involves a favoured 180° scattering event), the dominant modulations in the spectrum can be attributed to this one scattering path. If these directions can therefore be identified, one has a direct directional probe of the nearest neighbour substrate atoms and thus of the adsorption site. One simple way to identify these directions is to collect PhD spectra in a wide range of emission directions and take a Fourier transform of each; the directions in which the Fourier transform shows the largest amplitude short effective pathlength difference should be the one recorded in a nearest neighbour backscattering direction [17]. An illustration of the results of such a procedure are shown in fig. 3 using C 1s PhD data from acetylene, C_2H_2, adsorbed on Cu(111) [18]. PhD spectra were recorded at 10° intervals in polar emission angle in the three principal azimuths, <110> and the two inequivalent <211> directions, and the Fourier transforms of these spectra showed a low effective pathlength difference peak (at a value of approximately 4 Å) which could be attributed to a nearest neighbour Cu backscatterer to the C emitter. The contour map in fig. 3 shows the amplitude of this peak in the Fourier transform as a function of polar and azimuthal angle with simple interpolation between the experimental data points along the principal azimuths. Six essentially equivalent peaks are seen at polar angles of approximately 40° in each of the <211> directions. The schematic plan view of the structure shows the interpretation of this map in terms of two differently-sited C atoms, one occupying an fcc hollow site directly above a third layer Cu atom, the other occupying the hcp hollow directly above a second layer Cu atom. The three nearest neighbour Cu atoms of each of these C emitters appear in the Fourier transform map, rotated relative to one another by 60°, to give the pseudo-six-fold pattern of fig. 3. The clear implication is that these two different C emitters are the two C atoms of the acetylene molecule which lies with its C-C axis parallel to the surface so as to straggle these two hollow sites.

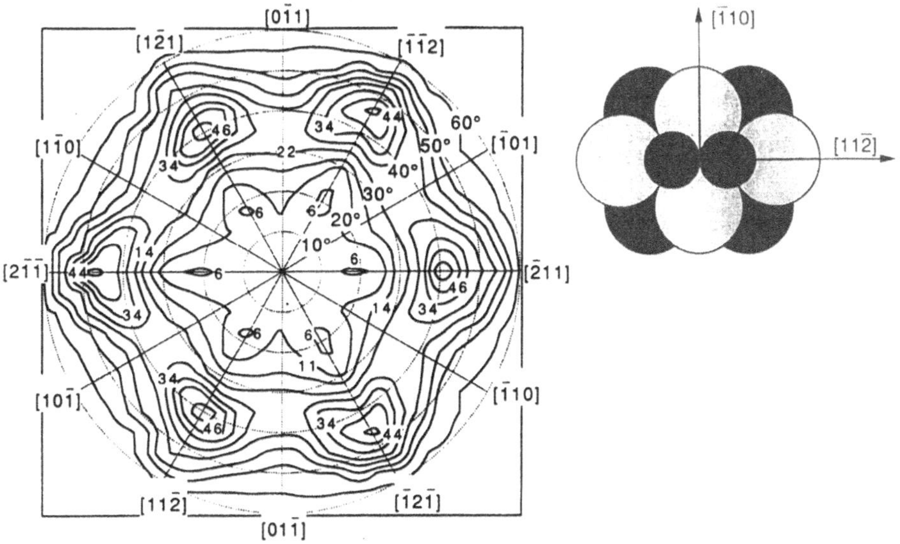

Figure 3. Contour map (in polar and azimuthal angle) of a 4 Å effective pathlength difference peak in Fourier transforms of C 1s PhD spectra recorded from adsorbed acetylene on Cu(111) [18]. Also shown is a plan view of the inferred structural model with C atoms in inequivalent hollow sites.

One limitation of this simple Fourier transform approach is that it gives only the approximate directions of the near neighbours, the effective pathlength difference differing from the true value due to the effects of scattering phase shifts. The method also requires quite a large data set (fig. 3 is derived from 24 PhD spectra). An alternative approach [19] based on the same underlying ideas involves replacing the complex harmonic function of the Fourier transform by a real modulation function based on a single scattering calculation for a substrate atom at different possible sites. We expect this projection integral to be maximised when the test site corresponds to the location of a substrate backscatterer and the frequency and phase of the dominant oscillations of the experimental and test spectra are matched. By summing the results of several such projection integrals using experimental spectra recorded in several (typically 4-8) emission directions, and weighting the results exponentially to strongly favour the very largest projection integrals, a map (or 'image') of the most probably locations of the backscatterers can be obtained. These 'images' appear to determine the emitter-backscatterer distances to within a few hundredths of an Ångstrom unit, as well as indicating the location in space.

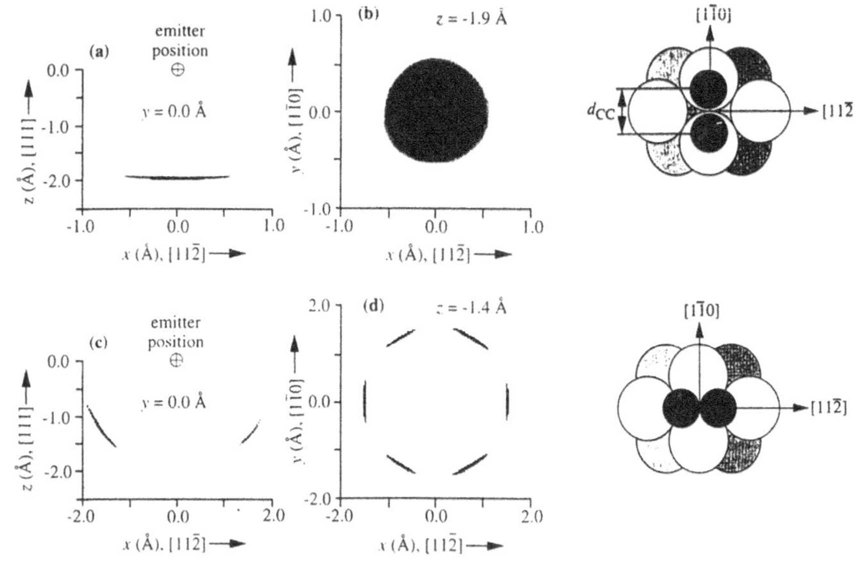

Figure 4. Results of the application of the projection method of direct inversion of PhD data for ethylene and acetylene adsorption on Ni(111) [20, 21], together with plan views of the implied structures. The two left-hand panels show grey-scale maps taken in cuts perpendicular to the surface in a <211> azimuth passing through the emitter at (0,0,0), while the centre panels show similar maps in cuts parallel to the surface at depths below the emitter chosen to cut the main features seen in the perpendicular cuts. The upper row relate to ethylene adsorption, the lower row to acetylene.

Fig. 4 shows the results of applying this projection method to C 1s PhD data for ethylene (C_2H_4) and acetylene adsorption on Ni(111). Because the projection method provides a three-dimensional map of a function related to the probability of finding a backscatterer at different locations, two-dimensional cuts of this space must be presented. The left-hand panels of fig. 4 show cuts perpendicular to the surface through the emitter in a <211> azimuth. In the case of the ethylene data this shows a characteristic arced feature directly below the emitter (by about 1.9 Å), while for acetylene one strong and one weak feature are seen about 1.4 Å below the emitter displaced laterally. The centre panels show cuts parallel to the surface at the depths of the features in the perpendicular cuts. For ethylene a single strong peak is shown implying that a Ni atoms lies (approximately) directly below the C atom emitter, suggesting the aligned bridge site for the molecule shown in the right-hand panel. For acetylene the parallel cut shows 6 symmetric features (c.f. fig 3) and is interpreted in terms of equal occupation of the two inequivalent hollow sites, with the molecule in a cross-bridge site, exactly as for acetylene on Cu(111) discussed above.

Although these two direct methods provide invaluable first-order structure indications from PhD, a proper quantitative structure determination still requires trial-and-error optimisation using full multiple scattering simulations in a process rather similar to that of LEED. The direct methods, on the other hand, narrow down the range of possible structures which need to be tested in this way, particularly for systems having high symmetry. In both the examples given above, full optimisation does confirm the qualitative site determination and semiquantitative bond-length determination of the direct methods. Some further comments on the significance of the results will be made in the next section. Before doing so, however, one further important development of the PhD technique is discussed, which is of especial significance for the elucidation of molecular adsorption structures.

In comparing PhD and SEXAFS on the one hand, and LEED and DLEED on the other, the advantage of elemental specificity in the structural information has been stressed. In LEED and DLEED, different atomic species are only distinguished by the differences in scattering amplitude and phase shifts, and for atoms close in atomic number these differences can be small. Furthermore, if we consider even a simple adsorbed molecule such as CO, a LEED structural search must include as primary search parameters the adsorption site, the molecular orientation, and the intra- and extra-molecular bondlengths. In PhD, on the other hand, one can collect separate O 1s and C 1s PhD data and, to a first approximation, determine the C and O sites separately; the results for the site of one of these atoms is only weakly correlated to the site of the other (through intramolecular scattering). Nevertheless, for many of the more complex problems encountered in molecular adsorption structures, even elemental specificity is less incisive than one might wish. For example, if a molecular adsorbate contains more than one atom of a given element, the PhD spectra from this element involve an incoherent sum of the diffraction at each of the atoms of this element. A similar problem arises if the surface has co-adsorbed species which contain the same element(s). Of course, in some cases the different atoms of the same element may be in symmetrically equivalent sites, so no added problem arises; for example, if one were to investigate benzene adsorption on a six-fold symmetric surface, one might expect that the molecule would lie flat on the surface with all six C atoms in equivalent sites. In reality, of course, most surfaces of interest have no more than threefold symmetry (e.g cubic (111) of hcp (0001)) so at least two inequivalent sites must be involved - as indeed was the case for acetylene on Cu(111) and Ni(111) in the cases described above.

A way to overcome this limitation in the case of the PhD technique is to exploit the fact that core level photoelectron binding energies are actually site dependent themselves; this basic fact is the basis of ESCA - now more commonly referred to as X-ray photoelectron spectroscopy (XPS) - as discussed in section 2.2. The actual source of these 'chemical shifts' in XPS, a complex mix of initial state changes in the one-electron binding energy and final state changes in the extra-atomic relaxation energy, both a consequence of a modification of the valence electron environment, need not concern us here. The key point is that such shifts do exist, so the chemically shifted components of a core level photoemission peak can be used to provide photoelectron diffraction spectra for each component and thus for each inequivalent environment of a given elemental species. Of course, the ability to exploit this idea depends on the size of the chemical shifts and the intrinsic and instrumental limitations on spectral resolution. The widespread use of ESCA or XPS for fingerprinting chemical states in surface chemistry bears witness to the significant range of potential application of this method, while the recent higher resolution studies mentioned earlier showing, for example, that the C and O 1s binding energies of CO adsorbed in different surface sites on Pt(111) [11] shows that there is still scope to extent this application further.

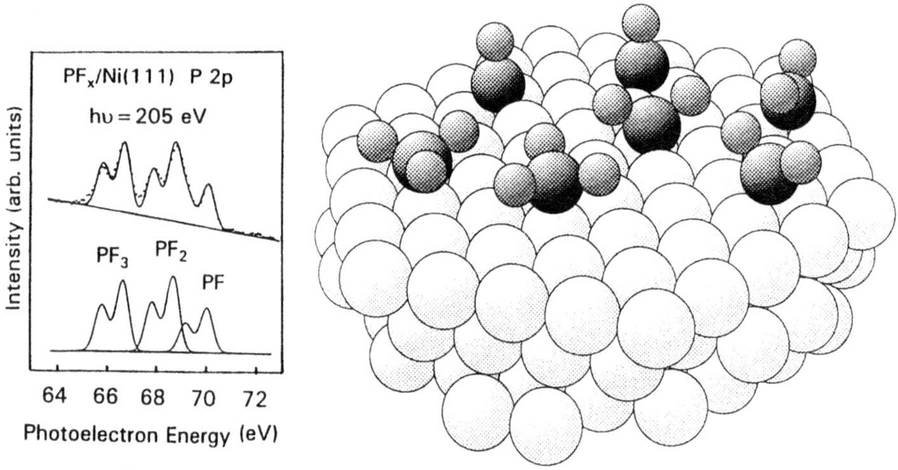

Figure 5. Photoelectron energy spectrum (left) recorded at a photon energy of 205 eV from a Ni(111) surface with coadsorbed PF_3, PF_2 and PF, showing the three chemically shifted components of the spin-orbit split P 2p photoemission peak. On the right is shown a schematic of the local geometry of each species as determined from photoelectron diffraction on each of these components [23].

Two specific demonstration experiments of chemical shift photoelectron diffraction (CSPhD) illustrate the potential of the method. One concerns the structure of the acetate species, CH_3COO-, on Cu(110); in this case the different photoelectron binding energies of the two C atoms in this species meant that it was possible to obtain independently, local structural information from the two (equivalently-sited) O atoms and separately from the methyl and carboxyl C atoms [22]. A second example concerns the coadsorbed species PF_3, PF_2 and PF on a Ni(111) surface produced by the partial electron beam dissociation of adsorbed PF_3 on this surface [23]. Fig. 5 shows the basis of the measurement at the outcome. A single photoelectron energy spectrum recorded at a photon energy of 205 eV shows a five-peaked structure which can readily be separated into three chemically-shifted P 2p spin-orbit split doublets attributable to the three coadsorbed molecular fragments. By recording similar such spectra over a range of different photon energies a set of PhD spectra for each component species was obtained and analysed. The resulting structure is shown in the schematic diagram of fig. 5. The PF_3 species occupies the same atop sites found without fragmentation while the PF_2 species occupy bridge sites and the PF are in the two inequivalent hollow sites. The result indicates that the P bonding coordination remains at a value of four throughout the fragmentation.

Clearly CSPhD has great potential as a technique for determining adsorbate sites of coadsorbed molecular fragments occurring in more industrially-relevant surface chemical processes, but to derive the full benefit of the method substantial improvements in the combination of high photon flux and high spectral resolution are required relative to those used in these demonstration experiments. In particular, chemical shifts on C 1s associated with different hydrocarbon states can be as little as a few tenths of an eV, compared with the shifts of approximately 2 eV seen in fig. 5. This potential already exists, however, in the form of undulator-based beamlines on third-generation synchrotron radiation sources, so further developments and applications can be anticipated.

4. Some results

While a complete and exhaustive review of the results of all quantitative determinations of molecular adsorbate structures lies beyond the scope of this brief review, it is instructive to try to pick out some trends in terms of the kinds of systems studied and the novelty or limitations of the results obtained so far. A recent update of the NIST Surface Structure Database [24] provides a reasonably up-to-date survey of structures solved. By far the largest number of molecular structures for which there exists proper quantitative solutions involve CO adsorption. In several cases these indicate the atop or bridge site occupation inferred from earlier vibrational spectroscopy, although, as already mentioned, for the case of CO on Ni(111) all the structural studies show occupation of hollow sites in contradiction to the interpretation of the C-O stretching frequency. In this context it is interesting to note that the same c(4x2) LEED pattern at the half-monolayer coverage seen on Ni(111) is also seen on Pt(111), but in this case LEED shows that the CO molecules occupy atop and bridge sites equally [25], rather than the hcp and fcc hollow sites seen on Ni(111). There are also a few studies of NO adsorption structures which yield results very similar to those for CO (indeed, it was the Ni(111)/NO adsorption system which first called into doubt the assignment of local sites on the basis of intramolecular vibrational frequencies [26]).

In addition to these investigations of adsorbed diatomic molecules, a few other nitrogen-containing molecules (NH_3, CH_3CN and C_5H_5N) have been studied, but the largest group of other molecules studied are simple hydrocarbons - principally C_2 species (ethylene and acetylene) and benzene (C_6H_6). One especially interesting aspect of these hydrocarbon structural studies is the extent to which the internal structure of the molecule is influenced by its adsorption on a surface; such an effect is not purely of academic interest, because significant distortions of an adsorbed molecule may represent a precursor to partial dissociation and thus to surface chemical reactions.

In the case of benzene adsorption the best-fit structures found in several systems do indicate the presence of distortions, either in the bondlengths within the ring (notably a Kekule type of distortion with alternating C-C distances around the ring), or in bending the planar ring into a boat conformation. However, a careful evaluation of the estimated precision of these structural studies leads to the rather disappointing conclusion that none of these distortions are formally statistically significant. To some extent this arises because both LEED and PhD are relatively less sensitive to atom movements parallel to the surface compared to perpendicular movements. Precisions in atomic displacements perpendicular to a surface can be as small as 0.02 Å, but parallel to the surface are often 0.10 Å or even larger. As benzene appears to bond with its molecular plane essentially parallel to the surface in all the cases studied, this leads to significant imprecision in the determination of the C-C distances. Although the same basic problem also applies to ethylene and acetylene bonded with their C-C axes parallel to the surface, large changes in the C-C distances found in these systems do appear to be chemically significant. In particular, the geometry adopted by acetylene on both Ni(111) and Cu(111) places the two carbon atoms in the inequivalent hollow sites, as indicated by the direct inversions of the PhD data shown in figs. 3 and 4. Full multiple scattering fitting of these data confirms these sites [18,21], thus giving C-C bond lengths of 1.44 Å and 1.47 Å respectively. By comparison the C-C distance in gas-phase acetylene is only 1.21 Å, so even with error estimates in the bondlength obtained from the PhD study of ± 0.10 Å, this bondlength extension is significant. Indeed, it implies a very considerable reduction in the bond-order of the C-C bond from a value of three in the free molecule to between 1 and 2 on the adsorbed species. (Note that the C-C distance in a gas-phase ethylene molecule, in which the bond order is two, is 1.34 Å). This conclusion is also consistent with the results of vibrational spectroscopy which shows a pronounced softening of the C-C stretching frequency.

Several other molecular structures which have been solved involve adsorbed species which are intermediates in some surface reaction. These include the methoxy (CH_3O-) and formate (HCOO-) species formed, respectively, by deprotonation of methyl alcohol and formic acid, and ethylidyne (C_2H_3). One further problem of specific note in the context of surface reactions is the study of ethylene and acetylene adsorption on Ni(111) from which the results of fig. 4 are taken. Probably the very first demonstration of the application of any surface spectroscopy to follow a surface reaction on a single crystal surface under UHV conditions was that of Demuth and Eastman in 1974 [27]. They showed that the valence band spectra provided by ultraviolet photoelectron spectroscopy (UPS) could be used to follow the dehydrogenation of adsorbed ethylene to adsorbed acetylene on Ni(111). In particular, at low temperatures (nominally 100 K) the UPS spectra of adsorbed ethylene and adsorbed acetylene showed the 'spectral fingerprint' to be expected of these molecules, but warming the surface to around 230 K after adsorbing ethylene at low temperature led to new spectrum identical to that of adsorbed acetylene (at both 100 K and 300 K). The reaction was thus

$$(C_2H_4)_a \rightarrow (C_2H_2)_a + 2H_a$$

with the adsorbed hydrogen atoms combining to desorb as molecular H_2 at a higher temperature.

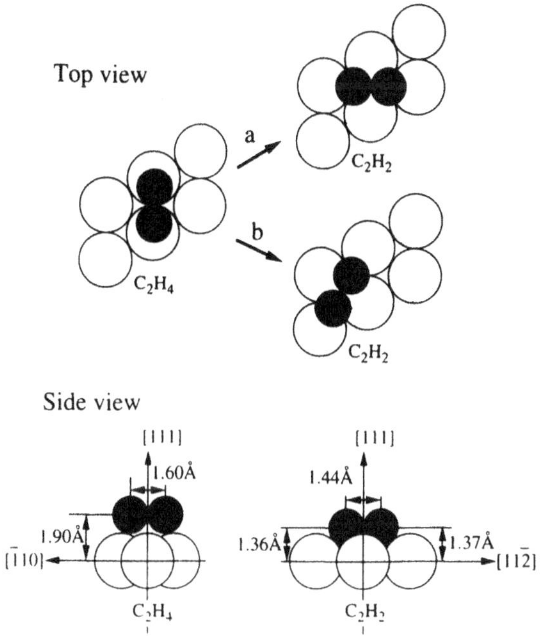

Figure 6 Schematic diagram showing two possible geometrical reaction paths involved in the dehydrogenation of adsorbed ethylene to adsorbed acetylene on Ni(111), based on the absorption geometries of the reactant and product molecules obtained from a PhD investigation [20,21].

The PhD study of adsorbed ethylene and adsorbed acetylene on Ni(111) also included a study of the coadsorbed acetylene and hydrogen produced by this surface reaction, and found the adsorption geometry of this coadsorbed acetylene to be identical to that of acetylene adsorbed onto a clean Ni(111) surface [20,21]. The results of this study in terms of the local adsorption geometries of the reactant and product species of this surface reaction are shown in the schematic diagram of fig. 6. Also shown are the two possible geometrical reaction paths which might be associated with the reaction on the basis of this structural study (assuming the loss of the two H atoms is a concerted reaction). One pathway involves a rotation of the C-C bond by 90° about the surface normal, the other a smaller rotation but also a translation parallel to the surface. In either route, of course, one H atom is lost from each carbon atom (the H atom sites are unknown and thus not shown), the C-C bondlength decreases, and the molecule moves closer to the surface. These two alternative routes provide a valuable starting point to understanding the reaction better through appropriate quantum chemical calculations.

One final class of surface structures which are being solved by proper quantitative methods is coadsorption phases. One specific example already mentioned is coadsorbed PF_3, PF_2 and PF on Ni(111) as the result of a partial fragmentation reaction. So far this is the only example of its type, but clearly coadsorbed reaction products form a potentially very important class of systems. Because of the lack of long-range order in such systems, and thus the enhanced need for chemical state specificity, it is likely that further work in this area will be dependent on the further development of the CSPhD technique. Other coadsorption systems involving long-range ordered phases have been studied, however. For example, there have been several CO/benzene coadsorption phases studied in cases where the pure benzene layer does not have such long-range order. Coadsorption systems of this type have an intrinsic interest in trying to understand how each adsorbate is modified by the presence of the other. In some cases, of course, such a coadsorption phase can be a precursor to a reaction between the coadsorbates. In other systems, especially alkali/molecule coadsorption [e.g. 28], there is interest from the point of view of the role of additives to catalysts which modify the activity or selectivity of the catalyst. This is another area in which increased activity can be anticipated in the next few years.

Acknowledgements

Although this short review attempts to cover, in general terms, the work of many authors, most of the specific examples are taken from work performed in collaborations involving young scientists in my own group and those in other institutions. The PhD work, in particular, is the result of a collaboration with the group of Prof. Alex Bradshaw of the Fritz Haber Institute in Berlin, together with invaluable theoretical support from Dr. Volker Fritzsche of the TU Dresden. I am happy to acknowledge the crucial role played by all the members of these groups, many of whom are listed in the individual authors lists of the references cited below. This work has also benefited from the financial support of the Engineering and Physical Science Research Council and from the European Commission under the Human Capital and Mobility programme.

References

1. A more complete summary of all of the techniques surveyed in section 2 is to be found in: Woodruff, D.P. and Delchar, T.A. (1994) *Modern Techniques of Surface Science - Second Edition*, Cambridge University Press, Cambridge.
2. Robinson, I.K., Ferrer, S., Torrelles, X., Alvarez, J., Van Silfhout, R., Schuster, R., Kuhnke, K. and Kern, K. (1995) Vibrational anisotropy of a CO monolayer on Ni(110), *Europhys.Lett.* **32**, 37-42
3. Kerkar, M., Woodruff, D.P., Avila, J., Asensio, M.C., Fernández-Garcia, M. and Conesa, J.C. (1992), Structure determination for PF_3 adsorption on Ni(111), *J.Phys.:Condens.Matter* **4**, 6509-6522
4. Kerkar, M., Hayden, A.B., Woodruff, D.P, Kadodwala, M. and Jones, R.G. (1992) An unusual

adsorption site for methoxy on Al(111) surfaces, *J.Phys.Condens.Matter* **4**, 5043-5052

5. Sheppard, N. and Nguyen, N.T. (1978) The vibrational spectra of carbon monoxide chemisorbed on the surfaces of metal catalysts - a suggested scheme of interpretation, *Adv.IR Raman Spectr.* **5**, 67-148
6. Schindler, K-M., Hofmann, Ph., Weiss, K-U., Dippel, R., Fritzsche, V., Bradshaw, A.M., Woodruff, D.P., Davila, M.E., Asensio, M.C., Conesa, J.C. and González-Elipe, A.R. (1993) Is the frequency of the internal mode of an adsorbed diatomic molecule a reliable guide to its local adsorption site? *J.Electr.Spectros.Rel.Phenom.* **64/65**, 75-83
7. Davila, M.E, Asensio, M.C., Woodruff, D.P, Schindler, K-M., Hofmann, Ph., Weiss, K-U., Dippel, R., Gardner, P., Fritzsche, V., Bradshaw, A.M., Conesa, J.C and González-Elipe, A.R. (1994) Structure determination of Ni(111)c(4x2)-CO and its implications for the interpretation of vibrational spectroscopic data *Surface Sci.* **311**, 337-348
8. Becker, L., Aminpirooz, S., Hillert, B., Pedio, M., Haase, J. and Adams, D.L. (1993) Three-fold coordinated hollow adsorption site for Ni(111)c(4x2)-CO: a surface extended X-ray absorption fine structure study, *Phys.Rev.B* **47**, 9710-9714
9. Mapeldoram, L.D., Bessent, M.P., Wander, A. and King, D.A. (1994) An automated tensor-LEED analysis of the Ni(111)c(4x2)-2CO structure, *Chem.Phys.Lett.* **228**, 527-532
10. Fernandez, V., Schindler, K-M., Schaff, O., Hofmann, Ph., Theobald, A., Bradshaw, A.M., Fritzsche, V., Davis, R. and Woodruff, D.P (1996) Structure determination of a CO/O coadsorption phase on Ni(111), *Surface Sci.*, **351**, 1-12
11. Björneholm, O., Nilsson, A., Tillborg, H., Bennich, P., Sandell, A., Hernnas, B., Puglia, C. and Mårtensson, N. (1994) Overlayer structure from adsorbate and substrate core-level binding energy shifts - CO, CCH_3 and O on Pt(111), *Surface Sci.* **315**, L983-989
12. Pendry, J.B. (1974) *Low Energy Electron Diffraction*, Academic Press, London
13. Van Hove, M.A. (1993) Automated determination of complex surface structures by LEED, *Surface Sci.Rep.* **19**, 191-229
14. Heinz, K. (1995) LEED and DLEED as modern tools for quantitative surface structure determination *Rep.Prog.Phys*, **58** 637-704
15. Starke, U., Barbieri, A., Materer, N., Van Hove, M.A. and Somorjai, G.A. (1993) Ethylidyne on Pt(111): Determination of adsorption site, substrate relaxation and coverage by automated tensor LEED, *Surface Sci.* **286**, 1-14
16. Woodruff, D.P. and Bradshaw, A.M. (1994) Adsorbate structure determination on surfaces using photoelectron diffraction, *Rep.Prog.Phys.* **57**, 1029-1080
17. Fritzsche, V. and Woodruff, D.P. (1992) A direct photoelectron diffraction method for adsorbate structure determination, *Phys.Rev.B* **46**, 16128-16134
18. Bao, S., Schindler, K-M., Hofmann, Ph., Fritzsche, V., Bradshaw, A.M. and Woodruff, D.P. (1993) The local adsorption site of acetylene on Cu(111), *Surface Sci.* **291**, 295-308
19. Hofmann, Ph., Schindler, K-M., Bao, S., Bradshaw, A.M. and Woodruff, D.P. (1994) Direct identification of atomic and molecular adsorption sites using photoelectron diffraction, *Nature* **368**, 131-2
20. Bao, S., Hofmann, Ph., Schindler, K-M., Fritzsche, V., Bradshaw, A.M., Woodruff, D.P., Casado, C. and Asensio, M.C. (1994) Following the changes in local geometry associated with a surface reaction: the dehydrogenation of ethylene, *J.Phys.Condens.Matter* **6**, L93-L98
21. Bao, S., Hofmann, Ph., Schindler, K-M., Fritzsche, V., Bradshaw, A.M., Woodruff, D.P., Casado, C. and Asensio, M.C. (1995), The local structure of reactant and product in a surface reaction: the dehydrogenation of adsorbed ethylene on Ni(111), *Surface Sci.* **323**, 19-29

22. Weiss, K-U., Dippel, R., Schindler, K-M., Gardner, P., Fritzsche, V., Bradshaw, A.M., Kilcoyne, A.L.D. and Woodruff, D.P. (1992), Chemical shift photoelectron diffraction from molecular adsorbates, *Phys.Rev.Lett.* **69**, 3196-3199
23. Weiss, K-U., Dippel, R., Schindler, K-M., Gardner, P., Fritzsche, V., Bradshaw, A.M., Woodruff, D.P., Asensio, M.C. and González-Elipe, A.R. (1993) Structure determination of coadsorbed molecular fragments using chemical shift photoelectron diffraction' *Phys.Rev.Letters* **71**, 581-4
24. Watson, P.R., Van Hove, M.A. and Hermann, K. (1996) *NIST Surface Structure Database Ver. 2.0, NIST Standard Reference Data Program*, Gaithersburg, MD.
25. Ogletree, D.F., Van Hove, M.A. and Somorjai, G.A. (1986) LEED intensity analysis of the structures of clean Pt(111) and of CO adsorbed on Pt(111) in the c(4x2) arrangement *Surface Sci.* **173**, 351-365
26. Asensio, M.C., Woodruff, D.P., Robinson, A.W., Schindler, K-M., Gardner, P., Ricken, D., Bradshaw, A.M., Conesa, J.C. and González-Elipe, A.R. (1992), Single local site structure for vibrationally distinct adsorption states: NO on Ni(111), *Chem.Phys.Letters*, **192**, 259-264
27. Demuth, J.E. and Eastman, D.E. (1974) Photoemission observations of the π-d bonding and surface reactions of adsorbed hydrocarbons on Ni(111), *Phys.Rev.Lett* **32**, 1123-1127
28. Davis, R., Woodruff, D.P., Schaff, O., Fernandez, V., Schindler, K-M., Hofmann, Ph., Weiss, K-U., Dippel, R., Fritzsche, V. and Bradshaw, A.M. (1995) Structure determination of an alkali metal-CO coadsorption phase: Ni(111)-K/CO, *Phys.Rev.Lett.* **74**, 1621-4

THE STRUCTURE AND REACTIVITY OF $TiO_2(110)$ SUPPORTED PALLADIUM AND RHODIUM

BRIAN E.HAYDEN
Department of Chemistry
The University
Southampton SO17 1BJ
United Kingdom

ABSTRACT Palladium and rhodium overlayers on $TiO_2(110)$ deposited by metal vapour deposition (MVD) are characterised using a combination of XPS, LEED and FT-RAIRS of adsorbed CO. The RAIRS is sensitive to the local dielectric response in the neighbourhood of the CO probe molecule, and together with its fingerprint properties allows a characterisation of the size and morphology of the metal phases, particularly when combined with the other complimentary techniques. Palladium and rhodium overlayers deposited by MVD exhibit quite different structural behaviour on $TiO_2(110)$. A range of small to larger particles in registry with the substrate are produced in the case of palladium, while rhodium produces exclusively a large particle size with no apparent registry. High dispersion of rhodium is only possible using techniques of metal organic chemical vapour deposition (MOCVD), and this is illustrated with studies using rhodium carbonyl chloride as a precursor. The thermal stability and re-generation of the highly dispersed $TiO_2(110)$ supported rhodium gem-dicarbonyl is described, together with its reactivity with hydrogen.

1. Introduction

The growth, stability and reactivity of metal overlayers on oxide and semi-conductor surfaces are important aspects of surface science because of a variety of important technological and commercial applications, one of these being the field of heterogeneous catalysis. This has provided the impetus for modelling and characterising various phases of metals supported on well characterised oxide surfaces. In addition such studies contribute to the development of new preparative techniques to produce active phases, providing alternatives to the traditional impregnation techniques. One of these, illustrated here, is the use of MOCVD to produce homogeneous distributions of highly dispersed phases. These precursors can be used to produce high area oxide supported metal catalysts under mild and controlled conditions to produce heterogeneous catalysts [1] with high activity and specificity.

Oxide supported palladium and rhodium catalyse a number of important reactions including hydrogenolysis [2-5], hydrogenation [6-8], syngas [9] and automotive exhaust [10-13]

R.M. Lambert and G. Pacchioni (eds.),
Chemisorption and Reactivity on Supported Clusters and Thin Films, 215-235.

conversion. In general, palladium exhibits structural insensitivity in a wide variety of reactions compared to platinum, iridium and rhodium [14-15], although a crystallographic face dependence has been observed for reaction rates on single crystal surfaces [16], and the support influences the activity and selectivity in a number of reactions [9]. In the case of rhodium the dispersion affects both the selectivity and activity of the catalyst [17], and because of its high cost, dispersion also has important economic consequences. The degree of dispersion is also influenced by the reacting gases, and inter-conversion of the metallic rhodium clusters and the gem-dicarbonyl $\{Rh(CO)_2\}$ is induced by CO and H_2 [18-23].

Since the pioneering work of Eichens and Pliskin on supported palladium [24], and Yang and Garland on supported rhodium [25], IR spectroscopy of CO as an adsorbed probe molecule has provided a characterisation of the crystallographic faces in metallic particles, and identification of dispersed species such as rhodium gem-dicarbonyl [26]. Indeed the usefulness of CO as a probe molecule in vibrational spectroscopy has increased as a result of RAIRS and HREELS studies of CO on a variety of single crystal surfaces. FT-RAIRS (Fourier Transform Reflection Absorption Infrared Spectroscopy) of CO has recently been applied to the study of $TiO_2(110)$ supported palladium [27] and rhodium [28,29] phases which have been deposited by both MVD and MOCVD techniques. When combined with XPS (X-ray Photoelectron Spectroscopy), LEED (Low Energy Electron Diffraction) and TPD (Temperature Programmed Desorption) measurements this has allowed the identification of crystal faces, various morphologies, and the thermal and chemical inter-conversion of various phases. Here, the results for MVD deposited layers of palladium and rhodium on $TiO_2(110)$ are compared and contrasted. In the case of rhodium, unique phases deposited through MOCVD are described, and the study of their stability, reactivity, and inter-conversion to the metallic phase described.

2. Experimental

Experiments were performed in a UHV chamber which is described in detail elsewhere [30], and is equipped with a quadrupole mass spectrometer, LEED, XPS and FT-RAIRS. The $TiO_2(110)$ crystal was cleaned by cycles of Ar^+ bombardment, annealing at 1000K in UHV and oxygen treatment at ca. 400K [29,30]. $[Rh(CO)_2Cl]_2$ was prepared and purified by vacuum sublimation [31]. It was subsequently transferred to a small glass vial (volume $2cm^3$) which was attached to the UHV chamber and evacuated to $1x10^{-8}$ Torr. Dosing was performed through a conventional leak valve and a resistively heated stainless steal tube (350K) positioned 10mm from the $TiO_2(110)$ surface. The compound was initially dosed for 10 minutes prior to exposure to the $TiO_2(110)$ surface allowing equilibrium to be attained with the internal walls of the dose line. Rhodium and palladium MVD overlayers were deposited by resistively heating a 0.5 mm diameter wire of the metal. Coverages were estimated by XPS, assuming uniform growth during deposition at 300K, and using published data on mean free path [32]. XPS data were obtained using Mg Kα radiation at an electron pass energy of 50eV. All XPS binding energies are referred to the $Ti(2p^{3/2})$ core level at 458.5eV. Polarised FT-RAIRS spectra were recorded on a Perkin Elmer 1720 using MCT and InSb detectors at 2 and 4 cm^{-1} resolution. The incident beam was p-polarised, and the angle of incidence of the beam to the surface normal was 83°.

3. Theory

The incident angle for FT-RAIRS was chosen to be below the Brewster angle for titania (Figure 1). Under these conditions reflectivity is high, and the normal (p_n) and tangential (p_t) components of the p-polarised radiation can couple to the normal and parallel components

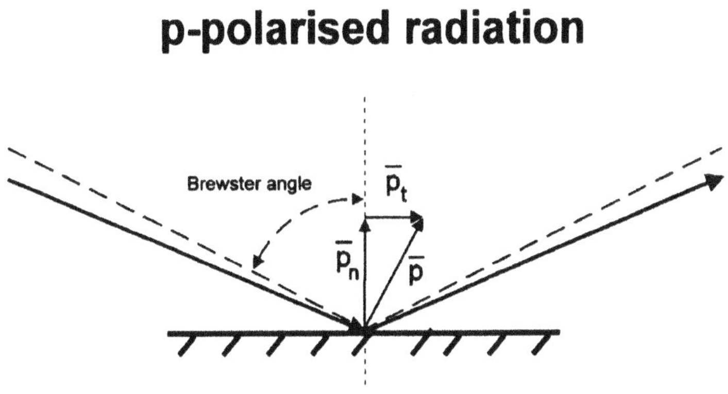

Figure 1 The optical configuration for FT-RAIRS

of a surface oscillator. Note that unlike reflection from a metal surface [33], the electric field generated parallel to the surface is not screened by a titania substrate [34] (the imaginary part of the refractive index is near zero), and both normal and parallel vibrational components will be observed. However, for a 2-dimensional layer of an isotropic absorber with resonant frequency ω_0, one may expect an increase in reflectivity for the coupling of p_n to the normal component of the oscillator, and a decrease in reflectivity for the coupling of p_t to the parallel component. These two bands will not be observed at ω_0, but will be shifted in frequency because of vibrational dipole coupling, resulting in a shift to either higher (normal component) or lower (parallel component) frequency depending on the relative orientation of the dynamic dipole moment. This is shown schematically in Figure 2, and spectra for such an isotropic layer of $Mo(CO)_6$ on $TiO_2(110)$ indeed show [35] that experimentally one observes exactly such behaviour for the ν(C-O) T_{1u} mode under these reflection conditions.

Molecules adsorbed and oriented on $TiO_2(110)$ in the monolayer may exhibit vibrational modes with components either normal or parallel to the surface, and as will be shown here, both can be observed in FT-RAIRS with an increase or decrease in reflectivity for the two components respectively. The situation becomes more complex in FT-RAIRS during the

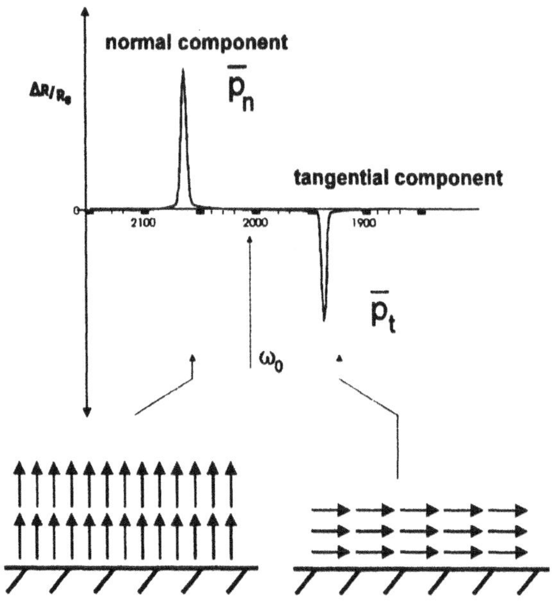

Figure 2 The coupling of normal and parallel components in an isotropic surface layer.

modification of the surface dielectric by a metallic overlayer. Eventually one may expect the local dielectric to reflect that of the metallic layer at which point the normal component of the vibrational modes of adsorbed molecules will result in a decrease in reflectivity, with complete screening of the parallel modes. The thickness and homogeneity of the metal layer will determine when such a point is attained and there is a complex region of metal coverage and dispersion where one may expect a transition from the behaviour expected for the titania to that of the metal as the local dielectric changes. It will become apparent that the size of metal particles on the surface clearly influences the observed behaviour in FT-RAIRS.

4. Results and Discussion

4.1. MVD phases of Palladium and Rhodium

The usefulness of CO as a probe molecule in vibrational spectroscopy has increased as a result of RAIRS and HREELS data for CO adsorbed on various crystal faces of, for example, palladium [36-41]. It is perhaps not surprising therefore that vibrational spectroscopy of CO adsorbed on model supported Pd catalysts is useful for identifying facets of Pd crystallite [40,42,43] and in following surface reactions on these surfaces [44-46]. Comparison with single crystal studies allows an identification of surface planes of the Pd [33,47], and the bridging species **B_1 (1910-1980cm^{-1}) and B_2 (1910-1945cm^{-1}) identify Pd(110)/Pd(100) and Pd(111) facets respectively** for CO adsorption at 300K. The FT-RAIRS spectra [27] for increasing coverages of CO adsorbed at 300K on a 1.2ML Pd MVD layer deposited at 300K on $TiO_2(110)$ are shown before and after annealing the surface to 550K in Figure 3.a. and Figure 3.b. respectively. The ν(C-O) band at 1980 cm^{-1} observed on the freshly deposited

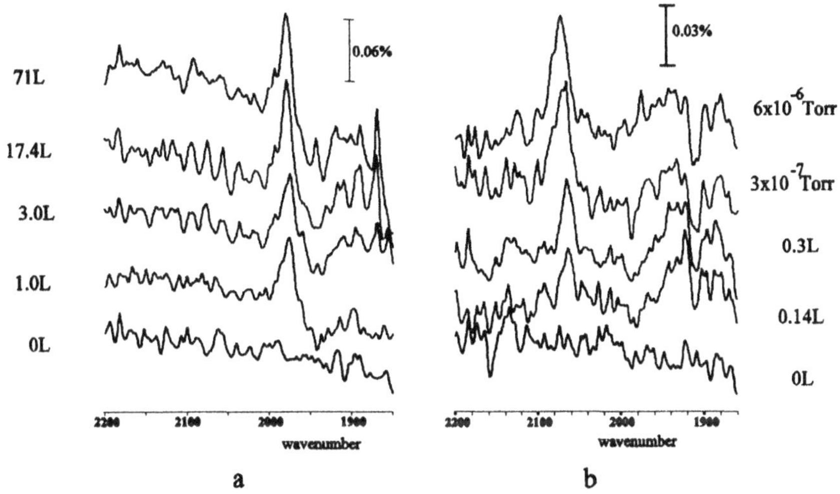

Figure 3 FT-RAIRS spectra of CO adsorbed at 300K on a 1.2ML MVD Pd layer on $TiO_2(110)$ following deposition at 300K (a) and annealing at 550K (b).

layer (Figure 3.a) is assigned to a B_1 bridged species characteristic of (100)/(110) facets of Pd. The extent of nucleation and particle growth during deposition at 300K is difficult to establish but it will become evident that nucleation to produce large Pd particles increases with increasing coverage and during annealing, with a concomitant change in morphology. The ν(C-O) band at 2063-2072 cm^{-1} observed on the annealed layer (Figure 3.b) is characteristic of a strongly bound linear species L* (it is not removed by pumping at 300K) which is associated with CO adsorbed at the edge sites of Pd particles [27]. A smaller contribution from a bridging ν(C-O) band at 1940cm^{-1} (Figure 3.b) is characteristic of the B_2 species associated with Pd(111) facets. The results are interpreted as indicating the agglomeration of the Pd deposited at 300K to form small particles with a high density of edge to close packed (111) faces. It is interesting to compare the FT-RAIRS results with IR results for adsorbed CO on small colloidal particles [48]. These show that the linear defect species totally dominates for particles of 10Å, and at 18Å the contribution of the bridged species appears similar to that observed for the annealed 1.2ML layer (Figure 3.b). We also note that the bridged CO contribution on the small colloidal particles is dominated by the B_2 species (1938 cm^{-1}) associated with Pd(111) facets. All FT-RAIRS modes (Figure 3) of ν(C-O) result in increases in reflectivity at the oscillator frequency, indicating that even after agglomeration the particles must be sufficiently small that the local dielectric remains that of the titania

Further evidence that annealing the Pd MVD layers results in the agglomeration of Pd into islands derives from observations of a sharpening of the substrate LEED spots, a decrease in the relative Pd(3d) XPS intensity, and a shift in Pd($3d^{5/2}$) binding energy fro 334.9eV to 335.1eV [27] as a result of heating the surface. The Pd($3d^{5/2}$) binding energy after annealing to 600K is shifted still further from the bulk value of 334.5eV.Since this shift of +0.6eV from the bulk expectation is dominated by final state effects in small particles [49], the Pd particles produced on annealing must be highly dispersed. Indeed similar final state shifts have been observed for supported Pd particles of ca. 20Å diameter [50]. Due to the possible influence of the substrate and particle morphology on the final state shift, we estimate a Pd particle size in the range 15-30Å for the annealed 1.2ML MVD Pd on TiO_2(110) surface.

Similar shifts were recently observed for Rh layers on TiO_2(110), but only when unusually high dispersion was achieved using an organometallic precursor [28]. Such dispersion of Rh could not be achieved using MVD [29]. Figure 4 shows FT-RAIRS spectra for various coverages of CO adsorbed at 300K on a 1ML Rh MVD layer on TiO_2(110) after deposition at 300K (Figure 4.a) and a 3ML Rh MVD layer after annealing at 700K (Figure 4.b). The frequency of ν(C-O) on the freshly deposited layer shifts in the range 2027-2079cm^{-1} which is the same as that observed for the linearly bound species on Rh(111). The sticking

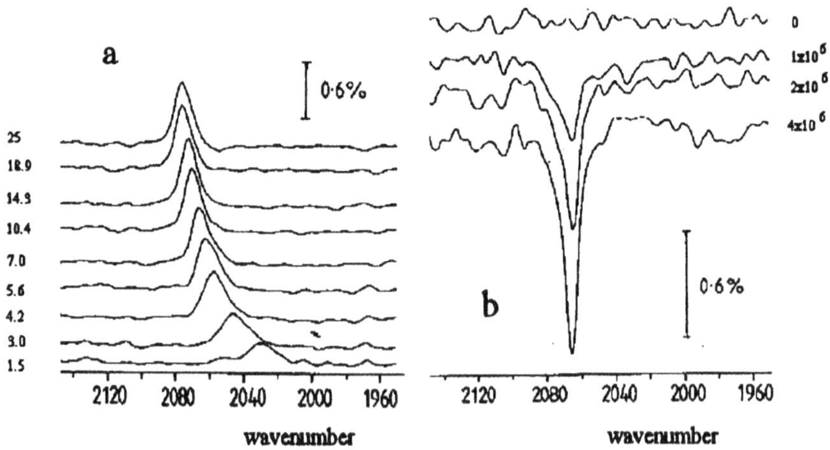

Figure 4 FT-RAIRS of CO adsorbed on a 1ML MVD Rh layer an TiO_2(110) deposited at 300K (a) and a 3ML MVD layer annealed at 700K (b). Exposures are given in Langmuirs.

probability is also similar to that found for the single crystal. The increase in reflectivity at ν(C-O) for the freshly deposited film (Figure 4.a) is indicative of a thin two dimensional Rh layer where the local dielectric is similar to titania. The frequency of ν(C-O) on a freshly deposited 20ML MVD Rh layer shifts in the same frequency range (2027-2079cm^{-1}) as that observed for the 1ML Rh layer (Figure 4.a), is observed at the same exposures, but with the band manifested as an absorption feature in FT-RAIRS. This indicates that the largely two dimensional layer now exhibits the dielectric behaviour of the metal. The absence of significant agglomeration is indicated by the absence of the behaviour observed for the annealed layer (below).

Annealing the layer to 700K (Figure 4.b) and adsorbing CO at 300K results in the observation of the same linear species but with a coverage independent frequency ν(C-O) at 2064 cm^{-1}. The band is also observed as an absorption in FT-RAIRS indicating the dominance of a metallic local dielectric. This is a result of extensive agglomeration during annealing and the generation of a Rh phase consisting of large particle sizes. Similar absorption bands are observed following annealing of MVD Rh layers of lower coverages [51]. The agglomeration is also evidenced by the reduction in Rh(3d) intensity and concomitant increase in Ti(2p) intensity in XPS during annealing [29]. Note that the linearly bound species is only observed at high CO exposures, unlike the un-annealed layers. This, and the constant frequency, is associated with the decoration of the Rh particles by oxygen after agglomeration, and the slow reaction of CO with oxygen at 300K to produce increasingly large islands of adsorbed CO of constant density [29].

FT-RAIRS spectra obtained for CO adsorbed at 300K on high coverage, ca. 100ML MVD layers of Pd, following deposition and subsequent annealing, are shown in Figure 5, and are characterised always by absorption bands [29]. During adsorption on the freshly deposited layer, CO adsorbed at 300K produces bridging bands shifting in the range 1913-1940cm^{-1} (B_2) and 1913-1994cm^{-1} (B_1), and a band at 2080cm^{-1} (L) which appears at highest coverage. The linear species observed here differs from that observed on the 1.2ML Pd layer (Figure 3.b) in that it is observed only at highest exposures and is removed by pumping. This weakly bound species is associated with densely packed CO layers on Pd (111) and (100)/(110) facets (indicated by the presence of the B_1 and B_2 species) by analogy to single crystal results [36-41]. At saturation coverage, the most intense bridging band is that associated with the B_1 species at 1994cm^{-1} associated with Pd(100)/(110) facets. A small amount of CO adsorbed on the freshly deposited film (appearing in the baseline spectrum) is responsible for the feature at 1906cm^{-1}. Annealing the surface to 700K and re-adsorbing CO at 300K again produces two bands in the bridging region (Figure 5.b) associated with the B_2 (1901-1943cm^{-1}) and B_1 species (1910-1991cm^{-1}), but with reduced intensity and a dominance of the B_2 band at saturation coverage. The intensity of the linear species 2078-2086cm^{-1} remains similar to that found at saturation on the un-annealed surface Conversion of Pd(110)/(100) to Pd(111) facets is therefore evident from FT-RAIRS for the 100ML Pd films, with a conversion of B_1 (Figure 5.a) to B_2 (Figure 5.b) absorption bands on annealing the surface to 700K.

At Pd coverages of 100ML, the LEED spots of the substrate were no longer observed, and a diffuse hexagonal Pd(111) pattern could now be observed in the absence of the $TiO_2(110)$-(1x1) spots [29]. Annealing the surface to 700K resulted in the appearance of a sharp

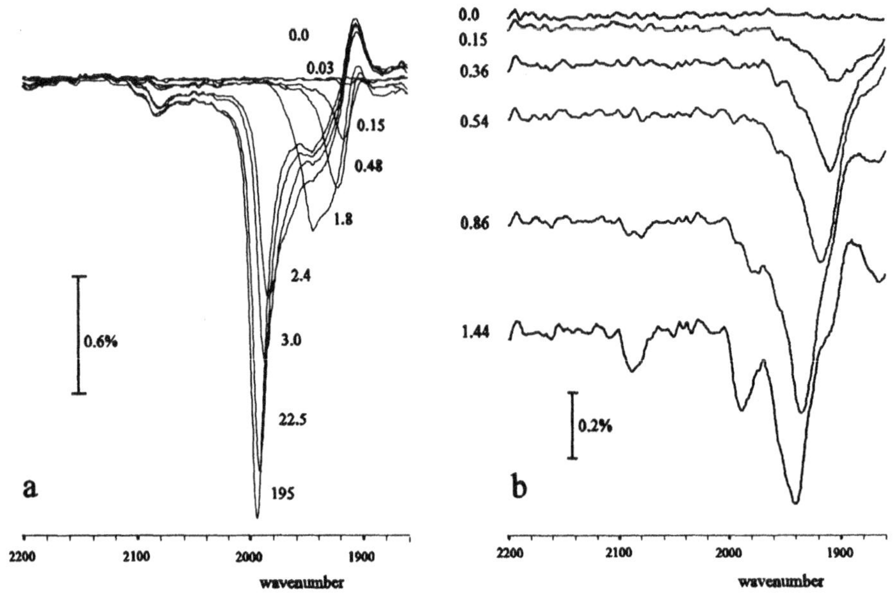

Figure 5 FT-RAIRS of CO adsorbed at 300K on a 100ML MVD Pd overlayer following deposition at 300K (a) and subsequent annealing at 700K (b). Exposures are given in Langmuirs.

hexagonal LEED pattern with distances corresponding to the Pd(111) face, and the substrate spots remained absent. The Pd($3d^{5/2}$) binding energy was 334.5eV for both the freshly deposited and annealed surfaces, although annealing the surface resulted in the appearance of Ti(2p) peaks which are totally absent following deposition of the 100ML Pd layer. The Pd(3d) peaks also reduce slightly in intensity as a result of annealing. At these coverages of Pd, agglomeration of the Pd phase takes place already at 300K to produce particles which are always sufficiently large to result in long range ordered domains of Pd(111), a dielectric response at the surface expected for the metallic phase, and a Pd($3d^{5/2}$) binding energy that corresponds to the bulk value. The large particle size is also reflected in the reduced contribution of the strongly bound linear (L*) band associated with CO adsorbed at the edge sites, and the smaller contribution of the weakly bound linear species which is observed (Figure 5) only at high coverages and exposures (or pressures) associated with CO adsorbed on the close packed surface [36-41].

The tendency to produce Pd(111) facets on the TiO_2(110) supported particles as coverage is increased, or the layer is annealed, is therefore reflected in the FT-RAIRS (Figure 5) and the LEED. Results for Pd MVD layers of intermediate coverages (10-20ML) [27] shows the superposition of the Pd(111) LEED spots on those of the titania (Figure 6) allowing a determination of the alignment of the Pd(111) with respect to the TiO_2(110) substrate. In order to observe diffraction from the Pd(111) facets, the particles must present Pd(111) facets parallel to the TiO_2(110) surface, be sufficiently large to allow coherent scattering (>ca.100Å), and must be oriented (or aligned) with respect to the TiO_2(110) surface. Such an alignment of the top layer Pd(111) with respect to the TiO_2(110) surface is depicted in Figure 6. This is similar to the influence of the anisotropic unit cell of TiO_2(110) on the growth of copper deposited by MVD [52] where the oxygen rows are important in establishing an orientated (111) super lattice in the same direction observed for Pd (Figure 6). Alignment of the Pd(111) structure in particles growing on the oxygen terminated face of ZnO(0001) is also observed

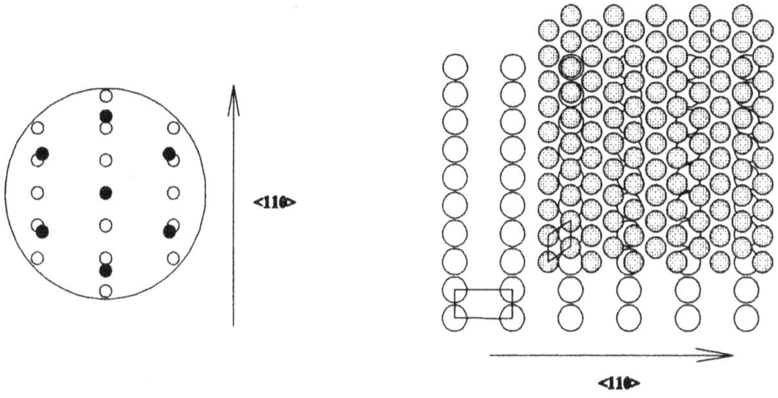

Figure 6 A schematic of the LEED pattern (○) substrate and (●) Pd (111) derived spots) for a 14ML MVD Pd overlayer on TiO_2(110) after annealing to 450K. A real space model shows the corresponding alignment of the Pd(111).

[53] at similar Pd coverages. In order to assess the extent and nature of this orientation and particle morphology we have carried out ex-situ AFM measurements [54] on the 100ML Pd film following annealing the overlayer. Figure 7 shows a typical AFM micrograph of the annealed layer which exhibited primarily Pd(111) facets in FT-RAIRS and Pd(111) diffraction in LEED. A relatively homogeneous distribution of large Pd particles is evident with a rectangular footprint of ca. 90x120 nm, and a height of ca. 25 nm. Note also the aligned orientation of the particles which is along the high symmetry directions of the TiO_2(110) face. The AFM results are therefore consistent with the LEED since the Pd particles present large

domains of what must be Pd(111) facets parallel to the $TiO_2(110)$ surface. In situ STM measurements are now under way to establish the morphology of the smaller Pd particles produced at lower coverages, and the conversion of this phase to that consisting of the larger Pd particles evident in the ex-situ AFM measurements.

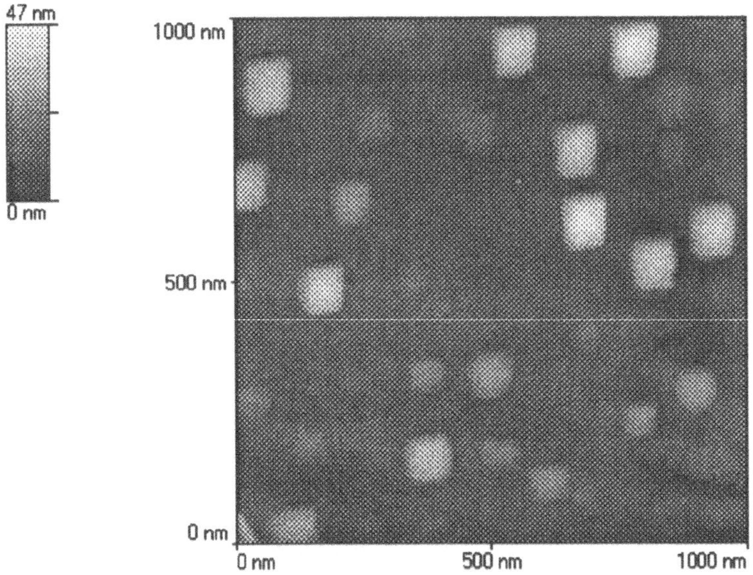

Figure 7 Ex-situ AFM of 100ML Pd MVD overlayer on $TiO_2(110)$ deposited at 300K and annealed at 700K.

4.2 MOCVD phases of Rhodium

Highly dispersed phases of rhodium can be formed on $TiO_2(110)$ using a metal organic precursor, $\{Rh(CO)_2Cl\}_2$. Evidence for the formation of the rhodium gem-dicarbonyl species $Rh(CO)_2$ is obtained from the FT-RAIRS spectra obtained [29] during the exposure of the surface to $[Rh(CO)_2Cl]_2$ at 300K, shown in Figure 8.a.1. A positive band (increase in reflectivity) shifting from 2104-2113cm^{-1} with increasing exposure is assigned to the symmetric stretch υ_{sym}(C-O) of $Rh(CO)_2$ arising from a coupling to the normal component (p_n) of the p-polarised radiation. υ_{sym}(C-O) appears at a slightly higher frequency than observed in transmission IR experiments (typically ~2095cm^{-1} [18-23,26]). The weaker absorption band at 2028 cm^{-1} is assigned to the asymmetric stretch υ_{asym}(C-O) of $Rh(CO)_2$ and arises through a coupling to the tangential component (p_t) of p-polarised radiation. It is in good

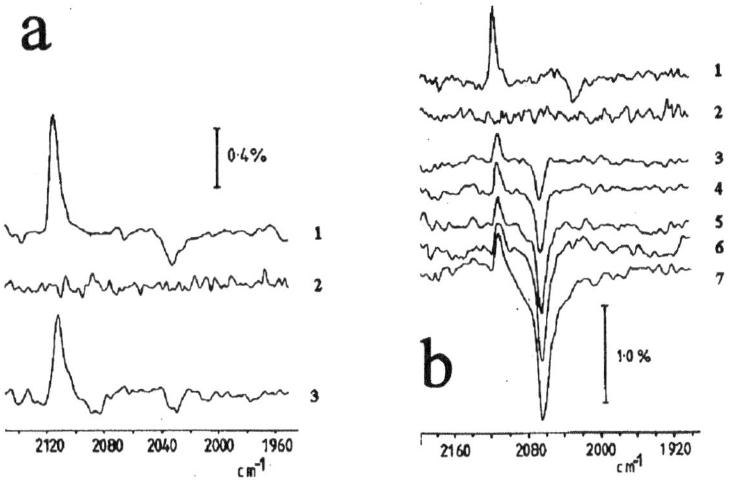

Figure 8 FT-RAIRS of the gem-dicarbonyl (**a**) produced by reaction of $\{Rh(CO)_2Cl\}_2$ with TiO_2(110) at 300K (1), following heating to 500K (2), and re-generation with 100L of CO. The gem-dicarbonyl in (**b**) is heated longer at 500K (2), and exposure to CO (3-7) only results in partial re-generation.

agreement with the expected value of ~2030 cm^{-1} [20,22]. For this highly dispersed layer of rhodium, the local dielectric remains dominated by that of the titania. If the molecule is preferentially aligned on the surface, the coupling arguments depend on the azimuthal arrangement for the FT-RAIRS. In principle the alignment and orientation of the gem-dicarbonyl can be deduced from the azimuthal dependence of the υ_{sym}(C-O) and υ_{asym}(C-O) bands [55].

A plot of the intensity of the two bands as a function of $[Rh(CO)_2Cl]_2$ exposure [29] clearly demonstrates that the two bands derive from the same surface species, and that dissociative adsorption of $[Rh(CO)_2Cl]_2$ follows Langmuir behaviour. The coverage independence of the $Rh(CO)_2$ IR vibrational frequencies has been presented as evidence for its existence as an isolated species on high area oxide supports [25,26]. The small 9 cm^{-1} frequency increase of the symmetric stretch vibrational frequency as $Rh(CO)_2$ coverage increases towards saturation in this work may arise from dynamic dipole and chemical interactions between CO atoms [33] in adjacent $Rh(CO)_2$ species similar to (but much smaller than) those observed in CO adsorption studies on Rh single crystals [56].

The gem-dicarbonyl species is characterised by a $Rh(3d^{5/2})$ binding energy of 309.1eV (Figure 10.1) and C(1s) binding energy of 287.8eV, in agreement with previous measurements

[18,57]. This is only slightly shifted from the value obtained [28] for $[Rh(CO)_2Cl]_2$ multi-layers (309.3eV) adsorbed at 150 K, suggesting that the electronic environment of the $Rh(CO)_2$ species bound to $TiO_2(110)$ via Rh-O linkages is similar to that of the parent molecule. The $Cl(2p^{3/2})$ binding energy is 198.3eV and is clearly shifted from that of the parent molecule at 199.1eV. The saturation coverage is estimated from XPS to be ~0.35ML (±0.05ML).

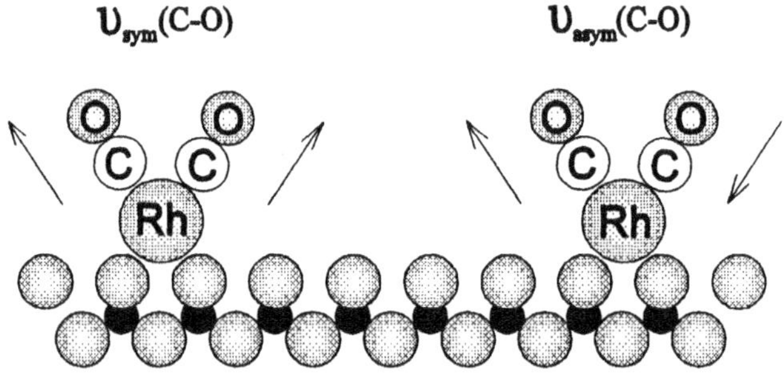

Figure 9 The symmetric and antisymmetric stretch of the $TiO_2(110)$ supported gem-dicarbonyl.

On the basis of the Cl(2p) binding energy alone, we are not able to deduce whether the chlorine ligand remains associated with the rhodium, or is adsorbed on the titania surface. However, we note that the Cl(2p) XPS spectrum is unchanged following $Rh(CO)_2$ decomposition at 450K, and secondly that $Rh(CO)_2$ can be regenerated in the absence of adsorbed chlorine following its reaction with water [29]. It is therefore unlikely that the $Rh(CO)_2$ species generated on $TiO_2(110)$ by dissociative adsorption of $[Rh(CO)_2Cl]_2$ contains a Rh-Cl bond. In addition, the reduction in symmetry from C_{2v} of $Rh(CO)_2$ to C_s of a chlorine containing species would still give rise to two IR active CO vibrational modes, but both would have components normal and parallel with the surface giving rise to a more complex coupling to the p-polarised IR radiation (Figure 2). Comparison of the Ti(2p) XPS spectrum before and after $Rh(CO)_2$ formation reveals a small reduction in intensity on the low binding energy side of the $Ti(2p_{3/2})$ peak at ~457.0eV which is indicative of Ti^{3+} defect sites [58]. This may arise from Cl chemisorption at Ti^{3+} defect sites as proposed by Jones and Hockey [59], or via a charge transfer interaction associated with the gem-dicarbonyl species.

The $Rh(CO)_2$ species decomposes via loss of CO at 450...550K which is observed in TPD [29], and in FT-RAIRS (Figure 8.a.2). where the CO band intensity is reduced in the range 450...550K, and no residual carbon is seen in XPS. $Rh(CO)_2$ decomposition is accompanied by a reduction in the $Rh(3d^{5/2})$ binding energy to 307.5eV (Figure 10.2). The Cl remains adsorbed on the $TiO_2(110)$ surface with no binding energy or intensity change observed in XPS. The adsorbed chlorine desorbs (or diffuses into the bulk) in the temperature range 600-850 K. Following the decarbonylation of the gem-dicarbonyl by heating a surface to 500 K, subsequent exposure to CO at 300 K results in relatively facile regeneration of the

$Rh(CO)_2$ species. This is evidenced in both FT-RAIRS (Figure 8.a.3) and XPS (Figure 10.3). The amount of the gem-dicarbonyl species which can be regenerated is critically dependent on both the annealing temperature, and time for which the surface is held at this temperature.

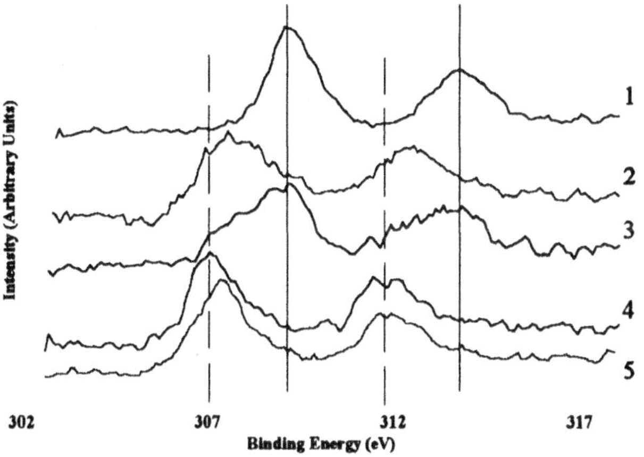

Figure 10 XPS of the Rh(3d) core level region (1) of freshly adsorbed gem-dicarbonyl, (2) following heating to 500K and (3) subsequent exposure to $6x10^6$L of CO at 300K. Heating the adsorbed gem-dicarbonyl to 800K is shown in (4) with exposure to $3x10^7$L of CO (5).

Figure 8.a.3 shows the FT-RAIRS spectrum of the gem-dicarbonyl species formed at 300K, the same surface heated briefly to 500K showing the desorption of the CO, and following exposure of the same surface to CO once re-cooled to 300K. It can be seen that CO exposure causes the re-appearance of the peaks associated with the gem-dicarbonyl at 2111cm^{-1} and 2028cm^{-1}, and about 75% of the original FT-RAIRS peak intensity is recovered.

Figure 8.b shows a similar sequence of FT-RAIRS spectra following the decomposition of the gem-dicarbonyl (Figure 8.b.1) at 500K for a prolonged period (Figure 8.b.2), and subsequently exposed to CO at 300K (Figure 8.b.3-7). The gem-dicarbonyl (transmission band at 2111cm^{-1}, the corresponding band expected at 2028cm^{-1} is too weak to be observed) is partly re-generated (ca. 35%) by exposure to CO at about 100L. At considerably higher exposures with a CO pressure of 10^{-2} Torr, an additional intense FT-RAIRS absorption band develops at ~2064cm^{-1}. This is assigned to a linear (terminal) CO species chemisorbed on Rh_x^0 particles (Rh_x^0-CO_t) and is in accordance with experimentally observed frequencies for linear CO on Rh(111) [56,60,61] of 1990...2070cm^{-1}, and on Rh/TiO_2 [62] of 2050...2063cm^{-1}. However, the initial ν_{linear}(CO) (2071cm^{-1}), and small downward frequency shift to 2064cm^{-1} with increasing exposure, together with a very low sticking probability, suggests that the CO adsorption behaviour is different to that observed on Rh single crystals [56,60,61] or the freshly deposited MVD Rh films on TiO_2(110) (Figure 4.a), but similar to CO adsorption an annealed MVD Rh layers (Figure 4.b).

The change in the Rh(3d) core level environment during the decomposition and regeneration sequence for which FT-RAIRS data are shown in Figure 8, is shown in the XPS spectra of Figure 10. The gem-dicarbonyl is characterised by a Rh($3d^{5/2}$) binding energy of 309.1eV (Figure 10.1). Following the desorption of the CO induced by heating the surface to 500K, the Rh($3d^{5/2}$) binding energy shifts to 307.5eV, and the rhodium peaks are considerably broadened (Figure 10.2). This binding energy is close to that expected for metallic rhodium at 307.1eV [63], however the small but significant difference is associated with a rhodium phase in the form of very small metallic particles probably as a result of final state effects. The broad peaks

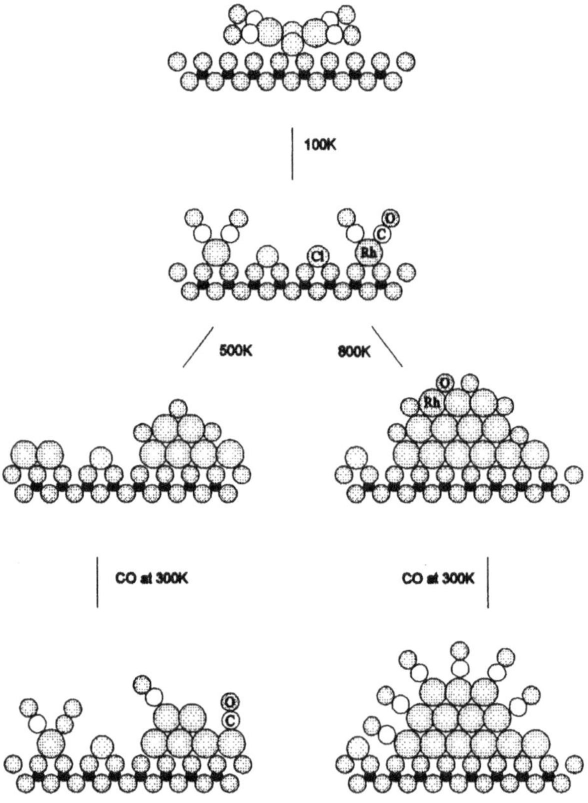

Figure 11 A schematic showing the phase changes associated with the MOCVD rhodium layers.

indicate that there is a wide distribution of particle sizes in the rhodium phase. Re-exposure of the surface to CO (Figure 10.3) is accompanied by the re-oxidation of the rhodium phase, concomitant with the observed re-generation of the gem-dicarbonyl species in FT-RAIRS (Figure 8.b.3-7). After 7.7×10^7 L CO exposure, the majority of the Rh has been re-oxidised (with the Rh($3d^{5/2}$) binding energy at 309.1eV) and is associated with the $Rh(CO)_2$ species. The linear CO is adsorbed on the remaining small metallic Rh_x^0 particles, with a relatively un-shifted Rh($3d^{5/2}$) binding energy (Figure 10.3). The rhodium phase changes are summarised schematically in Figure 11. The C(1s) XPS intensity increases quickly during the formation of the gem-dicarbonyl [29], however, no additional C(1s) peak intensity can be observed during the slow adsorption of the linear CO species. Hence the surface coverage of this species is therefore estimated to be substantially less than 0.04ML (detectivity limit of the XPS in the C(1s) spectral region). The intense FT-RAIRS absorption band associated with this low surface concentration (Figure 8.b.3-7) must be due to the fact that the small metallic rhodium particles are nevertheless sufficiently large to exhibit the dielectric characteristics of a bulk metallic phase. Comparison of the band intensities for CO on the metal phase to that associated with the gem-dicarbonyl simply highlights the low sensitivity of an isolated species dominated in the FT-RAIRS measurement by the dielectric behaviour of the oxide surface.

Heating the rhodium gem-dicarbonyl phase to 800K produces larger rhodium particles. The Rh(3d) core level environment following heating to 800K (Figure 10.4) is more homogeneous (narrower peaks) relative to the spectrum obtained after heating to 500K, with the Rh($3d^{5/2}$) binding energy shifted to that expected [63] of the metallic phase at 307.1eV.The FT-RAIRS indicates [29] that re-exposure of the surface to CO is unable to re-generate the gem-dicarbonyl with CO adsorption taking place only on the metal particles. The XPS spectrum (Figure 10.5) confirms that during exposure to CO, the rhodium remains in the reduced phase during CO adsorption, although at high coverages of CO on the metal particles, a very small shift in Rh($3d^{5/2}$) binding energy to 307.3eV is evident, which is a result of the electron withdrawing effect (Rh(3d) ---> CO(2π)) of the CO [64] on the small particle. The phase changes associated with the higher temperature 800K annealing of the surface are also summarised in Figure 11.

The variation in the ν(C-O) Rh_x^0-CO_t peak area in FT-RAIRS is linear with respect to CO exposure [29] and the peak does not saturate within the exposure range investigated. As in the case of the annealed Rh MVD phase (section 4.1), the rate of growth of this linear CO_t species on the metal particles produced by the MOCVD route is too slow (the sticking probability too low) to be accounted for by the direct adsorption on metallic rhodium, where the sticking probability is significantly higher (ca. 0.75 on Rh(111) [63]). The small coverage dependent frequency shift for CO on these phases (Figure 4b, 8b) is also unlike the large upward shift observed on the rhodium single crystal [56,62], and the un-annealed rhodium layers deposited by MVD (Figure 4.a) which is associated with static and dynamic lateral interactions. This behaviour is not consistent with a reduction in the number of adsorption sites arising from Rh agglomeration or with direct site blocking by a mobile TiO_x species [65]. A mechanism involving migration of mobile oxygen atoms during heating is proposed [29]. Oxygen migration requires a lower activation energy than oxygen desorption from TiO_2 (110). This correlates with the presence of Ti defects detected by XPS after sample heating to 800 K. We also note similar behaviour in the suppression of CO adsorption previously observed on oxidised Rh/Al_2O_3 catalysts [66]. These oxygen atoms then block CO adsorption sites until they are removed by reaction with CO to form CO_2. Subsequently, facile CO chemisorption can occur giving rise to the observed CO FT-RAIRS frequencies which are indicative of either

CO adsorbed in islands of high local coverage or CO adsorbed on a partially oxygen covered surface [56]. We observe a larger **reduction** in υ_{linear}(C-O) frequency during the **desorption** of the CO from the Rh^0_x particles (not shown), consistent with desorption from oxygen free areas of a clean surface (Figure 4.a). Oxygen migration and reaction with CO has also been observed on Fe/TiO_2(110) [67] and more recently on Rh/CeO_2 catalysts after annealing at temperatures as low as 400 K [68]. The observed slow rate of reaction at 300K is consistent with that found [69] for CO oxidation on Rh single crystals.

It has been proposed that CO induced oxidative disruption of Rh_x^0 to $Rh(CO)_2$ follows either a mechanism involving Rh oxidation by oxygen atoms from dissociated CO molecules [18,19,70,71] or oxidation by surface hydroxyl groups [20-23,62]. We observe regeneration of the gem-dicarbonyl species at similar pressures to those required on high area powder TiO_2 [62] without prior observation of the Rh_x^0-CO_t species. There is no evidence that surface OH groups are required for the regeneration. The regeneration of the $Rh(CO)_2$ species is fast compared to other studies on TiO_2 [62] and Al_2O_3 [20,21,72] in which linear CO is detected prior to $Rh(CO)_2$ formation and a mechanism involving gradual disruption of large Rh_x^0 crystallites has been proposed. $Rh(CO)_2$ regeneration is complete after the first CO exposure of 1.2×10^6L in Figure 8.b.3; it is therefore likely that facile CO adsorption takes place on relatively small Rh_x^0 particles whilst larger particles above a critical size are not disrupted by CO at 300 K. The absence of any surface carbon in XPS, and the observation of clean molecular CO desorption in this work, rules out any mechanism involving CO dissociation on TiO_2(110). It has been proposed [29] that a mechanism involving the stabilisation of the dispersed rhodium by surface defects (oxygen vacancies) [73] on the TiO_2(110) surface. The relatively high mobility of oxygen atoms and known redox behaviour of Ti^{4+} could provide sites which can accommodate either Rh^I or Rh^0. Van't Blik et al [18,19] have recently proposed that a titanium redox coupling is important in the oxidative disruption of Rh particles on high area titania. Since chlorine was present on the TiO_2(110) surface during re-generation of the Rh gem-dicarbonyl species, it was important also to establish [29] that similar re-generation on TiO_2(110) takes place in the absence of chlorine.

4.3 The reaction of Rh gem-dicarbonyl with hydrogen.

The reaction of hydrogen with chemisorbed CO on supported rhodium is of particular significance from the standpoint of a variety of catalytic reactions. For example, oxide supported rhodium exhibits a high activity towards CO and CO_2 hydrogenation, particularly on titania [74]. It is not surprising that a number of papers have focused on this aspect [8,20,21,23,75,76].

Exposure of the gem-dicarbonyl species to hydrogen at 300K leads to the loss of the CO. This process can be followed by both FT-RAIRS and XPS [29]. Exposure to ca. 10^3L of hydrogen leads to a significant loss in the v_{sym}(C-O) and v_{asym}(C-O) band intensities of the gem-dicarbonyl, and following an exposure of ca. 10^6L, the bands can no longer be detected. XPS indicates that during the hydrogen exposure of 10^6L, a concomitant shift in the $Rh(3d^{5/2})$ binding energy from 309.1eV to 307.4eV takes place, producing a peak with half width between those found following reduction by heating to 500K (Figure 10.3) and 800K (Figure 10.4). Unlike the thermolysis, the process giving rise to the hydrogen induced removal of CO, and reduction of the rhodium, is not associated with a simple decarbonylation of the gem-dicarbonyl [29]. The XPS of the C(1s) core level during the hydrogen reaction at 300K indicates that the CO ligands have been removed, but a residual C(1s) peak at a binding energy of 284.5eV indicates that between 35 ... 50% of the original CO has been reduced to a graphitic or carbidic form which remains on the surface. The precise composition of the residual carbon is unknown, but it clearly results from a disproportionation of a significant

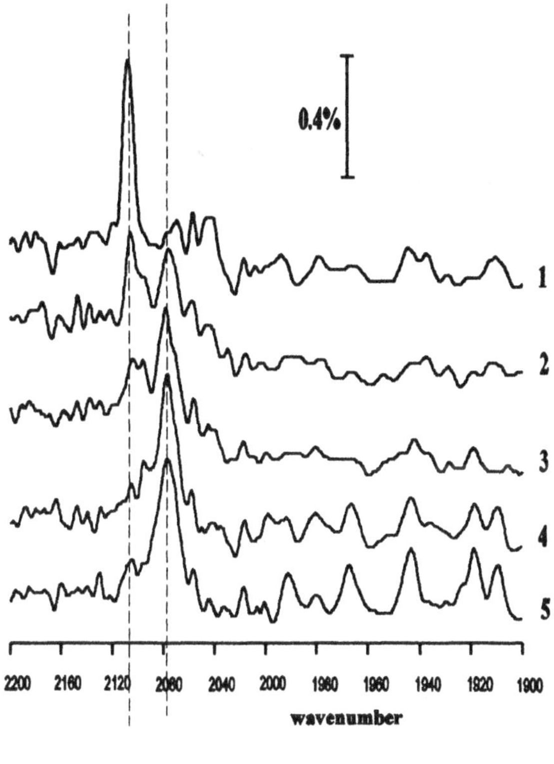

Figure 12 FT-RAIRS of the reaction of the $TiO_2(110)$ supported Rh gem-dicarbonyl with hydrogen: (2) $6x10^6$L; (3) $1.2x10^7$L; (4) $2.4x10^7$L; (5) $3.6x10^7$L.

fraction of the $TiO_2(110)$ supported gem-dicarbonyl CO ligands. CO exposure at 300 K of the hydrogen reduced surface leads to only partial regeneration of the gem-dicarbonyl species (ca. 30%), as evidenced in FT-RAIRS and XPS [29].

Although metallic rhodium particles are produced during the reaction with hydrogen, the route to carbon production does not appear to involve CO dissociation on Rh_x^0. Such a process was never observed on $TiO_2(110)$ supported particles of any size [29]. A competing reaction to the hydrogen induced displacement of CO appears to take place at the gem-dicarbonyl which can result in residual carbon. Originally [29] no intermediate in the reaction of hydrogen with the $TiO_2(110)$ supported Rh gem-dicarbonyl could be identified at 300K, such as RhH(CO) [21,23,75,77] or $RhH_2(CO)$ [21]. More recently , however, we have been able to carry out the same reaction and observe an intermediate species in FT-RAIRS [78]. Figure 12 shows a sequence of FT-RAIRS spectra recorded during the reaction. The spectrum characteristic of the gem-dicarbonyl (Figure 12.1) is replaced by the appearance of a single positive (increase in reflectivity) band at 2078cm^{-1}. This is associated with the intermediate RhH(CO) or $RhH_2(CO)$ on the basis of the ν(C-O) frequency and XPS [78]. The retention of the local titania dielectric is consistent with the suggestion of this highly dispersed intermediate.

The facile reaction of the gem-dicarbonyl with hydrogen on $TiO_2(110)$ suggests a possible direct mechanism to hydrogenated products involving the dispersed phase of rhodium in the form of the gem-dicarbonyl. Carbon is also a product of the hydrogenation (see [74] and references therein). In addition the regeneration of the Rh gem-dicarbonyl demonstrates a feasible catalytic cycle involving this highly dispersed phase. Whilst $Rh(CO)_2$ is observed to disappear at lower temperatures in the presence of H_2 on high area catalysts [8,20,21,27,75], some authors [8,75,79] conclude that it is not directly hydrogenated, in contrast to the linear and bridge bound CO species adsorbed on the metal particles. On $TiO_2(110)$, we clearly see that $Rh(CO)_2$ reacts to produce an intermediate [78] which subsequently decomposes directly to produce a phase of carbon which could be an intermediate to hydrogenation products at slightly elevated temperatures. We observe no reaction of the hydrogen with linear Rh_x^0-CO_t species under identical conditions [29].

5. Conclusions

FT-RAIRS of the probe molecule CO (particularly when combined with XPS, TPD and LEED) is a powerful tool in characterising the phases of metals such as Pd and Rh, and studying their inter-conversion, when supported on oxide single crystal surfaces. The local dielectric of the titania dominates for small particles and thin films resulting in both positive (increased reflectivity) and negative (absorption) bands being observed in p-polarised FT-RAIRS measurements through a coupling of the normal and tangential components of the radiation. Large metal particles or thick films result in a local dielectric which is that of the metal, resulting in strong absorption bands. The CO fingerprint also enables the characterisation of the facets and edge site of the particles.

MVD Pd layers produce a thin film at low coverages which exhibits primarily (100)/(110) facets. Annealing the layer produces small particles with a high density of edge sites and a small proportion of (111) facets. High Pd coverages result in the agglomeration of the film at 300K to produce larger particles with (111) but mainly (100)/(110) facets, but annealing results in a conversion to large particles exhibiting primarily (111) facets, observed by FT-RAIRS and LEED. MVD Rh layers produce films exhibiting less agglomeration at 300K, and a morphology independent of Rh coverage. Annealing Rh MVD layers of all coverages produces large Rh particles which are decorated by oxygen derived from the $TiO_2(110)$ support.

High dispersion of Rh can only be achieved using MOCVD. The precursor $\{Rh(CO)_2Cl\}_2$ reacts with the $TiO_2(110)$ surface to produce the Rh gem-dicarbonyl species $Rh(CO)_2$. This is characterised in FT-RAIRS through observation of both ν_{sym}(C-O) and ν_{asym}(C-O). Heating to 500K results in decarbonylation, producing particles small enough to allow regeneration of the gem-dicarbonyl through exposure to CO at 300K. Heating to higher temperature results in the production of larger particles analogous to those produced by MVD which cannot be converted to the gem-dicarbonyl species. Once produced, the reactivity of the $TiO_2(111)$ supported gem-dicarbonyl can be investigated, and reaction intermediates identified in FT-RAIRS.

6. References

[1] M.Ichikawa; Advan.in Catalysis **38** (1994) 283.
[2] J.H.Sinfelt.; Adv.Catal. **23** (1973) 91.
[3] R.Pitchai, S.S.Wong, N.Takahashi, J.B.Butt, R.L.Burwell Jr. and J.B.Cohen; J.Catal., **94** (1985) 478.
[4] Z.Karpinski, J.T.Butt and W.M.H.Sachtler; J.Catal., **119** (1989) 478.
[5] S.Gao and L.S.Schmit; J.Catal., **115** (1989) 356.
[6] A.Sarkany, A.H.Weisse and L.Guczi; J.Catal **98** (1986) 550.
[7] G.A.DelAngel, B.Coq, G.Ferrat, F.Figueras and S.Fuentes; Surf.Sci. **156** (1985) 943.
[8] F. Solymosi, M. Tobacz, M.J. Koscis; J. Catal. **75** (1982) 78.
[9] Yu.A.Ryndin, R.F.Hicks and A.T.Bell; J.Catal., **70** (1981) 287.
[10] F.R.Hartley Ed., "Chemistry of the Platinum Group Metals"; Elsevier Science Publishers (1991).
[11] F. Solymosi, J. Starkany; Appl. Surf. Sci. **3** (1979) 68.
[12] D.K. Paul, T.H. Ballinger, J.T. Yates Jr.; J. Phys. Chem. **95** (1991) 1699.
[13] D.K. Paul, T.H. Ballinger, J.T. Yates Jr.; J. Phys. Chem. **94** (1990) 4617.
[14] F.Fajula, R.G.Anthony and J.H.Lunsford; J.Catal., **73** (1982) 237.
[15] Z.Karpinski; Adv.Catal., **37** (1990) 45.
[16] T.G.Rucker, M.A.Logan, T.M.Gentle, E.L.Muttertles and G.A.Somorjai; J.Phys.Chem., **90** (1986) 2703.
[17] H.C. Yao, Y.F. Yu, K. Otto; J. Catal. **56** (1979) 21.
[18] H.F.J.Van't Blik, J.B. van Zon,, T. Huizinga, J.C. Vis, D.C. Koningsberger, R. Prins; J. Am. Chem. Soc. **107** (1983) 3139.
[19] H.F.J. Van't Blik, J.B. van Zon, T. Huizinga, J.C. Vis, R. Prins; J. Phys. Chem. **87** (1983) 2264.
[20] F. Solymosi, M. Pasztar; J. Phys. Chem. **89** (1985) 4789.
[21] F. Solymosi, M. Pasztar; J. Phys. Chem. **90** (1986) 5312.
[22] P. Basu, D. Panayotov, J.T. Yates Jr.; J. Phys. Chem. **91** (1987) 3133.
[23] P. Basu, D. Panayotov, J.T. Yates Jr.; J. Am. Chem. Soc. **110** (1988) 2074.
[24] R.P.Eichens and W.A.Pliskin; Advan.Catal., **10** (1958) 1.

[25] A.C. Yang, C.W. Garland; J. Phys. Chem. **612** (1957) 1504.
[26] J.T. Yates Jr., K. Kolanski; J. Phys. Chem. **79** (1983) 1026.
[27] J.Evans, B.E.Hayden and G.Lu; Surf.Sci. in press (1996).
[28] J. Evans, B.E. Hayden, F. Mosselmans, A. Murray; J. Am. Chem. Soc. **114** (1992) 6912.
[29] J. Evans, B.E. Hayden, F. Mosselmans, A. Murray; Surf. Sci. Lett. **279** (1992) L159.
J. Evans, B.E. Hayden, F. Mosselmans, A. Murray; Surf. Sci.**301** (1994) 61.
[30] J.Evans, B.E. Hayden, F. Mosselmans, A. Murray; NATO ASI Series, "Elementary Reaction Steps in Heterogeneous Catalysis", Ed. R.W.Joyner and R.A.van Santen, (1993) 179.
[31] M.Volmer and A.Weber; Z.Phys.Chem., **119** (1926) 277.
[32] M.P.Sear amd W.A.Dench; Surf.Interface Anal., **1** (1979) 2.
[33] B.E.Hayden; "Methods of Surface Characterisation", Plenum, Ed. J.T.Yates and T.E.Madey, **4** (1987) 267.
[34] Y.Chabal; Springer Proc.in Phys., **22** (1987) 301.
Y.Chabal; Surf.Sci.Rep., **8** (1988) 211.
[35] J.Evans, B.E.Hayden and G.Lu; J.Chem.Soc.Faraday Trans. in press (1996).
[36] A.M.Bradshaw and F.M.Hoffmann; Suf.Sci., **72** (1978) 513.
[37] A.Ortega, F.M.Hoffmenn and A.M.Bradshaw; Surf.Sci., **119** (1982) 79.
[38] M.A.Chesters, G.S.McDougall, M.E.Pemble and N.Sheppard; **164** (1985) 425.
[39] R.Raval, S.Haq, M.A.Harrison, G.Blyholder and D.A.King; Chem.Phys.Letts., **167** (1990) 391.
[40] W.K.Kuhn,J.Szanyi and D.W.Goodman; Surf.Sci., **274** (1992) L611.
[41] J.Szanyi, W.K.Kuhn and D.W.Goodman; J.Vac.Sci.Tech., **11** (1993) 1969.
[42] R.F.Hicks and A.T.Bell; J.Catal., **90** (1984) 205.
[43] P.Gelin, A.R.Siedel and J.T.Yates; J.Phys.Chem. **88** (1984) 2978.
[44] V.Matolin and E.Gillet; Surf.Sci., **238** (1990) 75.
[45] J.Szanyi and D.W.Goodman; J.Phys.Chem., **98** (1994) 2972.
[46] J.Szanyi, W.K.Kuhn and D.W.Goodman; J.Phys.Chem., **98** (1994) 2978.
[47] F.M.Hoffmann; Surf.Sci.Reports, **3** (1983) 107.
[48] J.S.Bradley, J.M.Millar, E.W.Hill, S.Behal, B.Chaudret and A.Duteil;Faraday Discuss., **92** (1991) 255.
[49] L.Hilaire, P.Legare, Y.Holl and G.Maire; Sol.Stat.Comm., **32** (1979) 157.
[50] Ch.Kuhrt and M.Harsdorff; Surf.Sci. **245** (1991) 173.
[51] A.Murray; PhD Thesis, Southampton (1993).
[52] P.J.Moller and M.-C.Wu; Surf.Sci., **224** (1989) 265.
[53] J.W.M.Jacobs and D.Schryvers; J.Catal., **103** (1987) 436.
[54] J.Evans, B.E,Hayden and I.Nandhakumar; Surf.Sci., to be submitted.
[55] B.E.Hayden and A.King; to be published.
[56] L.H. Dubois, G.A. Samorjai; Surf. Sci. **91** (1980) 514.
[57] B.G. Frederick, G. Apai, T.N. Rhodin; J. Am. Chem. Soc. **109** (1987) 4797.
[58] N. Beatham, A.F. Orchard, G. Thornton; Phys. Chem. Solids **42** (1981) 1051.
[59] P. Jones, J.A. Hockey; Trans. Faraday Soc. **67** (1971) 2669.
[60] J.E. Crowell, G.A. Samorjai; Appl. Surf. Sci. **19** (1984) 73.
[61] F.T. Wagner, T.E. Moylan, S.J. Schmieg; Surf. Sci. **195** (1988) 403.

[62] D.A. Buchanan, M.E. Hernandez, F. Solymosi, J.M. White; J. Catal. **125** (1990) 456.
[63] L.A. Delouise, J. White, E. Winograd; Surf. Sci. **147** (1984) 252.
[64] G. Apai, S.T. Lee, M.G. Mason, L.J. Gerenser, S.A. Gardner; J. Am. Chem. Soc. **101** (1980) 6880.
[65] D.N. Belton, Y.M. Sun, J.M. White; J. Catal. **102** (1986) 338.
[66] R.R. Cavanagh, J.T. Yates; J. Chem. Phys. **74** (1981) 4150.
[67] J. Deng, D. Wang, R. Zhai, H. Wang; Surf. Sci. **249** (1991) 213.
[68] G.S. Zafris, R.J. Gorte; J. Catal. **139** (1993) 561.
[69] M. Bowker, Q. Guo, R.W. Joyner; Surf. Sci. **280** (1983) 50.
[70] M. Primet; J. Chem. Soc. Faraday Trans. **74** (1979) 2570.
[71] G. Bergeret, G. Gallezot, P. Gellin, B. Taarit, Y. Fefebre, C. Naccache, R.D. Shannon; J. Catal. **104** (1987) 279.
[72] F. Solymosi, H. Knozinger; J. Chem. Soc. Faraday Trans. **86** (1990) 389.
[73] W.Gopel, J.A.Anderson, D.Frankel, M.Jaehnig, K.Phillips, J.A.Schafer, G.Rocker; Surf. Sci. **139** (194) 333.
[74] A. Erdoheyl, F. Solymosi; J. Catal. **84** (1984)446.
[75] S.D. Worley, G.A. Matteson, R.J. Caudhill; J. Chem. Phys. **87** (1983) 1671.
[76] J.C. Conesa, M.T. Sainz, J. Soria, G. Muneura, V. Rives Arnau, A.J. Munoz, J; Mol. Catal. **17** (1992) 231.
[77] J.T. Kiss, R.D. Gonzalez; J. Phys. Chem. **92** (1988) 1589.
[78] B.E.Hayden and A.King; to be published.
[79] K. Fujimoto, M. Kameyama, T. Kunugi; J. Catal. **61** (1980) 7.

ANGLE-SCANNED PHOTOELECTRON DIFFRACTION

A structural probe for near-surface atomic layers

GAETANO GRANOZZI AND MAURO SAMBI
Dipartimento di Chimica Inorganica, Metallorganica ed Analitica, Università di Padova, via Loredan 4, 35131 Padova (Italy)

Abstract

Electrons emitted from the core levels of a photon-irradiated crystalline sample undergo scattering by atoms in the vicinity of the emitting species. Subsequent interference phenomena between the electron wave portions propagating to the detector produce intensity modulations as a function of the direction of detection. This process constitutes the physical basis of the angle-scanned X-ray Photoelectron Diffraction (XPD) technique. The resulting modulations, properly interpreted, are rich in structural information concerning the near-surface atomic layers.

In the present notes, after an introduction to the principles of XPD, some selected results in the field of chemisorption and ultrathin epitaxial films will be reported in order to outline the merits of the technique. Qualitative structural information (e.g. growth modes and lattice distortions, orientation of chemisorbed molecules) can be directly obtained from the experimental raw data without any need of theoretical simulation. On the other hand, quantitative structural parameters and chemisorption sites may be deduced by using a trial-and-error fitting procedure based on simple single-scattering models.

R.M. Lambert and G. Pacchioni (eds.),
Chemisorption and Reactivity on Supported Clusters and Thin Films, 237-266.

1. Introduction

The knowledge of the structure of the near-surface atomic layers is a crucial step towards the full comprehension of the phenomena occurring at the surfaces of solids. This goal is of strategic importance in fields such heterogeneous catalysis, material science and for developing nanodevice technologies.

In the case of chemisorption, the relevant structural information is related with the geometry of both the chemisorption site and the chemisorbed molecule. Evidence is accumulating in the literature which points out that the structural perturbations induced by chemisorption are not limited to the outmost atomic layer of the crystalline substrate. On the other hand, passing to the field of ultrathin epitaxial growth, it is now well established that small structural changes within the outmost 10-20 monolayers (MLs), with respect to the standard structural parameters of the corresponding bulk phase, are effective in promoting surprisingly new chemical, electrical and magnetic properties of the new material. Hence, it is evident that for a full understanding of both chemisorption and epitaxial systems it is important to have at hand a tool capable of detecting the structural parameters of a system formed by stacking atoms up to 10 MLs in thickness.

This is not a trivial task if one considers the most common surface structure experimental techniques. Low Energy Electron Diffraction (LEED) can provide very detailed quantitative structural information concerning the outermost surface atomic layers, but it is limited by the requirement of long-range translational order along the surface, which is not always fulfilled in epitaxial and chemisorption systems. Scanning Tunneling Microscopy (STM), is mostly sensitive to the structure of the topmost layer and it is not atom-specific, so it can hardly provide the desired information. Electron Microscopies (e.g. TEM and STEM) can provide the structural data, but they are strongly limited by the sample preparation requirements which restrict the number of the investigable samples. Standard X-ray Diffraction (XRD) is a bulk-sensitive technique, although with a glancing-angle geometry the structural information from the outer atomic layers can be obtained.

The ideal technique would be a non-destructive one, sensitive to short-range order and possibly atom- and site-specific in order to probe the environment of each component of the surface layers. To provide the structural information this ideal technique should rely on some elastic scattering phenomenon. The electron-atom scattering possesses the required demands: it is non-destructive and, due to concurrent inelastic attenuation, it is a short-range phenomenon.

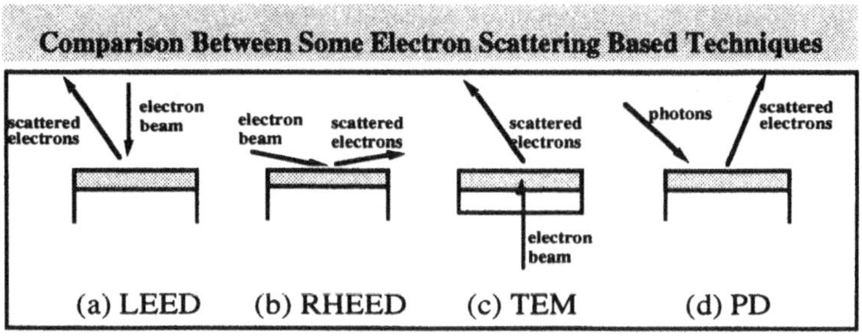

Figure 1. (a) Low Energy Electron Diffraction (LEED), (b) Reflection High Energy Electron Diffraction (RHEED) and (c) Transmission Electron Microscopy (TEM) differ in the energy of the primary electron beam and in the geometry of the experimental set-up. In (d) Photoelectron Diffraction (PD) the source of electrons is originated *in situ* by the photoemission process.

In a standard electron diffraction experiment, both the source of electrons and the detector are positioned outside the sample: in LEED a beam of external electrons in the 20-300 eV range is used and the diffraction pattern detected is generated by the back-scattered components of the diffused electron waves (Figure 1a); in the closely related Reflection High Energy Diffraction (RHEED) experiments, the energies of the electron beam are in the order of some tens of keV, and the incident and exit angles are grazing (Figure 1b). In TEM a beam of high energy electrons (in most cases in the range 10-120 keV) is transmitted through a very thin sample (< 200 nm thick) (Figure 1c). Alternatively, the source of electrons can be internal to the sample, generated *in situ* by some excitation process. This is the case of photoemission, and the technique based on the diffraction phenomena originated by such photoelectrons is called Photoelectron Diffraction (PD) (Figure 1d). As will be clear in the next paragraph, PD is related to the well known X-ray Photoelectron Spectroscopy (XPS) and has the advantage of satisfying the mentioned requirements of atom- and site-specificity. In fact, PD has been applied with success to follow the epitaxial growth of thin and ultrathin films and to study the chemisorption processes on single crystals [1-3].

In the present notes, we first introduce the principles of the angle-scanned version of PD, and then report some selected results in the field of chemisorption and ultrathin epitaxial films in order to outline the merits of the technique. Due to the tutorial character of this notes, the focus will be on the information obtainable by PD and details on the scattering model formalisms will be omitted. However, the interested student can refer to some excellent reviews on PD which cover all the aspects of the theory and applications [1-5]. At the end we will also briefly discuss some future trends and perspectives.

2. Basic principles of Photoelectron Diffraction

Electrons emitted from the core levels of a photon-irradiated crystalline sample (either photoelectron or Auger electrons) undergo scattering by atoms close to the emitting species. Subsequent interference (Figure 2) between the primary electron wave Ψ_0, which originates at the emitter and propagates to the detector along the **k** direction without undergoing any elastic scattering, and the secondary or scattered electron waves Ψ_j, which are caused by the scattering of the primary wavefront at atoms (scatterers) surrounding the emitter, produces strong intensity modulations as a function of either the direction of detection **k** or the kinetic energy (E_k) of the emitted electron. These modulations can be as high as 70% of the signal, and an overall anisotropy is commonly defined as $\chi=(I_{max}-I_{min})/I_{max}$. This fundamental process constitutes the physical basis of the PD technique. The resulting modulations, properly interpreted, are rich of structural information concerning the near-surface atomic layers.

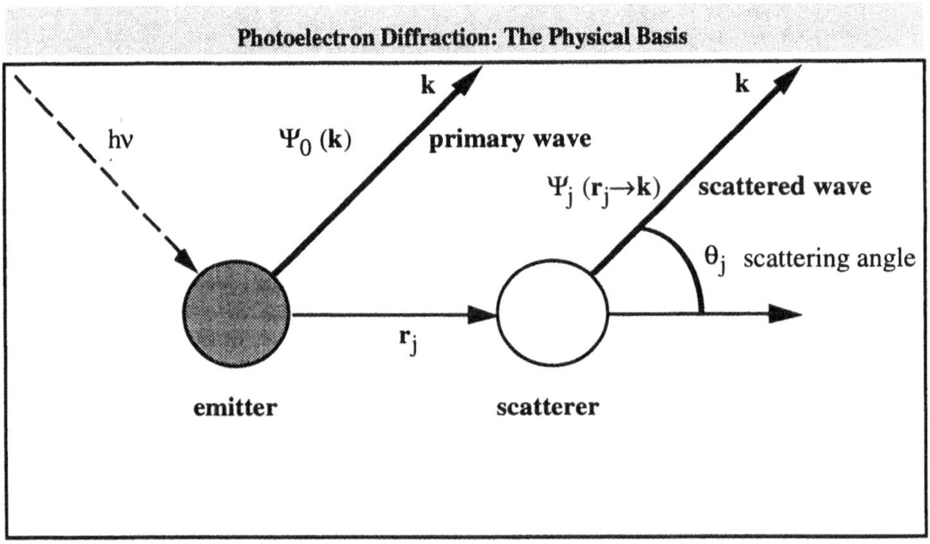

Figure 2. The primary wave Ψ_0, originated at the emitter by the photoemission process, reaches the detector without undergoing any scattering event. The secondary wave Ψ_j is produced by scattering of the primary wavefront at the jth scatterer at a distance r_j from the emitter. θ_j is the scattering angle. The interference between the primary and secondary waves at the detector (positioned along the direction **k**) gives rise to the intensity oscillations seen in a PD pattern.

Traditionally, PD has been carried out by two different methodologies, either scanning the direction of detection **k**, referenced as angle-scanned PD (or X-ray Photoelectron Diffraction, XPD), or scanning the E_k of the emitted electron by using a tunable photon source, referenced as energy-scanned PD. The latter is often cited as PhotoElectron Diffraction (PED) or Angle Resolved PhotoEmission Fine Structure (ARPEFS) in order to emphasize its similarity to the more familiar Extended X-ray Absorption Fine Structure (EXAFS) technique. When the electrons involved in the process are photoexcited Auger electrons, the technique is cited as Auger Electron Diffraction (AED). Nowadays the distinction between the two different approaches is fading away since most of the PD experiments carried out on synchrotron beamlines combine the two angle-scanned and energy-scanned methodologies, especially in connection with the new field of photoelectron holography [6, 7]. It has been, in fact, demonstrated that intensity distributions of Auger and photoelectrons can be considered as holograms, and that such holograms can be inverted using Fourier-transform methods to yield images of atomic positions. Further discussion on this innovative and highly exciting aspects of PD is out of the scope of this lecture and the interested student is directed to the pertinent literature [1d, 2b, 8].

In the following we will only review very briefly the basic principles and the main peculiarities of the angle-scanned version of the PD technique. The discussion will be kept as far as possible on qualitative grounds, and it will be restricted only to the concepts which are needed to understand the reported examples. Principles and applications of energy-scanned PD are described in another chapter of this book.

PD is a surface structural probe sensitive to short-range order, with atomic, chemical state and site specificity. The short-range order surface sensitivity is due to the lack of coherence between photoelectron waves emitted at different atomic sites because of the random nature of the emission process [2b], coupled with the 1/r decay of the outgoing photoelectron wave and with the relatively short electron inelastic mean free paths at the energies typical for a PD experiment. On the other hand, to obtain an XPD pattern a long-range ordered sample is needed in order to sum up the contributions from many equivalent emitters. If the local structural environment varies from emitter to emitter, the final modulation would be averaged to zero. In this respect XPD has a different application field with respect to the closely related technique of Extended X-ray Absorption Fine Structure (EXAFS), which is especially useful to study the short-range order in amorphous samples.

The atomic and chemical state specificity is easily understood if one considers that XPD is an implementation of the conventional X-ray Photoelectron Spectroscopy (XPS). In fact, an XPD curve is obtained experimentally simply by singling out any spectral feature from an XPS spectrum of a crystalline sample, and by monitoring its intensity as a function of the emission direction **k** (Figure 3). Ordinarily, the angle-scanned experiments are carried out by moving the sample on a suitable manipulator. The geometry and the notation usually adopted for the angles are reported in Figure 3. Two basic types of measurements are possible, in which either the polar angle (Θ) or the azimuthal angle (ϕ) is scanned while keeping fixed all the other parameters; polar scans and azimuthal scans are thus obtained, respectively. XPD curves can therefore be obtained for different core levels of the same element, of various elements present on the surface, and, at higher energy resolution, even of different chemical states of the same element. Coupled with such chemical specificity of the structural information, there is also the capability of observing the very surface and near surface region of the sample (up to thicknesses of 20-30 Å), due to the limited escape depths of photoelectrons at the kinetic energies usually employed (typically ~400-1500 eV). Going to grazing angles of photoelectron emission yields even higher surface sensitivity [1].

One inherent advantage of XPD with respect to other surface structural techniques is that, in the 400-1500 eV E_k range, the electron-atom scattering process is a relatively simple phenomenon, dominated by forward-scattering (FS) processes [1]. This is well illustrated by Figure 4, where the typical behaviour of the scattering amplitude $|f(\theta)|$ vs the scattering angle θ for different values of E_k of the incident electron is shown. It is evident that, for E_k above a few hundred eV, the only significant scattering amplitude is in the forward direction ($\theta \approx 0°$). As a consequence, polar and/or azimuthal scans for sufficiently energetic electrons furnish the information on bond orientations directly, since in this case the intensity is enhanced when the angle between the emitter-scatterer and the emitter-detector directions is close to 0°, i. e. whenever a scatterer intersects the emitter-detector direction. This 0th order interference effect, often called *forward-focusing,* is a very effective tool for the determination of internuclear directions and it has been extensively and successfully exploited for the determination of the orientation of chemisorbed molecules and in studies of epitaxial growth [1-3]. The examples discussed in the next section will be mainly based on the FS-XPD technique and the relevant structural information is mostly obtained from the raw XPD data.

On the other hand, such 0th order FS effects do not give any information on interatomic distances, because when $\theta \approx 0°$ the phase shift between the primary and the scattered wavefronts is due entirely to the interaction of the photoelectron with the atomic scattering potential and not to pathlength differences between the wave portions.

Angle-scanned Photoelectron Diffraction: The Experimental Procedure

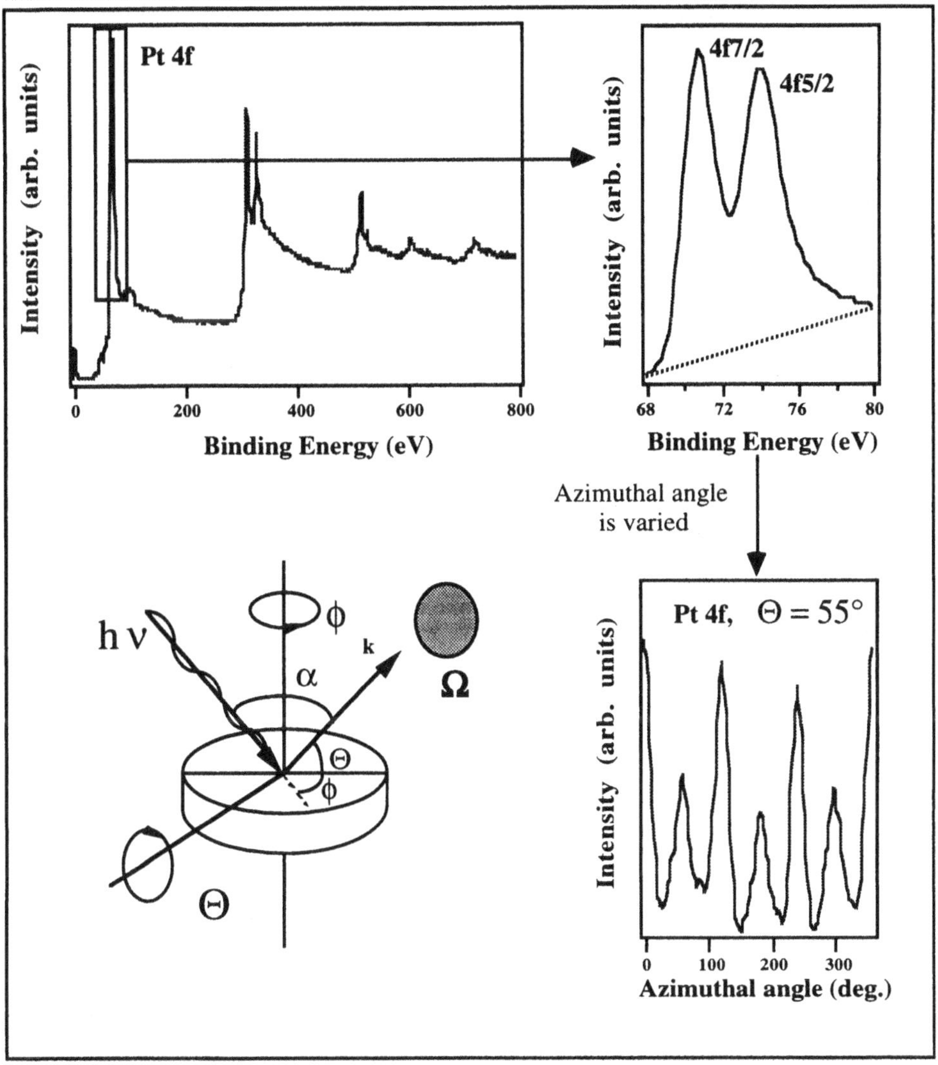

Figure 3. Schematic representation of the relation between XPS spectra and angle-scanned PD (XPD) curves. The example shows how an XPD scan is obtained from XPS data of a Pt(111) single crystal surface. Once chosen the pertinent XPS peak and after the background subtraction, the area of the peak is monitored as a function of the emission direction **k**. The sample is mounted on a manipulator with five degrees of freedom (three translations and two rotations). The geometry usually adopted for an XPD experiment, the definition of polar (Θ) and azimuthal (ϕ) angles are also reported. The polar angle is usually measured with respect to the crystal surface (Θ=90° is perpendicular to the surface) while the azimuthal angle is referred with respect to a low index direction on the crystal surface. α represents the angle between the photon source and the detector and it is usually fixed by the experimental set-up of the spectrometer.

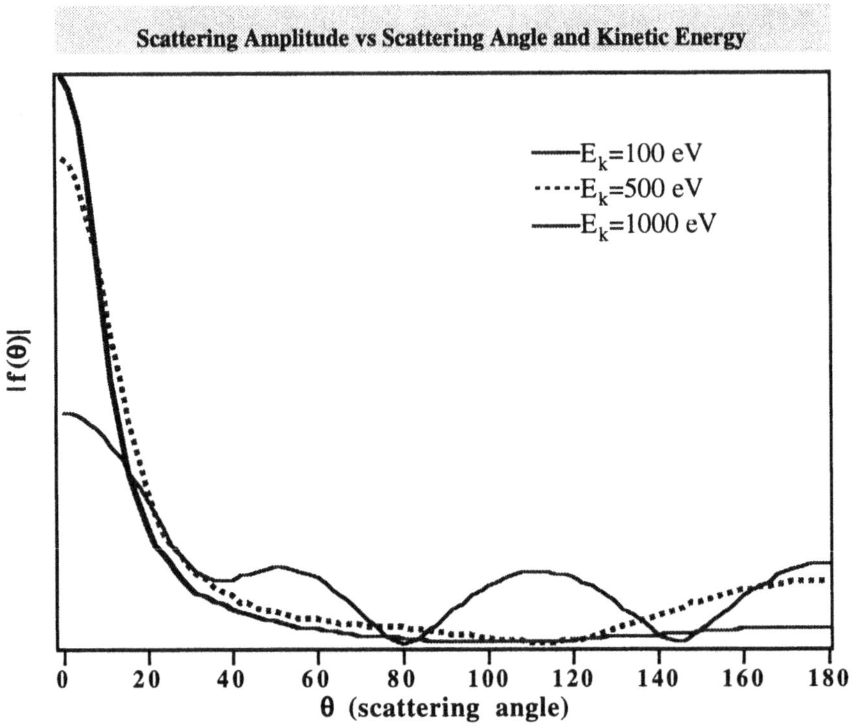

Figure 4. Scattering amplitude $|f(\theta)|$ for an electron plane wave incident on a Ni atom as a function of the scattering angle θ and for different values of the kinetic energy E_k of the colliding electron. At high E_k the scattering amplitude falls off rapidly with increasing scattering angle, giving rise to the *forward-focusing effect.*

The information on bond lengths is contained in the first order and higher orders constructive interference peaks, and associated with scattering angles higher than 0°. Unfortunately, at high energies, where FS dominates, the probability of large-angle scattering is small. Hence, higher order interference fringes are seen usually only as a rather weak fine structure accompanying the strong FS features. It has been demonstrated, however, that some distinct maxima in XPD curves may be originated by higher order interference effects, especially at intermediate energies, and these can contain information on bond distances. For these reasons and in order to exploit the whole structural information present in an XPD pattern while minimising misinterpretations of FS features, the correct approach to XPD analysis relies on a trial-and-error fitting procedure of the experimental data to theoretical simulations of the XPD curves, carried out assuming a structural model for the system under investigation. In this context, the possibility of using a simple kinematical approach, which considers only single scattering (SS) events between the emission and the detection of the photoelectron, has proven to be successful in much of the work performed by XPD [1]. This simplifying opportunity arises from the aforementioned peaked nature of the scattering amplitudes in the forward direction at energies exceeding some hundred eV, which makes the probability of large angle multiple scattering (MS) events exceedingly small. Summation of photoelectron wave amplitudes over all SS paths within a cluster of finite size, which models the surface under investigation, leads to the most widely used theoretical approach to XPD: the Single Scattering Cluster (SSC) theory. In its simplest form, the photoelectron wave, that undergoes the scattering events at the atomic sites in the cluster, is regarded as a simple plane wave and the model is referenced to as the SSC-Plane Wave (SSC-PW) model [1a-c], and it allows calculating the electron wave intensity I(**k**) along the direction of detection **k**. As an example, the expression for photoelectrons emitted from an s-level, in its simplest form, looks like:

$$I(\mathbf{k}) = \left| \hat{\boldsymbol{\varepsilon}} \cdot \hat{\mathbf{k}} + \sum_j \frac{\hat{\boldsymbol{\varepsilon}} \cdot \hat{\mathbf{r}}_j}{r_j} \left| f_j(\theta_j) \right| \exp\left\{ i \left[k r_j (1 - \cos\theta_j) + \psi_j(\theta_j) \right] \right\} \right|^2$$

Here, $\hat{\boldsymbol{\varepsilon}}$ is the polarization vector of the incident x-rays, $\mathbf{k} = k\hat{\mathbf{k}}$ is the photoelectron wave vector and $\mathbf{r}_j = r_j \hat{\mathbf{r}}_j$ is the position of the jth scatterer. The exponential factors account for the phase shifts of the scattered waves with respect to the primary wave due to the pathlength difference (by means of the term $kr_j(1-\cos\theta_j)$) and to the interaction with the scattering potential (by means of $\psi_j(\theta_j)$). At the heart of the model lies the plane wave complex scattering factor for the jth atom $f_j(\theta_j)$, which quantifies the probability of an electron wave being scattered at θ_j degrees from the direction of incidence. The full expression for I(**k**) [1] can account for the random orientation of the polarization vector if non-polarized radiation is used, for the attenuation of the photoelectron wave amplitude due to inelastic scattering, for the attenuation produced by atomic vibrational motion (by means of simple Debye-Waller factors), for the small surface refraction of the photoelectron wave due to the effect of the surface potential V_0, and for the finite angular width of the analyzer aperture (denoted as Ω in Figure 3).

An improvement to this simple SSC-PW theory has also been proposed, which essentially takes into account the correct spherical wave nature of the photoelectron final state, and this leads to the SSC-Spherical Wave (SSC-SW) model [3, 9, 10]. In the Rehr-Albers method

[10], the one employed in most of the simulations reported in the following section, the simple PW picture is recovered in terms of site-dependent effective scattering factors, which reduce to the plane wave scattering amplitudes in the PW limit (large emitter-scatterer distances). A further improvement of the theory involves the inclusion of MS events. It has been demonstrated that in the higher energy regime of XPD (above ca. 600-700 eV), MS plays a major role only for directions coinciding with dense atomic rows in a semiinfinite substrate, for which cases a defocusing effect that strongly reduces the FS amplitudes becomes evident [11]. It is profitable, however, that MS events along rows of atoms affect the intensity but not the position of the peaks. Thus, SS theory overestimates intensities along dense atomic rows, but otherwise has been found to yield a good description of experimental data. Anyhow, several methods have been suggested for the inclusion of MS corrections in the simulations when necessary, either based on a LEED-type approach that requires full long-range translational symmetry along the surface [11], or on more appropriate cluster approaches (MS Cluster, MSC) [12, 13].

3. Examples of applications

In the preceding section we have briefly outlined the principles of PD and we have distinguished between the angle-scanned and energy-scanned ways of producing the interference conditions. In addition, PD experiments can be carried out in two different energy regimes, either at high photoelectron kinetic energy (> 500 eV), where FS processes are dominant, or at low kinetic energy, where back-scattering (BS) events play a relevant role. In the following we will confine ourselves to reporting examples of using the high energy FS angle-scanned PD technique which has the remarkable advantage of being easily interpreted using the simplified single scattering approach.

3.1 Chemisorption systems

The different ways of obtaining structural information by PD in the case of chemisorption systems is schematically reported in Figure 5. Direct information on the adsorbate alignment with respect to either the surface normal (by polar scans) or azimuthal directions (by azimuthal scans) [14-22] can be obtained by XPD experiments when the adsorbate intramolecular FS (Figure 5a) is exploited: by measuring the angular distribution of the photoelectrons which originate from the core levels of the atoms directly bound to the surface, whose total intensity is modulated by the scattering from the outer atoms, the orientation of chemisorbed molecules is directly obtained from angular positions of the intense FS peaks. In the following, we will show an example of this methodology to obtain the azimuthal orientation of the formate anion at the Cu(100) surface.

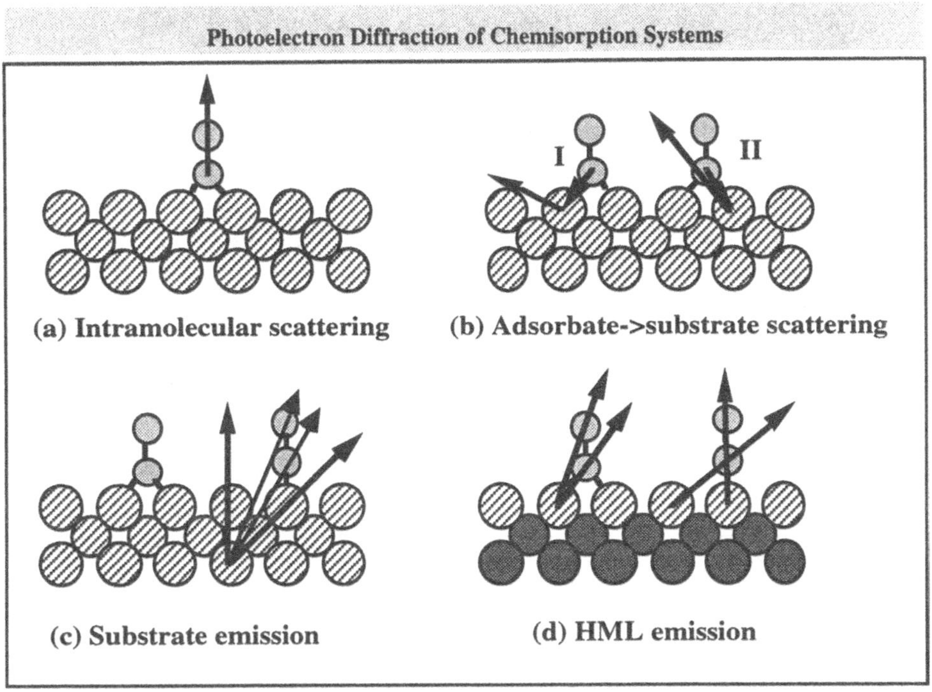

Figure 5. Schematic diagrams showing the principal scattering events in PD experiments on chemisorption systems: (a) intramolecular adsorbate scattering; (b) adsorbate→substrate scattering detected at grazing emission angles in the forward-scattering regime (I) or at higher angles in the back-scattering regime (II); (c) strong substrate→substrate forward-scattering arising from sub-surface atomic layers and weak substrate→adsorbate forward-scattering; (d) forward-scattering of the primary photoelectron wave emitted from a heteroepitaxial monolayer (HML) by adsorbed molecules occupying two different chemisorption sites.

However, in general intramolecular scattering cannot give insight into the nature of the chemisorption site, since this information is contained in the adsorbate→substrate scattering pattern, which is usually very weak and hard to be detected. In fact, as the emitting atom lies above the substrate, the adsorbate→substrate scattering which originates oscillations at the detector implies large scattering angles, which are associated to low scattering amplitudes at high kinetic energy (see Figure 4). Only at grazing emission angles (event I in Figure 5b), where the scattering angles involving the substrate are small enough to give rise to sufficient amplitudes of the secondary waves, significant FS-XPD patterns can be detected [23]. For this reason energy-scanned PD applied to adsorbate emission has proven to be of much more general use in the field of adsorption site determinations for small molecules on single-crystal surfaces [4]. Its merit lies in the exploitation of the low energy regime where BS events show large scattering amplitudes (event II in Figure 5b). Thus a high sensitivity is gained to the arrangement and distances of the nearest neighbours which lie *below* the emitter [4].

On the other hand, FS-XPD experiments based on the emission from the substrate cannot be used in general for the purpose of adsorption site determination, because the strong substrate→substrate FS modulations arising from sub-surface atomic layers dominates over the weaker components due to the substrate→adsorbate one (Figure 5c). We have recently proposed a new method to obtain information on the chemisorption site by using the FS-XPD approach [24]. The method is based on the adsorption of an atomic or molecular species on a heteroepitaxial monolayer (HML) deposited on a suitable single-crystal substrate. The HML is used as the source of the primary photoelectron wave, which undergoes FS by the atoms constituting the adsorbate (Figure 5d). Since the ideal HML is bidimensional, the problem of the substrate→substrate FS is removed. An example of application of the HML method to determine the site of chemisorption of the systems c(2x2)-O on Ni(100), a system extensively studied by means of several techniques, will be reported in the following. We have demonstrated that the direct structural information gained by the HML method, without any theoretical simulation, is comparable with that obtained by other techniques after extensive theoretical analysis of the data.

3.1.1 *Formate anion on Cu(100)*

Surface formate is an intermediate in methanol synthesis using copper-derived catalysts, and for this reason, many studies have been performed on formate adsorbed on Cu single crystals by means of different surface science techniques, in order to assess its electronic and structural properties. In particular the chemisorption site (see Figure 6 for the definition of the different sites) has been the object of long debate [25-29].

We have carried out experiments to try to assess the formate chemisorption site on Cu(100) by means of FS-XPD [22]. The experimental work has been accompanied by rigorous first principle quantum mechanical calculations in the framework of LDF (Local Density Functional) theory [22, 30].

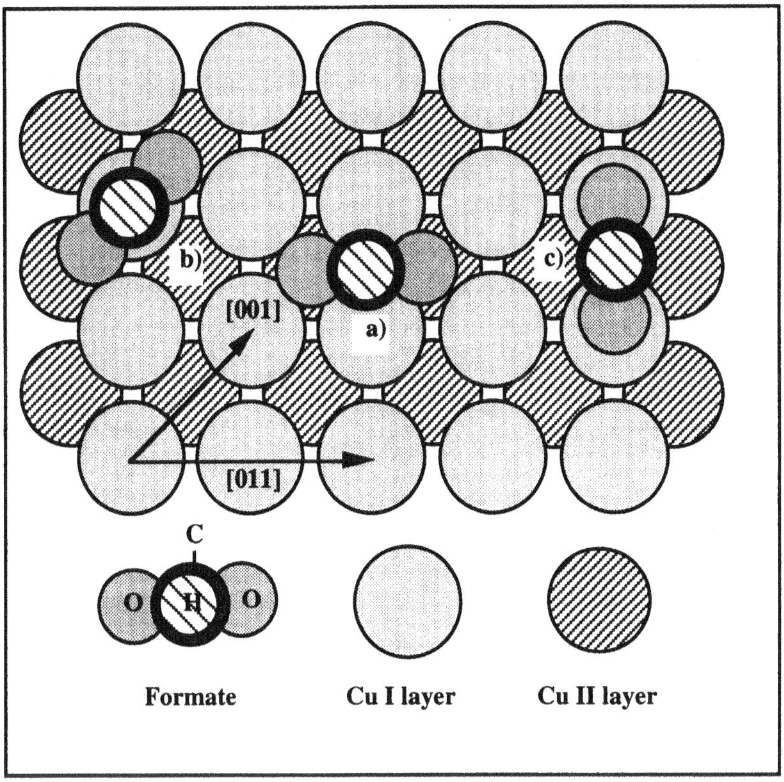

Figure 6. Top view of the different adsorption sites proposed for formate on Cu(100): (a) cross bridge (CB); (b) diagonal atop (DA); (c) short bridge (SB).

Azimuthal orientation of the adsorbed species was determined by recording O1s photoelectron intensity as a function of azimuthal angle at a polar angle of 30° with respect to the surface, in order to scan through the two FS directions corresponding to O-C bonds. The fourfold averaged data (Figure 7b) show directly, and with no need of theoretical simulations, that formate is chemisorbed with its molecular plane parallel to the (0 1 1) and (0 $\bar{1}$ 1) planes of the copper crystal, whose alignment was identified by assigning the FS maxima in Figure 7a. The presence of four equivalent peaks in a 360°-scan demonstrates that there are two sets of adsorbates perpendicular to each other. This result rules out the DA site, which was proposed after extensive and rather refined SEXAFS data analysis [28], but does not allow to exclude either the SB or the CB one, both possible candidates from a preliminary inspection of the experimental data. In this case, however, a careful analysis of the PD data, with the aid of SSC-SW simulations, was capable of extracting experimental evidence for the preference of the SB site.

In Figure 7b and 7c the results of the SSC-SW simulations for the two SB and CB sites are compared with the experimental results. A good agreement between the fine structure of the experimental and theoretical curves can be found if the low frequency hindered rotational modes of the adsorbate (often referred to as librations) are included in the simulations. Such vibrations produce substantial displacements from the atomic equilibrium positions and may have a more important effect on XPD curves than higher frequency internal modes, which can be taken into account rather satisfactorily by means of simple Debye-Waller factors. A simple model has been assumed to simulate the low frequency τ_z libration around the adsorbate C_2-axis (the axis parallel to the C-H bond and perpendicular to the surface) [22]. According to this model the adsorbate has been treated as a harmonic torsional oscillator: simulated azimuthal curves have been obtained as harmonic oscillator envelopes of several displacements from equilibrium position, for various values of decreasing vibrational frequency ω and, thus, of increasing γ_{rms} (root mean square torsional displacement), with a progressive occupation of the oscillator first excited state. R-factor curves *vs* γ_{rms} show a better agreement with the experimental data for the SB site, with the minimum lying around $\gamma_{rms}= 11°$. The best-fit frequency is thus approximately 70 cm^{-1}. At our best knowledge, this is the first experimental determination of such libration frequency.

A simple visual inspection of the best-fit simulations (corresponding to R-factor minima) for both sites shows clearly that the simulation for the SB arrangement gives the best agreement with the experimental data (Figures 7c and 7d).

The body of the results induced us to conclude that there is a PD experimental evidence of the SB site occupation for formate on Cu(100), in accordance with ARPEFS [29] results and with LDF calculations [22, 30].

Also the Cu-O distance has been determined with good accuracy by means of the R-factor analysis. Because the fine structure in the PD curves arises primarily from the scattering of the O1s photoelectron from the substrate, it is reasonable to expect that it is very sensitive to the adsorbate-substrate distance. The simulated curves relative to different adsorbate-substrate distances have been then compared with the experimental scan, and the best fit was obtained for a Cu-O bond length of 1.95 Å, with an estimated error somewhat between 0.05 and 0.1 Å. This Cu-O distance is coincident with the one calculated by the LDF [22, 30] theory, and in excellent agreement with ARPEFS [29] experimental data reported in the literature (1.98±0.04 Å).

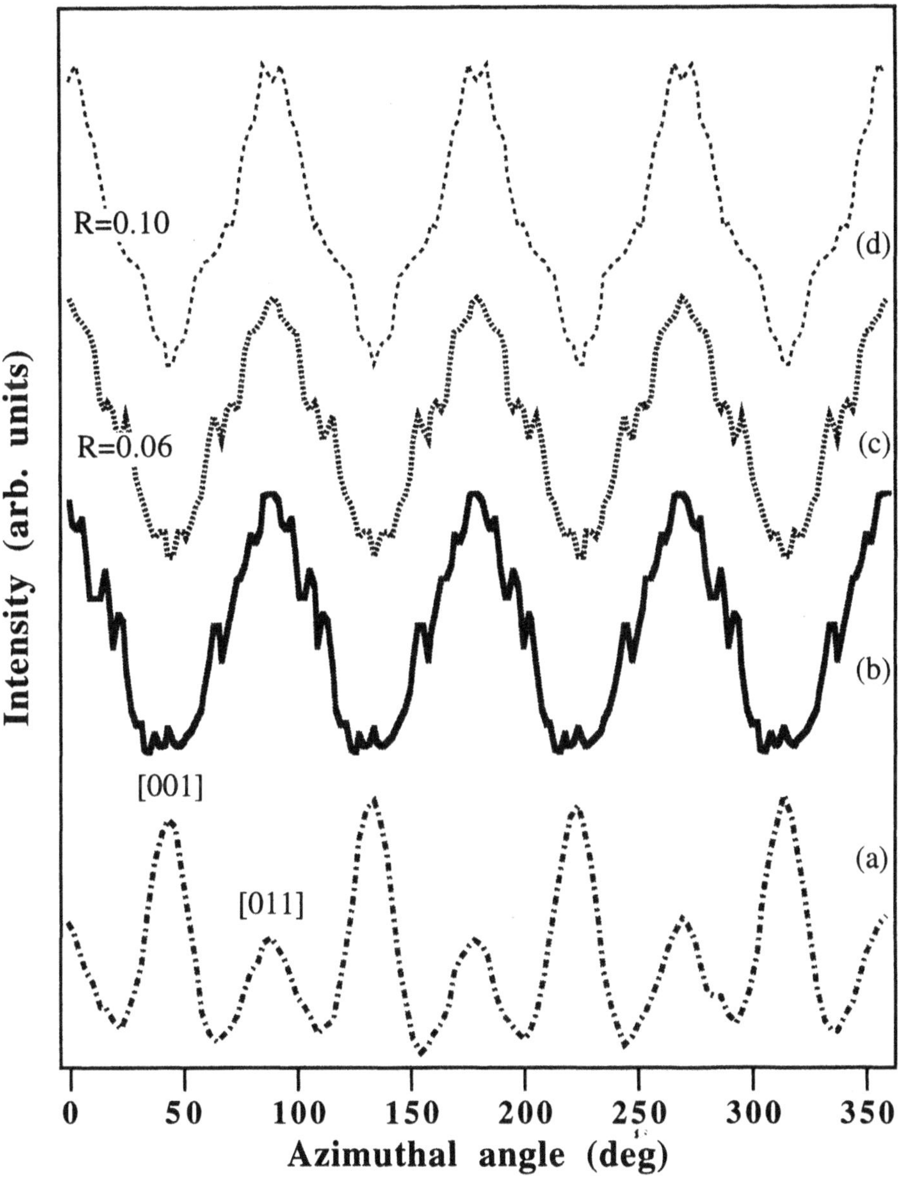

Figure 7. (a) $Cu2p_{3/2}$ azimuthal scan at a polar angle Θ=45° for a clean Cu(100) surface; (b) Fourfold averaged O_{1s} azimuthal scan for formate on Cu(100) taken at a polar angle Θ=30° with respect to the surface; (c) and (d) Best fit SSC-SW simulations for SB and CB site respectively (R-factors are reported) (see ref. [[22]).

3.1.2 *c(2x2)-O on Ni(1ML)/Cu(100)*

The c(2x2)-O on Ni(100) system is currently adopted as a benchmark for surface science techniques. To obtain the structural information (i.e. the site and the Ni-O distance) extensive theoretical simulations of the spectroscopic data are usually required. In all cases reported in literature the fourfold hollow site has been demonstrated (see Figure 8), with a Ni-O bonding distance varying from 1.92 to 1.98 Å, depending on the used experimental technique. Accordingly, the ψ angle, defined in Figure 8 with respect to the surface, ranges from 23.6° to 27.6 °.

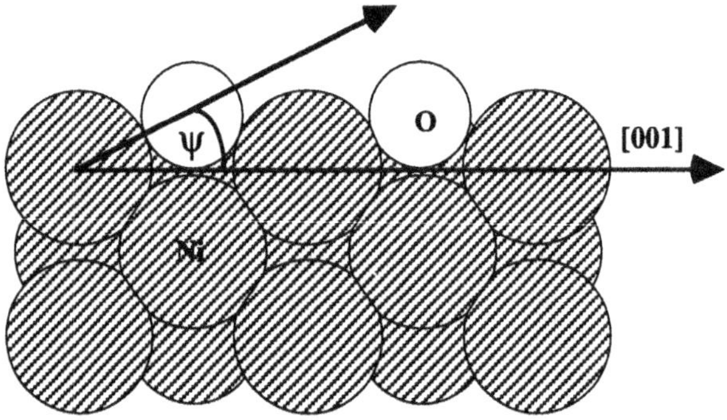

Figure 8. The structure of c(2x2)-O on Ni(100), where the oxygen atoms occupy the fourfold hollow sites.

We have adopted a similar system, i.e. oxygen chemisorbed over 1 ML of Ni on Cu(100), in order to test the capability of the HML method in providing the structural information directly from the raw XPD data without theoretical simulation [31].

The growth mode of Ni on Cu surfaces is somewhat controversial. Some authors claim a layer-by-layer growth, which would furnish the ideal heteroepitaxial ML, but some years ago Egelhoff [3] gave unambiguous evidence, based on FS-XPD measurements, for the fact that the growth proceeds *via* agglomeration of the Ni overlayer and partial segregation of Cu onto the Ni, and furnished a simple thermodynamic rationale for this behaviour. On the other hand, Egelhoff demonstrated that the Ni layer can be forced to lie flat on the Cu surface if oxygen is used as surfactant [3]. Actually, oxygen tends to float out to the growing Ni surface and keeps the Ni layer lying flat. Thus, the system cut out for our purposes is easily prepared simply by evaporating 1 ML equivalent of Ni atoms over a Cu(100) surface which was predosed with oxygen to give a c(2x2) LEED pattern.

To confirm the occupation of the fourfold hollow sites, Figure 9 (top) reports an azimuthal scan on Ni2$p_{3/2}$ at the polar angle Θ=20°.

The c(2x2)-O on Ni (1 ML)/Cu(100) system

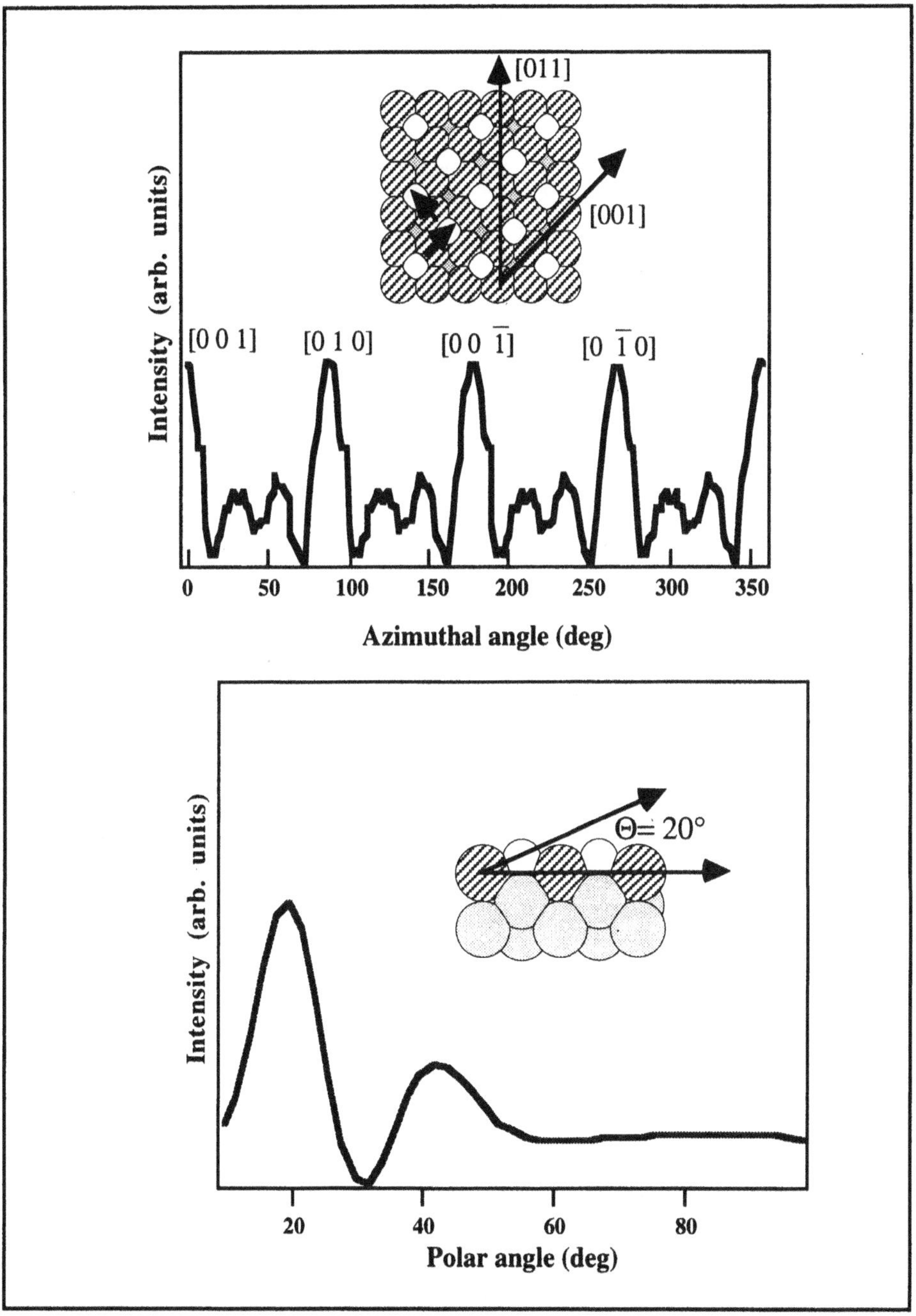

Figure 9. Top: $Ni2p_{3/2}$ azimuthal scan at a polar angle of Θ=20°; *bottom:* $Ni2p_{3/2}$ polar scans along the [0 0 1] azimuth for (2x2)-O on Ni (1 ML)/Cu(100) (see ref. [31]).

As expected for a fourfold site, the curve shows four maxima along the <001> directions to be related with the FS events originated by the four nearest neighbours. All maxima have the same FWHM, and are separated exactly by 90° intervals. Thus the demonstration of the occupied site is only a matter of recording an azimuthal scan.

Since oxygen is expected to lie slightly above alternate fourfold hollows, $Ni2p_{3/2}$ polar scans along the diagonal azimuths on the surface have been performed, in order to scan through the Ni→O directions. The curve, obtained averaging the four polar scans along the four equivalent <001> azimuths, is reported in Figure 9 (bottom). The strong peak at $\Theta=20°$ is to be attributed to Ni→O forward-focusing and furnishes a direct measure of ψ angle, as defined in Figure 8.

Assuming a pseudomorphic growth of the Ni ML on Cu(100), and having determined the angle of the Ni-O bond with respect to the surface, the Ni-O distance is easily obtained *by simple trigonometric calculations*: d(Ni-O)=1.92±0.04 Å. The error is estimated simply by considering the uncertainty in the determination of the angle (±1°). The value of the Ni-O bond distance is within the interval reported in literature, as obtained by other surface science techniques. The smaller angle detected in our case ($\Theta=20°$) allows to maintain approximately the same Ni-O distance on the strained pseudomorphic Ni overlayer as on bulk Ni(100). We want to emphasize that the Ni-O distance is here obtained from the rough data without any theoretical simulation.

3.2 Ultrathin epitaxial systems

Ultrathin epitaxial films grown on single-crystal substrates are nowadays becoming important new subjects for research in the field of solid-state chemistry and physics, material science and device technology. The innovation coming from ultrathin films consists in the possibility of creating new systems with uncommon structural properties. Actually, from a structural point of view, the ultrathin layers may present peculiarities that make them quite distinct from the standard bulk phases: overlayers with distorted geometries and metastable phases (stable only at high pressure and/or high temperature when bulk phases are concerned) can be grown due to the lattice mismatch between film and substrate. When the growth of the film is extended (after ca. 10-20 MLs) the memory of the substrate gets lost, causing the film to recover the bulk structure of the material, so that the whole drama is played out within the first ten MLs.

By now, various studies have been reported in the literature [1-3] where XPD has been used to evaluate:

Establishing Growth Modes and Interlayer Spacing Contractions/Expansions in Epitaxial Layers by XPD

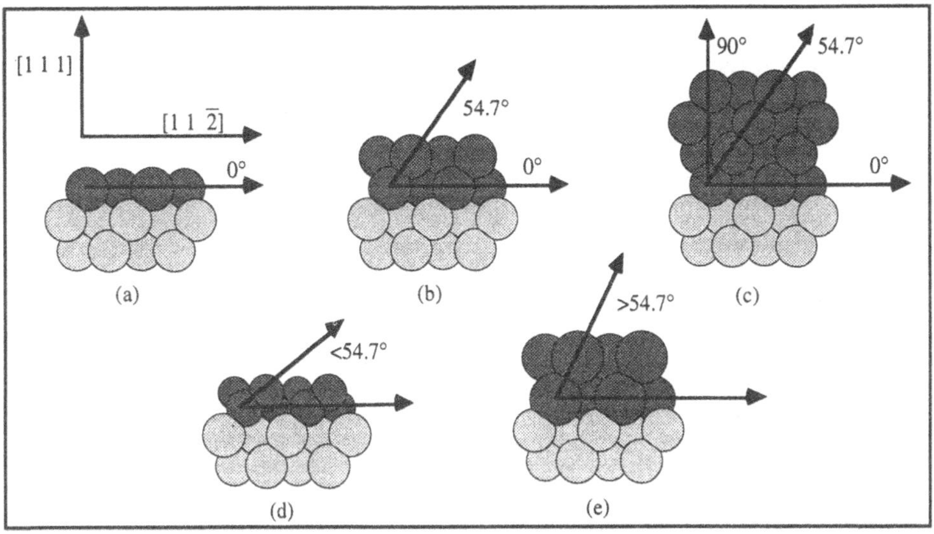

Figure 10. Cross-section along the [1 1 $\bar{2}$] direction of a fcc (111) epitaxial growth. By doing an XPD polar scan on photoemission lines from atoms of the growing layer, and by looking at the most prominent peaks originated by the *forward-focusing effect* from the nearest-neighbours, it is easy to see that intensity enhancement will be present (a) for 1ML only at very grazing angles, (b) for two MLs at Θ=54.7° and (c) for four MLs also at Θ=90°(even if less intense due to the larger emitter-scatterer distance). Thence by inspection of an XPD polar scan, it is possible to derive the number of layers grown, and, knowing the number of equivalent MLs deposited, to put forward conjectures regarding the growing mechanism. Similarly, if the actual Θ value associated in the polar scan to the most intense peak is compared with the expected value of Θ=54.7°, it is possible to detect a contraction (d) or expansion (e) in the vertical atomic layer spacing consequent to the fact that ultrathin epitaxial films are often strained in-plane to match the substrate pseudomorphically.

-the growth mode of epitaxial layers;

-lattice expansions and contractions with respect to bulk phases;

-the structural parameters (distances and bond angles) of the epilayers;

-structural defects, such as stacking faults.

Concerning the first issue, one question that XPD can answer is related to the discrimination between the types of film growth, i.e. layer-by-layer (Frank-van der Merwe, FM), three-dimensional (3-D) crystallite growth (Volmer-Weber, VW) and mixed-mode (Stranski-Krastanov, SK). It is quite simple from geometrical considerations to predict the angle at which intense FS peaks should be present in a polar scan as soon as a chain of two or more atoms forms in a growing epitaxial film. As an example of metal-on-metal epitaxy, Figure 10a-c shows that for the epitaxial growth on a fcc (111) surface a strong intensity enhancement at the polar angle of Θ=54.7° is expected only when a second atomic layer has been deposited and that a peak at Θ=90° will appear when the fourth layer is grown. From the actual angular position of the prominent peaks in a polar scan it is also easy to evaluate possible lattice expansions and contractions of the epilayer (see Figure 10d-e). Obtaining information on the structural parameters and on structural defects is somewhat more complex and requires more detailed comparisons with the results of SSC, or perhaps MSC simulations.

In thc following, examples of XPD studies of metal-on-oxide and oxide-on-oxide epitaxial growths will be briefly discussed. The focus will be on the unique information obtainable by XPD and not on the details of the experimental data.

3.2.1 *Metal-on-oxide growth: V on $TiO_2(110)$*

In the last few years much effort has been devoted to studying metal/oxide systems mainly because of their importance in electronic and sensor devices and heterogeneous catalysis. One key point to be clarified is the relation between the metal/oxide interaction at the interface and the growth mechanisms of metals at the oxide surfaces. Among the various oxide surfaces, $TiO_2(110)$ is the most studied system and an overview paper has recently appeared wherein the results on $M/TiO_2(110)$ (M=alkaline, transition and noble metals) were critically discussed [32]. Metals having an oxygen affinity higher than or comparable to that of Ti in TiO_2 strongly interact with the substrate. In particular, strong interactions, accompanied by electronic charge transfer from the metal to TiO_2, are reported for alkaline metals, aluminium, and metals at the left of iron in the periodic table. On the contrary, weak interactions were observed for Ni, Cu, Rh, Pd and Pt. In the same review article, some structural conclusions on the $M/TiO_2(110)$ film growth have been also reported [32]. The wetting ability of the overlayer, its growth morphology, and the tendency for developing ordered or disordered structures scale with the reactivity of the metal. For metals with high reactivity towards oxygen, where strong oxidation-reduction reactions occur at the interface, flat metal overlayers can be grown, whereas agglomeration in 3-D islands takes place for less reactive metals. On the other hand it has been reported that the low reactive overlayers tend to produce ordered films, with a weak degree of long-range order (according to LEED measurements) but with a well developed short-range order, as probed by XPD. For very reactive overlayers (metals at the left of Cr), evidence for the growth of ordered overlayers was lacking in the literature, and in the case of Hf and Ti disordered structures have been found [32].

In this context, we recently used XPD to demonstrate [33] that few monolayers of V, a highly reactive metal, deposited at the $TiO_2(110)$

surface, grow forming islands with a bcc structure, with the [100] direction normal to the substrate surface. In addition, the comparison between the V2p and Ti3p azimuthal curves have shown that the [001] azimuth of the overlayer is aligned with the $[\bar{1}\,1\,0]$ direction of the substrate, which clearly affects the overlayer orientation; however there is no long-range coherence between the two lattices, as demonstrated by the absence of a LEED pattern. We will now briefly discuss these experimental data to demonstrate how straightforward it is to extract the structural information from them.

In Figure 11a, we report an azimuthal XPD scan on Ti3p from clean $TiO_2(110)$ at a polar angle $\Theta=65°$, taken for the purpose of crystal alignment. The curve reflects the twofold symmetry of the surface, with the two intensity maxima at $\phi=90°$ and $\phi=270°$ being due to FS of the Ti photoelectrons emitted from each atomic layer and forward-focused by Ti atoms of the layer above. These peaks identify the $[0\,0\,1]$ and $[0\,0\,\bar{1}]$ directions on the surface, while the two features at $\phi=0°$ and $\phi=180°$ are related to the $[1\,\bar{1}\,0]$ and $[\bar{1}\,1\,0]$ azimuths. Figure 11b was obtained with V2p photoelectrons at a polar angle $\Theta=35°$ for a ~5 ML thick V film deposited at the TiO_2 surface. The scan has a good anisotropy ($\chi=19\%$), although no LEED pattern was detectable after V deposition, and it displays an evident fourfold symmetry, with intensity maxima at $\phi=45°$, $\phi=135°$, $\phi=225°$ and $\phi=315°$ and with less intense features between them.

Polar scans measured along the directions identified by the four main FS maxima in Figure 11b are all equivalent. These results prove that the V film deposited on $TiO_2(110)$ is short-range ordered and it has a cubic structure exposing the (100) surface. In fact, a (110) surface should produce a twofold symmetric azimuthal XPD curve, while a (111) orientation should lead to threefold symmetry. In addition, it is interesting to determine whether the bcc bulk structure of V is maintained in the ultrathin film or whether it grows in a metastable fcc stacking. To answer this question, two polar scans on V2p were obtained for two principal azimuths of the overlayer (determined from the azimuthal scan of Figure 11b) at $\Delta\phi=45°$ from each other. The resulting curves are reported in Figure 11c and 11d. In spite of the fact that the curves reported represent raw data, i.e. no removal of the $1/\sin\Theta$-like contribution to the photoelectron intensity typical of a thin layer [1] has been performed, and therefore the intensity at grazing angles is overestimated with respect to data at higher polar angles, a clear-cut conclusion can be drawn from these results.

Simple geometrical considerations allow predicting that, for a bcc structure, the most intense FS peak (due to the nearest-neighbours) should be located at $\Theta=35°$ in a polar scan along the [011] azimuth, while, for an fcc structure, the strongest peak should be located at Θ=45° in a polar scan along the [001] azimuth. It is evident that the strongest feature (stronger even than the normal emission FS) is detected in curve 11c at Θ=35°. The data reported in Figure 11c and 11d thus unequivocally demonstrate the bcc stacking of the vanadium thin film.

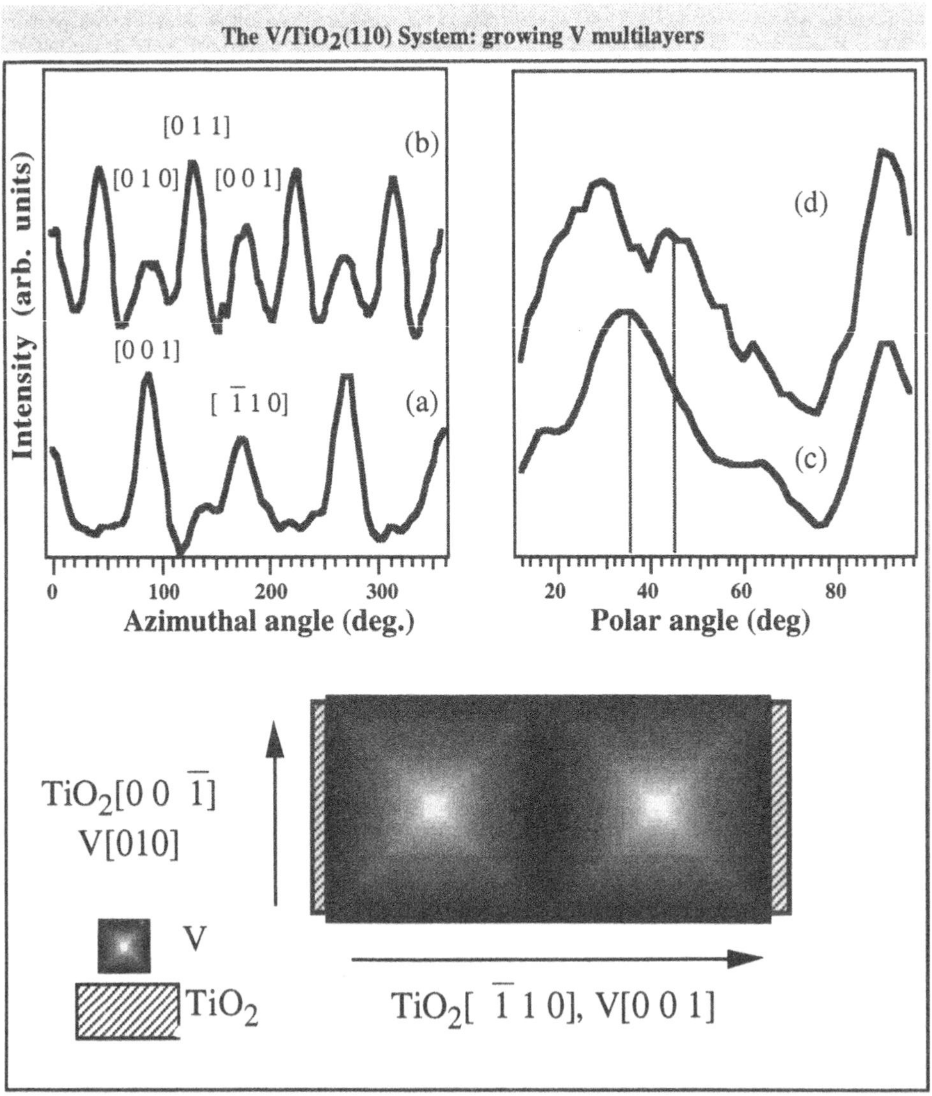

Figure 11. XPD results on the V/TiO_2(110) system [33]. Curve (a) is an azimuthal XPD scan on Ti3p from clean TiO_2(110) at a polar angle Θ=65°, taken to the purpose of crystal alignment. Curve (b) is a V2p azimuthal scan at a polar angle Θ=35° for a ~5 ML thick V film. Curves (c) and (d) are V2p polar scans on the V multilayer along the [011] and [001] azimuths of the overlayer, respectively. An analysis of the FS features in these polar scans allows to conclude that the V film has a bcc structure.

It should be also pointed out that the FS maxima in both polar scans of Figure 11 are detected at polar angles expected for an ideal bcc structure. Therefore, there is no experimental evidence for a tetragonal distortion of the V lattice such as interlayer spacing contraction and/or expansion in order to compensate some possible in-plane strain between substrate and overlayer.

Moreover, the comparison between the V2p and Ti3p azimuthal curves of Figure 11a and 11b shows that the [001] azimuth of the overlayer is aligned with the [$\bar{1}$ 1 0] direction of the substrate, as reported in the inset of Figure 11. These findings support the idea, already proposed by other authors for the growth on $TiO_2(110)$ of less reactive metals (as Fe and Cr [34]), that the substrate affects the orientation of the V layer by matching two V square surface unit cells to the $TiO_2(110)$ rectangular surface cell of the substrate. In this case, the interface lattice mismatch is +2% along the shorter side of the $TiO_2(110)$ surface cell (3.02 Å vs 2.96 Å) and -7% along the longer side (6.04 Å vs 6.49 Å). The lack of a definite LEED pattern, however, suggest that long-range order is lacking.

3.2.2 *Oxide-on-oxide growth: VO_2 on $TiO_2(110)$*

Epitaxial growth of oxide layers at oxide surfaces is nowadays of primary importance for many technological and catalytic applications as well as for studying the physical properties of low dimensional systems.

As a continuation of the study of vanadium on $TiO_2(110)$, we have investigated the structural properties of the mixed V-O-Ti interfacial layer. It has been shown [35] that the interface, prepared by depositing vanadium atoms at room temperature at the $TiO_2(110)$ surface in a sub-monolayer regime (0.2-0.3 ML) and annealing it at 473 K in UHV, presents V atoms occupying Ti substitutional sites in the topmost layer of the substrate without underneath diffusion.

The experiment was performed by dosing 0.2-0.3 ML of V on the clean, stoichiometric and well ordered $TiO_2(110)$ surface. After the deposition, the initially sharp (1x1) LEED pattern of the substrate became extremely faint and barely detectable. For this reason, a mild annealing (T=473 K for 1') was carried out to promote the surface order and to reoxidise the Ti atoms at the surface, which were reduced by the deposited V atoms. After annealing, a sharp (1x1) LEED pattern was recovered. All the reported measurements were performed on the surface thus prepared.

The middle curve in the top left panel of Figure 12 represents an experimental V2p azimuthal scan at $\Theta=34°$ for 0.2 ML of V deposited at the TiO_2 surface. It reflects the intrinsic symmetry of the surface, with

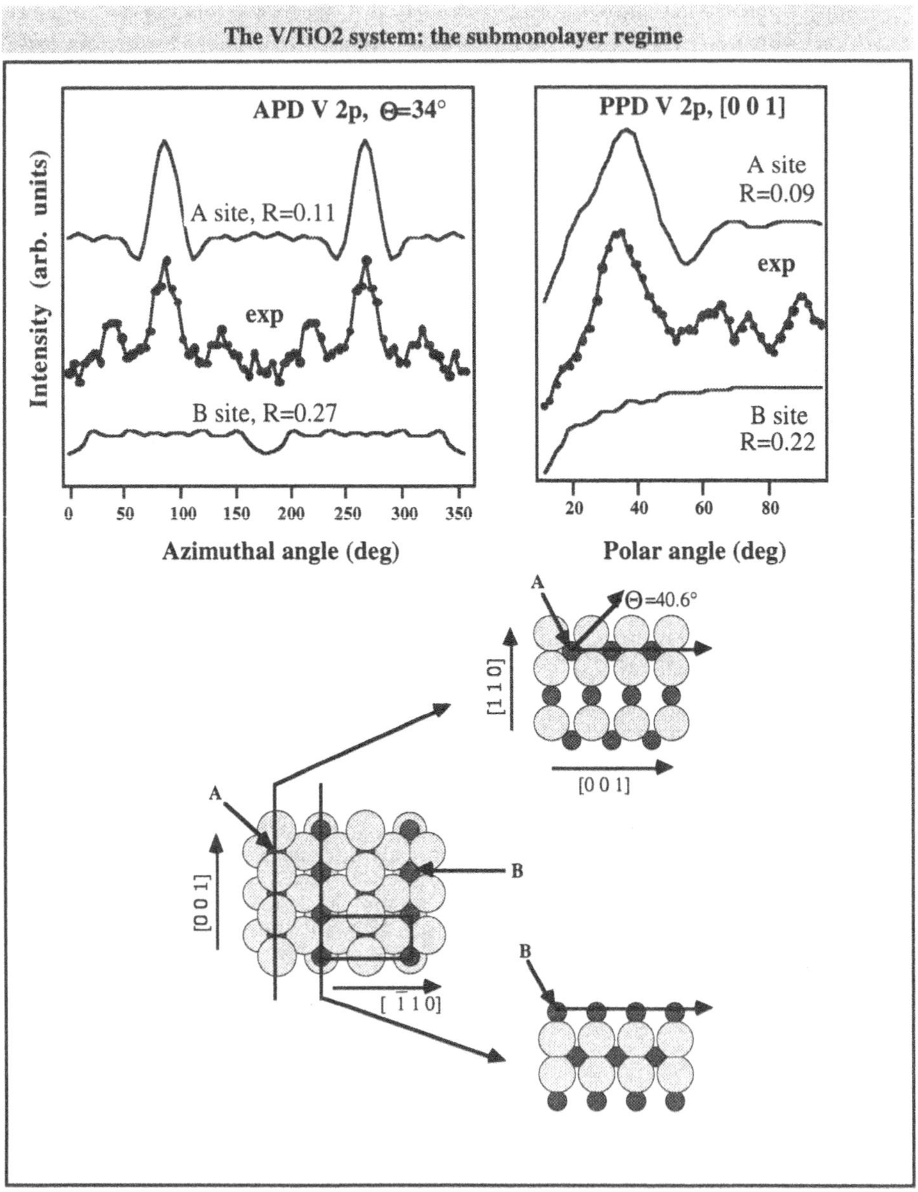

Figure 12. Top left: Experimental twofold-averaged V2p XPD azimuthal scan at Θ=34° for 0.2 ML of V deposited at the TiO_2 surface (center), compared to SSC-SW simulations assuming either A or B substitutional sites for V atoms in the first TiO_2 layer. R-factors are quoted. *Top right:* Experimental V2p XPD polar scan along the [0 0 1] azimuth for 0.2 ML of V deposited at the TiO_2 surface (center), compared to SSC-SW simulations assuming either A or B substitutional sites for V atoms in the first TiO_2 layer. *Bottom:* top view and two inequivalent cuts normal to the surface plane along the [0 0 1] azimuth for the (1x1) TiO_2(110) surface. It consists of alternate rows of sixfold-coordinated (A sites) and fivefold-coordinated (B sites) Ti cations parallel to the [0 0 1] direction (see ref. [35]).

the two most intense peaks occurring along the [0 0 1] and [0 0 $\bar{1}$] main azimuths of the substrate. The anisotropy of the curve is very good (χ=24%), which suggests that V atoms sit in a precise site at the surface: the hypothesis was made that V atoms occupy Ti substitutional sites of the first surface layer in the early stages of epitaxy. The model generally accepted for the (1x1) surface termination of a clean stoichiometric $TiO_2(110)$ surface is reported in Figure 12 (bottom). It consists of alternate rows of sixfold- and fivefold-coordinated Ti cations parallel to the [0 0 1] direction. The sixfold-coordinated cations, occupying A-sites in Figure 12, support a row of bridging O atoms protruding from the surface, while fivefold-coordinated cations (B-sites) are almost coplanar with four O atoms of the surface layer. A more detailed analysis of Figure 12 in terms of possible FS events for a V2p photoelectron originating from V atoms positioned at substitutional Ti sites of either A or B type in the first surface layer is useful. In the bottom panel of Figure 12 two non-equivalent cuts of the surface region along the [0 0 1] direction are reported as side-views. It can be easily seen that a V atom substituting an A-type Ti atom, if the surface is thought to be unrelaxed, would produce only one main FS feature at a polar angle of approximately 40° from the surface, due to the presence of a bridging O atom. Conversely, a V atom in a B-site would not see any scatterer above the surface, giving rise to a poorly modulated XPD scan. The same arguments may be applied for the [$\bar{1}$ 1 0] direction (side views not reported): this time a V atom in an A-site would produce no prominent 0th order scattering features, while a V photoelectron originated at a B-site would give rise to a strong FS event at Θ~20°, due to the bridging oxygen on the adjacent row.

In Figure 12 (top left), the experimental V2p azimuthal scan at Θ=34° is compared to SSC-SW simulations assuming either A- or B-site occupancy for deposited V atoms. It is evident that the overall shape, the twofold symmetry and the azimuthal orientation of the curve are very well reproduced if V atoms are assumed to occupy A-type substitutional sites in the first layer of the substrate. The B-site assumption leads to a poorly modulated curve which produces a much worse agreement with the experiment, as argued both by the R-factor values and by a simple visual comparison of the data. The nature of the two main features seen along the [0 0 1] direction in the experimental azimuthal scan may now be attributed to FS from V to bridging O.

In order to confirm the site assignment and to shed light on the possible relaxation of bridging O atoms, a V2p polar scan along the [0 0 1] main azimuth was acquired, and compared to SSC-SW simulations for V atoms in A or B sites. The best fit results are reported in Figure 12 (top right). The only important feature found at Θ=35° in the experimental scan is well reproduced by theoretical simulations if an A-type site is assumed for substitutional V atoms. As expected, the B-type site produces no prominent features along this azimuth. An R-factor analysis has been performed for different values of the V-O distance. The best fit between the experiment and the A-site simulation is obtained for a downward relaxation of the bridging oxygens by an amount of 0.2 Å, in very good agreement with quantum mechanical calculations for the clean substrate reported in the literature [36]. The experimental polar scan along the $[\bar{1}\ 1\ 0]$ direction (not reported herein) is practically featureless (χ~3%), and this consideration alone again favours the A-site assignment on the basis of the above-mentioned FS analysis. As a general remark, it should be stressed that the diffusion of V atoms underneath the first layer would produce additional peaks in polar scans both along the [001] and the $[\bar{1}\ 1\ 0]$ directions, which are not seen in the experimental curves.

These results are consistent with an incipient epitaxial growth of vanadium oxide at the $TiO_2(110)$ surface; furthermore, the reappearance of the LEED pattern after a mild annealing hints at the possibility of growing oxide layers which preserve the long-range translational order of the surface. In fact, the attempt of growing epitaxial vanadium oxide layers as thick as several MLs, which show clearly detectable LEED patterns, has been successful [37]. The procedure devised to grow an ordered vanadium oxide layer consists in alternating V depositions and mild annealing cycles in 10^{-6} mbar of oxygen.

Figure 13 displays full linear 2π Θ/φ intensity plots for Ti2p emission from the clean TiO_2 substrate, and for V2p from a ~5 ML thick vanadia overlayer. In this kind of plot the centre corresponds to the surface normal, a radial section represents a polar scan, a circular section is an azimuthal scan and the photoelectron intensity is given by the colour scale. The overall symmetry and the main intensity maxima for the Ti2p and V2p plots are very much similar, although the relative intensity of the features may be different. Apart from the spot along the surface normal, there are two maxima along the $[0\ \underline{0}\ 1]$ direction at Θ=36° and Θ=65°, and another two peaks along the $[\bar{1}\ 1\ 0]$ azimuth at Θ=45° and Θ=74°.

As a whole, the results obtained indicate very clearly the possibility of growing long-range-ordered ultrathin films of vanadium oxide pseudomorphic to the $TiO_2(110)$ substrate. The pseudomorphism of the epilayer is demonstrated partly by the permanence of the 1x1 LEED pattern characteristic of the substrate, at least up to 5 ML-thick deposits, with no lateral expansion or contraction of the surface lattice parameters, as strongly supported by the 2π Ti2p and V2p XPD plots reported in Figure 13. The coincidence of the two plots, as far as the overall symmetry and the exact position of all the main FS features are concerned, indicates that the crystal structure of the surface layers of clean $TiO_2(110)$ and of the thin film, grown by successive V deposition/O annealing cycles, is very much the same. The differences detected in the relative intensities of the FS features on passing from the clean substrate to the overlayer may be reasonably explained by the different weight that Multiple Scattering is expected to have in a bulk crystal and in a thin film obtained by stacking just a few MLs.

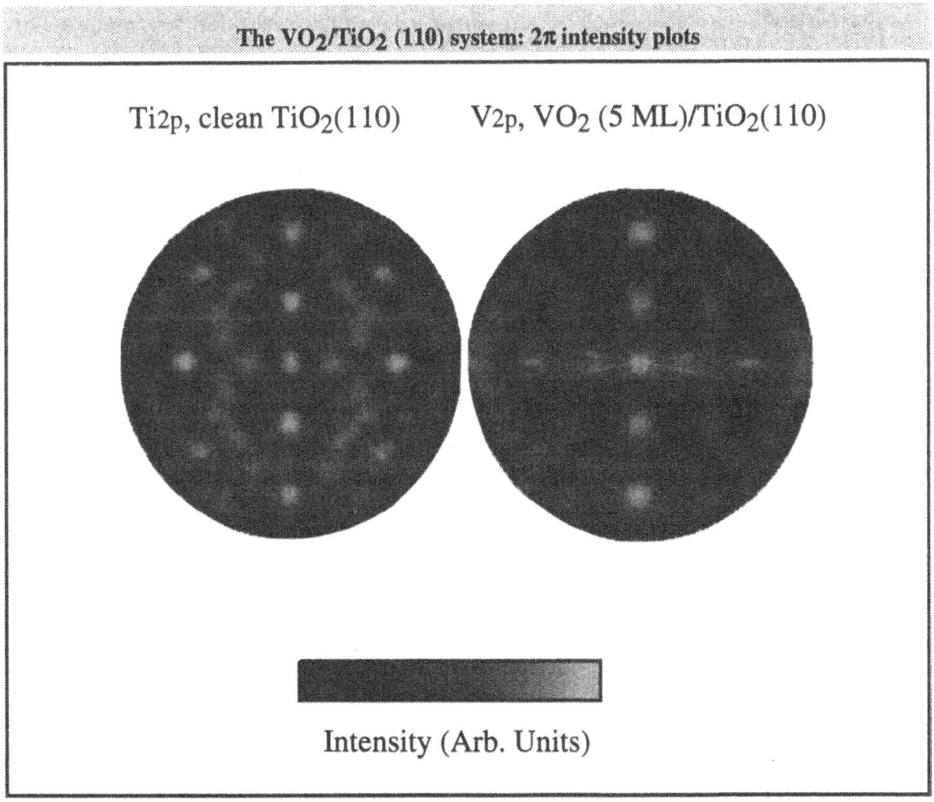

Figure 13. Full linear 2π Θ/ϕ intensity plots for Ti2p emission from the clean TiO_2 substrate, and for V2p from a ~5 ML thick vanadia overlayer (see ref. [37]). The overall symmetry and the main intensity maxima for the Ti2p and V2p plots are exactly the same, although the relative intensity of the features may be different.

4. Perspectives and future directions

We hope that this short introduction to the principles of angle-scanned photoelectron diffraction and the few examples herein reported have been able to demonstrate the capabilities and the inherent simplicity of the technique. From an experimental point of view, it only requires a standard XPS spectrometer equipped with a sample manipulator capable of polar and/or azimuthal motions. In many cases, qualitative structural information (e.g. growth modes and lattice distortions) is directly obtained from the experimental raw data without any need of theoretical simulation. Thence, we expect that in the future it will become a standard characterization technique for monitoring the quality of an epitaxial growth, not limited to specialists of PD. On the other hand, extracting quantitative and detailed structural information (e.g. quantitative structural parameters, stacking faults and other structural defects) may require a bit of theoretical simulation in a trial-and-error fitting procedure, which implies some expertise in the electron scattering theory. However, user friendly SSC computer programs for this purpose are currently available [10c].

Passing now to the expectations for the future, we would say that two main breakthroughs can be anticipated. The first is a methodological one and refers to the holographic 3-D reconstruction of surface structures. The analogy between photoelectron intensity maps and optical holography was first suggested in 1986 [6], and, in 1988, it was shown theoretically that PD patterns can be Fourier transformed to generate 3-D surface site images [7]. The advantage of the holographic method over the traditional trial-and-error method is in the direct 3-D imaging by means of a computer transformation that has no model-dependent assumption. The state of the art in this field has been recently reviewed [1d, 2b, 8].

The second novelty is related to the use of synchrotron radiation sources for XPD experiments. The third generation synchrotron facilities, provide photon fluxes several orders of magnitude higher that the conventional laboratory x-ray sources, with much better energy and spatial resolution. We would expect, then, that studies on sub-monolayer molecular or atomic deposits and low-concentration components will be possible due to the higher sensitivity. Moreover, for the same reason time-dependent studies will also be achievable. Finally, the high spatial resolution will make possible the exploration of irradiated areas as small as a few hundred Å square, allowing a selective investigation of epitaxial systems which do not have uniform surface composition.

Acknowledgement

Thanks are due also to the whole group of the Surface Science and Thin Film Laboratory of the University of Padova, and in particular to prof. Eugenio Tondello for his encouragement in the implementation of the XPD technique.

References

[1] (a) C. S. Fadley, Prog. Surf. Sci. 16 (1984) 275; (b) C. S. Fadley, Phys. Scr. T17 (1987) 39; (c) C. S. Fadley, in: Synchrotron Radiation Research: Advances in Surface Science, Ed. R. Z. Bachrach (Plenum, New York, 1992); (d) C. S. Fadley, Surf. Sci. Rep. 19 (1993) 231.

[2] (a) S. A. Chambers, Adv. Phys. 40 (1990) 357; (b) S. A. Chambers, Surf. Sci. Rep. 16 (1992) 261.

[3] (a) W. F. Egelhoff, Jr., Crit. Rev. Solid State Mater. Sci. 16 (1990) 213; (b) W. F. Egelhoff, Jr., in: Ultrathin Magnetic Structures I, An Introduction to Electronic, Magnetic and Structural Properties, Ed. J. A. C. Bland and B. Heinrich (Springer-Verlag, Berlin, 1994).

[4] D. P. Woodruff, Surf. Sci. 299/300 (1994) 183.

[5] J. Osterwalder, P. Aebi, R. Fasel, D. Naumovic, P. Schwaller, T. Kreuz, L. Schlapbach, T. Abukawa and S. Kono, Surf. Sci. 331-333 (1995) 1002.

[6] A. Szoeke, in "Short Wavelength Coherent Radiation: Generation and Applications", Ed. D. T. Attwood and J. Bokor, AIP Conf. Proc. No 147, American Institute of Physics, New York (1986).

[7] J. J. Barton, Phys. Rev. Lett. 61 (1988) 1356.

[8] (a) J. G. Tobin and D. Waddill, Surf. Rev. Lett., Vol. 1, Nos. 2 & 3 (1994) 297; (b) S. Y. Tong, Hua Li and H. Huang, Surf. Rev. Lett., Vol. 1, Nos. 2 & 3 (1994) 303; (c) K. Heinz and H. Wedler, Surf. Rev. Lett., Vol. 1, Nos. 2 & 3 (1994) 319; (d) C. M. Wei, I.H. Hong and Y. C. Chou, Surf. Rev. Lett., Vol. 1, Nos. 2 & 3 (1994) 335.

[9] (a) J. J. Barton and D. A. Shirley, Phys. Rev. B32 (1985) 1892; (b) J. J. Barton and D. A. Shirley, Phys. Rev. B32 (1985) 1906; (c) J. J. Barton, S. W. Robey and D. A. Shirley, Phys. Rev. B34 (1986) 778.

[10] (a) J. J. Rehr, R. C. Albers, C. R. Natoli, E. A. Stern, Phys. Rev. B34, 4350 (1986); (b) J. Mustre de Leon, J. J. Rehr, C. R. Natoli, C. S. Fadley, J. Osterwalder, Phys. Rev. B39, 5632 (1989); (c) D. J. Friedman and C. S. Fadley, J. Electron Spectrosc. Relat. Phenom. 51 (1990) 689.

[11] (a) H. C. Poon and S. Y. Tong, Phys. Rev. B30 (1984) 6211; (b) S. Y. Tong, H. C. Poon, D. R. Snider, Phys. Rev. B32 (1985) 2096.

[12] (a) J. J. Barton, M.-L. Xu and M. A. van Hove, Phys. Rev. B37 (1988) 10475; (b) M.-L. Xu, J. J. Barton and M. A. van Hove, J. Vac. Sci. Technol. A6 (1988) 2093; (c) M.-L. Xu, J. J. Barton and M. A. van Hove, Phys. Rev. B39 (1989) 8275.

[13] (a) J. J. Rehr and E. A. Albers, Phys. Rev. B41 (1990) 8139; (b) A. P. Kaduwela, G. S. Herman, D. J. Friedman, C. S. Fadley and J. J.Rehr, Phys. Scr. 41 (1990) 948; (c) A. P. Kaduwela, G. S. Herman, D. J. Friedman and C. S. Fadley, J. Electron Spectrosc. Relat. Phenom. 57 (1991) 223.

[14] (a) L.-G. Petersson, S. Kono, N. F, T. Hall, C.S. Fadley and J. B. Pendry, Phys. Rev. Lett. 42 (1979) 1545; (b) P. J. Orders, S. Kono, C. S. Fadley, R. Trehan and J. T. Lloyd, Surf. Sci. 119 (1981) 371.

[15] D. A. Wesner, F. P. Coenen and H. P. Bonzel, Phys. Rev. Lett. 60 (1988) 1045.
[16] R. S. Saiki, G. S. Herman, M. Yamada, J. Osterwalder and C. S. Fadley, Phys. Rev. Lett. 63 (1989) 283.
[17] A. V. de Carvalho, M. C. Asensio and D. PP. Woodruff, Surf. Sci. 273 (1992) 381.
[18] L. S. Caputi, R. G. Agostino, A. Amoddeo, E. Colavita and A. Santaniello, Surf. Sci. 282 (1993) 62.
[19] L. S. Caputi, G. Chiarello, M. G. Lancellotti, G. A. Rizzi, M. Sambi and G. Granozzi, Surf. Sci 291 (1993) L756.
[20] M. C. Asensio, A. V. de Carvalho and D. P. Woodruff, J. Phys.: Condensed Matter 3 (1991) S111.
[21] O. Knauff, U. Grosche, H. P. Bonzel and V. Fritzsche, Mol. Phys. 76 (1992) 787.
[22] M. Sambi, G. Granozzi, M. Casarin, G. A. Rizzi, A. Vittadini, L. S. Caputi and G. Chiarello, Surf. Sci. 315 (1994) 309.
[23] S. Kono, S. M. Goldberg, N. F. T. Hall and C. S. Fadley, Phys. Rev. Lett. 41 (1978) 1130.
[24] G. Granozzi, M. Sambi, L. Zaratin, G.A. Rizzi, E. Tondello and A. Santaniello, Surf. Sci., 331-333 (1995) 35.
[25] A. Outka R. J. Madix and J. Stöhr, Surf. Sci. 164 (1985) 235.
[26] J. Stöhr, D. Outka, R. J. Madix and U. Dobler, Phys. Rev. Lett. 54 (1985) 1256.
[27] M. D. Crapper, C. E. Riley and D. P. Woodruff, Phys. Rev. Lett. 57 (1986) 2598.
[28] M. D. Crapper, C. E. Riley, D. P. Woodruff, Surf. Sci. 184 (1987) 125.
[29] D. P. Woodruff, C. F. McConville, A. L. D. Kilcoyne, T. Lindner, J. Somers, M. Surman, G. Paolucci and A. M. Bradshaw, Surf. Sci. 201 (1988) 228.
[30] M. Casarin, G. Granozzi, M. Sambi, E. Tondello and A. Vittadini, Surf. Sci. 307-309 (1994) 95.
[31] M. Sambi, L. Zaratin, G. Granozzi, G. A. Rizzi, E. Tondello and A. Santaniello, Surf. Sci. 321 (1994) L214.
[32] U. Diebold, J.-M. Pan and Th.E. Madey, Surf. Sci. 331-333 (1995) 845, and references therein.
[33] M. Sambi, E. Pin, G. Sangiovanni, L. Zaratin, G. Granozzi and F. Parmigiani, Surf. Sci., 349 (1996) L169.
[34] J.-M. Pan, B. L. Maschhoff, U. Diebold and Th. E. Madey, Surf. Sci. 291 (1993) 381.
[35] M. Sambi, G. Sangiovanni, G. Granozzi and F. Parmigiani, submitted.
[36] M. Ramamoorthy, R. D. King-Smith and D. Vanderbilt, Phys. Rev. B49 (1994) 7709.
[37] M. Sambi, G. Sangiovanni, G. Granozzi and F. Parmigiani, manuscript in preparation.

CO-ADSORPTION ON METAL-OXIDE CRYSTAL SURFACES

Cases of CO/Cu/ZnO(0001) and CO_2/Na/TiO_2(110)

P.J. MØLLER
University of Copenhagen, Department of Chemistry
Universitetsparken 5, DK-2100 Copenhagen, Denmark

Abstract

Coadsorption on metal-oxide crystal surfaces is discussed for two model catalyst examples in which a (CO_2^-) complex is involved: 1) Adsorption of CO onto Cu clusters and thin films supported by a thermally etched (0001) surface of a ZnO single crystal and 2) Adsorption of CO_2 onto Na clusters and thin films supported by 1x1- and 1x2-terminated (110) surfaces of a TiO_2 single crystal.

In the former case we will apply electron spectroscopy results from adsorption of CO on clean and Cu-deposited ZnO(0001) surfaces to discuss evidence for formation of a (CO_2^-) chemisorption complex and for Cu oxidation during CO exposure.

In the latter case we will apply synchrotron-radiation-induced core-level-, valence-band and resonant photoemission results from adsorption of CO_2 on a 1x1 truncated and on a 1x2 reconstructed TiO_2(110) surface to discuss the difference in Na adsorption behavior of the two surfaces (adsorption onto two different sites) and the strong reactivity towards CO_2 when submonolayer quantities of Na are preadsorbed. Valence-band data shows formation of carbonate. At low coverage the presence of a (CO_2^-) complex suggests a $2CO_2^- \rightarrow CO_3^{2-} + CO$ reaction.

1. Introduction

The system consisting of particles and ultrathin films of Cu deposited onto ZnO surfaces is an important component of the Cu/ZnO/Al_2O_3 heterogeneous catalyst used in synthesis of the industrially important product methanol, CH_3OH, through the process $CO + 2H_2 \rightarrow CH_3OH$ [1]. We will here discuss the behavior of deposited Cu and its effect on adsorption of CO.

Also important in heterogeneous catalysis is the use of promotors in many processes. The alkali metals are strong promotors [2]. We will here discuss the behavior of Na deposited as particles and ultrathin films onto TiO_2 and its effect on adsorption of CO_2. TiO_2 rutile has found widespread use in catalysis and gas-sensor applications [3,4].For both systems we have used low-index surfaces of single crystals of the metal oxides: the terminated ZnO(0001) and the TiO_2(1$\dot{1}$0) surfaces, respectively. A main question

R.M. Lambert and G. Pacchioni (eds.),
Chemisorption and Reactivity on Supported Clusters and Thin Films, 267-284.

concerning these systems is the respective roles of the metal particles and their oxide substrate. There are several reasons for the variation of the catalytic activity with metal concentration on the oxide substrate. At the sub-monolayer stage of deposition, the metal adatoms may be considered as new active adsorption and reactivity sites. On the other hand, islands or clusters may exhibit different adsorption and reactivity properties in comparison with the substrate [5,6].

We have chosen primarily to work with bulk single crystals to get our model studies as close as possible to the real catalysts used in industrial application. Since the metal oxides most often are wide-gap semiconductors or insulators [7] it is, however, often difficult to carry out experiments using electron spectroscopy and diffraction due to charging effects. It has proven possible in many cases, though, to obtain sufficient conductivity through induced oxygen vacancies [8].

2. Cu/ZnO(0001)

2.1 LEED

2.1.1 *Clean ZnO(0001)*

Firstly we will, of course, make sure that we start out the experiment with an atomically clean and ordered ZnO(0001) surface. In this case this was obtained by

Figure 1. LEED from a 800K-annealed ZnO(0001) surface. Primary energy 63 eV [10].

argon ion bombardment and annealing cycles at 750 K, and a sharp LEED pattern appeared showing pronounced pure singlets with six-fold symmetry, which is referred to as a 1x1 pattern [9]. Further annealing at 800 K yields an ordered array of sub-sextets around each of the main diffraction spots (Fig. 1).This latter behavior may be understood in terms of diffracted reflection from a ZnO(0001) surface containing hexagonal pits created by thermal etching during the annealing [10].

2.1.2 *Cu/ZnO(0001)*

Deposition of 0.3 ML of Cu causes elimination of the sub-sextet structure and with Cu deposition to 2 ML a new hexagonal pattern gradually appears. With 6 ML of Cu it fully covers the surface, showing an (111)-ordered formation of Cu islands.

2.2 ELECTRONIC STRUCTURE

2.2.1 *TCS*

The electronic structure was investigated by total (target) current spectroscopy (TCS), and by electron energy-loss spectroscopy in the electr-onic regime (EELS). Figure 2 shows the total current spectrum, $S(E)=\mathrm{d}J(E)/\mathrm{d}E$ versus incident energy E, measured at normal incidence.

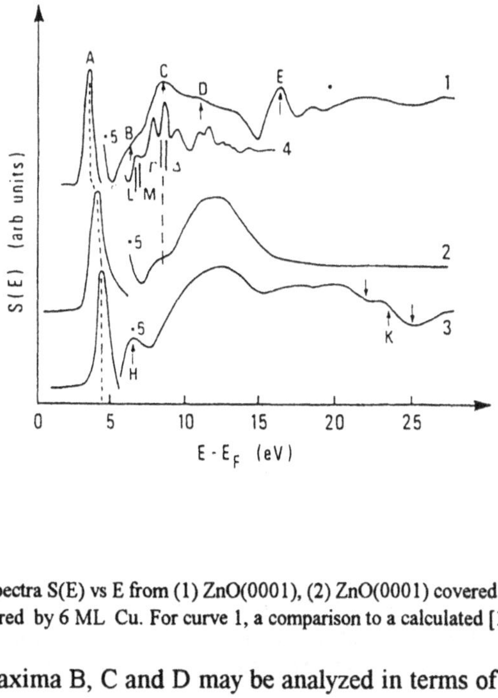

Figure 2. TCS spectra S(E) vs E from (1) ZnO(0001), (2) ZnO(0001) covered by 0.3 ML Cu and (3) Z nO(0001) covered by 6 ML Cu. For curve 1, a comparison to a calculated [11] DOS is made (4).

The origin of maxima B, C and D may be analyzed in terms of the matching method for determining elastic reflection coefficients [12-14]. It has also been shown that the energy location of extrema in TCS coincides with the location of band-structure critical points in the Brillouin zone for empty electronic states [15,16].

However, the energy position of the density-of-states (DOS) maxima corresponds to the critical points of the band structure whence we may expect a correlation between the energy structure of the DOS and TCS curves. Figure 2 curve 4 shows good agreement between the experimental maxima B, C and D and the DOS structure in corresponding critical points (L, M, G and D). E is connected with reflection intensity variation due to diffraction from thermo-etched pits [10]. The TCS primary peak A shows a vacuum-level position in the energy diagram and indicates that the work function is about 3.5 eV, in good accordance with literature values for annealed and relaxed ZnO(0001) surfaces [17,18].

Work-function change. The shift of this peak with Cu deposition shows a work-function change to ϕ=4.0 eV for 0.3 ML Cu and 4.5 eV for 6 ML Cu. The obtained value of ϕ is smaller than f for bulk Cu(111), indicating that the surface is covered with Cu(111) islands without complete surface coverage. The broadening of the primary peak peak also reflects formation of patchwise Cu coverage with Cu deposition.

Fine structure. The TCS fine structure from the ZnO surface is strongly attenuated only the strong peak C is still seen after 0.3 ML Cu. The Cu related peaks H and K are related to the band edges in the Cu band structure. The wide hump in the 9-14 eV energy range may be caused by copper-induced pertiubation of the ZnO surface band structure since it is more pronounced at submonolayer coverage and gradually disappears with increased Cu thickness.

2.2.2 *EELS*

Similarly, Figure 3 gives the EELS results (Fig. 3), with information on the changes in the electronic structure. Peak P is assigned to a bulk plasmon, and B,C and D to interband transitions between valence and conduction bands. Peak A is connected with electron transition from defect states in the ZnO gap. A set of vertical lines (curve 3) indicates the energy positions for the losses for the Cu(111) surface while vertical lines below curve 6 shows the losses obtained for a Cu_2O surfaces [8a].

2.3. ADSORPTION OF CO

Curves 4-6 of figure 3 illustrates the influence of CO adsorption. A set of losses F_1 to pronounced CO-adsorption-induced losses are F_1 and F_2, indicating the CO_2^- chemi-sorption-complex formation on the open ZnO area [22].

For the clean surface the amount of desorbed CO_2 is about one order of magnitude higher than that of CO, demonstrating a practically complete CO oxidation. Immediately upon switching on the illumination, a jump in the CO_2 partial pressure takes place (curve 1), obviously caused by CO_2 photodesorption since the sample surface temperature at this moment is equal to room temperature. This result is justified by previous research [19,23,24]. The two pronounced humps (marked by dotted lines in Figure 4) may be connected with the a (125 ^{0}C) and b (190 ^{0}C) phases of CO_2 thermodesorption from the ZnO surface. The shape of the CO desorption curve (curve 1a) corresponds to the origen of the observed desorption corresponding to the a (140 ^{0}C) of CO thermodesorption from ZnO [25].

Desorption from a 0.3 ML Cu surface exhibits a higher adsorption capacity and the desorption starts from a lower temperature (70 ^{0}C) and the amount of CO is about 3 times higher than for a clean ZnO surface.

Desorption from a 6-ML-Cu surface exhibits predominantly desorption of CO and it is clearly seen (curve 3) that there is no photodesorption component in the CO_2 desorption flow. Thus the influence of copper particles on electronic properties of ZnO restricts the validity of the CO_2 molecule photodesorption mechanism. At the same time a gradual elimination of the CO catalytic oxidation takes place.

2.4. CO AND CO_2 DESORPTION PROCESSES

Finally, we will investigate photoinduced and thermoinduced processes of CO and CO_2 desorption from the above described surfaces. The processes were caused by a beam of

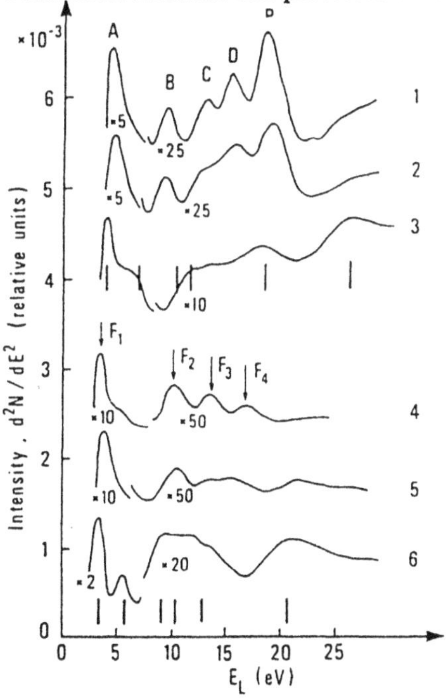

Figure 3. EELS, primary energy 95 eV. (1) Clean ZnO(0001); (2,3) ZnO(0001) covered by 0.3 ML Cu and 6 ML Cu, respectively; (4-6) Difference spectra representing CO adsorption influence on the loss spectra [22].

light from a high-pressure xenon lamp which was externally focused through a window onto the sample mounted in UHV. The desorption was measured by quadrupole mass spectrometry.

Figure 4 shows the development of the CO_2 (curves 1-3) and CO (curves 1a-3a) desorption processes from the clean ZnO(0001) (curves 1 and 1a), from the 0.3 ML Cu (curves 2 and 2a) and from the 6 Ml Cu (curves 3 and 3a) surfaces. Curve 4 shows the temperature variation with time [22]. P is assigned to a bulk plasmon, and B, C and D to interband transitions between valence and conduction bands. Peak A will be connected with transition from defect states in the ZnO gap.

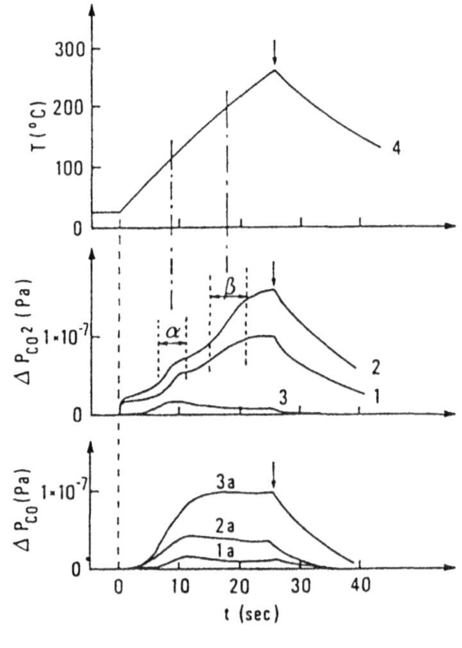

Figure 4. CO_2 (curves 1-3) and CO (curves 1a-3a) photoinduced desorption curves from the clean ZnO(0001) surface (curves 1 and 1a), and from the surfaces covered by 0.3 ML Cu (curves 2 and 2a) and by 6 ML Cu (curves 3 and 3a); Curve (4) shows temperature variation with time. Note the photoinduced pressure rise in the first section [22].

The deposition of 6 ML Cu (Figure 3 curve 3) shows a general increase in intensity by a factor 5 with respect to the clean Zno surface and the shape is more similar to the one for Cu(111). A set of vertical lines indicates the loss positions for the Cu(111) surface. The observed EELS signal at primary energy 95 eV thus originates from the copper overlayer.

The CO desorption results are in accord with temperature-programmed desorption (TPD) results from the CO/Cu on nonpolar ZnO surfaces, where the main increase in the CO desorption flow takes place in the same range (70-100 ^{0}C), and sub-monolayer Cu deposition is accompanied by a significant increase in amount of desorption [26].

3. CO/Cu/ZnO(0001) DISCUSSION

3.1. CLEAN ZnO(0001) EXPOSED TO CO WITHOUT LIGHT

We may thus conclude, that at the clean ZnO(0001) exposed to CO *without illumination* the oxidation takes place on defect sites, bonding to substrate oxygen atoms and creating CO_2^- chemisorption complexes. The observation of the CO_2 can be explained using the photoelectron mechanism of desorption [23-25]. Since the xenon lamp produces a large UV fraction (the quantum energy exeeds the the ZnO gap width of 3.3 eV [21]), the absorption of light quanta in the subsurface region leads to creation of free electrons and holes. A capture of a photoexited hole to the CO_2^- complex is accompanied by its energy excitation and its neutralization. After neutralization the particle is in an antibonding state and is being acted upon by forces of repulsion from the surface, causing desorption [19,22]. Thus the particular properties of the clean ZnO(0001) surface which contains hexagonal-pit defects (of atomic-scale size) are: (i) a high reactivity for CO oxidation followed by CO_2^- chemisorption-complex formation [24], (ii) validity of a photo-electron mechanism for CO_2 photodesorption.

3.2. 0.3-ML-Cu/ZnO(0001)

In the low-coverage range below 0.3 ML of Cu the Cu atoms are bonded preferentially to defect sites, filling etch pits that directly lead to elimination of the sub-sextet structure in LEED. It was earlier shown [10] that this sub-sextet structure was originated from the diffraction reflection from a ZnO(0001) surface containing hexagonal etch pits. Simultaneously oxidation of Cu atoms occurs, as observed by EELS. Thus we see Cu islands, perhaps containing Cu_xO fragments, at the ZnO surface. It has been shown [27] that CO adsorbs more readily on Cu than on ZnO and have a higher sticking probability, particularly to Cu clusters [28]. We observe, as a result, CO_2 molecular desorption from ZnO areas (and photodesorption is still seen) while have CO molecular desorption from Cu (or Cu_xO).

3.3. 6-ML-Cu/ZnO(0001)

At the high coverage of Cu the Cu atoms merge into (111) oriented epitaxial clusters that are preferentially oriented parallel to the ZnO(0001) surface, as seen by LEED. Following the themally activated CO adsorption, partly Cu oxidation occurs (EELS indicated formation of Cu_2O fragments). Predominantly, the CO desorbs; a surface CO oxidation was strongly eliminated, influenced by the Cu particles. A photodesorption

CO_2 component is absent in this case, hence CO_2 molecules appearing in the flow may be associated with oxidation of CO either on the Cu_xO fragments or on the open ZnO areas.

4. CO/Cu/ZnO(0001). Conclusion

For the Cu/ZnO(0001) system we reach the conclusion that the nearly fully Cu-covered ZnO surface loses its catalytic activity with respect to CO oxidation and does not show validity of a photoelectron mechanism for particle desorption.

Let us now turn to the alkali-promoted system:

5. Na/TiO_2(110)

Only very few single-crystal studies on the alkali-metal/metal-oxide interface have been carried out [7] in spite of their interest due to the strong enhacement in surface basicity [2]. However, rutile TiO_2 is an exeption since several alkali-metal studies have been carried on it [29-34]. Of the different surfaces of crystalline rutile, TiO_2(110) is considered to be the most stable. Annealing at elevated temperatures in UHV causes reconstruction to occur. Annealing to 600^{0}C thus yields a sharp 1x1 LEED pattern is obtained and annealing further to 800^{0}C yields a sharp 1x2 reconstructed surface. The structure was attributed to a missing-row structure originating from removal of every second of the bridging surface-oxygen rows [35]. This has recently been confirmed in several STM experiments [36-39] from which some modifications, particularly with regard to lateral displacement of surface Ti atoms, were introduced to the original model.

The surface oxygen vacancies are of general high interest in relation to the reactivity of metal oxides whence the rutile reconstruction to a 1x2 surface is of particular interest. One may consider the surface an array of ordered defects which can varied over the above temperature range [35] and thus give possibility for detailed studies of the effect of a group of common defects.

Resonant photoemission (ResPES) on the 1x1 and 1x2 surfaces showed clear differences in the Ti-O hybridization which was related by Nerlov *et al.* [40] to elongation of the Ti-O surface bonds upon reconstruction. This was evidenced in the most recent STM study [36].

ResPES. A few words should be in order on the basic principle of ResPES. It has in recent years proved quite a useful tool in studies of trans-ition metal oxides [7,41]. The resonance effect occurs when sweeping the photon energy through the 3p absorption edge of the transition metal. Autoionization of the excited $3p^5 3d^{n+1}$ state yields a final state $3p^6 3d^{n-1} + e^-$, which is degenerate to the direct valence-band excitation. In the case of TiO_2, which is nominally a $3d^0$ state, de-excitation of the excited $3p^5 3d^1$, $3p^5 4s^1$, ... states occurs through interatomic channels [41], i.e.:

$$\text{Ti } 3p^5 3d^1 + \text{O } 2p^6 \rightarrow \text{Ti } 3p^6 3d^0 + \text{O } 2p^5 + e^-.$$

The resonant emission will then be dominated by states with relatively large Ti-O hybridization, and by separating the resonant from the non-resonant part valuable information on the Ti-O hybridization is obtained. Due to the energetic coincidence of the 3p absorbtion edges in transition metals and the minimum of the mean free path of electrons in solids, ResPES data obtained at these edges obviously are very surface sensitive [40].

Also, the absence of "in resonance" bandgap emission for both the 1x1 and the 1x2 surface indicated that for both surfaces all surface-Ti should be considered as Ti^{4+} [40]. Surface Ti^{3+} states may be created, however, by photo-induced desorption [35].

Onishi *et al.* [29] found a c(4x2) structure on the 1x1 surface near 0.5 ML Na from which a model consisting of "Na_2O dimers" was proposed while an STM study by Murray *et al.* [32] on the 1x2 surface found a p(4x2) structure of single Na atoms.

The former study proposes bonding of Na with two bridging O atoms while the latter study found the Na to adsorb adjacent to the O rows in a 3-fold arrangement. Hence, a comparative study of the two surfaces is called for. Onishi *et al.* also found that submonolayer amounts of Na deposited on the 1x1 surface dramatically enhance the reactivity of the 1x1 surface toward CO_2 at RT, leading to what was interpreted as carbonate species. The onset of the CO_2 uptake was found to coincide with the formation of "Na_2O dimers" while saturation of the CO_2 uptake occurred near 1 ML, corresponding to full dimerisation of the surface. Nerlov *et al.* [42] show, however, that due to the higher resolution in synchrotron-radiation-induced photoemission (SRPES) and ResPES an observed shift can be related to different adsorption sits on the two surfaces while the Na-induced valence-band signal from the 1x2 surface is quite similar to what would be expected from Na_2O compounds.

Finally, the reactivity of Na/TiO_2 1x1 and 1x2 toward CO_2 is discussed, and it is found possible to distinguish the active from the non-active Na sites [42].

5.1. Na ADSORPTION ON TiO_2(110)-1x2

5.1.1. *Core-level measurements*

Figure 5 shows the changes in the Ti 3p and Na 2p levels with deposition of Na. A small shift of about 0.1 eV toward higher binding energy (BE) was detected and attributed to additional band bending. The Na 2p shows a total BE shift of 1.15 eV (including the band bending) towards lower BE.

A continuous shift towards lower BE of the Na2p emission is observed, the total shift amounting to 1.15 eV within the coverage range studied. This value includes the 0.1 eV correction due to band bending.

Figure 6 (lower graph, upper curve) shows the growth curve for Na on the 1x2 surface. We observe two breakpoints within the coverage investigated. A clear breakpoint is seen at deposition time d_t= 21.1 min. The position of the break point practically coincides (within 5%) with what was observed for the 1x1 interface. In the following this point is used as a reference point, coverage θ=1 ML, for the 1x2 surface. total change is 2.5 eV (2.2 eV at θ=0.5 ML), reaching an assymptotic value at θ=1 ML. The accuracy of the coverage is estimated to be within 0.05 ML. Simultaneously, a monotonous decrease of the work function is observed (Figure 6, upper graph).

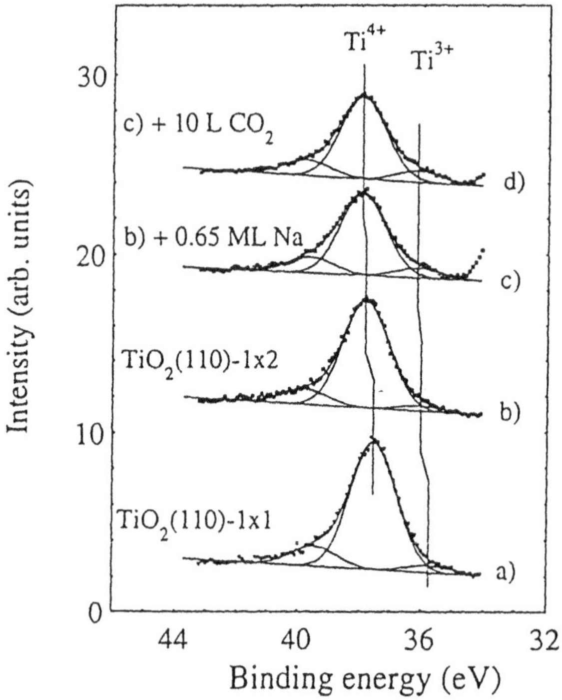

Figure 5. Ti 3p core-level spectra obtained from (a) clean TiO_2(110)-1x1 and (b) -1x2 surfaces as function of Na deposition time d_t. BE of Ti^{3+} and Ti^{4+} indicates the band bending which is not included [42].

The 1x2 LEED pattern was gradually attenuated without appearence of additional spots as Na was deposited.

We observe an additional breakpoint at 8 min (θ = 0.35 ML). This coincides with a relatively large change in BE of the Na 2p level. From the photoemission signal obtained at θ= 0.25 ML we further observe a prominent asymmetry in the shape of the Na 2p peak (Figure 7, lower graph). This suggests that the observed BE shift and the large increase in the width of the Na 2p emission observed at low θ (Figure 8, lower graph) are due to the presence of two separate contributions to the Na 2p emission. At θ= 0.4 ML the asymmetry of the Na 2p peak is reversed, the low-binding-energy feature being the most intense (Figure 7). The spectra obtained at θ=0.4 ML are well described by two Gaussians of equal width. The resulting BE are given by BE= 32.32 ± 0.05 eV for the high-BE feature (a), while the low-BE feature (b) is positioned at 31.37 ± 0.05 eV. Thus, at relatively low θ, the Na adsorption is characterized by a BE of 32.32 eV, while adsorption in the range 0.2 ML < θ< 0.4 ML is dominated by a BE of 31.37 for Na 2p.

Upon further Na adsorption we observe a continuous shift towards lower BE. At θ> 1 ML we also note an increased asymmetry of the Na 2p emission. These are all indications of the presence of a third feature in the Na 2p emission at higher θ. A curve-fitting procedure yields BE= 30.41 ± 0.10 eV for the high-θ Na feature (c). The variation with θ for the three Na 2p features is shown in Figure 8 (upper graph). From this it is obvious that the breakpoint observed at 1 ML in Figure 6 coincides with what seems to be saturation of feature (b). We assign the breakpoint position to the completion of the Na adlayer involved in bonding to the substrate. The presence of the feature (c) with noticeable intensity indicates that strict monolayer formation does not occur at this surface.

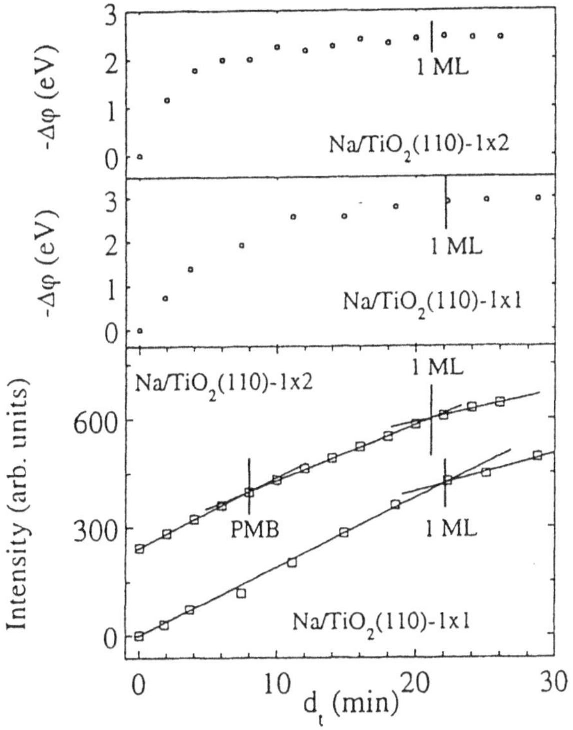

Figure 6. Lower graph: Growth curves obtained for Na deposition on the 1x1 (lower curve) and 1x2 (upper curve) based upon integration of the Na 2p signal obtained at 125 photon energy as a function of deposition time. The size of the data points corresponds to the estimated error. PMB denotes the premonolayer breakpoint on the 1x2 surface. Middle graph: Change in θ with Na deposition for the 1x1 surface. Upper graph: Change in θ with Na deposition for the 1x2 surface[42].

5.1.2. *Valence-band emission*

The difference spectrum of thevalence band of the 1x2 surface obtained at θ= 0.35 ML exhibits a pronounced peak in the band-gap region (BE= 0.7 eV), and broad features around BE= 5.9-6.5 eV, 8.0 eV and 9.5-11.5 eV. We also note a substantial decrease in emission around BE= 4.0 eV.

5.2. Na ADSORPTION ON TiO_2(110)-1x1

The growth curve, the BE changes and VB emission for the 1x1 are given in the curves together with the 1x2 data.

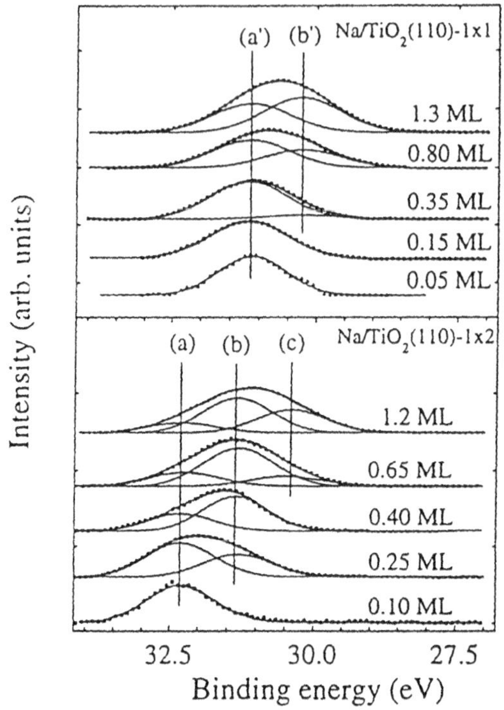

Figure 7. Lower graph: Emission from the Na 2p level obtained rom the 1x2 surface at various θ as well as the Gaussians fits to the peak shape. At θ< 0.4 ML, the Na emission is fitted to two Gaussians centered at BE= 32.32 eV (a) and BE= 31.37 eV (b), while the emission obtained at higher θ are fitted to one additional Gaussian centered at BE= 30.42 eV (c). Upper graph: Emission from the Na 2p level obtained from the 1x1 surface at various θ as well as the fits. The data fitted at and below 0.15 ML are fitted to a single Gaussian centered at BE=31.15 eV while the emission obtained at higher θ are fitted to a Gaussian centered at BE= 30.25 eV. All curves are scaled to same intensity [42].

5.3. COADSORPTION OF CO_2 AND Na ON $TiO_2(110)$

Figure 9 gives the EDC's at photon energy 36 eV at various CO_2 exposures for 0.8-ML-Na covered surfaces. The dosing of CO_2 was carried out at 160 K while the Na deposition was performed at RT. As CO_2 is adsorbed we observe additional peaks at BE= 13.0 eV (i) and 10.8 eV (ii).

(The CO_2 uptake (insert, fig. 9) is judged from the intensity of the dominating peak ii). Hence, the uptake on the 1x2 surface at these coverages far exceeds the uptake at the 1x1 surface. Also, core-level spectra shows no change in the Ti 3p with CO_2 adsorption, whence the Na-to-substrate charge transfer indicates that the Ti^{3+} states are unaffected by the CO_2.

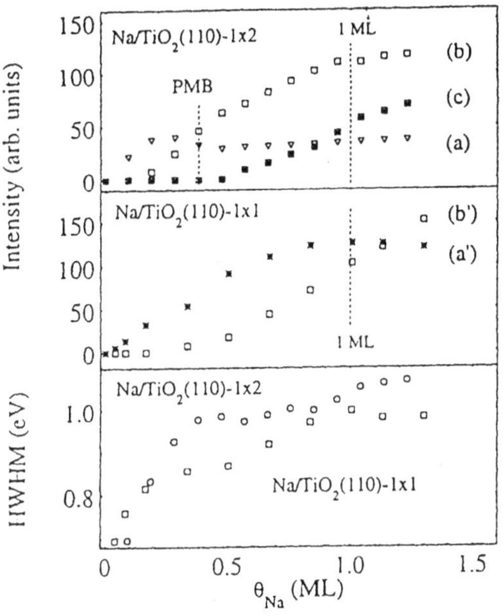

Figure 8. Lower graph: The variation of the half-width-half-maximum (HWHM) of the Na 2p emission with θ on the 1x1 (squares) and 1x2 (circles) obtained by fitting the Na 2p emission to one Gaussean (background subtracted). Middle graph: Variation with θ of the intensity of features (a') and (b') contributing to the overall Na 2p. Upper graph: Variation with θ of the intensity of features (a), (b) and (c) contributing to the overall Na 2p. PMB denotes the premonolayer breakpoint observed in Figure 6 [42].

An increased broadening and asymmetry of the Na 2p peak was found, indicating that the (small amount of) adsorbed CO_2 induces a negative shift in Na binding energy involved in bonding CO_2. A overall shift towards lower binding energy of the Na 2p emission at 0.4 ML was seen, and the asymmetry of the Na 2p peak is still for present, indicating two separate contributions. A CO_2 induced shift of -0.25 eV was seen both components. Further, not all of the components within the Na 2p were shifted; the third feature (c) was unshifted.

5.4. DISCUSSION CO/Na/TiO_2(110)

5.4.1. *Na adsorption*

The presence of two Na peaks in the monolayer region for the 1x1 surface is not consistent with the Na_2O-dimerisation model applied earlier [29] since that deals only with one submonolayer site. Our finding of two separate contributions is consistent [42], however, with their earlier Auger data. The higher resolution of our SRPES core-level results enables a separation of the various contributions in the signal by analyzing the asymmetries and thus demonstrates the advantage of SRPES in this case. The rather large 1.5 eV difference in binding energy between the two surfaces suggests different adsorption geometries. We would intuitively expect the Na atoms to maximize their oxygen contribution during adsorption, i.e. 3-fold for both surfaces [42].

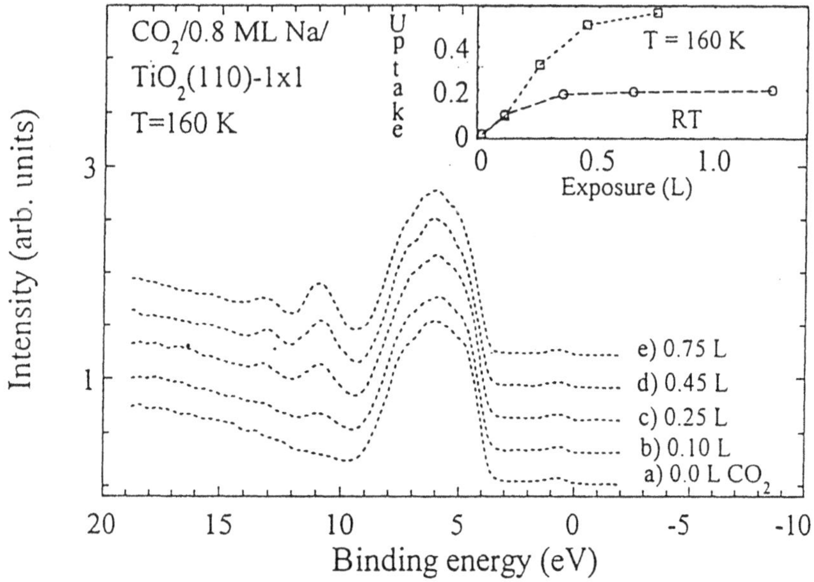

Figure 9. VB spectra obtained at photon energy 36 eV for increasing exposures of CO_2 for a 0.8 ML-Na covered 1x1 surafe at 160 K (band bending not included). Insert: The CO_2 uptake judged by the intensity of the peak observed at 10.8 eV for CO_2 adsorption on 0.8-ML-Na/TiO_2(110) at 160 K and RT, respectively[42].

However, we may have two types of 3-fold -coordinated sites: Either Na bonding to two planar O atoms an one bridging (adjacent to the O rows) or the reverse (in between the bridging atoms). Earlier calculations [43-45] propose a relaxation of about 0.02 nm for the 1x1 surface, but recent SRPES results [40] found that specific surface-related resonant PES, presumably arising from surface relaxation, vanishes with Na adsorption. This was attributed [40] to surface derelaxation, giving support to the assumption of bulk truncation. Hence, we assume bulklike positions for the 1x1 substrate surface atoms with Na positioned in hollow sites. By further assuming the Na to O distances equal to the sum (0.229 nm) of the respective ionic radii we obtain the Na position given in Table Ia [42].

Murray *et al.* [36] found that the removal of O rows in the reconstruction is accompanied by ca 0.05 nm lateral displacement of the 5-fold Ti surface atoms toward the missing rows. Even though it was not imaged in the STM data [36] we would expect a displacement of the same order for the neighbouring planar O atoms, giving the positions shown in Table Ib together with the positions obtained for the Na in the optimum 3-fold coordinated sites [42].

Adsorption adjacent to the oxygen rows favours sp^3 hybridization of the bridging O atoms, while no O(bridge)-hybridization favouring Na bonding exists when Na is adsorbed in between the rows. In the latter case the stongest Na-O interaction is probably through the unhybridized O $2p_y$ orbital of the planar O atoms perpendicularly directed toward to the surface. But these orbitals are also involved in π to Ti(t_{2g}) orbitals [46]. Therefore, considering the smaller Ti coordination for O(bridge), we believe the N-O (planar) bonding, in the case of adsorption in between the rows, to be weaker than

TABLE Ia. The positions (in Å) within the $TiO_2(110)$-1x1surface unit cell of surface Ti and O atoms and the proposed positions of Na for the two adsorption sites with maximum oxygen co-ordination together with the Na-Ti distances obtained for these adsorption sites.

	X	Y	Z	d(Na,Ti-5-fold)	d(Na,Ti-6-fold)
O(top)	0	0	1.267		
Ti 6-fold	0	± 1.479	0		
Ti 5-fold	± 3.248	0	0		
O(planar)	± 1.981	± 1.479	0		
Na(adjacent)	2.25	0	1.74	2.01	3.20
Na(between)	1.46	1.479	2.25	3.23	2.68

The Na-O(bridge) bonding found for Na adsorption adjacent to the O rows [42]. This leads us to suggest that the shift observed when comparing the two surfaces may be explained by a change in the initial Na adsorption site from "in between" on the 1x1 to 'adjacent to" the rows on the 1x2 surface, as further discussed in ref. 42. That (a), in the relative intensities of the high Na 2p binding energies (a) and (a′), is approximately half of the intensity of (a`) is in accordance with the removal of half the oxygen rows upon reconstruction and gives support to the model of initial adsorption in connection with the O rows [42].

TABLE Ib. The positions (in Å) within the 1x2 surface unit cell of surface Ti and O atoms and the proposed positions for the two adsorption sites with maximum oxygen co-ordination together with the Na-Ti distances d obtained for these adsorption sites. The positions for O(top) and Ti(6-fold) are as shown in Table Ia.

	X	Y	Z	d(Na,Ti-5-fold)	d(Na,Ti-6-fold)
Ti 5-fold	± 3.75	0	0		
O(planar)	± 2.48	± 1.479	0		
Na(adjacent)	2.25	0	1.75	2.30	3.51
Na(between)	1.55	1.479	2.10	3.38	2.61

5.4.2. *CO_2 and Na adsorption*

Several indications suggest that CO_2 adsorption on the Na/$TiO_2(110)$ leads to carbonate-like species [47,48], but Freund and Messmer [49] emphasise the large similarities between valence-band (VB) PES from CO_2^- and CO_3^{2-}species [42]. From a comparison of experimental results for CO_2^-/CO_3^{2-} systems [42], differences in the upper part of the VB are expected when comparing what have been assigned to CO_2^- and CO_3^{2-}. For carbonate, we would expect three peaks in the upper part of the VB, derived from $1a_2{}'$, $1e''$ and $4e'$ orbitals, respectively. These show up experimentally, however, as one broad feature in the VB. In case of CO_2^- we expect two separated features in this region: a high-energy peak from $6a_1$ states and a high-energy peak by combined $1a_2$ and $4b_2$ states [42].

From these considerations we suggest that the difference spectra (fig. 12) should be divided into two distinct parts, the lower being CO_2^- related and the two upper ones to CO_3^{2-}.

There are, essentially, two cases which could lead to the observed emission: Two types of adsorption sites (one for CO_2^- and one for carbonate formation), but due to the adsorption energies involved in converting CO_3^{2-} to CO_2^-, which is higher than the thermal energies, we would have a dominating CO_3^{2-}. Alternatively, carbonate may be formed by the reaction $2CO_2^- \rightarrow CO_3^{2-} + CO$, where CO desorbs (it was found that CO would not even at 160 K be adsorbed). This reaction has been proposed for metal surfaces by Freund and Messmer [49] who propose a precursor consisting of a CO_2^- dimer.

5.5. CONCLUSION ON CO_2/Na/TiO_2(110)

SRPES Na 2p core-level spectra from the Na/TiO_2(110)-1x1 interface show two features in the submonolayer region, and the 1x2 interface shows three features. The initially adsorbed Na shows a 1.15 eV higher binding energy for the 1x2 interface. The data agrees with a simple adsorption model in which the 1x2 reconstruction facilitates Na adjacent to the protruded O rows rather than in between the bridging O atoms as for the 1x1. Thereby will earlier contradicting models [29,32] for core-level data from both surfaces suggest that simultaneous occupation of various Na adsorp-tion sites occurs, but relating to the density of O rows present.

CO_2 does not adsorb on clean 1x1 and 1x2 surfaces (not even at 160 K), but it does so on Na-precovered 1x1 and 1x2 surfaces at RT. The CO_2 uptake relates to the density of O rows. The reactivity towards CO_2 is strongly enhanced upon reconstruction to 1x2, indicating that bridging O atoms are not involved in bonding of CO_2. It is implied that the initial adsorbate should be considered to be CO_2^- which at higher uptake forms carbonate through the reaction $2CO_2^- \rightarrow CO_3^{2-} + CO$.

6. Acknowledgements

The author acknowledges the valuable contributions of many collaborators in this work, including S.A. Komolov, J. Nerlov, Q. Ge, E.F. Lazneva, E. Hjelm Pedersen, S.Vronning Christensen and S. Weichel, and The Danish Natural Science Research Council, The Commission of Science and Technology (E.U.) and the NATO Advanced Study Institute for financial support.

References

1. Klier, K. (1982), *Adv. Catal.* **31**, 243.
2. Tanabe, K. (1984) in B. Imelike, C. Naccache, G. Coudurier, Y.B. Taarit and J.C. Vedrine (eds.), *Catalysis by Acids and Bases*, Elsevier Publishers, Amsterdam.
3. Anderson, J.R. and Boudart, M. eds. (1982) *Catalysis: Science and Technology*, Springer, Berlin.
4. Göpel, W., Hesse, J. and Zeinel, J.N.,eds. (1991) *Sensors - A Comprehensive Survey* vol. 2, VCH,Weinheim.
5. (a) Chinchen, G.C., Spencer, M.S., Waugh, K.C. and Whan, D.A. (1987), *J. Chem. Soc. Farad. Trans. I* **83**, 2193-2212.
   (b) Rasmussen, P.B., Holmblad, P.M., Askgård, C.V., Ovesen, C.V., Stolze, P., Nørskov, J.K. and Chorkendorff, I. (1994), *Catal. Lett.* **26**, 373.
   (c) Nakamura, J., Uchijima, T., Kanai, Y., and Fujitani, T. (1996), The role of ZnO in Cu/ZnO methanol synthesis catalysts, *Catal. Today* **28**, 223-30.
6. Møller, P.J. and Nerlov, J. (1994) Ultrathin films of Cu on ZnO(11-20): growth and electronic structure, *Surf. Sci.* **307-309**, 591-596.
7 Henrich, V.E. and Cox, P.A. (1994) *The Surface Science of Metal Oxides*, Cambridge University Press, Cambridge.
8. (a) Møller, P.J. (1994), Copper and nickel ultrathin films on ceramic oxide crystal surfaces, in J. Nowotny (ed.) *Science of Ceramic Interfaces* , vol II, Elsevier Publishers, Amsterdam, pp. 473-526.
   (b) Møller, P.J., Komolov, S.A. and Lazneva, E.F. (1993) Structural and electronic properties of ultrathin Cu-layers on some crystalline metal-oxides, in H.-J. Freund and E. Umbach (eds), *Adsorption on Ordered Surfaces of Ionic Solids and Thin Films*, Springer-Verlag, Berlin, pp. 156-166.
9. Chang, S.C. and Mark, P. (1974), The chrystallography of the polar (0001)Zn and (000-1)O surfaces of zinc oxide, *Surf. Sci.* **46**, 293-300.
10. Møller, P.J., Komolov, S.A. and Lazneva, E.F. (1994) VLEED from a ZnO(0001) substructure, *Surf. Sci.* **307-309**, 1177-1181.
11. Bloom, S. and Ortenburger, I. (1973), *Phys. Status Solidi (b)* **58**, 561.
12. Jaklevic, R.C. and Davis, L.C. (1982), Band signatures in the low-energy-electron reflectance spectra of fcc metals, *Phys. Rev. B* **26**, 5391.
13. Herlt, H.-J., Feder, R., Meister, G. and Bauer, E.G. (1981), Experiment and theory of the elastic electron reflection coefficient from tungsten, *Solid State Commun.* **38**, 973-76.
14. Komolov, S.A. (1992) *Total Current Spectroscopy of Surfaces*, Gordon and Breach, Philadelphia.
15. Komolov, S.A. and Strocov, V.N. (1991), *Phys. Status Solidi (b)*,**167**, 605.
16. Schäfer, I., Schlüter, M. and Skibowski, M. (1987), Conduction-band structure of graphite studied by angle-resolved inverse photoemission and target current spectroscopy, *Phys. Rev.B* **35**, 7663.
17. Jacobi, K., Zwicker, G. And Gutmann, A. (1984), *Surf. Sci.* **141**, 109.
18. Jacobi, K. (1984), Clean ZnO surfaces and their interaction with gases, in H.-J. Freund and E. Umbach (eds.) *Adsorption on Ordered Surfaces of Ionic Solids and Thin Films*, Springer Verlag, Berlin, p. 103-114.
19. Møller, P.J., Komolov, S.A. Lazneva, E.F. and Pedersen, E.H. (1995) $CO_2^{\cdot}$ intermediates in the CO/ZnO(0001) interface, *Surf. Sci.* **323**, 102-108.
20. Au, C.T., Hirsch, W. and Hirschwald, W. (1988), Adsorption of CO and CO_2 on annealed and defect zinc oxide (000-1) surface studied by photoelectron spectroscopy (XPS and UPS), *Surf. Sci.* **197**, 391-401.
21. Göpel, W., Bauer, R.S. and Hansson, G. (1980), Ultraviolet photoemission studies of chemisorption and point defect formation on ZnO nonpolar surfaces, *Surf. Sci.* **99**, 138-158.
22. Møller, P.J., Komolov, S.A. and Lazneva, E.F. (1994) Influence of atomic Cu-layer epitaxy on CO_2 and CO photoinduced from ZnO(0001), *Appl. Surf. Sci.* **82/83**, 569-575.
23. Shapira, Y., Cox, S.M. and Lichtman, D. (1976), Chemisorption, photodesorption and conductivity measurements on ZnO surfaces, *Surf. Sci.* **54**, 43-59.
24. Møller, P.J., Komolov, S.A. and Lazneva, E.F. (1993), Laser-induced CO_2 desorption from a CO/Cu/ZnO(0001) surface, *Surf. Sci. Lett.* **290**, L677.
25. Cheng, W.H. and Kung, H.H. (1982), Interaction of CO, CO_2 and O_2 with nonpolar, stepped and polar surfaces, *Surf. Sci.* **122**, 21-39.
26. Ge, Q. and Møller, P.J. (1994) CO adsorption on clean and atomic-layer-Cu-covered ZnO(10-10) surfaces,

Appl. Surf. Sci. **82/83**, 305.
27. Fu, S.S. and Somorjai, G.A. (1991), Zinc oxide and overlayers on Cu(110): a model for Cu-Zn-O catalysts, *Appl. Surf. Sci.* **48/49**, 93-103.
28. Nygren, M.A. and Siegbahn, P.E.M. (1992), Theoretical study of chemisorption of CO on copper clusters, *J. Phys. Chem.* **96**, 7579.
29. Onishi, H., Aruga, T. Egawa, C. and Iwasawa, Y. (1988), Modification of surface-electronic structure on TiO_2(110) and TiO_2(441) by Na deposition, *Surf. Sci.* **199**, 54-66.
30. Onishi, H., Aruga, T., Egawa, C. and Iwasawa, Y. (1989), Active structures and electronic states for adsorption of CO_2 and NO on Na/TiO_2(110) surface, *J.Chem. Soc. Far. Trans.* **85**, 2597-2604.
31. Praphakaran, K., Purdie, D., Casanowa, R. Muryn, C.A., Hardman, P.J., Willcott, P.L. and Thornton, G. (1992), Alkali-metal-to-substrate charge transfer in TiO_2(110)c(2x2)K, *Phys. Rev. B.* **45**, 6969.
32. Murray, P.W., Condon, N.G. and Thornton, G. (1995), Na adsorption sites on TiO_2(110)-1x2 and its 2x2 superlattice, *Surf. Sci. Lett.* **323**, L281-L286.
33. Thomas, A.G., Hardman, P.J., Muryn, C.A., Dhariwal, H.S., Prime, A.F.,Thornton, G., Roman, E. and de Segovia, J.L. (1995), *J. Chem. Soc. Far. Trans.* **91**, 3569.
34. Heise, R. and Courth, R. (1995), *Surf. Sci.* **331-335**, 1460.
35. Møller, P.J. and Wu, M.-C. (1989) Surface geometrical structure and incommensurate growth: ultrathin Cu films on TiO_2(110), *Surf. Sci.* **224**, 265-276.
36. Murray, P.W., Condon, N.G. and Thornton, G. (1995), Effect of stoichiometry on the structure of TiO_2(110), *Phys. Rev.B* **51**, 10989.
37. Sander, M. and Engel, T. (1994), Atomic level structure of TiO_2(110) as a function of surface oxygen coverage, *Surf. Sci. Lett.* **302**, L263-L268.
38. Onishi, H. and Iwasawa, Y. (1994), Reconstruction of TiO_2(110) surface: STM study with atomic-scale resolution, *Surf. Sci. Lett.* **313**, L783-L789.
39. Szabo, A. and Engel, T. (1995), Structural studies of TiO_2(110) using STM, *Surf. Sci.* **329**, 241-254.
40. Nerlov, J., Ge, Q. and Møller, P.J. (1996) Resonant photoemission from TiO_2(110) surfaces: implications on surface bonding and hybridization, *Surf. Sci.* **348**, 28.
41. a) Bertel, E., Stockbauer, R. and Madey, T.E. (1983), Resonant electron emission in Ti and TiO_2,*Phys. Rev.B* **27**, 1939-42.
b) Davis, L.C. (1986), Photoemission from transition metals and their compounds, *J. Appl. Phys.* **59**, R25-63.
c) Zhang, Z., Jeng, S.-P. and Henrich, V.E. (1991), Cation-ligand hybridization for stoichiometric and reduced TiO_2(110) surfaces determined by resonant photoemission, *Phys. Rev. B* **43**, 12004.
42. Nerlov, J., Christensen, S.V., Weichel, S., Pedersen, E.H. and Møller, P.J. (1996) A photoemission study of the adsorption of CO_2 and Na on TiO_2(110)-1x1 and -1x2 surfaces: adsorption geometry and reactivity, *Surf. Sci.*, in press.
43. Vogtenhuber, D., Podloucky, R. Neckel, A., Steinemann, S.G. and Freeman, A.J. (1994), Electronic structure and relaxed geometry of the TiO_2(110) rutile surface, *Phys. Rev. B* **49**, 2099.
44. Ramamoorthy, M., King-Smith, R.D. and Vanderbilt, D. (1994), Defects on TiO_2(110) surfaces, *Phys. Rev. B* **49**, 7709.
45. Reinhardt, P. and Heß, B.A. (1994), Electronic and geometrical structure of rutile surfaces, *Phys. Rev. B* **50**, 12015.
46. Sorantin, P.I. and Schwarz, K. (1992), Chemical bonding in rutile-type compounds, *Inorg. Chem.* **31**, 567-76.
47. Au, C.T., Hirsch, W. and Hirschwald, W. (1988), Adsorption and interaction of carbon dioxide, formic acid and hydrogen/carbon dioxide mixtures on (10-10) zinc oxide surfaces studied by XPS and UPS, *Surf. Sci.* **199**, 507-517.
48. Conner, J.A., Considine, M. and Hillier, I.H. (1978), Low energy photoelectron spectroscopy of solids: photoelectron spectra of carbonate, phosphate, sulphate, triborate and ethanoate ions, *J. Chem. Soc. Far. Trans. II* **74**, 1285-1291.
49. Freund, H.-J. and Messmer, R.P. (1986), On the bonding and reactivity of CO_2 on metal surfaces, *Surf. Sci.* **172**, 1.

THEORY OF ADSORPTION AND SURFACE REACTIONS

B. HAMMER AND J. K. NØRSKOV
CAMP and Department of Physics
Technical University of Denmark
DK-2800 Lyngby
Denmark

1. Introduction

The adsorption and further reaction of molecules from the gas or liquid phase on a solid surface is a very common phenomenon. All solid surfaces have varying amounts of adsorbed gases in air, and adsorption and chemical reactions at surfaces form the basis of a large number of technically important processes ranging from corrosion to heterogeneous catalysis.

Adsorption and reactions at surfaces are intriguing processes that are not simply described in the usual vocabulary of chemistry or that of solid state physics. To understand reactions at surface it is important both to understand the gas phase and the solid, but there are a number of phenomena that are unique to processes at the surface. A chemical reaction at a surface requires that the reactants adsorb, that is, bind to the surface. This often requires the breaking of the bonds of the molecule, and therefore requires a complicated process where energy is exchanged between the adsorbate and surface degrees of freedom. The adsorbates will then diffuse on the surface before they eventually meet other adsorbates and react.

It is only within the last 25 years that experimental techniques have been developed allowing a quantitative assessment of even the most basic questions like the geometrical structure of the surfaces and the adsorbates on the surface or the rates of individual processes on the surface. For an introduction into the field of surface science and surface reactivity we refer to the recent book by Somorjai[1].

The development of theoretical tools to describe and understand adsorption and reactions at surfaces has been quite parallel to the experimental one. A number of useful concepts have been introduced to describe the electronic structure, the interaction energies and the dynamics of ad-

R.M. Lambert and G. Pacchioni (eds.),
Chemisorption and Reactivity on Supported Clusters and Thin Films, 285-351.

sorbates on surfaces. It is now possible to describe at a semi-quantitative level the electronic structure and energetics of adsorption on metal surfaces[2, 3, 4, 5, 6]. The accuracy is still not sufficient to calculate rates of chemical reactions, but it is sufficient for a semi-quantitative description of adsorption and reaction processes, and in particular for comparing different systems. The latter is particularly important if one wants theoretical input into a search for surfaces with a desired activity or selectivity for a given chemical reaction.

Even if we are at a point where bond energies and activation energies can be calculated for the simplest systems with a reasonable accuracy, there is a strong need for models and simplified theories of bonding. The large scale calculations can be viewed as computer experiments and just as for real experiments we need a conceptional framework for an understanding or rationalization of the results. It is not possible or even desirable to have to do a full calculation for each new system to be considered. If we are, for instance interested in finding new alloy surfaces with a particular reactivity or bond strength towards a certain molecule it would be extremely helpful with an understanding of the most important factors determining the ability of a surface to bind or react with the molecule in question. Traditionally, these factors have been divided into "electronic" and "geometrical", but in order to be really useful concepts like that need to be come quantitative: which surface and molecule properties is it that enters into the bond strength and similar properties of the combined system?

In the present lecture notes it is the intention to give a detailed introduction to the theoretical aspects of adsorption and surface chemical reactions. There are several reviews and books describing various aspects of the theory of adsorption and reactions. For a simple introduction to some of the key aspects of adsorption on metal surfaces we refer to the book by Zangwill[7], Masel[8] or the lecture notes by Holloway and Nørskov[9]. A more quantum chemical view of adsorption has been published by Hoffmann[10] or most recently by Van Santen and Neurock[11].

The present lecture notes will concentrate on the theory of adsorption and reactions. This means that experimental results will be included to show to what extent the theory is able to describe the physical reality, but no attempt will be given to overview experimental methods or results. The approach we shall take is the following. First, in Chapter 2, we illustrate the capabilities of large scale self-consistent calculations of adsorption properties, by discussing in detail three examples where there has been detailed experimental investigations as well as extensive calculations. Having shown that the calculational methods are at a level where they describe experiments quite accurately, we then go on to an extensive discussion of the theory.

The basis for our discussion of adsorbate-surface reactions will be the density functional theory. We use density functional calculations in accurate self-consistent calculations of adsorption properties, but also as a basis for the development of models which bring out the important physical quantities determining the reactivity of a given surface. We therefore start in Chapter 3 by giving an introduction to density functional theory. In the following Chapters we then discuss adsorption and reactions on metal surfaces in some detail. We take the simplest case, adsorption of a single atom, first and then go on to discuss molecular adsorption and finally simple reactions where bonds between adsorbates are broken or formed. We have chosen to concentrate in these lecture notes on metal surfaces. Many of the concepts developed will, however, be more generally applicable to semi-conductors and insulators.

2. Three case stories.

In this Chapter, we discuss three case stories of adsorption systems where there has been extensive experimental and theoretical work. The aim is to show the strengths and weaknesses of the theoretical methods that will be used throughout this text. We consider atomic chemisorption, molecular chemisorption and a simple dissociation reaction to illustrate to which extent the calculations are able to reproduce the experimentally observed adsorption geometries, vibrational properties, adsorption energies, and activation barriers. The details of the calculation method used in these calculations will be described later. Here we just mention that they are all based on the density functional theory (DFT) using the generalized gradient approximation to treat exchange and correlation effects[12, 13].

2.1. OXYGEN ADSORPTION ON PT(111).

Consider as a first example oxygen adsorption on the Pt(111) surface. This system has been studied in detail experimentally[14]. We concentrate on the ordered p(2×2) oxygen overlayer structure observed at a quarter of a monolayer of oxygen atoms. Fig. 1 shows the geometry of this structure including relaxations of the Pt atoms at the surface and three layers down as deduced from low energy electron diffraction (LEED) experiments. We also show the result of a DFT calculation. The equilibrium structure is determined by allowing all the coordinates of the atoms in the oxygen adlayer and in the two first layers to vary untill the lowest energy geometry is found. Clearly, all bond lengths and the relaxation pattern are reproduced by the calculation down to the 1% level.

The chemisorption energy, including relaxations of the Pt substrate is calculated to be ΔE_O =4.29 eV per O atom relative to an atom in the gas

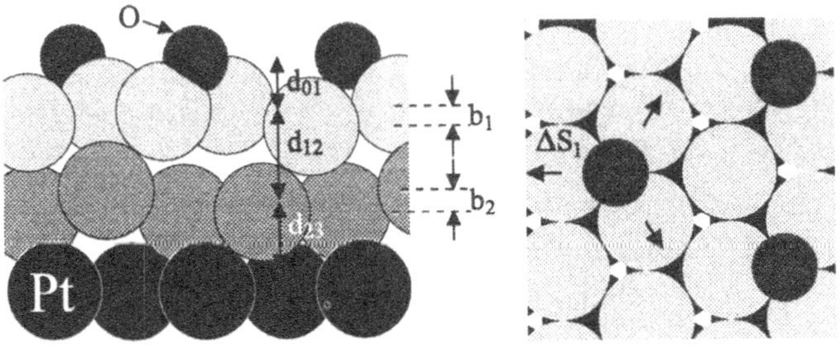

Sideview Pt (111)-p(2x2)-O Topview

	Experiment	Theory
d_{01}	1.18 Å	1.21 Å
$(d_{12} - d)/d$	1.3 %	0.6 %
$(d_{23} - d)/d$	0.0 %	-0.1 %
b_1/d	3.1 %	3.5 %
b_2/d	4.0 %	3.9 %
$\Delta S_1/S_1$	1.6 %	2.9 %

Figure 1. The experimental and theoretical equilibrium structure for the Pt(111)-p(2×2)-O system. The experimental data is from Ref. [14].

phase. The binding energy per O atom in O_2 is 2.95 eV (in the DFT-GGA calculations), so the heat of adsorption is 2.68 eV per molecule. This is in reasonable agreement with results from micro-calorimetry which gives an integral heat of adsorption for a quarter of a monolayer of oxygen of 2.34 eV/O_2[15].

We can also deduce the vibrational frequency for the oxygen vibration perpendicular to the surface. The experimental value is 58 meV[16] and the value calculated from the curvature of the total energy as a function of the height of the O atoms above the surface is 59 meV.

2.2. CO AND N_2 ADSORPTION ON RU(001).

As an example of molecular adsorption consider CO and N_2 adsorption on the Ru(001) surface. Here a $(\sqrt{3} \times \sqrt{3})$ structure has been observed for both adsorbates, and the geometrical, vibrational and bonding characteristics are collected in Table 1. In the table we also show the calculated values[17]. Again geometrical data are quite well reproduced, the only difference being the larger N_2-induced surface relaxation in the calculated results. Shifts in the molecular vibrational frequencies are also well described and the chemisorption energies are within a few tenths of an eV

TABLE 1. Comparison of calculated and experimental (in parenthesis) adsorption properties of CO and N_2 on Ru(001). All results are for the $(\sqrt{3} \times \sqrt{3})$ overlayer structure. The lowest energy site is found in the calculations to be ontop of the Ru atoms. The molecules are upright on the surface with the C end down in the case of CO. The Ru-molecule bond length is measured from the Ru atom underneath the molecule to the nearest atom (C or N). The buckling is the lifting of the Ru atoms under the molecule relative to its nearest neighbors in the surface layer. Underneath, we show calculated occupancies of the molecular states. This will be used in the discussion in Chapter 5. From Ref. [17]

	CO	N_2
Molecular bond length (Å)	1.17 (1.10)	1.13 (1.10)
Ru-molecule bond length (Å)	1.92 (1.93)	2.00 (2.00)
Buckling (Å)	0.13 (0.07)	0.15 (0.0)
Chemisorption energy (eV)	1.90 (1.66)	0.61 (0.44)
$N_{2\pi}$	0.030	0.175
$N_{5\sigma}$	0.916	0.970
$N_{1\pi}$	0.997	0.997
$N_{4\sigma}$	0.997	0.990

of experiment. The large difference in chemical activity of CO and N_2 is clearly borne out by the calculation.

2.3. H_2 DISSOCIATION ON NI, CU, PT, AND AU(111).

As our final example, we will consider the simplest kind of surface reaction, the dissociation of a diatomic molecule. Figure 2 summarizes the calculated reaction energetics for H_2 outside the (111) surface of the four metals Ni, Cu, Pt, and Au[18]. The figure shows the energy of the system as a function of the height of the molecule above the surface. For each height the H-H bond length has been allowed to relax to give the lowest energy configuration, and in a similar way the configuration of the molecule relative to the surface unit cell has been allowed to relax on the way towards the surface. The insert shows the calculated reaction path for H_2 dissociating over the Au(111) surface. It can be seen that H_2 dissociation is activated on both Cu and Au and essentially non-activated on Ni and Pt in complete agreement with experimental evidence[19, 20, 21, 22, 23, 24, 25]. It is also seen that Au stands out as having both the highest barrier for dissociation and the least stable chemisorption state. The lack of reactivity of gold is thus clearly borne out by the calculations.

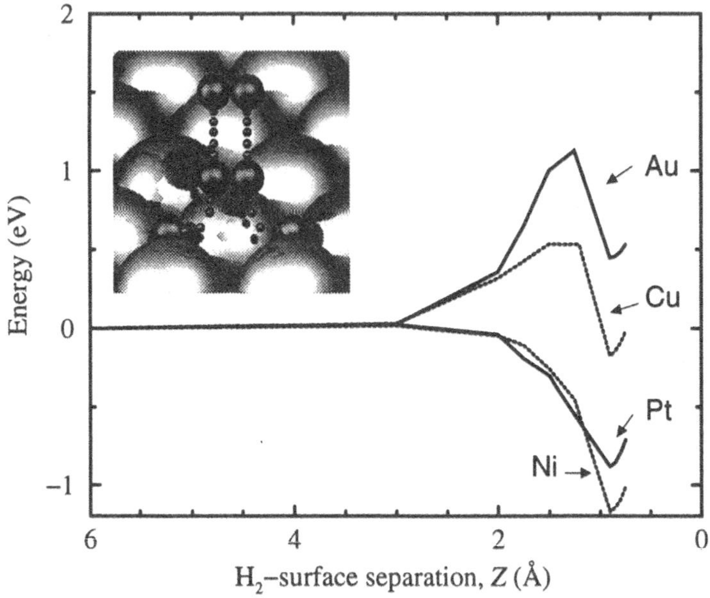

Figure 2. The calculated energy along the minimum energy reaction path for H_2 dissociating (in a $\sqrt{3} \times \sqrt{3}$ surface unit cell) over the (111) surface of Ni, Cu, Pt, and Au. Z is the height of the molecular center of mass above the plane of the surface atoms. The path for H_2 over Au(111) is illustrated in the insert. The reaction paths over the other metals also involve an orientation of the molecular axis parallel to the surface. From Ref. [18]

We conclude that the DFT calculations give equilibrium geometries to within a few percent of the experimental results. Vibrational frequencies are equally well described The absolute accuracy on chemisorption energies and activation barriers is of the order 0.25 eV, but differences from one surface to the next or one adsorbate to another seem even more accurate than that. This is a reasonable starting point for a theoretical description and understanding of adsorption and reactions on metal surfaces.

3. Density functional theory

In this Chapter we discuss in some detail the theoretical background of the calculational scheme used to give the results discussed in the preceding section. We start by discussing the basic principles of density functional theory before we discuss the detailed implementation to surface problems in the following Chapter. We end this Chapter by a discussion of approximate schemes for obtaining the electronic structure and bonding at surfaces.

The reader who is not interested in the detailed theoretical basis can skip this chapter an go directly to Chapter 4 and may still be able to follow the rest of the text at a certain level of understanding.

3.1. THE BASIC PRINCIPLES OF DENSITY FUNCTIONAL THEORY

The total energy of a system of interacting electrons is uniquely determined by the external potential. In the case of a surface with adsorbates, the external potential is solely given by the position and charges of the nuclei. If we know the external potential we should then solve the Schrödinger equation to get the many-electron wave function and from this the ground state energy and other observables of the system. In quantum Monte Carlo (QMC) calculations[26, 27] it is possible to calculate statistically the observables (e.g. the total energy) determined by this many-body wave function for the simplest few-electron systems. Being extremely computer intensive, however, the QMC is not generally applicable for systems of interest in surface physics and chemistry though recent developments show the great potential of this method[28, 29, 30]. An alternative route to the calculation of the total energy of such many-electron systems is to first do a calculation within the Hartree-Fock approximation and then include correlation effects by the Configuration Interaction (CI) method[6, 11]. Such calculations have been done for a number of adsorption systems, and are presently the most accurate in terms of the treatment of the electronic structure problem. To make the problem solvable the surface is always modeled by a cluster of metal atoms[6]. We will be returning to these accurate calculations in the following when we evaluate the accuracy of the more approximate, but computationally much faster density functional calculations.

Density functional theory (DFT) provides an extremely useful view of the problem of many interacting electrons in an external potential. The basis for DFT is the proof by Hohenberg and Kohn[31] that the total energy of a system of interacting electrons is a unique functional $E[n(\mathbf{r})]$ of the electron density $n(\mathbf{r})$ of the system. The proof is quite simple and the reader is referred to e.g. the original paper of Hohenberg and Kohn[31] for the details.

Given that the total energy for the system is a unique functional of the electron density, the usual Rayleigh-Ritz variational principle of quantum mechanics implies that the total energy has a minimum E_0 for the ground state density $n_0(\mathbf{r})$. The ground state energy and electron density can therefore in principle be obtained by minimizing the total energy with respect to the electron density. The only problem is that it is not possible to write down explicitly the energy functional $E[n(\mathbf{r})]$ to be minimized. Hohenberg and Kohn proposed to write the total energy functional formally as[31]

$$E_{HK}[n] = T_{HK}[n] + F[n] , \tag{1}$$

where $T_{HK}[n]$ is the kinetic energy of a noninteracting electron gas with density n and $F[n]$ is the sum

$$F[n] = E_{es}[n] + E_{xc}[n] \,, \tag{2}$$

of the average electrostatic energy

$$E_{es}[n] = \frac{1}{2}\int\int \frac{\rho(\mathbf{r})\rho(\mathbf{r}')}{\mid \mathbf{r}-\mathbf{r}' \mid} d\mathbf{r}d\mathbf{r}' \tag{3}$$

(ρ is the sum of the electron density n and the nuclear charge; omission of the nuclear self-interaction is understood) and the rest, which is called the exchange-correlation energy $E_{xc}[n]$. We still do not know how to write the explicit dependence of the exchange-correlation energy on the electron density, but traditionally $E_{xc}[n]$ has been described using the local density approximation (LDA)[32, 33, 34]

$$E_{xc}[n] = \int f_{xc}(n(\mathbf{r}))d\mathbf{r} = \int n(\mathbf{r})\varepsilon(n(\mathbf{r}))d\mathbf{r} \,. \tag{4}$$

Lately better, non-local corrections also depending on the gradient of the density have been introduced[13]. We will return to this question later.

Since $T_{HK}[n]$ is the kinetic energy of a system of non-interacting electrons with the electron density equal to the density $n(\mathbf{r})$ of the interacting electron system, we can get $T_{HK}[n]$ in the following way. Define an effective one-electron potential v with the property that the solutions to the one-electron Schrödinger equation

$$(-\frac{1}{2}\nabla^2 + v)\psi_\alpha = \varepsilon_\alpha \psi_\alpha \tag{5}$$

satisfy the condition

$$\sum_{\alpha=1}^{N} \mid \psi_\alpha(\mathbf{r}) \mid^2 = n(\mathbf{r}) \,. \tag{6}$$

The solutions now give the kinetic energy as

$$T_{HK}[n] = \sum_{\alpha=1}^{N} < \psi_\alpha | -\frac{1}{2}\nabla^2 | \psi_\alpha > = \sum_{\alpha=1}^{N} \varepsilon_\alpha[v[n]] - \int v[n] n d\mathbf{r} \,. \tag{7}$$

The effective one electron Schrödinger equations Eq. 5 are called the Kohn-Sham equations[32].

The potential v_0 corresponding to the ground state is related to the density (up to an additive constant) through the relation

$$v_0[n_0] = \frac{\delta F}{\delta n}[n_0] = \phi[n_0] + v_{xc}[n_0] \,, \tag{8}$$

where $\phi[n_0] = \frac{\delta E_{es}}{\delta n}[n_0]$ is the Hartree potential

$$\phi(\mathbf{r}) = \int \frac{\rho_0(\mathbf{r}')}{|\mathbf{r}-\mathbf{r}'|} d\mathbf{r}' , \tag{9}$$

and $v_{xc}[n_0] = \frac{\delta E_{xc}}{\delta n}[n_0]$ the exchange-correlation potential, which in the LDA takes the form:

$$v_{xc}(n_0(\mathbf{r})) = \frac{df_{xc}}{dn}(n_0(\mathbf{r})) . \tag{10}$$

This follows by considering a first order change in E_{HK} from Eq. 1 and 7:

$$\delta E_{HK} = \delta T_{HK} + \delta F = -\int v\delta n d\mathbf{r} + \int \frac{\delta F}{\delta n}\delta n d\mathbf{r} \tag{11}$$

Due to the variational property of $E_{HK}[n]$ we must have the first order change $\delta E_{HK} = 0$ for the ground state which directly gives Eq. 8.

In general the variational principle implies that

$$E_{HK}[n_0 + \delta n] = E_0 + O(\delta n^2) . \tag{12}$$

We now make an extra, very useful step[35, 36]. Instead of regarding v in the Kohn-Sham equations 5 as given by the density, we may regard it as an independent variable[35, 36]:

$$T[n, v] = \sum_{\alpha=1}^{N} \varepsilon_\alpha[v] - \int v n d\mathbf{r} . \tag{13}$$

For a variation around the ground state configuration (n_0, v_0) we then have

$$\begin{aligned} \delta T[n,v]|_0 &= \int n_0 \delta v d\mathbf{r} - \int n_0 \delta v d\mathbf{r} - \int v_0 \delta n d\mathbf{r} = -\int v_0 \delta n d\mathbf{r} \\ &= \delta T_{HK}[n]|_0 , \end{aligned} \tag{14}$$

so the generalized energy functional

$$E[n, v] = T[n, v] + F[n] , \tag{15}$$

is stationary in the ground state with respect to independent variations in the density *and* the potential.

Due to the variational property we immediately get

$$E[n_0 + \delta n, v_0 + \delta v] = E_0 + O(\delta n^2, \delta v^2, \delta n\delta v) = E_0 + O^2(\delta n, \delta v) , \tag{16}$$

i.e. first order errors in the density *and* the potential give rise to only second- and higher order errors in the total energy.

A number of energy functionals with variational properties can be derived from Eq. 15. If the potential is considered as a functional of the density as given implicitly by Eqs. 5 and 6 we of course reobtain the Hohenberg-Kohn functional. Another possibility is to choose the potential as the density functional given explicitly by a generalization (away from n_0) of Eq. 8 into: $v[n] = \int(\rho(\mathbf{r}')/\mid \mathbf{r}-\mathbf{r}'\mid)d\mathbf{r}' + v_{xc}(n(\mathbf{r}))$, in which case the total energy is treated as only a density functional, $E'[n]$, through:

$$E'[n] = E[n, v[n]] \,. \tag{17}$$

Yet another choice would be to regard the potential as the only independent variable and let the density be a potential functional defined by solving the Kohn-Sham equation 5 and use Eq. 6 to get the density. This is the approach that has been taken by Andersen[37] and others[38] when discussing the so-called *force theorem*. The most general functional Eq. 15 has been used extensively to generate approximate total energy expressions in the framework of the effective medium theory[35, 36]. We will discuss this in detail later. Recently, Methfessel has also used this approach to get approximate total energies[39]

3.2. DENSITY FUNCTIONAL CALCULATIONAL METHODS

Density functional theory can be implemented in numerous ways to describe the electronic structure and total energy of atomic configurations. Mathematically elaborate, but numerically accurate and physically transparent are the liniear muffin tin orbital (LMTO) and linear augmented plane wave (LAPW) methods in which all electrons of every atomic species are described (as implicitly prescribed by the Hohenberg-Kohn formalism). Almost as accurate are approaches in which the localized electrons in the closed shells around the nuclei – the core electrons – are assumed independent of the environment of the individual nucleus. In the frozen core appoximation the core electrons are simply fixed in their atomic orbitals, and in the pseudopotential approximation, they are even projected out of the problem by replacing the $-Z/r$ potential of the nuclei by so-called pseudopotentials[40, 41]. These are potentials that are constructed to bind only electrons from a given energy shell and up (in energy). The corresponding pseudowave functions mimic the true atomic wave functions at large distances from the nucleus, while at close proximities the rapid oscilations required for the orthogonalization to the core states have been removed (see Fig. 3). This in turn means that only slowly varying basis functions are required for an expansion of the one-electron (pseudo)wave functions and the use of plane waves as basis functions becomes possible.

Ideally, single crystalline surfaces represent at the atomic level systems of semi-infinite extension. While Greens function methods[42, 43, 44] can

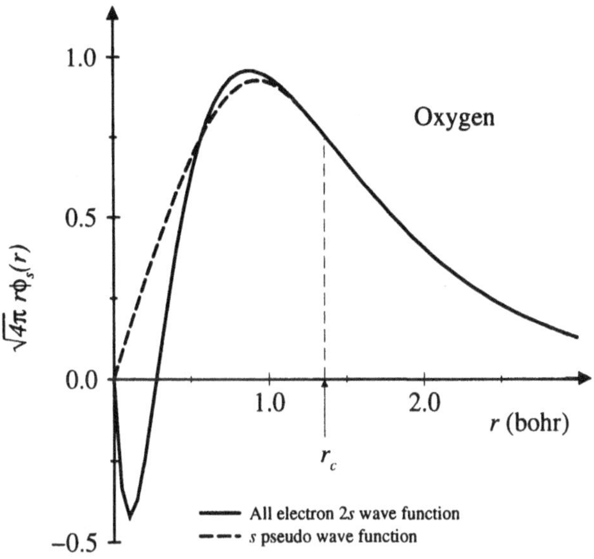

Figure 3. The Oxygen 2s radial wave function from an all-electron calculation compared with the nodeless s state bound by the pseudopotential describing oxygen. At large distances away from the nuclei (at $r = 0$), the two wave functions are identical by construction. This is the region important for bonding in molecules and solids.

represent them as such even within density functional theory, it is often more practical to resort to more approximate descriptions, some of which are illustrated in Fig. 4. One possibility is to represent the surface by a cluster of atoms in an arrangement similar to that in the semi-infinite surface. Such clusters may have to be rather large and to be truncated (passivated) in particular ways in order to behave electronically as the surfaces one wants to model. Alternatively, a surface may be represented by a slab geometry, in which case a cluster constituting a so-called super cell is replicated in the two dimensions of space defined by the surface plane. This in fact is just a particular choice for the boundary conditions in the cluster approach, which however is generally found to describe the surface properties better than the cluster approach for a given number of atoms in the super cell or the cluster respectively. In both methods, only a finite amount of atomic layers may be included and this will always represent an approximation. When representing a surface by a slab, any surface defects or adsorbates that are introduced will automatically be replicated in the two dimensions as well. While this is highly desireable for ordered overlayers for instance, it complicates to some extend the study of isolated defects and adsorbates, in which case care must be taken that the surface dimensions of the super cell are large enough that e.g. such adsorbates do not interact appreciably

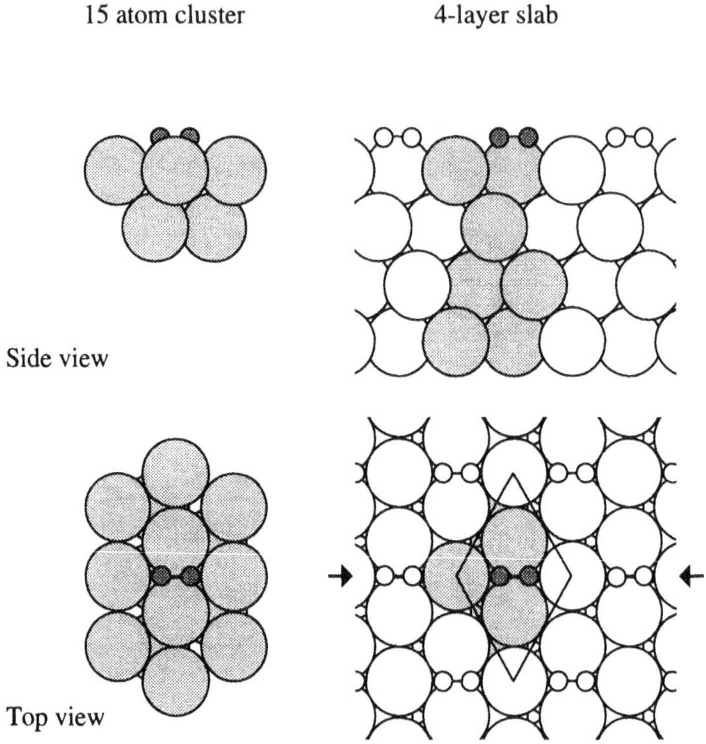

Figure 4. Illustration of possible representations of an H_2 molecule (small circles) interacting with a semi-infinite surface. The cluster shown has 15 subtrate atoms, while the slab shown has 12 atoms in the super cell. Replicas of the super cell atoms are shown with open circles. The side view for the slab is taken along the line indicated by the arrows in its top view. In terms of the number of independent substrate atoms, the slab method usually converge faster than the cluster method towards behaving electronically as the semi-infinite surface.

through the boundary conditions. Finally, when using plane waves for the basis set, the slab systems must be replicated in the dimension perpendicular to the surface as well. When this is done some amount of vacuum between the slabs is included to decouple the slabs.

While the above mentioned approximations (e.g. the use of pseudopotentials, finite basis sets and super cells) can be checked systematically, the main source of uncertainty in density functional theory calculations typically stems from the approximation of the exchange-correlation energy, $E_{xc}[n]$. Based on Ceperley and Alders quantum Monte Carlo calculation for the homogeneous electron gas[33] analytic parametrizations of the local density exchange correlation functionals have been made by Perdew and Zunger (unspin polarized, LDA)[34] and by Wosko, Wilk and Nusair (spin polarized, LSDA)[45]. These functionals have proven highly successfull for producing properties for rather homogeneous systems as bulk metals, semi conductors and insulators and as the surfaces of these even when involving simple adsorbates (e.g. positions, heights and vibrational frequencies). The remarkable success of the Ceperley-Alder LDA has been attributed the fact that it describes at least one *physically existing* system *exactly*[46], i.e. the homogeneous electron gas. The LDA description of less homogeneous systems e.g. isolated atoms and molecules (including their chemisorption energies to surfaces and their reaction barriers over such) is less good and extensions beyond the LDA are required for these systems. Much work has therefore been aimed at improving the exchange-correlation functionals. The most straight forward way is to include in f_{xc} a dependence on the local gradient of the density in what is called the gradient expansion approximation (GEA):

$$E_{xc}[n] = \int f_{xc}(n(\mathbf{r}), |\nabla n(\mathbf{r})|) d\mathbf{r} \,. \tag{18}$$

However, Perdew and others[46] recognized that as such functionals no longer describe any physically existing system exactly they will perform less well than the LDA functional unless the functional form of $f_{xc}(n(\mathbf{r}), |\nabla n(\mathbf{r})|)$ is constrained to ensure that certain physical properties are reproduced by f_{xc}. One such property is e.g. that the exchange-correlation hole contains exactly (minus) one electron indicating that if one 'rides' on one particular electron (in a many-body wave function) the probability of finding other electrons in the nearby region will be lowered by one. With the constraints the GEA is termed the generalized gradient approximation (GGA). The GGA[13] has proven succesfull in describing both rather homogeneous and more inhomogeneous systems. For instance gas-phase[47] and surface[2] reactions of molecules have become feasible with the GGA and many interesting studies are now being carried out[3, 5, 11, 48, 49, 50, 51, 52, 53, 54, 55, 56, 57, 58, 59, 60, 61].

Finally, as an illustration of the accuracy of the pseudopotential plane wave GGA density functional theory slab approach that we take in these lecture notes, we consider the value of the barrier for the dissociative adsorption of CH_4 into CH_3 and H on Ni(111). As we review in Chapter 6, DFT calculations[61] yield a barrier of about 1.0 eV compared to roughly 0.7 eV in elaborate CI cluster calculations[62]. The discrepancy is due to the use of a slab versus a cluster geometry and of the approximate exchange-correlation description within the density functional study and sets a scale on the absolute accuracy of the calculations. Differences in barrier heights from surface to surface or even from one reaction to the next are typically much more accurately reproduced rendering the DFT approach extremely usefull as we will illustrate in the following. For more direct comparisons of DFT and CI calculations we refer to the extensive work of Nacthigal *et al*[63] for H_2 dissociation on Si clusters.

3.3. DENSITY FUNCTIONAL THEORY AS A BASIS FOR SIMPLIFIED THEORIES OF BONDING

As a supplement to the accurate, self-consistent solutions described above it is important to establish simpler models which enables an understanding of the large-scale calculations and experiments. Density functional theory also provides a theoretical basis for such models.

Most intuitive arguments about bonding in chemistry are based on changes in the one-electron spectrum – the formation of bonding and antibonding states etc. It is clear from Eq. 1 that while the sum of the one-electron eigenvalues does enter the expression for the total energy through the kinetic energy term Eq. 7 there are other equally important terms. Direct calculations show that usually the total energy of a system is very far from given by the one-electron energy sum alone. Never the less it turns out that by using the variational properties of the total energy functional one can get small total energy differences directly from suitably calculated one-electron energy differences. This result is important enough for us to discuss it in some detail. This discussion naturally gives an introduction to the approximate effective medium theory.

One way of simplifying the calculation of the one-electron spectrum is to introduce a simplified basis set instead of the extremely large sets used in obtaining the accurate solutions. We discuss the simplest tight-binding or LCAO basis sets and the models that this gives rise to. In particular, we briefly discuss the surface version of the tight binding model, the so-called Newns-Anderson model.

3.4. ESTIMATING SMALL ENERGY DIFFERENCES

In this section we concentrate on evaluating small total energy differences directly without a full self-consistent calculation in each case[35, 36, 64]. Even though the arguments are perfectly general, we will consider explicitly an adsorbate a outside a metal surface m, and we want to estimate the change in adsorption energy of the adsorbate when the metal is modified slightly to $\tilde{m}$. The modification could for instance be that another atom or molecule was adsorbed on m close to a or that m was exchanged for another metal close to m in the periodic table.

We are interested in the difference in adsorption energy in the two cases:

$$\delta E_{ads} = E_{ads}[\tilde{m}] - E_{ads}[m]$$

$$
\begin{aligned}
&= (E_{tot}[\tilde{m}+a] - E_{tot}[\tilde{m}] - E_{tot}[a]) \\
&\quad -(E_{tot}[m+a] - E_{tot}[m] - E_{tot}[a]) \\
&= E_{tot}[\tilde{m}+a] - E_{tot}[\tilde{m}] - (E_{tot}[m+a] - E_{tot}[m]). \qquad (19)
\end{aligned}
$$

We now divide space up into two regions, A and M, as illustrated in Fig. 5. In the near adsorbate region, A, the density and one-electron potential will only be affected slightly by the change of m to $\tilde{m}$. Likewise, in the near metal region, M, the effect of the adsorbate is only weak. We can exploit this in connection with the generalized variational principle Eq. 16 in the following way. We choose "frozen" electron densities and one-electron potentials when evaluating δE_{ads}. Because region A is dominated by the adsorbate we choose the same density and potential in this region irrespective of the metal. And likewise, we let the density and potential in region M be independent of the presence of the adsorbate. We can do this with only second order errors in the total energies.

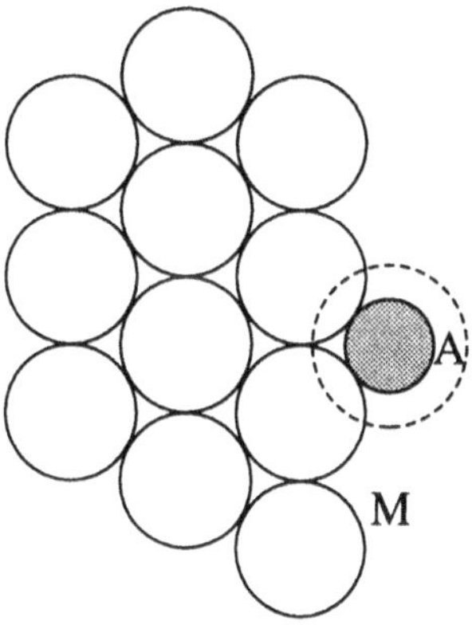

Figure 5. Division of space.

Let us now consider the contribution of F[n] , Eq. 2, to δE_{ads}. In the local density approximation, Eq. 4, the exchange correlation energy is a local function of the position in space. Even in the GGA, the functional dependence is quasi-local, i.e. when in region A the exchange-correlation energy density $f_{xc}[n]$, Eq. 4 does not depend on the properties of the system in region M and vice versa. This is not generally true of the electrostatic energy contribution, Eq. 3. If e.g. the adsorbate has a dipole moment this will give rise to an electrostatic potential in region M. For the moment we neglect such non-local electrostatic interactions between regions A and M. Then E_{es} is also local in the sense outlined above. We can then divide F

into contributions from the two regions $F = F_A + F_M$ and write

$$\begin{aligned}\delta F_{local} &= F_A[\tilde{m}+a] - F_A[\tilde{m}] - (F_A[m+a] - F_A[m]) + \\ &\quad F_M[\tilde{m}+a] - F_M[\tilde{m}] - (F_M[m+a] - F_M[m]) \\ &= (F_A[\tilde{m}+a] - F_A[m+a]) - (F_A[\tilde{m}] - F_A[m]) + \\ &\quad (F_M[\tilde{m}+a] - F_M[\tilde{m}]) - (F_M[m+a] - F_M[m]) \qquad (20)\end{aligned}$$

With the frozen density ansatz above, all terms in parenthesis () are zero. The only contribution from F to the adsorption energy difference is therefore the non-local electrostatic energy

$$\delta E_{es,A-M} = \int_A \int_M \frac{\rho(\mathbf{r})\rho(\mathbf{r}')}{|\mathbf{r}-\mathbf{r}'|} d\mathbf{r} d\mathbf{r}'. \qquad (21)$$

In the same way, the ansatz renders the net contribution from the nv integrals in the kinetic energy difference, Eq. 7, zero, and only the difference in the one-electron energies calculated with the frozen potentials outlined above contribute:

$$\begin{aligned}\delta T_{HK} = \delta E_{1el} &= \sum \varepsilon\{v_M[\tilde{m}], v_A[a]\} - \sum \varepsilon\{v_{M+A}[\tilde{m}]\} \\ &\quad -(\sum \varepsilon\{v_M[m], v_A[a]\} - \sum \varepsilon\{v_{M+A}[m]\}) \qquad (22)\end{aligned}$$

The difference in adsorption energy is therefore given by the one-electron energy difference plus the difference in electrostatic interaction between the surface and the adsorbate in the two situations:

$$\delta E_{ads} = \delta E_{1el} + \delta E_{es,A-M}. \qquad (23)$$

This is a very interesting result. It shows that energy *differences* are given by an inter-atomic electrostatic energy and the sum of the one-electron energies, provided they are calculated in the right way, using frozen potentials. The self-consistent total energy on the other hand is given by Eq. 1, 2, and 13 and even though the one-electron energy sum enters in Eq. 13 it is far from equal to the total energy. Eq. 23 gives a nice theoretical background for the use of arguments based on the one-electron spectra and shows how and when we can expect one-electron energies to give meaning. They are particularly useful for making comparisons between similar systems, not for calculating the total binding energy.

The formalism outlined above is a general theory for comparing the total energy in two similar systems. It is easy to see that it also shows how to calculate the total energy of an atom in a given set of surroundings $\tilde{m}$ by first calculating the energy in another set of surroundings m and then estimate the energy difference. This is a useful procedure if system

m is much simpler to handle computationally, for instance due to a higher symmetry, than system $\tilde{m}$. The simpler system m is often called the effective medium, and the theory the effective medium theory[35, 36].

3.5. THE SIMPLEST EFFECTIVE MEDIUM THEORY OF BONDING

The simplest effective medium one can imagine is a homogeneous electron gas. This can be used to get the adsorption energy at a metal surface in the following way[35, 36, 65]. Consider an atom interacting with a metal surface. The atom can be thought of as being embedded in an (inhomogeneous) electron gas set up by the neighbors. Imagine now, as a lowest order approximation, that the electron density from the neighbors is averaged in the vicinity of the atom, and the embedding energy is estimated as the embedding energy of the atom in a homogeneous electron gas:

$$\Delta E = E_c(\bar{n}) + \Delta E_{corrections} \tag{24}$$

where

$$\bar{n} = < \sum_i n_i(\mathbf{r}) > . \tag{25}$$

The sum is over the surrounding atoms with electron densities $n_i(\mathbf{r})$ and $<>$ stands for an appropriate average. $E_c(n)$ is the energy of the atom in a homogeneous electron gas as a function of the electron gas density n.

Using the frozen density and potential formalism outlined above, it is clear that the correction term can be written as the sum of an electrostatic and a one-electron energy difference for the atom in the two different surroundings:

$$\Delta E_{corrections} = \Delta E_{es} + \Delta E_{1el}. \tag{26}$$

If the effective medium is a reasonable approximation to the real surroundings of the atom in question the correction terms will be small, and we can get a reasonable qualitative description just considering the density dependent term. It should for instance be possible to estimate the binding energy of an atom outside a surface from a knowledge of the electron density of the metal outside the surface and the function $E_c(n)$, which can be calculated once and for all for each atom. This has been done in a number of cases and some examples are shown in Fig. 6. The $E_c(n)$ function for reactive atoms like O, C, S, Cu, Al, etc. has a single minimum. The depth of the minimum is a measure of a typical bond strength for the atom in question to a metal and the density at the minimum reflects the electron density that the atom prefers from its surroundings. The fact that the minimum is deeper for O than for S in Fig. 6 thus means that oxygen binds stronger to a free electron metal, and the fact that sulfur has its minimum

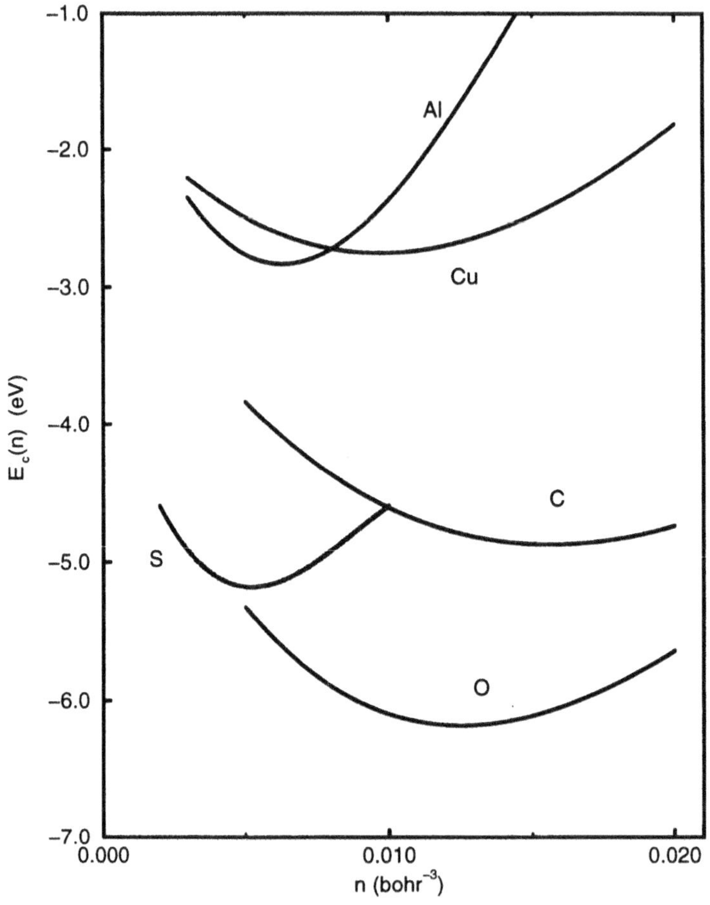

Figure 6. The cohesive function $E_c(n)$ for Aluminum, copper, oxygen, sulfur and carbon in a homogeneous electron gas calculated within the GGA approximation. The curves are similar for other reactive atoms.

at a lower density than oxygen, suggests that the metal-sulfur bond length should be larger than the metal-oxygen bond length. We illustrate this in Fig. 7 where we show the $E_c(n(r))$ for O and S outside a Ni surface. The (averaged) electron density for the Ni surface is taken from Stoltze[66]. We will return to these questions in Chapter 4.

When an adsorbate moves around on a metal surface, it will tend to stay at a constant electron density. This means, for instance, that the bond length between the nearest metal atoms and the adsorbate will depend on the number of nearest neighbor metal atoms. The more atoms that contribute to the electron density in the sum Eq. 25 the longer the bond length. This simple argument predicts, for instance, that the O-Ni bond

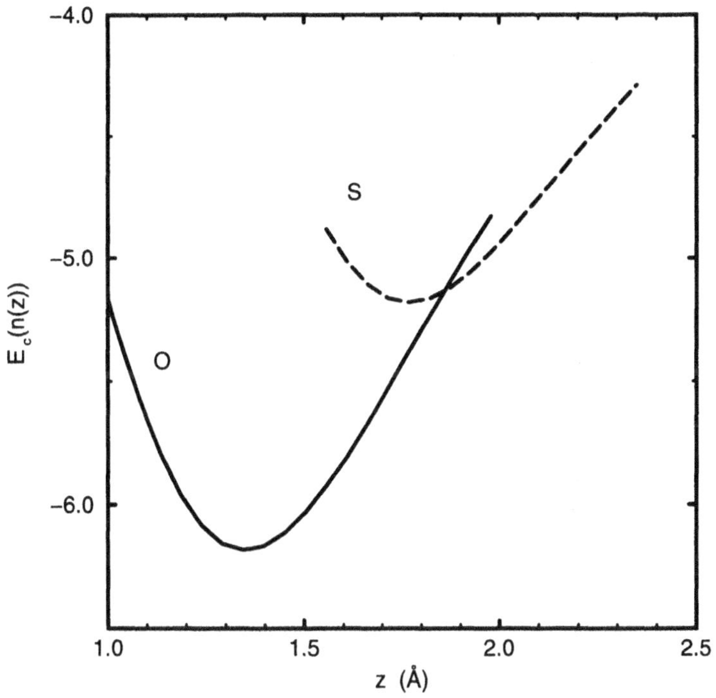

Figure 7. The contribution from the $E_c(n)$ function to the chemisorption energy for O and S outside a Ni(111) surface. The Ni electron densities entering into Eq. 25 are from Ref. [66].

length should be larger on the Ni(100) surface where the O atoms have 4 Ni neighbors than on the (111) surface where there are only three. This is in good agreement with experimental evidence[36]. Arguments like this have also been derived from quite different viewpoints, and are sometimes called bond order conservation[67].

The correction terms are in general more difficult to treat. It is found generally [35, 36, 65] that the electrostatic term favours large coordination numbers for atomic adsorbates. The reason is the following. Imagine defining a sphere around each atom in the system, and each sphere is defined to be neutral. If these spheres do not overlap, and we define the sphere around the adsorbate to be equal to the region A in Fig. 5, then the interatomic electrostatic interaction, Eq. 21, is small, there is at least no direct, zero'th moment Coulomb interaction. In reality, the spheres will overlap, and the repulsion between the positively charged nuclei will not be completely compensated by the electron charge. The electrostatic interaction between the adsorbate and the metal atoms in Eq. 26 will therefore be repulsive for small distances. The equilibrium site is therefore the one where the electron density from the surface atoms is close to the optimum value where the E_c function has its minimum and the adsorbate is as far from the metal atoms as possible. Since the metal-adsorbate distance, with a constant density is furthest from the metal atoms when the coordination number is high, this immediately gives the high coordination number rule. We saw that in the examples discussed above this holds well, and it is generally found that atomic adsorbates find high coordination sites on metal surfaces[1]. There are exceptions, though, where the one-electron energy difference in Eq. 26 becomes important and the atomic adsorbate prefers a lower coordinated site[2, 68].

An adsorbate can generally find its uptimum embedding density and consequently gain the same E_c contribution to the chemisorption energy over all metal surfaces simply by adjusting its height above these. We therefore note that the trends in chemisorption energy from one metal to the next must lie in the correction terms. As it is found emperically that for first row atoms and molecules at metal surfaces the one-electron energy term is the dominant, we shall concentrate on this in the following two sections and use extensively in the rest of these notes.

3.6. THE TIGHT BINDING APPROXIMATION

We consider in this section the simplest description of the quantum mechanics of interacting atoms and molecules, using the minimal basis set. The idea is not to give an accurate description, this was discussed in Section 3.3, but to bring out the essential physics. For an introduction into the tight binding theory, see e.g. the recent book by Pettifor[69].

If we consider two atoms A and B with one atomic state ψ_A and ψ_B on each, the simplest possible basis set we can use is the one consisting of these two states only:

$$\psi_{AB} = c_A\psi_A + c_B\psi_B = \mathbf{c}\left\{ \begin{array}{c} \psi_A \\ \psi_B \end{array} \right\}. \tag{27}$$

Each of the atomic orbitals are solutions to an atomic one-electron Kohn-Sham equation

$$(-\frac{1}{2}\nabla^2 + v_A)\psi_A = \varepsilon_A\psi_A \tag{28}$$

We now assume that the effective one-electron potential for the AB molecule is the sum of the atomic potentials:

$$v_{AB} = v_A + v_B, \tag{29}$$

and thus have for the one-electron Hamiltonian of the AB molecule:

$$H_{AB} = -\frac{1}{2}\nabla^2 + v_{AB} = H_A + v_B = H_B + v_A. \tag{30}$$

If we work in the basis $\{\psi_A, \psi_B\}$ and multiply the effective one-electron Schrödinger equation for the molecule

$$H_{AB}\psi_{AB} = \varepsilon\psi_{AB} \tag{31}$$

by ψ_A^* and ψ_B^* in turn and integrate over space we get it in matrix form:

$$(\mathbf{H} - \mathbf{S}\varepsilon)\mathbf{c} = 0, \tag{32}$$

where the vector c was defined in Eq. 27. The Hamiltonian, H, has matrix elements

$$H_{AA} =< \psi_A|H_A + v_B|\psi_A > \tag{33}$$

and

$$H_{AB} =< \psi_A| -\frac{1}{2}\nabla^2 + v_{AB}|\psi_B >= V \tag{34}$$

If we neglect small terms like $< \psi_A|v_B|\psi_A >$ we have

$$H_{AA} = \varepsilon_A. \tag{35}$$

The overlap matrix $\mathbf{S}$ has elements

$$S_{AA} =< \psi_A|\psi_A >= 1 \tag{36}$$

and

$$S_{AB} =< \psi_A|\psi_B >= S. \tag{37}$$

In the following we shall assume that the phases of ψ_A and ψ_B are chosen so that $S > 0$, whereby follows for realistic situations that $V < 0$, as v_{AB} is negative (attractive for an electron) everywhere in space.

Eq. 32 is easily solved to give new one-electron energy solutions

$$\varepsilon_{\pm} = \frac{(\varepsilon_A + \varepsilon_B) \mp \sqrt{4V^2 + (1 - S^2)(\varepsilon_A - \varepsilon_B)^2}}{2(1 - S^2)} \tag{38}$$

Typically S will be small and to first order in S we have

$$\varepsilon_{\pm} = \frac{(\varepsilon_A + \varepsilon_B)}{2} \mp \sqrt{V^2 + (\frac{\varepsilon_A - \varepsilon_B}{2})^2} - VS. \tag{39}$$

Two special cases of this expression are particularly instructive. One is that of a homo-nuclear diatomic molecule, where $\varepsilon_A = \varepsilon_B$. In this case Eq. 39 reduces to

$$\varepsilon_{\pm} = \varepsilon_A \mp |V| - VS. \tag{40}$$

The other case is that of the eigen levels being well separated compared to the coupling matrix element ($|V| \ll |\varepsilon_B - \varepsilon_A|$) where a Taylor expanssion gives the familiar second order perturbation expression:

$$\varepsilon_{\pm} = \begin{cases} \varepsilon_A - \frac{V^2}{\varepsilon_B - \varepsilon_A} - VS \\ \varepsilon_B + \frac{V^2}{\varepsilon_B - \varepsilon_A} - VS \end{cases} \quad \text{for } \varepsilon_B > \varepsilon_A. \tag{41}$$

In all cases above, two new *molecular* solutions appear below (bonding) and above (anti-bonding) the two atomic solutions, as illustrated in Fig. 8. Apart from the bonding/anti-bonding shifts, there is a general shift of the states by $-VS$ which is positive since S and V have opposite sign. This up-shift in the one electron energy is due to the overlap of the two atomic states (Pauli repulsion).

In the perturbation theory limit, we can write the molecular levels correct to first order in V:

$$\begin{aligned} \psi_+ &= \psi_A + \frac{V}{\varepsilon_A - \varepsilon_B}\psi_B, \\ \psi_- &= \psi_B + \frac{V}{\varepsilon_B - \varepsilon_A}\psi_A, \end{aligned}$$

which shows that the weight (the wavefunction squared) of the upper (lower) atomic state in the lower (upper) molecular state becomes proportional to $V^2/(\varepsilon_A - \varepsilon_B)^2$.

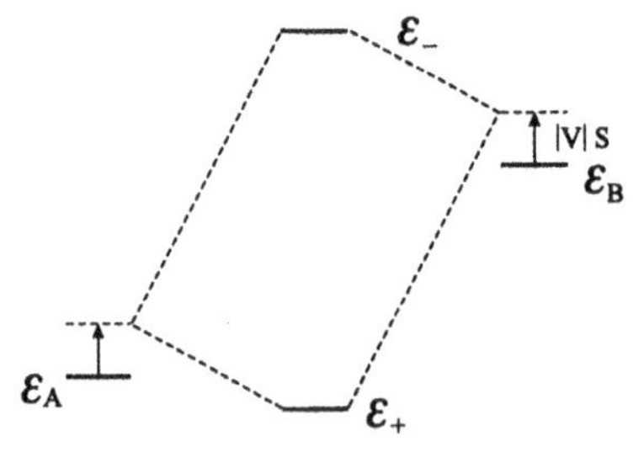

Figure 8. Schematics of bonding in a diatomic heteronuclear molecule.

The question is now how the changes in electronic structure when the two atoms A and B are brought into contact to form the molecule AB affects the total energy of the system. The total energy difference

$$\Delta E = E_{AB} - E_A - E_B \tag{42}$$

measures the strength of the molecular bond. We shall term it the bond energy. If ΔE is negative the molecule is stable, and $E_{diss} = -\Delta E$ is the dissociation energy. If ΔE is positive for all inter-nuclear distances the molecule is not stable at all.

In the following we consider only the sum of the one-electron energies

$$E = \sum_{i\ occ} \varepsilon_i, \tag{43}$$

where the sum is over the occupied levels only and also includes the two spin directions. This is clearly NOT the total energy of the system, c.f. Eq. 2, but we can get some qualitative insight from considering this simple approximation, and it is clear from Section 3.4, that if, for instance, we are considering small *changes* in the total energy from one atom to a similar one then these differences are governed to a large extent by the properly calculated one-electron energy sum.

In this very simple approximation, the bond energy is

$$\Delta E = \sum_{i\ occ} \varepsilon_i^{AB} - \sum_{i\ occ} \varepsilon_i^{A} - \sum_{i\ occ} \varepsilon_i^{B}. \tag{44}$$

In the following we first consider bonding in a simple homo-nuclear diatomic molecule. We then briefly discuss hetero-nuclear molecules. In the following section we use the same framework to discuss bonding in a solid.

Consider first a homo-nuclear diatomic molecule with one chemically important (valence) state on each atom. We assume that each atomic state has one electron, like in the H_2 molecule. We have discussed above how in this case the interaction between the two atomic states gives rise to two molecular states with energy ε_+ and ε_-. Only the bonding state with energy ε_+ will be filled with two electrons while the free atom states with energy ε_0 have one electron each. This gives the bond energy

$$\Delta E = 2\varepsilon_+ - 2\varepsilon_0 \tag{45}$$

Using the expression for the energy of the bonding level in the case where we can work to first order in the overlap, this gives

$$\Delta E = 2S|V| - 2|V|. \tag{46}$$

The first term on the right hand side of Eq. 46 is positive while the second term is negative. From the definition of the overlap and hopping matrix elements it is clear that both V(R) and S(R) must decay exponentially with the interatomic distance R. This follows from the fact that the wave functions decay exponentially. The decay constant must therefore be similar, and it is not unreasonable to assume that S and V are proportional: $S = -\alpha V$. We can then write Eq. 46 as

$$\Delta E = 2\alpha V^2 - 2|V|. \tag{47}$$

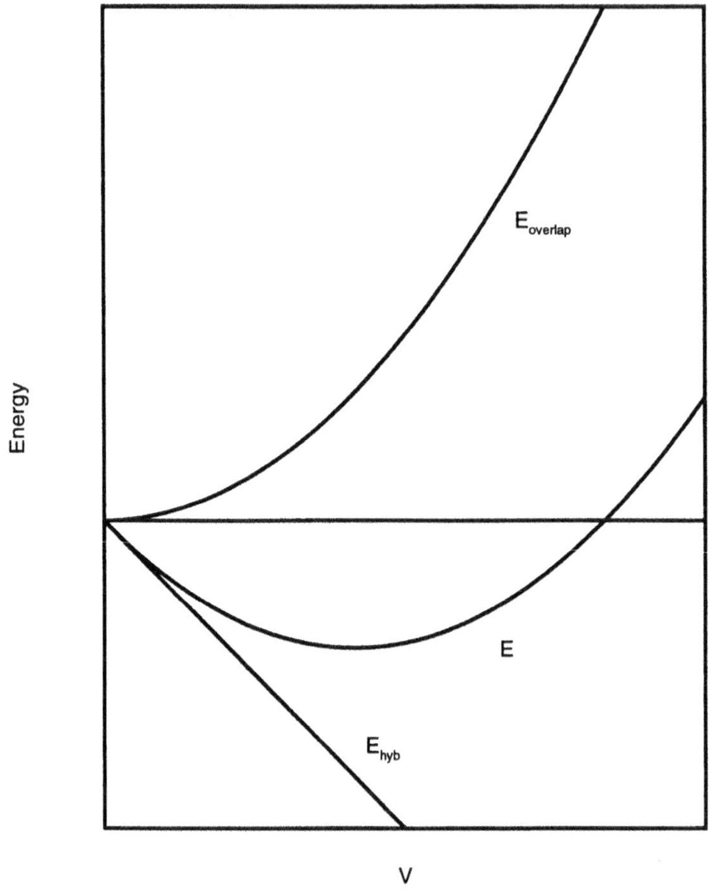

Figure 9. The binding energy in a H_2-like molecule as a function of the coupling matrix element V. The total energy is split up into the orthogonalization and hybridization parts.

In figure 9 ΔE and the two components are shown as a function of V. Since V depends monotonically on the inter-nuclear separation R, this figure can be directly converted into a $\Delta E(R)$ plot. The minimum in energy signifying the most stable configuration of the system comes about as a competition between the repulsion due to the overlap term $\Delta E_{overlap} = 2\alpha V^2$, and the attraction $\Delta E_{hyb} = -2|V|$ due to the quantum mechanical *hybridization* between the two atomic states.

If there is not one but two electrons in each of the atomic states (like in He), then both the bonding and the anti-bonding states will be filled and the bond energy is

$$\Delta E = 2\varepsilon_+ + 2\varepsilon_- - 4\varepsilon_0 = 4S|V|. \tag{48}$$

In this case the interaction is purely repulsive (He_2 is not a stable molecule, as you know). The bonding down shift is exactly canceled by the anti-bonding up-shift of the molecular states.

For a hetero-nuclear diatomic molecule with one valence electron on each atom, the bond energy is:

$$\begin{aligned}\Delta E &= 2\varepsilon_+ - \varepsilon_A - \varepsilon_B \\ &= -2SV - W_{AB}. \end{aligned} \tag{49}$$

This expression is very similar to the one for the homo-nuclear diatomic except that now the hybridization is given by $W = \sqrt{4V^2 + \Delta^2}$, where $\Delta = \varepsilon_B - \varepsilon_A$. It can be seen that in the case where Δ becomes small, the expression approaches the homo-nuclear limit. For the same hopping matrix element V, the interaction is stronger the larger the energy separation Δ is. Even when V is zero the system can gain energy by transferring the electron from the highest lying state to the lowest lying state. In this case the system becomes *ionic.*

The tight binding or LCAO description discussed above can be generalized readily to larger molecules. The basis set becomes larger, since there will be (at least) one state per atom, but the principle is the same. Again there will be a repulsive energy term proportional to SV for each pair of interacting atoms and there may be some attraction due to the hybridization between the atomic states.

For a solid with perfect periodicity, the problem is greatly reduced by using the translational symmetry of the system. The tight binding model is exactly solvable in simple cases and many of the basic features of the electronic structure in a solid can be understood on this basis[69]. The spectrum of one-electron eigenvalues is continuous, and usually it is simpler to consider the density of states rather than the individual one-electron levels.

Figure 10 shows a typical density of states for a transition metal. The broad s-band is half filled – all the transition metals have one s-electron – and the d-band is seen as a much narrower band. Its occupancy varies along the transition metals as the d-band shifts through the Fermi level. The narrow d-bands are a consequence of the small coupling matrix element V between the localized d states; it is one of the important conclusions from tight binding theory that the band width is proportional to V.

In the simple independent electron approximation, the bond energy per atom from a band with density of states $n(\varepsilon)$ is

$$\Delta E = 2\sum_{\mathbf{k},occ} \varepsilon(\mathbf{k}) - \varepsilon_0 \tag{50}$$

$$= \int_{-\infty}^{\varepsilon_F} n(\varepsilon)(\varepsilon - \varepsilon_0)d\varepsilon. \tag{51}$$

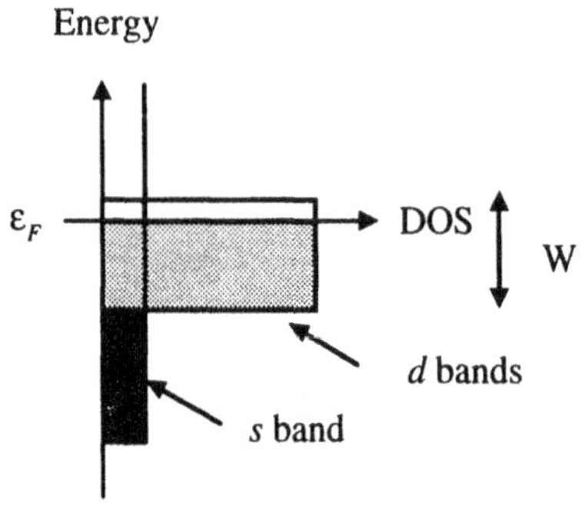

Figure 10. Schematic illustration of the density of states of a transition metal, showing the broad *s* band and the narrow *d* band around the fermi level.

If we calculate the density of states $n(\varepsilon)$ in a tight binding model where the overlap is neglected, then we must remember to add a term

$$\Delta E_{overlap} = -N_{nn}SV \tag{52}$$

where N_{nn} is the number of nearest neighbors of each of the atoms.

It is clear that states in the band that are below ε_0 are bonding – that is, they contribute to increase the bond energy (i.e. to make ΔE more negative) in Eq. 51, while the states above ε_0 are anti-bonding.

We can take the discussion a little further by making a simple assumption about the *shape* of the band. Assume that the density of states is rectangular as in Figure 10: $n(\varepsilon) = 2/W$, for $\varepsilon_0 - W/2 < \varepsilon < \varepsilon_0 + W/2$, and zero otherwise. W is the band width, which is proportional to the hopping matrix element V. The density of states is normalized so that the total number of electrons in the band is two; one for each spin. By centering the band at atomic energy level ε_0, we have implicitly neglected the overlap repulsion Eq. 52. For a rectangular band we can do the integral in Eq. 51 and we get the hybridization part of the bond energy as

$$\Delta E_{hyb} = \frac{2}{W}\int_{-W/2}^{\varepsilon_F} \varepsilon d\varepsilon \tag{53}$$

$$= \frac{(\varepsilon_F^2 - (W/2)^2)}{W}. \tag{54}$$

Now use that the fraction f of the band which is filled, is given by

$$f = \frac{1}{W}\int_{-W/2}^{\varepsilon_F} d\varepsilon \tag{55}$$

$$= \frac{\varepsilon_F}{W} + \frac{1}{2}, \tag{56}$$

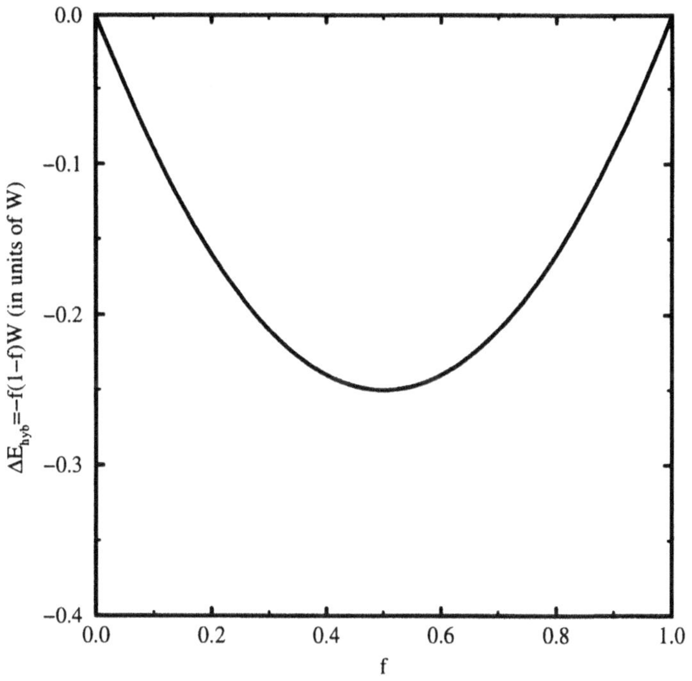

Figure 11. The hybridization energy in the rectangular band model as a function of the fractional filling of the band.

and get

$$\Delta E_{hyb} = -f(1-f)W. \tag{57}$$

Eq. 57 shows first of all that the hybridization energy is proportional to the band width W, which in turn is proportional to the hopping matrix element V. The picture of bonding in a solid is therefore very much the same as in the case of a simple diatomic discussed above. The equilibrium bond length and bond energy is given by a competition between a repulsive term $E_{overlap}$ increasing with the matrix element as V^2 and an attractive hybridization term proportional to V, as illustrated in figure 9.

Eq. 57 also shows how the bonding in a solid depends on the degree of filling of the band. Fig. 11 shows the bond energy as a function of the filling factor f. It has a minimum (maximum bond strength) for a half filled band. If we now return to the bonding in a transition metal, we can to a first approximation consider the bonding due to the s and the d bands separately. The bonding from the s band is maximum for all the transition metals, since they all have a half filled s band. This on the other hand does not give rise to largely different bonding for the different metals. These differences are much more pronounced for the d band contribution. Here the filling varies from zero to the left in the transition metal series to one to the far right. The simple model, which is originally due to Friedel[70, 71, 69], says that the bonding should be strongest for the transition metals with five d electrons, a half filled d band. This is exactly what is found experimentally. Figure 12 shows the cohesive energy ($= -\Delta E$) for all the transition metals as a function of the number of d electrons. Except for the magnetic metals in the $3d$ series, the cohesive energy pretty much shows the parabolic behavior displayed in Eq. 57.

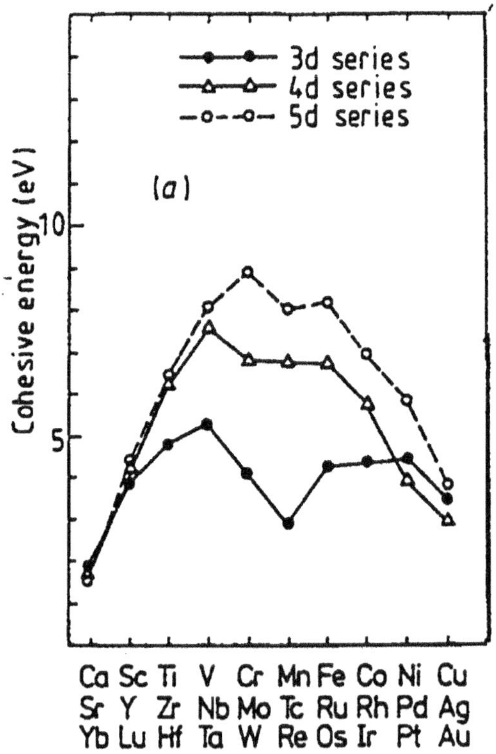

Figure 12. The measured cohesive energies for all the transition metals as a function of the number of d-electrons. From Ref. [72].

3.7. THE NEWNS-ANDERSON MODEL

The simple two level problem discussed above must be modified slightly at a surface. Here an adsorbate state will not just interact with a single surface state, but rather with a continuum of states. Consider a metal surface with one-electron states $|k>$ with energies ε_k, and an adsorbate with a single valence state $|a>$ with energy ε_a. When the adsorbate is taken from far away from the surface to a position just outside, the two sets of states are coupled by matrix elements $V_{ak} = <a|\mathbf{H}|k>$, where $\mathbf{H}$ is the Hamiltonian of the combined system. If we expand the solutions $|i>$ of $\mathbf{H}$ in terms of the free adsorbate and surface solutions:

$$|i> = c_{ai}|a> + \sum_k c_{ki}|k> \tag{58}$$

and neglect the overlap $<a|k>$, then the Schrödinger equation can be written

$$\mathbf{H}\mathbf{c}_i = \varepsilon_i \mathbf{c}_i \tag{59}$$

where $H_{aa} = \varepsilon_a$, $H_{kk} = \varepsilon_k$, and $H_{ak} = V_{ak}$. This is the simplest form of the so-called Newns-Anderson model [73, 74] . Other important contributions to this theory comes from Grimley[75] and Gadzuk[76]

There are infinitely many metal states and it is impossible to keep track of all of them when the coupling is switched on. It is therefore instructive to follow the projection of the density of states on the adsorbate state:

$$n_a(\varepsilon) = \sum_i |<i|a>|^2 \delta(\varepsilon - \varepsilon_i) \tag{60}$$

where the sum is over the eigenstates of the full Hamiltonian Eq.(59). We can write this as

$$n_a(\varepsilon) = -\frac{1}{\pi}\mathrm{Im}\sum_i \frac{<a|i><i|a>}{\varepsilon - \varepsilon_i + i\delta} = -\frac{1}{\pi}\mathrm{Im}G_{aa}(\varepsilon) \tag{61}$$

with $\delta = 0^+$. Here $G_{aa}(\varepsilon)$ is the projection on the adsorbate state of the so-called single particle Green function

$$\mathbf{G}(\varepsilon) = \sum_i \frac{|i><i|}{\varepsilon - \varepsilon_i + i\delta} \tag{62}$$

which is then defined by the formal equation

$$(\varepsilon - \mathbf{H} + i\delta)\mathbf{G}(\varepsilon) = 1 \tag{63}$$

To get $n_a(\varepsilon)$ we therefore just need the imaginary part of the $|a>$ projection of $\mathbf{G}(\varepsilon)$. Eq.(63) can be written in matrix form:

$$\begin{pmatrix} (\varepsilon + i\delta - \varepsilon_a) & -\mathbf{V}_{aM}- \\ | & \\ \mathbf{V}_{Ma} & \mathbf{1}(\varepsilon + i\delta) - \mathbf{H}_{MM} \\ | & \end{pmatrix} \begin{pmatrix} G_{aa} & -\mathbf{G}_{aM}- \\ | & \\ \mathbf{G}_{Ma} & \mathbf{G}_{MM} \\ | & \end{pmatrix} = \begin{pmatrix} 1 & -0- \\ | & \\ 0 & \mathbf{1} \\ | & \end{pmatrix} \tag{64}$$

Here we immediately get two equations:

$$\begin{aligned} (\varepsilon - \varepsilon_a + i\delta)G_{aa} + \sum_k V_{ak}G_{ka} &= 1 \\ V_{ka}G_{aa} + (\varepsilon - \varepsilon_k + i\delta)G_{ka} &= 0 \end{aligned} \tag{65}$$

and eliminating G_{ka} this gives

$$G_{aa}(\varepsilon) = \frac{1}{\varepsilon - \varepsilon_a - q(\varepsilon)} \tag{66}$$

where the self-energy

$$q(\varepsilon) = \Lambda(\varepsilon) - i\Delta(\varepsilon) \tag{67}$$

is given by real and imaginative parts

$$\Delta(\varepsilon) = \pi \sum_k |V_{ak}|^2 \delta(\varepsilon - \varepsilon_k) \tag{68}$$

and

$$\Lambda(\varepsilon) = \frac{P}{\pi} \int \frac{\Delta(\varepsilon')}{\varepsilon - \varepsilon'} d\varepsilon'. \tag{69}$$

Eq.(61) then immediately gives the projected density of states

$$n_a(\varepsilon) = \frac{1}{\pi} \frac{\Delta(\varepsilon)}{(\varepsilon - \varepsilon_a - \Lambda(\varepsilon))^2 + \Delta(\varepsilon)^2} \tag{70}$$

From Eq.(68) it can be seen that $\Delta(\varepsilon)$ can be regarded as a local projection of the metal density of states around the adsorbate.

There are two simple limits that can easily be worked out. The first is the case where $\Delta(\varepsilon) = \Delta_0$ is independent of energy. In this case Λ from Eq.(69) is zero. $n_a(\varepsilon)$ is then a simple Lorentzian of width Δ_0 centered around ε_a. Such a structure is usually encountered for adsorbate states interacting with broad sp-bands of metals. Here the surface density of states is slowly varying, and a single resonance is formed.

To see the physical meaning of the width Δ_0 of the resonance, let us assume that at $t = 0$ there is an electron in level $|a>$. The probability that there is an electron in $|a>$ at a later time t is then

$$P_a(t) = |<a|\exp(-i\mathbf{H}t)|a>|^2 \tag{71}$$

By Fourier transforming $\mathbf{G}(\varepsilon)$ in Eq. (63), one sees that

$$\begin{aligned} P_a(t) &= |<a|\mathbf{G}(t)|a>|^2 = |G_{aa}(t)|^2 = |\exp(-i(\varepsilon_a - i\Delta_0)t)|^2 \\ &= \exp(-2\Delta_0 t) \end{aligned} \quad (72)$$

This small calculation immediately confirms our intuition, that the adsorbate level is simply broadened due to the finite lifetime of an electron in the adsorbate state. The lifetime is $\tau = (2\Delta_0)^{-1}$ and Δ_0 is given by a simple Golden Rule type expression Eq.(68).

The other limit that can be treated off hand, is the case where the bandwidth W of the metal states is much smaller than the coupling matrix elements V_{ak}. Then the problem effectively degenerates into the two level problem discussed in the preceding section. Sharp bonding and anti-bonding states are formed above and below the two states. Such a case is frequently encountered when one is considering metal surfaces with d-bands. d-bands are usually narrow compared to the hopping matrix elements V_{ak} and therefore gives rise to bonding adsorbate - metal-d states below the d-bands (these will then be broadened out into resonances due to the s-p bands) and anti-bonding states (resonances) above the d-bands.

The two limiting behaviors are illustrated in Fig. 13. The broad band limit with a single resonance is often called "weak chemisorption" while the case where there are split off bonding and anti-bonding states is called "strong chemisorption"[77].

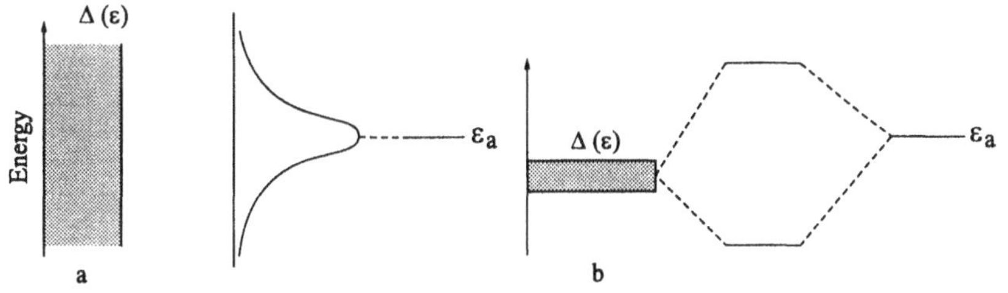

Figure 13. The local density of states at an adsorbate in the Newns-Anderson model in two limiting cases: a) The $\Delta(\varepsilon)$ independent of ε, and b) for a narrow metal band.

4. Atomic chemisorption on metal surfaces

In this Chapter we discuss in some detail the simplest case of chemisorption, where a single atom binds to a metal surface. In Chapter 2 we established that the self-consistent DFT calculations are able to account for a number of experimental observations regarding atomic chemisorption. Here we will go into a deeper discussion of the underlying physics.

4.1. THE ELECTRONIC STRUCTURE

As a starting point, consider the adsorption of atomic hydrogen on the four metals discussed above. In Fig. 2 the atomically adsorbed state is the final state minimum for the four reaction paths. In all cases the lowest energy configuration on these close packed fcc(111) surfaces is with the hydrogen atoms in the three fold site (see the insert). It is seen that the hydrogen-metal chemisorption bond is much weaker on the noble metals Cu and Au than on the transition metals Ni and Pt, and that the bond is weakest on Au. This is very general for atomic chemisorption of many of the constituents in the simplest gas molecules like H, O, S, C, or N. It is illustrated for O and S chemisorption on a number of noble and transition metals in Fig. 14. For both adsorbates it is seen that the noble metals bind weakest and the unique nobleness of Au is clearly borne out. The O-Au surface bond is weaker than the O-O bond (2.95 eV per O atom in the GGA) so oxygen molecules will not readily dissociate on Au at all. On the other hand, the Au-S bond is marginally stronger than the S bond in H_2S (the energy to split H_2S into H_2 and S is 3.37 eV in the GGA, neglecting zero-point energy contributions).

To understand the differences in chemisorption energy from one metal to the next, consider in Fig. 15 the density of states projected onto the O 2p states for O chemisorbed on Cu, Ag, Au, Ni, Pd, Pt, and Ru. In all cases the density of states have a similar structure, in Fig. 16 we show how one can think about the origin of the main peaks in the density of states. We imagine the bond formation takes place in two steps[77, 36]. First, we let the adsorbate valence p states interact with the metal s and p electrons. As discussed in connection with our introduction of the Newns-Anderson model, this will give rise to a single resonance. The O p derived resonances are well below the Fermi level and hence completely filled. We then turn on the coupling to the d-electrons. Since the d bands are narrow, this gives rise to strong interaction in the sense of the Newns-Anderson model, that is, it gives rise to a splitting of the O resonance into two: one state which is bonding with respect to the adsorbate and metal d states and another above the d-bands which is anti-bonding.

It is seen that for oxygen on Ag the bonding and anti-bonding states have essentially the same weight on the O state (as measured by the area under the peaks in the O projected density of states in Fig. 15. This means that the uncoupled adsorbate and metal d states are essentially in resonance for this system (only when $\varepsilon_A = \varepsilon_B$ will the problem be symmetrical and the two solutions ε_+, ε_- in Eq 39 will have the same weight on the two components). For the other metals the d band center is higher, and the bonding state has more O character, while the anti-bonding state has more

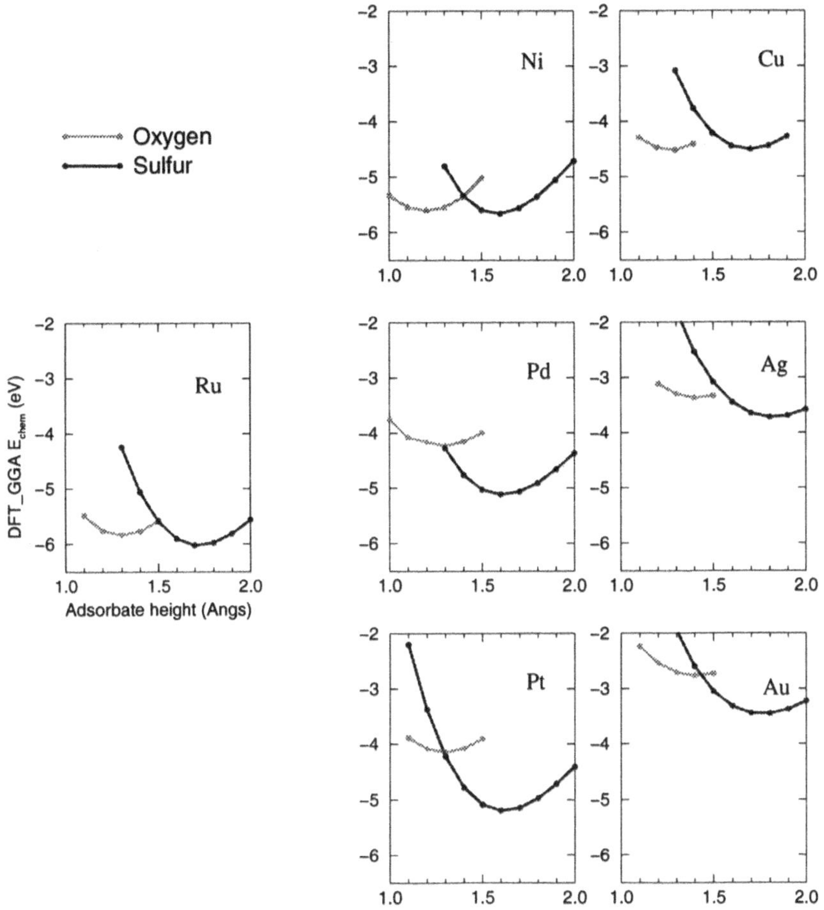

Figure 14. O and S chemisorption at a number of close-packed late transition and noble metal surfaces (hcp(0001) for Ru, while fcc(111) for Ni, Cu, Pd, Ag, Pt, and Au). A quarter monolayer of either O or S is adsorbed at the 3-fold fcc site (the 3-fold hcp for Ru) in a p(2×2) pattern. The adsorbate height is varied, while the metal ions are fixed at the truncated bulk positions.

metal *d* character. The picture is the same for the other metals and for other adsorbates, see Fig. 17.

4.2. UNDERSTANDING TRENDS IN ATOMIC CHEMISORPTION ENERGIES

The interpretation of the changes in electronic structure given above has some immediate consequences for our understanding of the trends in the binding energies. We can also think of the binding energy as having two

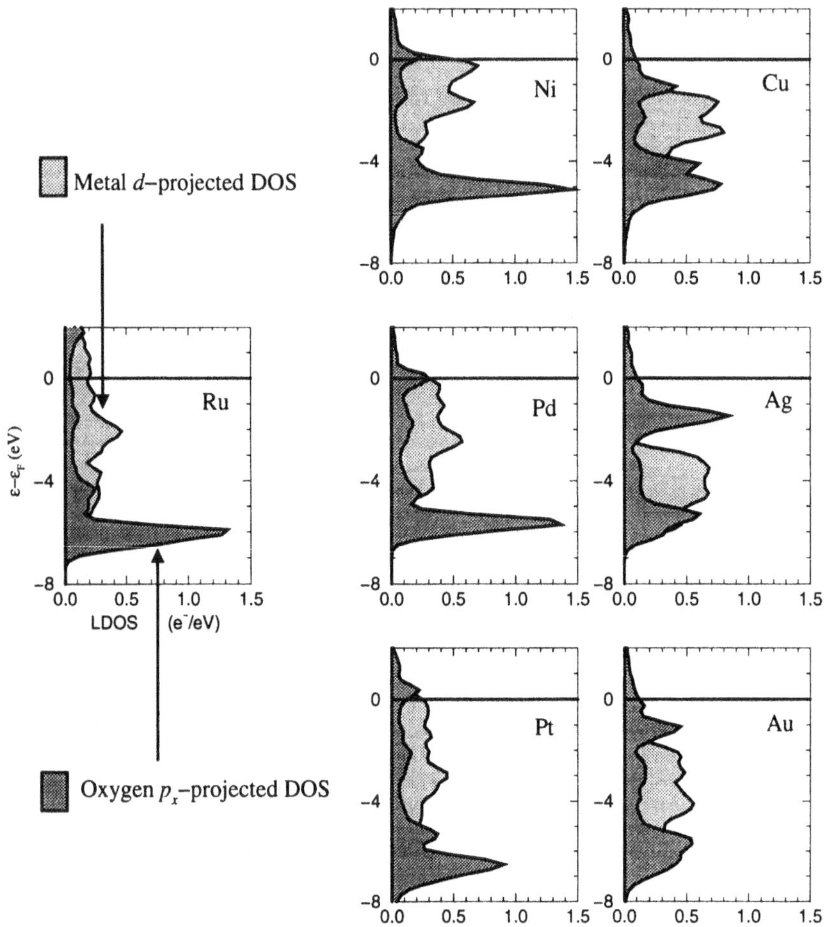

Figure 15. Local density of states projected onto the O p_x state (dark shaded area) for oxygen 1.3 Å above close-packed surfaces of late transition and noble metals (cf. 14. The light shaded area gives the metal d projected DOS for the respective metal surfaces before the oxygen chemisorption.

components, one from the coupling to the metal s-p states, and one due to the extra coupling to the d states[77, 64, 36, 53, 55]. If we assume that the introduction of the d-state coupling is only a weak perturbation on the adsorbate, then we can treat the energy difference due to the d-coupling using the frozen density and potential approximation. This means that the energy due to the coupling to the d electrons is simply given by the difference in the sum of the one-electron eigenvalues due to the inclusion of the coupling to the d electrons. Judging from the calculated densities of states in Fig. 15 and 17 this coupling is essentially a two level problem

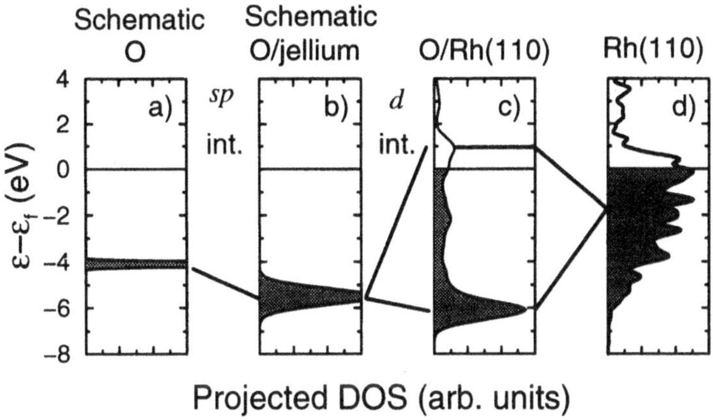

Figure 16. Schematic illustration of the change in local electronic structure at an oxygen atom upon adsorption on simple and transition/noble metal surfaces (in this case Rh(110)). First the sharp atomic states of the gas phase are broadened into ressonances and shifted down due to the interaction with the metal *sp* states. Next these renormalized states interact with the narrow *d* bands at the transition and noble metal surfaces forming covalent bonding and anti-bonding states below and above the initial adsorbate and surface states. From Ref. [78].

giving rise to a bonding and an anti-bonding state, and the d band is characterized only by the band center ε_d and its filling f. If we neglect the influence of the d-band width, the energy difference is given by a simple extension of Eq. 49 to the case where one of the states has occupancy 1 and the other (the effective metal d-state that the adsorbate state interacts with) has occupancy f[55]:

$$E_{d-hyb} = -(1-f)(W_{ad} - \Delta_{ad}) - 2(1+f)VS. \tag{73}$$

Here $W_{ad} = \sqrt{4V^2 + \Delta_{ad}^2}$ and $\Delta_{ad} = \varepsilon_d - \varepsilon_a$.

In Eq. 73 the first term is attractive if the d band is not completely full, while the second is the usual Pauli repulsion due to overlap. As usual we assume that $S = -\alpha V$. The constant α can be considered independent of the metal surface in question if the adsorbate is a sufficiently weak perturbation. In case the coupling is weak relative to the energy difference between the two states we can write Eq. 73 as:

$$E_{d-hyb} = -2(1-f)\frac{V^2}{|\varepsilon_d - \varepsilon_a|} + 2(1+f)\alpha V^2. \tag{74}$$

A large part of the variation of the d-coupling from on metal to the next is related to the variations in the coupling matrix element V. The absolute magnitude of V will also depend on the position of the adsorbate, but for

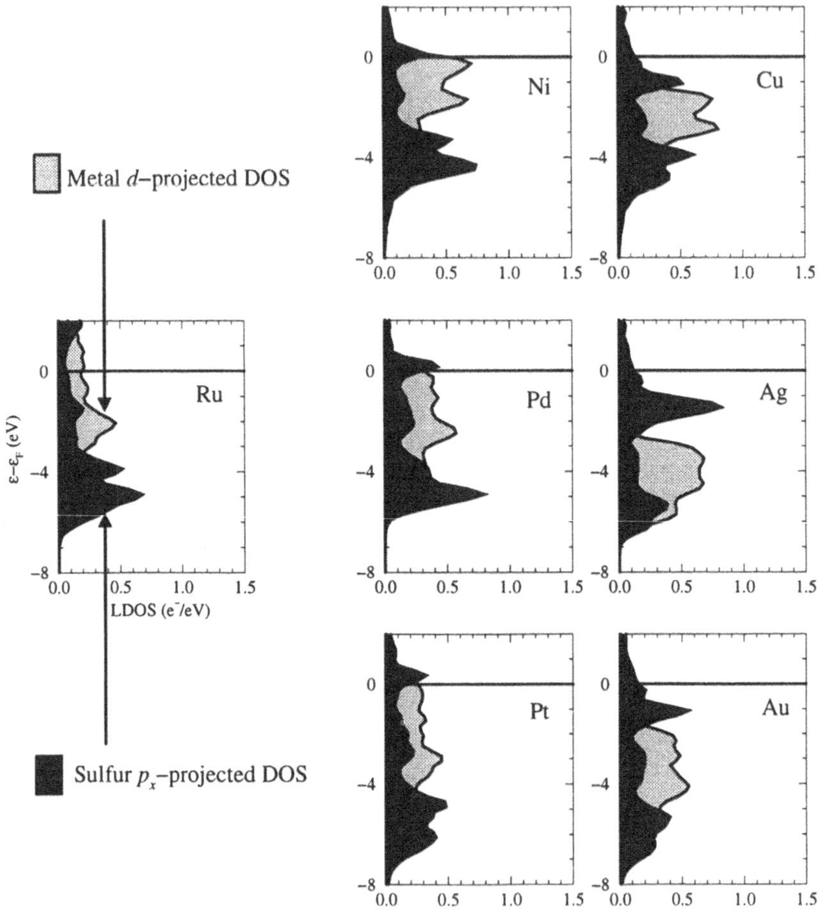

Figure 17. Same as figure 15, but for sulfur at a distance 1.7 Å from the close-packed transition and noble metal surfaces. Dark shaded area is the S p_x projected DOS, while light shaded area is the metal d projected DOS.

a fixed adsorbate geometry the variations of the matrix element depend on the metal one-electron potential and the extend of the metal d-states. This can for instance be seen if the matrix elements are calculated from LMTO theory. In appendix A we show in detail how the matrix elements can be calculated. Here it suffices to say that V can be written as a product of three terms, one depending only on the potential around the metal atoms, one depending only on the potential around the adsorbate and one depending only on the relative positions of the interacting atoms. Further, we have to use "frozen potentials" in the calculation of the one electron energy differences. This means that the potential around the metal atoms must be independent of the presence of the adsorbate. The variation in V from one metal to the next for a fixed adsorbate and adsorbate geometry is therefore a property of the metal alone. For each element we know the *relative* strength of the coupling matrix element directly from tabulated values of the LMTO potential functions [79, 80, 81]. The values are shown in Fig. 18 for a large section of the periodic table.

It is simple to include the effect of the width of the d band. We have to solve the Newns-Anderson problem, taking into account the fact that the calculated d projected density of states for the clean surface has a certain width. The width is known, however, from the calculations for the clean surfaces, and this makes only a slight difference in the calculation. In the calculations shown in the rest of this section we have solved the full Newns-Anderson problem assuming a semieliptic form of the metal DOS.

Idealized d- band filling

V_{ad}^2 [Relative to Cu]

0.1 20.8 Ca 4.12	0.2 7.90 Sc 3.43	0.3 4.65 Ti 1.50 3.05	0.4 3.15 V 1.06 2.82	0.5 2.35 Cr 0.16 2.68	0.6 1.94 Mn 0.07 2.70	0.7 1.59 Fe -0.92 2.66	0.8 1.34 Co -1.17 2.62	0.9 1.16 Ni -1.29 2.60	1.0 1.0 Cu -2.67 2.67	1.0 0.46 Zn 2.65
0.1 36.5 Sr 4.49	0.2 17.3 Y 3.76	0.3 10.9 Zr 1.95 3.35	0.4 7.73 Nb 1.41 3.07	0.5 6.62 Mo 0.35 2.99	0.6 4.71 Tc -0.60 2.84	0.7 3.87 Ru -1.41 2.79	0.8 3.32 Rh -1.73 2.81	0.9 2.78 Pd -1.83 2.87	1.0 2.26 Ag -4.30 3.01	1.0 1.58 Cd 3.1
0.1 41.5 Ba 4.65	0.2 17.1 Lu 3.62	0.3 11.9 Hf 2.47 3.30	0.4 9.05 Ta 2.00 3.07	0.5 7.27 W 0.77 2.95	0.6 6.04 Re -0.51 2.87	0.7 5.13 Os 2.83	0.8 4.45 Ir -2.11 2.84	0.9 3.90 Pt -2.25 2.90	1.0 3.35 Au -3.56 3.00	1.0 2.64 Hg 3.1

Bulk Wigner-Seitz radius, s [au]

ε_d [eV]

Figure 18. Section of the Periodic Table with the $3d$, $4d$, and $5d$ transition metals and the noble metals. Shown in lower right corner: Bulk Wigner-Seitz radius, s. Lower left corner: The center of the d band calculated for the most close packed surface for each of the metals ((111) for fcc, (001) for hcp and (110) for bcc). In upper right corner: Behavior of the adsorbate-(s or p)-metal d coupling matrix element squared, V_{ad}^2 . The V_{ad}^2's generally decrease for increasing nuclear charge within a row and increase down the groups. All above numbers except for the properties of Zn, Cd, and Hg, have been compiled from Ref. [79] Finally, upper left corner: the idealized d band fillings. These are found to be close to the actual, calculated bulk d band fillings considering the uncertainties in interpreting these[79].

In Fig. 18 we also include values of the two other metal d band properties entering into the description, the center of the d band ε_d and the filling of the d band f. The centers have been calculated for the most close packed surface for the metals considered.

If the contribution from the coupling to the metal sp states is approximately the same for the metals considered then the main trends in

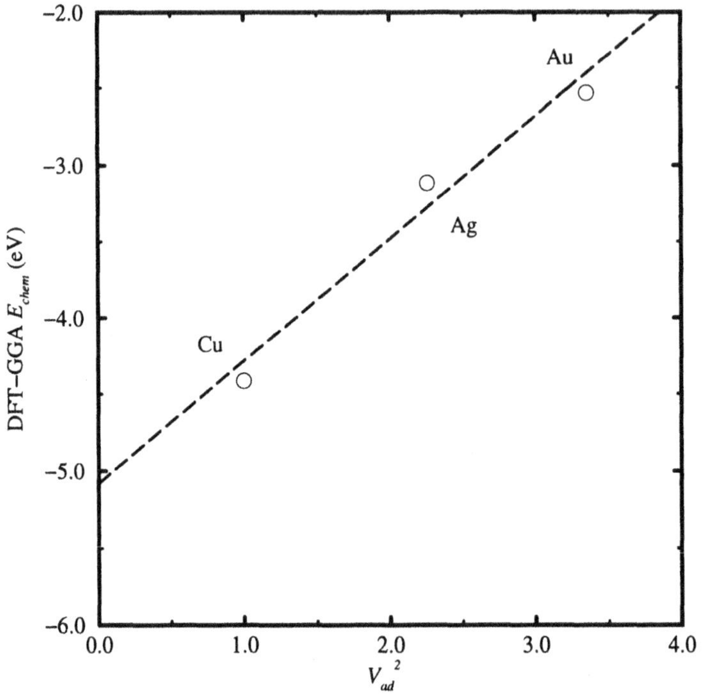

Figure 19. The calculated chemisorption energy, E_{GGA}, for 1/4 ML Oxygen over Cu(111), Ag(111), and Au(111) (cf. Fig. 14) plotted versus the coupling matrix elements squarred, V_{ad}^2. The oxygen-noble metal interaction is purely repulsive as predicted by Eq. 75.

the chemisorption energies should be given by the coupling to the *d* electrons[55]. For the noble metals, where $f = 1$, the model is particularly simple. In that case:

$$E_{bond} = E_0 + 3{\cdot}4\alpha V^2 \equiv E_0 + 3{\cdot}4\alpha\beta V_{ad}^2 \,, \tag{75}$$

where E_0 is the energy contribution from the *sp* electrons, where the factor of three accounts for the degeneracy of the adsorbate level (oxygen $2p$), and where β is the adsorbate and geometry dependent, but metal independent ratio V^2/V_{ad}^2 of the actual coupling matrix elements to the relative matrix elements, V_{ad}^2, given in Fig. 18. In Fig. 19 we plot the calculated bond energies as a function of these matrix elements and see how the simple model clearly captures the main trends. The model also shows that the contribution from the *sp* electrons is by far the largest (E_0 is of the order -5 eV) and that the contribution from the *d* electrons is purely repulsive. Au has the most extended *d* states, the largest V_{ad}^2 and therefore the largest repulsion. This makes O least stable on Au and thus Au the most noble metal. The

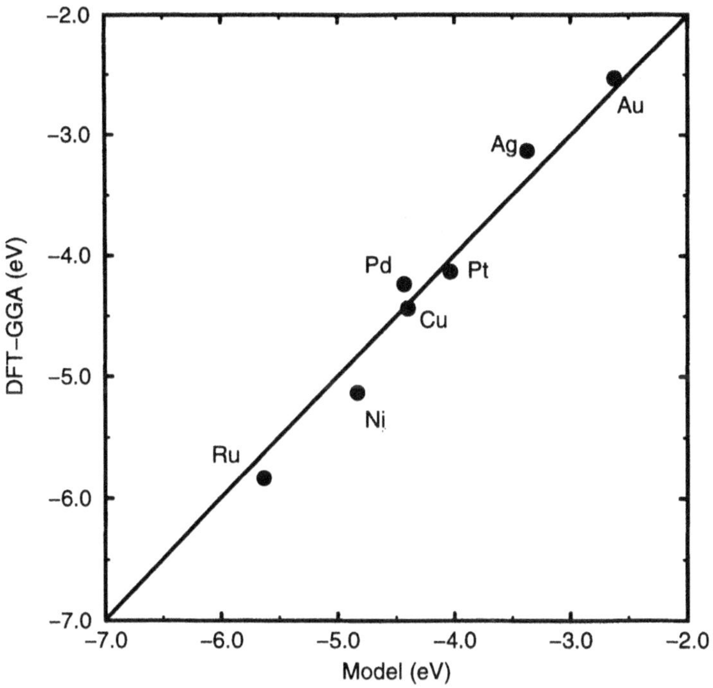

Figure 20. The DFT-GGA based oxygen chemisorption energies correlated with the model estimates of the bond strengths over the various close-packed transition and noble metal surfaces. The full Newns-Anderson model is used with $\alpha_O = 0.036$, $\beta_O = 2.1$, $\varepsilon_O = -5$ eV, $\Delta_0 = 1$ eV, $r_{M-O} = 2.05$ Å and $E_{0,O} = 5.18$ eV.

same is true for S and H (see Fig. 14 and 2) and should, according to the model, hold for other electronegative adsorbates like N, C or the halogens with deep lying valence states after coupling with the metal *sp* states.

A view at the values of V_{ad}^2 for the other metals with $f = 1$ in the periodic table in Fig. 18 shows that the smallest value is for Zn. The model therefore suggests that oxygen and the other electronegative adsorbates should bind most strongly to Zn in good agreement with experimental experience.

To test the model for cases where $f < 1$ we show in Fig. 20 and 21 an interpretation of the full DFT calculations for for a fixed adsorbate-metal bond length using the full Newns-Anderson model. Three parameters have been fixed; the contribution E_0 of the *sp* coupling to the chemisorption bond energy, the ratio $\beta = V^2/V_{ad}^2$, and the value of α. All three parameters are assumed independent of the metal. The model clearly describes the full results remarkably well. The parameters are given in Fig. 20 and 21. It is seen that the value of E_0 is large and negative for both O and S, showing, as

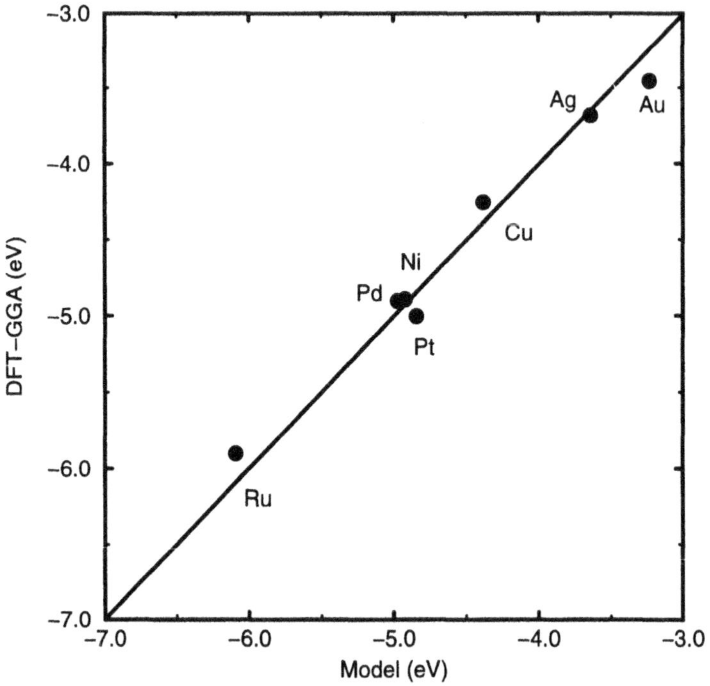

Figure 21. The DFT-GGA based sulfur chemisorption energies correlated with the model estimates of the bond strengths over the various close-packed transition and noble metal surfaces. The full Newns-Anderson model is used with $\alpha_S = 0.038$, $\beta_S = 1.5$, $\epsilon_S = -3$ eV, $\Delta_0 = 1$ eV, $r_{M-S} = 2.4$ Å and $E_{0,S} = 4.9$ eV.

discussed above, that while most of the differences in chemisorption energy from one metal to the next is given by the coupling to the *d* electrons, the coupling to the sp electrons gives most of the absolute binding energy. We also note that β is larger for O than for S. This is a simple consequence of the fact that the S atoms are further from the surface (see, Fig.14) giving rise to a smaller overlap to the localized metal *d* states. The value of α is essentially independent of the adsorbate.

We can now use the model to understand some of the trends in the O and S bond strengths. While Fig. 20 and 21 are based on the full Newns-Anderson description (Eqs. 70 and 44) it is most convenient to neglect the *d* band width in the qualitative discussion and base this on Eq. 73, which is what we will do in the following. As a starting point, we note that in Fig. 14 it is seen that the O bond strength on Ni is much larger than on Pd or Pt. The reason is that the attractive term in Eq. 73 is still smaller than the repulsive one for f=0.9. This makes the bonding of atomic adsorbates weak on Pd and Pt. The tendency that adsorbates bind weaker on the 4d and 5d

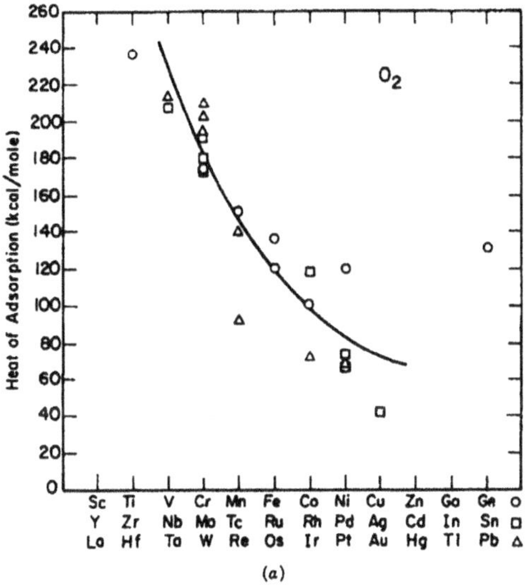

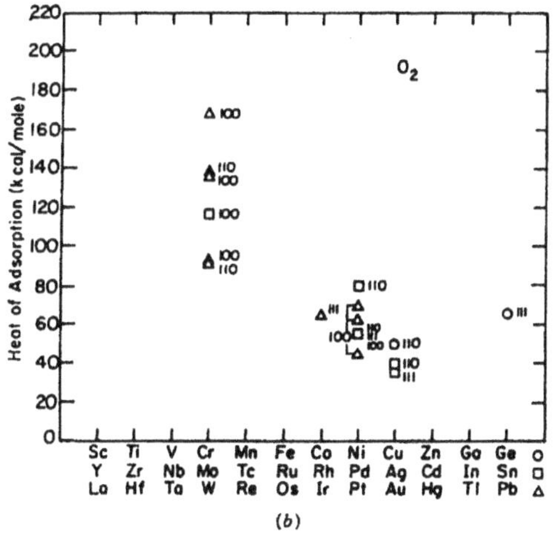

Figure 22. Experimental heats of oxygen adsorption (kcal/mole O_2): (a) over polycrystaline films, (b) over single crystal surfaces. The O-metal surface bond strength generally increase towards the left in the transition metal series. From Ref. [82].

than on the 3d metals is less pronounced for S than for O. The reason is that the S atoms are larger (the S-metal distance is larger) and the (repulsive) coupling to the localized d states is therefore weaker ($\beta_S < \beta_O$).

For Ru we have f=0.7 and here the O and S bonds are about the same and much stronger than for the metals further to the right. Generally, the model predicts that the bond strengths increase towards the left in the transition metal series, in agreement with experimental evidence, Fig. 22.

4.3. BOND LENGTHS AND BINDING SITES

Finally, we briefly return to the question what determines the bond length. Obviously, the adsorbate-*d* coupling will have some influence here, but the main effect comes from the coupling to the *sp* electrons which also contribute mostly to the absolute bond strength. One can use the simple effective medium picture to get an idea of the coupling between an atom and the metal *sp* electrons. It is clear from a comparison of Fig. 7 and 14 that the bond length is given mainly by the E_c function. We also note that the larger binding of O relative to S due to the *sp* coupling in Fig. 7 is in good agreement with the fact that we find E_0 for S smaller than for O in the fits in Fig. 20 and 21.

According to the model of the adsorbate-*d* coupling, it should be a general trend that the smaller the overlap to the *d* states, the less noble the 5d metal will appear relative to the 3d and 4d metals. This means that while e.g. atomic O binds more weakly to Pt than to Ni, the OH or H_2O molecule does not, because the O atoms in OH or H_2O are further away from the surface (this is simple to understand in the simple EMT picture: When the H atom(s) are close to the O atoms they will also contribute electron density to the surroundings of the O atoms which then do not need to be as close to the surface in order to be close to the optimum electron density). The same should be true of atomic C and the methyl molecule CH_3, for instance.

5. Molecular chemisorption on metal surfaces

As our prime example of molecular chemisorption on metals, we will be discussing CO chemisorption on a number of metals. We saw already in Chapter 2 that CO and N_2 chemisorption is reasonably well described by the DFT calculations. We can therefore again use the self-consistent calculations to guide us into an understanding of the molecular adsorption properties.

5.1. ELECTRONIC STRUCTURE OF ADSORBED MOLECULES

For CO, adsorption experiments[84] as well as many theoretical studies[10, 85, 86, 11] suggest that the filled 5σ and the doubly degenerate, empty $2\pi^*$ electronic states are the ones mainly responsible for the bonding to metal surfaces (see Fig. 23 for an illustration of the states of the free CO and N_2 molecule). The calculations for CO and N_2 adsorption on Ru(0001) discussed in Chapter 2 also shows very clearly that only the 2π and 5σ states are important in the coupling to the metal[17]. In Table 1 we have included the calculated occupancies of the four molecular states closest to

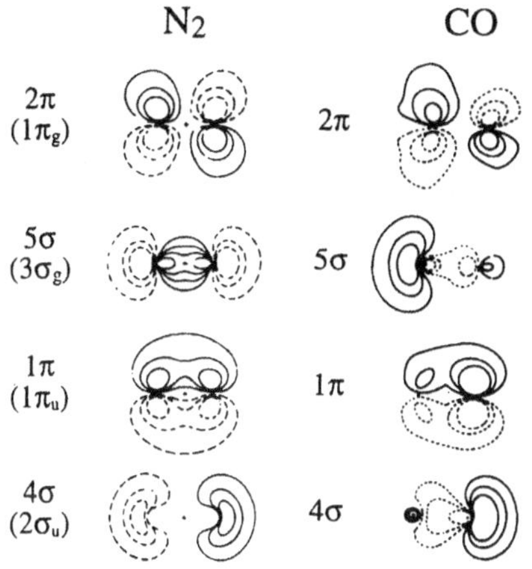

Figure 23. Contours of constant one-electron wave function amplitudes (solid and dashed lines represent opposite signs) for gas-phase N_2 and CO. Adapted from Ref. [83].

the Fermi level. The 2π state which is empty ($N_{2\pi} = 0$) and the 5σ state which is completely filled ($N_{5\sigma} = 1$) in the gas phase have changed their occupancy during adsorption, while the initially filled 1π and 4σ states remain essentially unchanged.

It is possible to understand the changes in electronic structure during CO chemisorption in exactly the same two-step process as for the atomic adsorbates. This is illustrated in Fig. 24. We consider here the change in the adsorbate density of states when a CO molecule interacts with a metal with only *sp* electrons and one where there are also *d* electrons that the CO states can interact with. It is seen that the interaction with the metallic *sp* electrons gives rise to a down shift and broadening of both the 2π and 5σ states, while the coupling to the metallic *d* states gives rise to bonding and anti-bonding states below and above the two original states.

5.2. TRENDS IN MOLECULAR CHEMISORPTION ENERGIES

We can use the model developed above for atomic chemisorption to treat the coupling of both the 2π and 5σ states to the *d* states. Due to different symmetry, they interact with different *d* orbitals, and the two interactions can be treated independently. For CO adsorption the *d*-coupling is fairly week, and we can use the perturbation expression Eq. 74, the only difference being that for the 2π state, which is initially empty, the prefactor is not

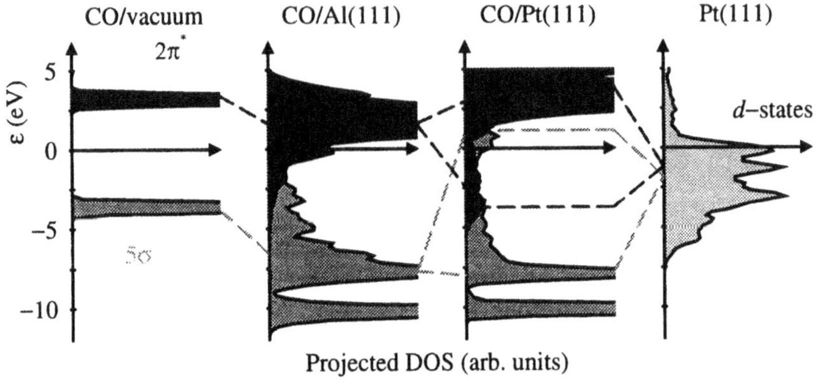

Figure 24. The self-consistent electronic density of states (DOS) projected onto the 5σ and $2\pi^*$ orbitals of CO: in vacuum and over Al(111) and Pt(111) surfaces. Also shown is the DOS from the d-bands in the Pt(111) surface. The sharp states of CO in vacuum are seen to broaden into resonances and shift down in energy over the simple metal surface (mixing with the 4σ state causes additional structure in the 5σ resonance). Over the transition metal surfaces the CO resonances further hybridize with the metal d states. This leads to shifts in the 5σ and $2\pi^*$ levels and to anti-bonding 5σ-d states at the top of the d bands and bonding $2\pi^*$-d states at the bottom. These states have low weight in the 5σ and $2\pi^*$ projections shown. After Ref. [56].

$(1-f)$ but f:

$$E_{d-hyb} \simeq -4\left[f\frac{V_\pi^2}{\varepsilon_{2\pi}-\varepsilon_d}+fS_\pi V_\pi\right] -2\left[(1-f)\frac{V_\sigma^2}{\varepsilon_d-\varepsilon_{5\sigma}}+(1+f)S_\sigma V_\sigma\right], \tag{76}$$

As before we take the d band centers and filling factors from Fig. 18. Here the relative matrix elements V_{ad}^2 are also listed and we take $V_\pi^2 = \beta_\pi V_{ad}^2$ and $V_\sigma^2 = \beta_\sigma V_{ad}^2$. From the DFT calculations we find that $\beta_\sigma/\beta_\pi \simeq 1.7$ so that there is only a single parameter $\beta = \beta_\pi = \beta_\sigma/1.7$ describing the absolute strength of the coupling to the metal d states. As for the atomic chemisorption case we assume $S_\sigma \simeq -\alpha V_\sigma$ and $S_\pi \simeq -\alpha V_\pi$. We also use that the transition metal surfaces have very similar half filled s-bands, so that the renormalization of the CO states by the delocalized metal states will be very alike. Guided by the results of DFT calculations for CO adsorption on Al(111) (Fig. 24) and on the transition metal surfaces with small coupling matrix elements, we use +2.5 eV and -7 eV (with respect to the Fermi level) for the renormalized $\varepsilon_{2\pi}$ and $\varepsilon_{5\sigma}$ positions, respectively.

Adjusting only α and β in the model, we obtain the excellent correlation in Fig. 25. α and β values of 0.063 eV^{-1} and 1.5 eV^2 respectively have been

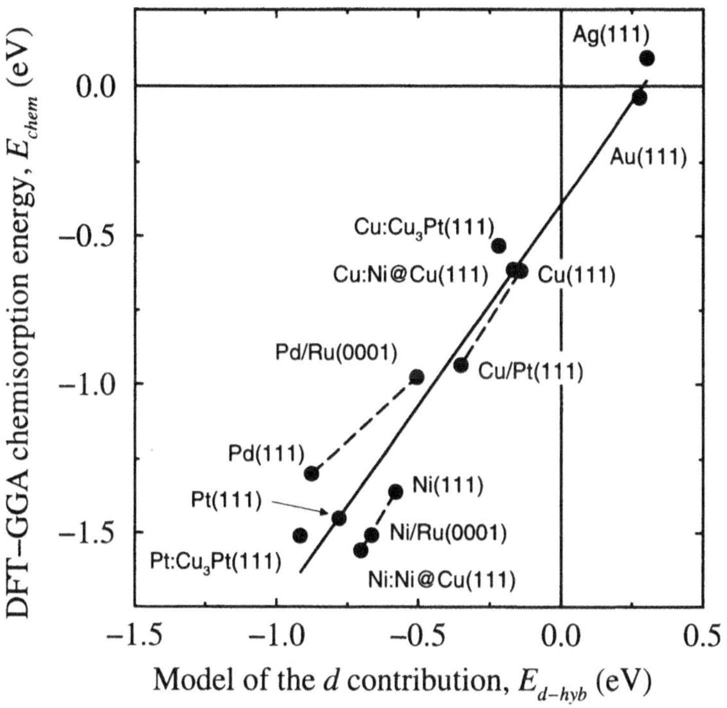

Figure 25. Comparison of the model and the full DFT-GGA chemisorption energies for a number of metal systems. From Ref. [56].

used. We note that in Fig. 25 the slope of the least square fitted curve guiding the eye is close to one, which means that the adsorbate-metal d interactions described by our model can account for the main trends in CO bonding from one surface to the next. The curve is offset by ~0.5 eV on the vertical axis for E_{d-hyb}=0, which fits with CO forming a bond of this strength on simple metal surfaces. Outside the scale in Fig. 25 is the result for CO/Ru(0001). Here the simple model (Eq. 76) estimates a d-contribution of −2.08 eV to the chemisorption energy which in the full calculation amounts to -1.80 eV. Both the model and the full calculation thus show that the CO-Ru bond is very strong. However, in this case (as for most of the metals to the left in the transition metal series) $\varepsilon_{2\pi} - \varepsilon_d$ is small, the d-band width is large, and the two level perturbation model of Eq. 76 leads to an overestimated E_{d-hyb}. If we use the full Newns-Anderson model this problem disappears[17].

As an important property of the model, we further note that it captures the shifts in the CO chemisorption energy from the single crystal surfaces to the overlayer structures, Ni/Ru(0001), Pd/Ru(0001), and Cu/Pt(111), where one type of metal is deposited pseudomorphically on another. This is apparent in Fig. 25 as seen by the dashed lines. For Cu_3Pt, which is a very stable covalently bonded alloy, the Cu sites become less reactive than predicted by the model, while the Pt sites behave roughly as expected. As the model only describes the coupling of the metal *d*-states to the renormalized CO $2\pi^*$ and 5σ states we conclude that it is this interaction, which is responsible not only for the gross trends in CO chemisorption energies over the wide range of the late transition metal surfaces considered but also for the details for metallic overlayers and alloy surfaces. We return to this below.

In Fig. 25 we have chosen to consider the DFT-GGA calculations as a computer experiment of CO adsorption on a number of metal surfaces with all ionic degrees of freedom kept fixed. Hereby we can concentrate on the ability of the model (Eq. 76) to capture the trends caused by the *electronic* factors of the CO-metal bonding. Substrate relaxations are therefore not considered, but rather the truncated bulk geometries are used. Further, for all surfaces, CO is put at the atop position with a fixed metal-carbon distance of 1.94 Å and a CO bond length of 1.14 Å as reported from calculations for CO/Pd(110)[87]. The use of fixed CO coordinates is actually a good approximation. For CO on Pt(111) we find that the relaxed values are r_{MC}=1.88 Å and r_{CO}=1.15 Å and that this relaxation influences E_{chem} by less than 0.05 eV. For the Cu/Pt(111) overlayer structure, we have also tried including relaxations, but the effect is minimal (less that 0.05 eV).

The present model of the CO bonding is in complete agreement with the theoretical interpretations developed by Blyholder[86], Bagus[85] and Van Santen and Neurock[11]. Usually the CO metal bond is described in terms of electron donation from the CO 5σ to the metal and back-donation from the metal to the CO $2\pi^*$. With the present division of the donation and back-donation into separate metal *sp* and *d* steps which follows the reasoning of Bagus and co-workers[85] we obtain a simple picture and a quantitative model of the electronic reason for the trends in the CO chemisorption energies over metal surfaces and overlayers.

One of the important results of Fig.25 is that the variations in the CO bond energy when one metal is deposited on another are well described. CO adsorbs more strongly on a monolayer of Cu on Pt(111) than on the clean Cu(111) surface. The same is true for Ni on Ru(0001) while the opposite holds for Pd on Ru(0001). This is also observed experimentally[88]. The variations can be traced simply to the variations in the *d* band centers. For CO adsorption on the late transition metals the coupling of the $2\pi^*$ level to the *d* bands is much stronger than the coupling to the 5σ level[56]. The change in the chemisorption energy due to a change in the *d*-band center by $\delta\varepsilon_d$ is therefore

$$\delta E_{d-hyb} \simeq -\frac{4fV_\pi^2}{(\varepsilon_{2\pi}-\varepsilon_d)^2}\delta\varepsilon_d \equiv \gamma\delta\varepsilon_d \tag{77}$$

This means that $\delta E_{d-hyb}/\gamma$ should be proportional to the change in the d band center $\delta\varepsilon_d$. This is illustrated in Fig. 26 for the three cases discussed above and for the more extreme case of a single Ni impurity atom in a Cu(111) surface, Ni@Cu(111). The self-consistently calculated change in the CO chemisorption energy is clearly given by $\delta\varepsilon_d$ as predicted by the model. Since the $2\pi^*$ level is above the Fermi level the interaction with the d-band becomes stronger when the d-band shifts up so that the energy denominator in the first term in Eq. 76 becomes smaller. When Ni is alloyed in Cu(111) or deposited on Ru(0001) the d-bands also shift up while the opposite is true for Pd.

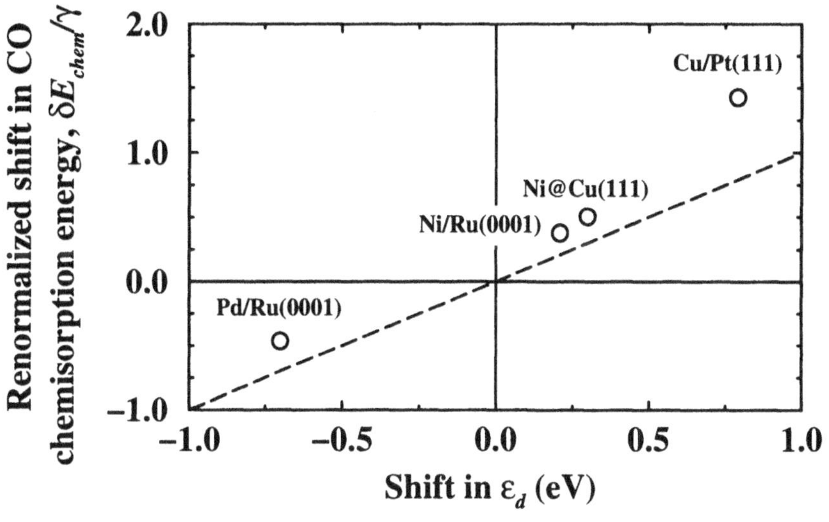

Figure 26. The scaled change in the CO chemisorption energy as a function of the change in the d-band center. From Ref. [81].

6. Bond breaking and making at metal surfaces

In the two preceding Chapters, we have examined the formation of bonds between an atomic or molecular adsorbate and a metal surface. Now we go one step further and examine the simultaneous breaking and formation of bonds at the surface. When a molecule dissociates on a surface one or more intra-molecular bonds must be broken and at the same time new bonds between the surface and the dissociation products are formed. It is this ability of the surface to break bonds and stabilize intermediates in the reaction which is the basis for the use of solid surfaces as catalysts. The presence of the surface makes processes that would have very low rates in the gas or liquid phase. Take ammonia synthesis as an example. To make ammonia from H_2 and N_2 one must break the H-H and N-N bonds. In the gas phase this requires energies of 4.75 eV and 9.76 eV respectively. This means that the concentration of N atoms is extremely low even at high temperatures and the rate of reaction negligible. Only in an arch with extremely high temperatures is it possible to perform the reaction homogeneously. On metallic Fe, on the other hand, N_2 dissociates readily and so does H_2. The ammonia synthesis can take place at a rate of about 10^{-4} molecules transformed per Fe site in the surface and second (corresponding to of the order 10^{-1}-1 molecules per *vacant* Fe site and second) at a pressure of 100 bar and a temperature of 700 K. This is the basis for the industrial ammonia production[89].

6.1. THE POTENTIAL ENERGY SURFACE.

When a molecule dissociates on a metal surface several interatomic coordinates usually vary simultaneously. To understand the process we therefore need to understand how the energy of the system varies with several coordinates. The total energy as a function of all the coordinates of the system defines a potential energy surface (PES) for the reaction. In the following we shall be assuming that only the ground state PES is important, that is, that electronic excitations are not important for an understanding of the reaction. This is far from always true, but the ground state PES is still usually a good starting point for a discussion of the reaction.

Consider as an example in Fig. 27 the PES for an H_2 molecule dissociating on a Cu(111) surface[3]. In the calculations the molecule is kept parallel to the surface and centered over the bridge as shown in the insert. The PES is displayed as a function of the molecular bond-length, b, and the molecular center of mass distance, Z, from the Cu surface. In such a PES the initial and final states are (local) minima – these states are locally stable forms of the system. In the case of H_2 dissociating on Cu(111), Fig. 27, the initial state is where the molecule is far from the surface and the final state is the minimum in the lower right corner in the figure where Z is small and b is large: the two atoms have separated and are bound close to the surface. The two minima are separated by a barrier, the height of which is the activation energy of the process.

The barrier for dissociation depends strongly on the point of impact in the unit cell and the orientation of the incoming molecule (the center of mass coordinates X and Y, along $[11\bar{2}]$ and $[\bar{1}10]$, the azimuthal angle, ϕ,

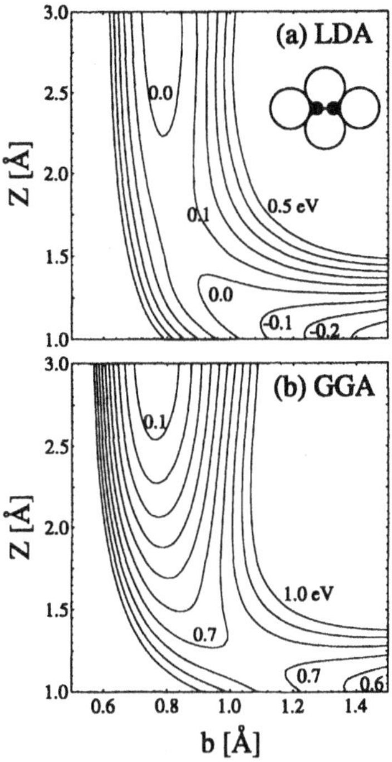

Figure 27. The PES for H_2 dissociation over Cu(111): (a) Calculated with the LDA and (b) with the GGA. Only (b) conforms with experiments. The insert shows the geometry. After Ref. [3].

and the polar angle, θ of the molecule relative to the surface) . Figure 28 shows the variation of the barrier in the (X,Y)-plane for two values of ϕ. It is clear that there is a strong variation of the barrier height in these three molecular degrees of freedom.

The strongest variation of the barrier is with the polar angle, θ, Turning the molecule from the optimum reaction configuration (Fig. 27) to the upright position at (b, Z)=(0.8 Å, 1.5 Å) requires already an energy of 0.4 eV and activating the H-H bond in this configuration means that one H will end up in the vacuum whereby a minimum of half the H_2 dissociation energy can be expected for the barrier height. This practically excludes large parts of configuration space from contributing signicantly to the reaction, except if the reactants get steered efficiently[90, 91].

The strong dependence of the PES on all degrees of freedom makes the dynamics quite complicated, and it is only recently that it has been understood in some detail[4, 90, 92]. For a detailed discussion of the dynamics of the H_2 dissociation process we refer to the review by Darling and Holloway[93].

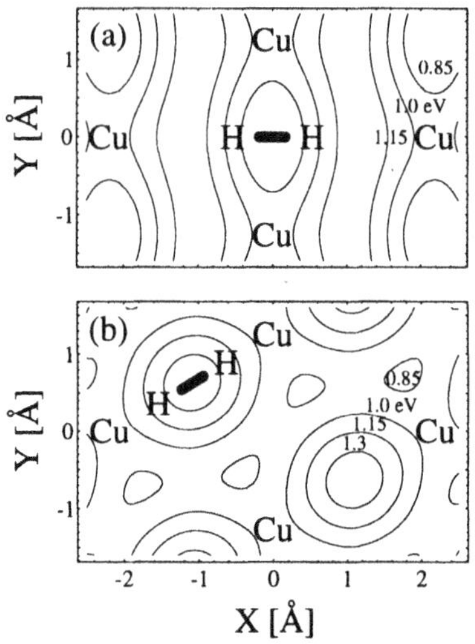

Figure 28. The barrier height as a function of X, Y. The 'Cu' marks the atomic positions in the Cu(111) surface. (a) ϕ=0. A molecule is depicted in the optimum reaction geometry, perpendicular to the Cu-Cu bridge. (b) $\phi=\pi/6$. A molecule is shown in the worst case, dissociation parallel with the bridge. From Ref. [3].

6.2. THE MINIMUM ENERGY PATH.

Instead of looking at the full PES it is often useful to define the minimum energy reaction path, which is the path connecting the initial and final states with the property that all derivatives of the energy perpendicular to the path are zero. This is the path of steepest descent from the saddle point separating the minima. In Fig. 27 the reaction path is indicated dashed, and in Fig. 2 the energy along the reaction path is shown for H_2 dissociating on Cu(111) as well as on the (111) surfaces of Ni, Pt and Au. The highest point along the path defines the minimum activation energy for the reaction (remember that there are many different barriers depending on the impact parameters and angles, but the definition of the minimum energy path ensures that it is the one with the lowest barrier).

If there is a gas of molecules in thermal equilibrium at a temperature T above the surface, then the probability that an incoming molecule has an energy higher than the activation energy E_a is given by the Boltzmann factor $e^{-E_a/kT}$. Because this factor depends so strongly on E_a only the paths with the lowest E_a's will contribute to the adsorption rate, and the E_a of the minimum energy path will be the one determining the thermal barrier. If there are only few impact parameters with barriers close to the minimum barrier, transition state theory says that the transition state will have a high entropy and the prefactor ν in an Arrhenius expression for the adsorption rate $r_{ads} = \nu \exp(-E_a/kT)$ will be small.

6.3. A REACTIVITY MEASURE.

The maximum in the minimum energy path is therefore a useful measure of the reactivity of a given surface. It is, however, quite cumbersome to calculate the minimum energy path. It is therefore more convenient to define a fixed path and even to consider only a single point on this path. In Fig. 29

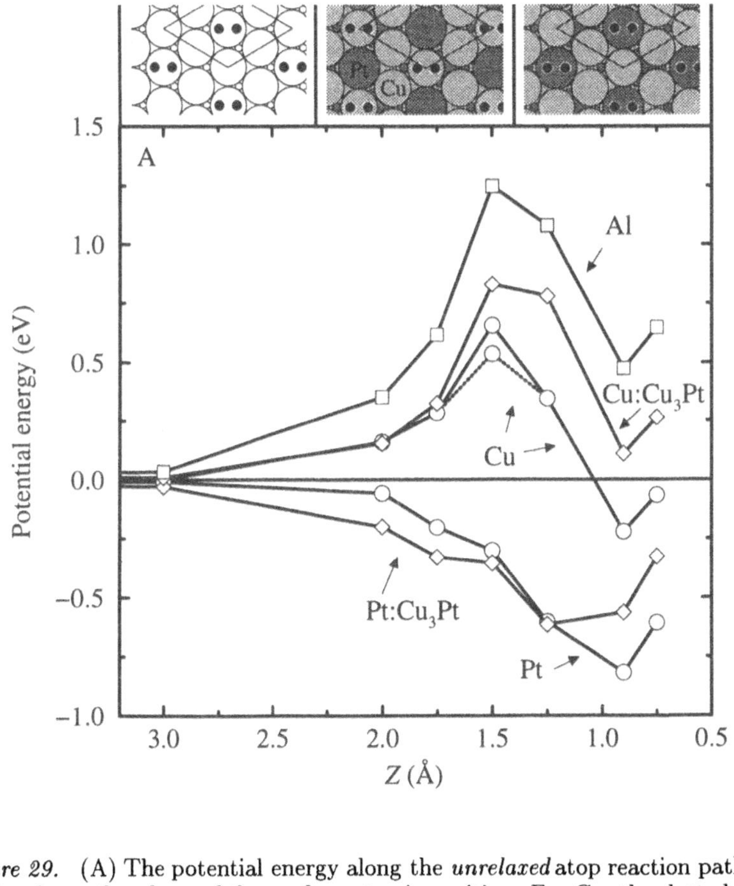

Figure 29. (A) The potential energy along the *unrelaxed* atop reaction path. Z is the H_2 height above the plane of the surface atomic position. For Cu, the dotted line indicates the minimum energy barrier (cf. 2). (B) The geometry over Al(111), Cu(111) and Pt(111). Large open circles represent the substrate atoms, small solid circles (•) the H atoms. The (2×2) cell is indicated. (C) and (D) The geometry for H_2 over the Cu (light gray) and Pt (dark gray) sites in $Cu_3Pt(111)$. From Ref. [55].

we show the energy along a *fixed* reaction path for H_2 dissociating outside Cu(111) and three other surfaces. The barrier used is not the minimum barrier, but it is quite close and still gives a measure of the relative reactivity of the metals considered. If we want a single number $\tilde{E}_a$ to measure the reactivity, we define this to be the energy at a fixed coordinate $\tilde{q}_{ts}$. It can be seen in Fig 29 and Fig 2 that if we chose it to be around the position of the maximum for Cu, then the trends in barrier for all the other metals are given from the energy value for the same reaction coordinate. A positive value for $\tilde{E}_a$ means that the process is activated, while a negative value means that it is non-activated or has a low barrier as for e.g. Pt(111) or Ni(111).

6.4. ELECTRONIC STRUCTURE AND TRENDS IN ACTIVATION BARRIERS.

To understand the coupling of a molecule at the transition state $\tilde{q}_{ts}$ to the metal we can apply exactly the same procedure as above for atomic and molecular chemisorption. In Fig. 30 we show the density of states at a common $\tilde{q}_{ts}$ for H_2 outside different surfaces corresponding to the cases shown in Fig. 29. Just like before, we see that outside the simple metal Al

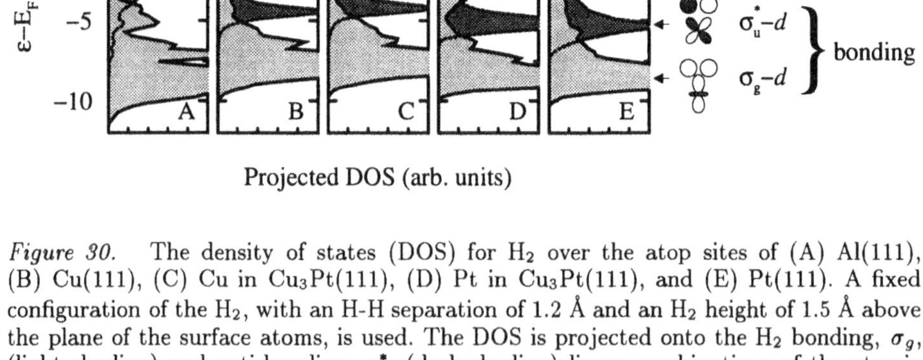

Figure 30. The density of states (DOS) for H_2 over the atop sites of (A) Al(111), (B) Cu(111), (C) Cu in $Cu_3Pt(111)$, (D) Pt in $Cu_3Pt(111)$, and (E) Pt(111). A fixed configuration of the H_2, with an H-H separation of 1.2 Å and an H_2 height of 1.5 Å above the plane of the surface atoms, is used. The DOS is projected onto the H_2 bonding, σ_g, (light shading) and anti-bonding, σ_u^*, (dark shading) linear combinations of the atomic H 1s orbitals. The arrows inside panels (B) through (E) point out the σ_g-d anti-bonding states. The character of the wave function at the different peaks in the DOS is depicted by the schematics at the right of the figure (the σ's – represented by two spheres – are shown atop the d's with the sign of the wave function indicated by the shading). From Ref. [55].

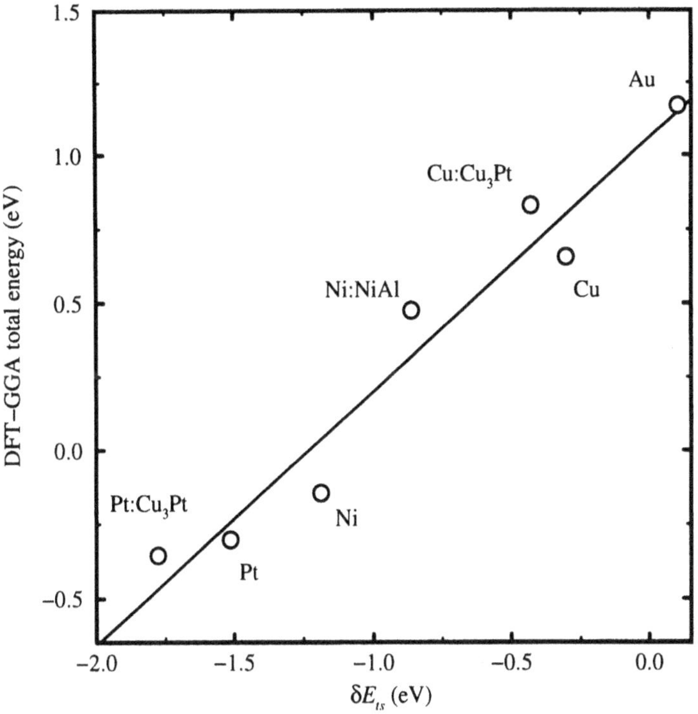

Figure 31. Correlation between the reactivity measure E_{ts} and the calculated activation barrier for adsorption (the total energy at (b, Z)=(1.2 Å, 1.5 Å) in Fig. 29). α in Eq. 79 has been used as a fitting parameter. After Ref. [55].

the two molecular states in question, the bonding σ_g and the anti-bonding σ_u^*, are just shifted and broadened, while the coupling to the metal d states in the case of a transition or noble metal surface gives rise to bonding and anti-bonding states with respect to the molecule metal d coupling.

Just like for CO adsorption, we can use the picture to give a quantitative measure δE_{ts} of the interaction between the adsorbate and the metal d states at the transition state:

$$\delta E_{ts} = -2\left[f\frac{V_u^2}{\varepsilon_{\sigma_u^*} - \varepsilon_d} + fS_uV_u\right] - 2\left[(1-f)\frac{V_g^2}{\varepsilon_d - \varepsilon_{\sigma_g}} + (1+f)S_gV_g\right] . \tag{78}$$

Here the first term is the hybridization and orthogonalization energy due to the d-σ_u state and the second is due to the coupling with the σ_g state. The σ_u state is initially empty like the CO $2\pi^*$ state. This means that there is a factor f in from of this term, while as usual there is a $(1-f)$ factor in front of the term describing the hybridization with the filled σ_g state. The parameters, ε_d, f and V_{ad}^2 are all listed in Fig. 18. Only the metal independent constants, the ratios $\beta_u = V_u^2/V_{ad}^2$, $\beta_g = V_g^2/V_{ad}^2$ and α are unknown. It turns out that $\beta_u \simeq \beta_g = \beta$. From the calculations for H_2 outside Al, we conclude that $\varepsilon_{\sigma_g} \simeq -7$ eV and $\varepsilon_{\sigma_u^*} \simeq 1$ eV relative to the Fermi level. The estimate, Eq. 78, can if we restrict ourselves to the late transition metal and noble metal surfaces thus be written:

$$\delta E_{ts} \simeq -2\frac{\beta V_{ad}^2}{\varepsilon_{\sigma_u^*} - \varepsilon_d} - 2(1-f)\frac{\beta V_{ad}^2}{\varepsilon_d - \varepsilon_{\sigma_g}} + \alpha\beta V_{ad}^2 , \qquad (79)$$

where we in the last term have lumped together the two orthogonalization terms. In Fig. 31 we show how, for reasonable values of α and β, the calculated barriers for adsorption from Fig. 29 correlates very well with the reactivity measure E_{ts} for a number of surfaces, including some alloys.

Just as for CO it is the interaction of the anti-bonding state with the d bands that dominate the trends. The coupling to the bonding σ_g state is weaker due to the $(1 - f)$ factor. A high lying d band and a large matrix element therefore give the lowest barrier. It is seen that Ni and Pt have about the same reactivity measure in good agreement with experiment. The origin of the high reactivity is quite different, though. Ni has high lying d bands and a fairly small V_{ad}^2, c.f. Fig. 18, while Pt has much lower lying d bands, but a larger matrix element.

We can illustrate the dependence of the activation barrier on the surface electronic structure for another reaction, the methane dissociation over a Ni(111) surface. Methane activation and conversion into synthesis gas, CO and H_2 is performed commercially over a Ni catalyst. The minimum energy path for the dissociation of CH_4 into chemisorbed H and CH_3 is found to be for the molecule right over one of the Ni atoms[51, 94], and after dissociation both the adsorbed H atom and the methyl group are found in three-fold sites. The calculated energy along the path is shown in figure 32. The process is highly activated, in agreement with experiments[95, 97].

We now keep the reaction path fixed and study the effect of alloying on the dissociation process by exchanging one or two of the neighboring Ni atoms by Au atoms. The dissociation still proceeds over a Ni atom and we are only studying the *change* in Ni properties due to alloying. The results are also included in figure 32. If one out of six Ni neighbors is substituted by a Au atom, the barrier for methane dissociation is increased by δE_1=0.17 eV (16 kJ/mole) and if the Ni atom has two Au neighbors the barrier is increased by δE_2=0.39 eV. More Au neighbors will increase the barrier even further. The density of states of the CH_3-H molecule dissociating over Ni(111) is shown in Fig. 33. Again the transition state is characterized by an empty anti-bonding molecular state which can interact with the metal d states.

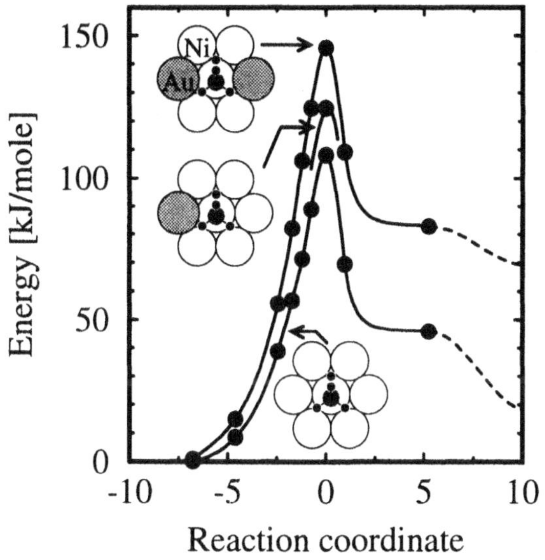

Figure 32. The calculated energy along the reaction path for CH_4 dissociating over a Ni atom in the Ni(111) surface and over a Ni atom that has either one or two gold nearest neighbors. The geometries of the CH_4 along the reaction path include full structural relaxation of the CH_3 group in each step. The dashed curves refer to infinite separation of the dissociated H and CH_3 group on the surface. The calculations are done in a super cell geometry with four Ni layer thick slabs having surface unit cells with four Ni atom, five Ni and one Au atom, and three Ni and one Au atom, respectively. Corrections due to the zero-point energy of the transition complex, which are not included in the presented curves, will tend to lower the calculated barrier heights. From Ref. [94].

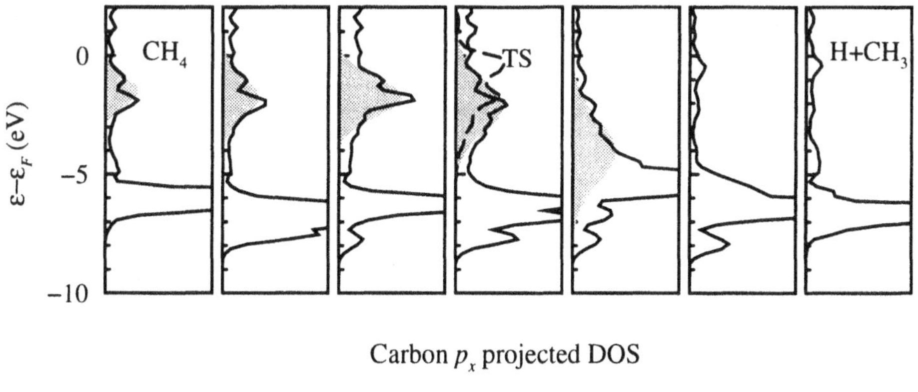

Figure 33. The Carbon p_x projected density of states (solid curve) along the reaction path of CH_4 dissociating over Ni(111). The left most panel is an almost undisturbed CH_4 molecule outside the Ni(111) surface. The panel marked TS is for the transition state configuration. The right most panel is that of CH_3 and H chemisorbed in individual 3-fold sites at the surface. The shaded regions highlight the anti-bonding (CH_4) molecular state, that builds up below the Ni *d* states (dotted curve) through hybridization with these. From Ref. [98].

If we only consider the coupling to the Ni atom right under the dissociating molecule (this is less well founded than for the H_2 molecule since the methane complex is more extended and will feel the neighbors to some extend), then the simple arguments given above would give that an upshift in the d states should lower the barrier and vice versa for a down shift. It turns out that the Ni atoms nearest neighbor to one Au experiences an downshift of the d band center of 0.02 eV and a Ni atom with two Au neighbors has an downshift of 0.06 eV. Again the correlation with the simple model is quite good.

The predictions of the calculations are quite well confirmed by a recent combination of STM and molecular beam experiments[94, 96]. From STM pictures it is found that at low Au coverages, θ_{Au} ML, the Au will be highly dispersed in substitutional sites on Ni(111). From these STM pictures the Au coverage dependent distributions of Ni sites with N_{Au} Au nearest neighbors can be determined. With the ansatz that above each Ni atop site will be a transition state for the CH_4 dissociation whose activation energy is solely determined by N_{Au} a subsequent molecular beam study lead to an experimental finding of a reduced sticking over Ni sites with gold neighbors. The amount of reduction in the sticking was shown to be in good agreement with the calculated barrier increases when using the experimentally determined reduction in sticking over Ni(111) versus the CH_4 translational energy as a reference[94]. Such good correlation between theory and experiment shows that the theoretical methods from density functional theory have reached a point where they can be used to predict trends in the chemical activity when a surface is modified by alloying. This opens up new possibilities in computer aided design of catalysts.

7. Electronic and geometrical effects in surface chemistry

It is one of the long term goals of surface science to establish an understanding of the important physical properties of a surface determining its chemical reactivity. Such an understanding could be the basis for an "atomistic engineering" of surfaces with desired catalytic activity or selectivity. Surface reactions are often divided into structure sensitive and structure insensitive reactions[1, 99]. The former depend on the surface structure or particle size in the case of a high surface area catalyst, while the latter does not. This very useful concept is somehow related to the concepts of geometrical end electronic effects in catalysis. There is no unique way of distinguishing between the effects of geometry and electronic structure – the electronic structure is a strong function of the geometrical arrangement of the atoms at the surface, and ultimately, it is the electronic structure which determines the geometry, by determining the total energy of the system.

Structure sensitivity of a reaction may thus be ascribed to electronic or geometrical effects but must ultimately be related to the electronic structure of the surface-adsorbate complex.

TABLE 2. Comparison of the barrier for H_2 dissociation on Cu(111) and (100). Both the barrier for the minimum energy path and for a fixed path directly over one Cu atom is shown. The *d* band centers are also compared.

	Cu(111)	Cu(100)
Minimum E_a (eV)	0.50	0.60
Atop E_a (eV)	0.66	0.63
ε_d (eV)	-2.67	-2.53

To illustrate the difficulties in distinguishing between the two effects consider as an example H_2 dissociation on Cu(111) and (100). Experimentally, the adsorption rate is slightly higher on the close-packed (111) surface than on the (100) surface[22]. The question is now if this is an electronic or a geometrical effect? The DFT calculations offer a very nice way of making this distinction[52]. We can start by noting that the barriers from the minimum energy paths for the two surfaces give the right result that the (111) barrier is lower than the (100) barrier, see Table 2. The two reaction paths are quite different as illustrated by the two very different minimum path transition states depicted in Fig. 34. On the (100) surface the H-H bond is much more stretched, and it is asymmetrical with respect to the Cu-Cu bridge[49, 52, 50].

We can then compare the dissociation barrier for a fixed reaction geometry. If we let the H_2 molecules dissociate over the on top site (see Fig. 34) then the local geometry is about the same for the two surfaces. In such a calculation the barrier is largest for the (111) surface as shown in Table 2. This is quite understandable in terms of the model developed above. The more open (100) surface has a higher lying *d* band (see Table 2) and the barrier should be lower here. Using Eq. 79 we get a difference in barrier for the two surfaces of 0.05 eV in good agreement with the difference from the full DFT calculation shown in Table 2.

The reason that the minimum barrier is lower on the (111) surface than on the (100) surface is therefore not to be found in a global property of the surface like the *d* band center. Rather it reflects the local properties like the possibility to have a good overlap with the metal states (the V^2 contribution). The bond in the H_2 molecule must be appreciably more stretched on

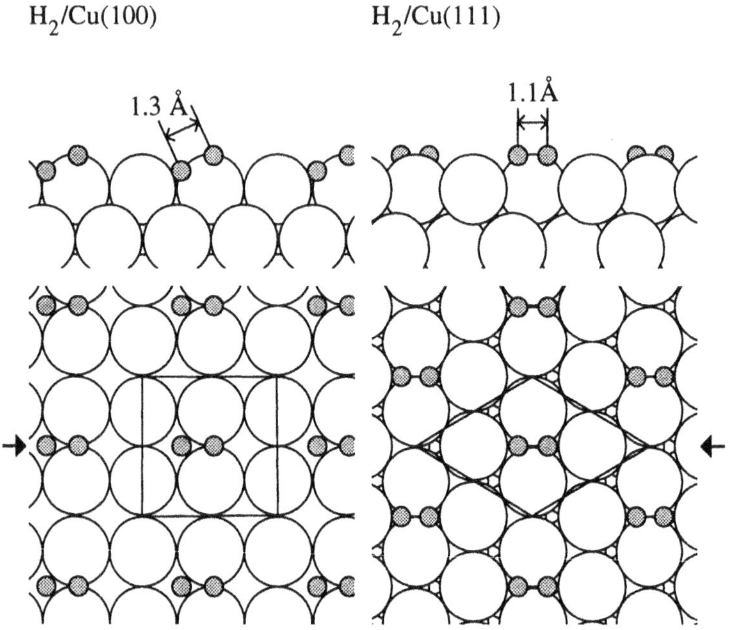

Figure 34. Transition state geometries for H_2 dissociation on Cu(100) and Cu(111). The lower panels show top views of the slabs with the surface unit cell indicated. The upper panels show cuts through the slabs where marked by the arrows in the lower panels. The numbers give the H–H bond length at the transition state. From Ref. [52].

Cu(100) before dissociation occurs, resulting in a higher barrier. A look at the dissociation geometry in Fig. 34 shows that this is natural because the center sites are much further apart on the (100) than on the (111) surface.

In general it may be useful to introduce the following distinction between electronic and geometrical effects in surface reactions: When we compare the reactivity of different surface structures we can compare activation energies or adsorption energies for a fixed local geometry – this will give a purely electronic effect. We can then investigate the different reaction probabilities that the different geometries and thereby different matrix elements give rise to. It must be stressed that the latter is also an electronic effect, but it may still be useful to separate the two electronic effects.

The short summary of the surface properties determining the ability of a surface to bond and react with adsorbates are the following:

1. The coupling of the adsorbate states to the metal *sp* electrons gives a strong contribution to the determination of
   - the over all bond energy (but not so much the trends from one system to the next)

- the bond length
- the adsorption geometry

and gives an important *renormalization* of the adsorbate electronic structure.

2. The coupling to metal d electrons gives trends in bond energies and activation energies and also contributes to the bonding geometry. Important d-band properties are:
   - the d band center
   - the coupling matrix element
   - the d band filling.

It is interesting to compare the surface properties singled out above as determining the atomic and molecular chemisorption energy and the energy barriers with other measures of reactivity proposed in the literature. Two simple properties of the surface have been successfully correlated with its reactivity: the local density of one-electron states at the Fermi level $LDOS(E_F)$[100, 101, 102, 103] and the number of holes N_h in the d-bands[104]. More general formulations of surface reactivity measures based on concepts from the theory of gas phase chemical reactivity[105, 106] have recently been developed[107], which give the two simple measures, $LDOS(E_F)$ and N_h, as limiting cases of a more general dependence of the reactivity on the surface local density of states[10].

We can use the data base generated by the DFT calculations to put these reactivity measures to a test. Consider in Fig. 35 the local density of states (DOS) projected on the d-orbitals of the Cu(111) and the Pt(111) surface atoms and the Pt and Cu atoms in the Cu_3Pt(111) surface. It can be seen that the d-bands of Cu_3Pt represent some intermediate between those of pure Cu and pure Pt; the main qualitative difference being that for the alloy, the Fermi level intersects right at the top of the d-bands. The d-bands of Cu_3Pt are therefore practically filled (i.e. there are essentially no d-holes!) and the DOS at the Fermi level is low compared to what it is for Pt. The simple measures therefore predict that Pt sites in Cu_3Pt(111) should be less reactive than Pt(111). This is not in agreement with the self-consistent calculation (cf. Fig. 29) or experiment[108] according to which, H_2 dissociation proceeds on the Cu_3Pt(111) surface without an appreciable barrier. On the other hand, the model proposed above which takes into account the d band centers correctly predicts that due to the effective upshift of these for the Pt sites in the alloy, these should still be able to dissociate hydrogen. For the Cu sites in the alloy, the d bands upshift does not manifest itself in an increased reactivity in the DFT calculations, presumably because of changes in the metal s-H_2 coupling, showing that the d band shifts are not the whole story. However, in conclusion, the shortcomming

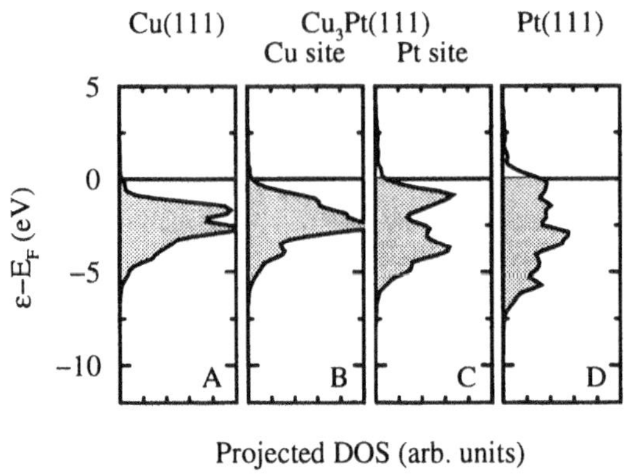

Figure 35. The density of states (DOS) for (A) Cu, (B-C) Cu_3Pt and (D) Pt. In (A) and (B) the solid curves are the DOS projected onto the five *d*-orbitals of atomic Cu positioned in the surface, while onto the *d*-orbitals of atomic Pt in (C) and (D). After Ref. [55].

of the simplest measures ($LDOS(E_F)$ and N_h) means that it is not just the states right around the Fermi level, but all the *d* states (as measured e.g. by the center of the *d* band) which determine the reactivity. There is clearly some correlation between the *d* band center and the DOS at the Fermi level, but this is not perfect.

8. Electronic structure and reactivity at alloy surfaces

It has long been realized that alloying can modify the catalytic activity and selectivity of a metal significantly[109, 110], and that alloy catalysts represent one of the most promising avenues towards new catalyst systems[1]. One important effect of alloying is to change the distribution of islands of a given metal which otherwise retain its reactivity (the ensemble effect)[109, 110, 111], but it may also change the local bonding geometry (the structure effect)[87], or directly modify the reactivity of the metal atoms (the electronic or ligand effect)[112, 88].

Based on the reasoning in the preceding Chapters, it is clear that the three factors ε_d, f and V^2 may all be affected by alloying. The smallest effect may be on the filling of the *d* bands, since the intra-atomic Coulomb repulsion is large in the localized *d* shells. The matrix elements will surely change when the structure and the identity of the surface atoms changes during alloying. And finally, alloying will change the centers of the *d* bands. Such changes are immediately measurable for e.g. adsorption on on-top sites where the coupling matrix elements are always the same. Yet we saw above that the CO bond energy or the barrier for H_2 dissociation may change

TABLE 3. Shifts in d-band centers of surface impurities and overlayers relative to the clean metal values (bold). The impurity/overlayer atoms are listed horizontally and the host entries are listed vertically. For each combination the two numbers listed is first the isolated surface impurity and next the overlayer. The surfaces considered are the most close packed and the overlayer structures are pseudomorphic. No relaxations from the host lattice positions are included. All values are in eV and the elemental d band centers are relative to the Fermi level.

	Fe	Co	Ni	Cu	Ru	Rh	Pd	Ag	Ir	Pt	Au
Fe	**-0.92**	0.05	-0.20	-0.13	-0.29	-0.54	-1.24	-0.83	-0.36	-1.09	-1.42
		0.14	-0.04	-0.05	-0.73	-0.72	-1.32	-1.25	-0.95	-1.48	-2.19
Co	0.01	**-1.17**	-0.28	-0.16	-0.24	-0.58	-1.37	-0.91	-0.36	-1.19	-1.56
	-0.01		-0.20	-0.06	-0.70	-0.95	-1.65	-1.36	-1.09	-1.89	-2.39
Ni	0.09	0.19	**-1.29**	0.19	-0.14	-0.31	-0.97	-0.53	-0.14	-0.80	-1.13
	0.96	0.11		0.12	-0.63	-0.74	-1.32	-1.14	-0.86	-1.53	-2.10
Cu	0.56	0.60	0.27	**-2.67**	0.58	0.32	-0.64	-0.70	0.58	-0.33	-1.09
	0.25	0.38	0.18		-0.22	-0.27	-1.04	-1.21	-0.32	-1.15	-1.96
Ru	0.21	0.26	0.01	0.12	**-1.41**	-0.17	-0.82	-0.27	0.02	-0.62	-0.84
	0.30	0.37	0.29	0.30		-0.12	-0.47	-0.40	-0.13	-0.61	-0.86
Rh	0.24	0.34	0.16	0.44	0.04	**-1.73**	-0.54	0.07	0.17	-0.35	-0.49
	0.31	0.41	0.34	0.22	0.03		-0.39	-0.08	0.03	-0.45	-0.57
Pd	0.37	0.54	0.50	0.94	0.24	0.36	**-1.83**	0.59	0.53	0.19	0.17
	0.36	0.54	0.54	0.80	-0.11	0.25		0.15	0.31	0.04	-0.14
Ag	0.72	0.84	0.67	0.47	0.84	0.86	0.14	**-4.30**	1.14	0.50	-0.15
	0.55	0.74	0.68	0.62	0.50	0.67	0.27		0.80	0.37	-0.21
Ir	0.21	0.27	0.05	0.21	0.09	-0.15	-0.73	-0.13	**-2.11**	-0.56	-0.74
	0.33	0.40	0.33	0.56	-0.01	-0.03	-0.42	-0.09		-0.49	-0.59
Pt	0.33	0.48	0.40	0.72	0.14	0.23	-0.17	0.44	0.38	**-2.25**	-0.05
	0.35	0.53	0.54	0.78	0.12	0.24	0.02	0.19	0.29		-0.08
Au	0.63	0.77	0.63	0.55	0.70	0.75	0.17	0.21	0.98	0.46	**-3.56**
	0.53	0.74	0.71	0.70	0.47	0.67	0.35	0.12	0.79	0.43	

dramatically.

It is therefore of some interest to have an overview of the way the d band centers change when one metal is forming a monolayer on top of or are alloyed into the first layer of another metal. These shifts can be calculated using the DFT and Table 3 summarizes results for a large number of combinations of catalytically interesting metals[81]. For each metal we have considered the equilibrium bulk metal structure (fcc, hcp or bcc) for the substrate, but we have found that variations of d-band centers for the close-packed surfaces of the different structures are small, of order 0.02 - 0.04 eV in the case of fcc(111) and bcc(110) and even smaller in the case of hcp(0001) and fcc(111) faces. The differences between different facets of the same bulk structure are generally larger, of order 0.2 eV in the case of fcc(111) and fcc(100) facets.

We point out, that the results are for overlayers and surface impurities occupying lattice positions of the substrate, and we have not here taken into account whether these structures are actually stable under experimental conditions. Yet, the table can be used to get an idea about possible ways of modifying the electronic structure of a surface. There is, however, a tendency that the inclusion of relaxations will decrease the shifts seen in table 3, but this will not in general change the sign or relative magnitude of the effect.

It is known from several theoretical studies that the variation in the center of gravity of the *d*-bands for metal overlayers is accompanied by a similar variation in the surface core level shifts – at least towards the right in the transition metal series[113, 114, 115]. It has also been shown that the *trends* in variations in the surface core level shifts for different overlayers are often given by the initial state shift, that is, by the changes in the electronic structure of the unperturbed surface. From this we conclude that we can view the variation in the surface core level shifts as a measure of the variation in the *d*-band center. This is fortuitous as it means that in the surface core level shifts, we have an *in situ* experimental measure of the valence *d*-band center shifts that are important for the chemical properties of the metal surface as discussed above. As an outstanding example of the information carried by the surface core level shifts, Rodriguez and Goodman have in their very detailed studies of metallic overlayers observed a strong correlation between the surface core level shift of such overlayers and the CO chemisorption energy at these[88].

In Fig. 36 we examine if the present theoretical analysis can be used on the original experimental data[88]. The figure strongly supports the model of chemisorption presented throughout these lecture notes. Identifying the shift in the *d*-band center with (minus) the core level shifts and assuming the CO chemisorption energy shift proportional to (minus) the TPD peak shift ($\Delta T/\gamma$), a linear relation between the renormalized TPD peak shift and the core level shift is expected(cf. Eq. 77). Such a relationship is clearly present in Fig. 36, however, slightly offset from the origin.

It must be noted that while there seems to be a good correlation between the *d* band center and the surface core level shift for *overlayers*, the same does not appear to hold for simple surface *impurities*[116]. This, unfortunately puts a limit to the use of core level shifts as a reactivity measure.

Finally we illustrate how Table 3 can be used to search for a better catalyst for a given reaction. Consider again the methane activation over Ni. It was shown in Chapter 7 that alloying of Au into the Ni surface shifts up the *d* band of the Ni atoms (this cannot be seen in Table 3 which

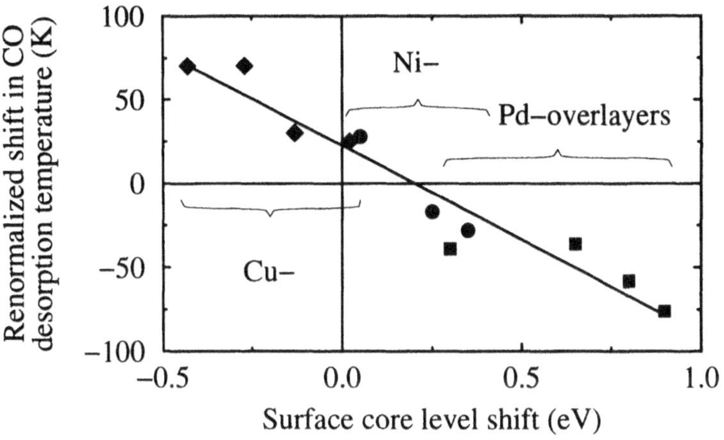

Figure 36. The experimental data of Rodriguez and Goodman[88] for CO adsorption on Ni-, Cu- and Pd- overlayers plotted as the renormalized TPD peak shifts ($\Delta T/\gamma$) versus the measured shifts in surface core level positions. The reference systems for the CO TPD peaks and the surface core level positions for the overlayer structures are the respective single crystal surfaces. From Ref. [56].

only includes shifts for the metal that is alloyed into the surface (in this case Au)), and increases the barrier. It can bee seen that if Ni is alloyed into a Cu rich surface, on the other hand, then the Ni *d* band should be shifted up and the barrier should descrease. Direct calculations confirm this prediction: Ni atoms surrounded by Cu atoms in the surface have a barrier for dissociation which is 0.05 eV lower than the clean Ni. This is in accordance with experience from experiments on alloy catalysts, where it is found that Cu can *promote* the steam reforming rate on Ni catalysts[117].

ACKNOWLEDGMENTS

We thank a large number of collaborators for contributing to the work described here including P. Kratzer, Y. Morikawa, K. Terakura, J. J. Mortensen, K. W. Jacobsen, P. Stoltze, K. Stokbro, S. Holloway, B. I. Lundqvist, F. Besenbacher, I. Steensgård, E. Lægsgaard, L. P. Nielsen, I. Chorckendorff, J. H. Larsen, M. Holmblad, J. Jacobsen, V. Ganduglia-Pirovano, M. Scheffler, H. L. Skriver, A. Ruban, E. Törnqvist, B. Clausen, and H. Topsøe. The present work was in part financed by the Danish Research Councils through The Center for Surface Reactivity and grant # 9501775. Center for Atomic-scale Materials Physics is sponsored by the Danish National Research Foundation.

A. Deriving the coupling matrix elements

For completeness we also discuss how variations in the coupling matrix elements V_{ad} from one metal to the next can be estimated from LMTO theory. As shown in Refs. [80, 35, 36, 79] the coupling matrix element between an adsorbate state and a metal d state a distance r away can be written

$$V_{ad} = \eta \frac{M_a M_d}{r^{l_a+l_d+1}} \tag{80}$$

where η is given by the structure constants in LMTO theory and l_a denotes the angular momentum quantum number of the adsorbate state and l_d of the metal d state. M_a and M_d are related to the LMTO potential parameters Δ by

$$M_l = (s^{2l+1}\Delta_l)^{1/2} \tag{81}$$

Here s denotes the neutral sphere radius for the atom in question. It is seen that for a given adsorbate at a given distance from a metal atom, the *trends* in the matrix element for the metals will depend only on M_d. Since the values for Δ for the bulk metals have been tabulated for all the metals [79] we can also tabulate the relative values of the matrix elements. In Fig 18 we present $V_{ad}^2 \propto M_d^2$ relative to the value for Cu for the relevant section of the Periodic Table. We have used the bulk Wigner-Seitz radii as well as the bulk metal values for Δ in calculating M_d. Direct calculations show that this is a good approximation even for the surface.

References

1. G. A. Somorjai, *On Introduction to Surface Chemistry and Catalysis*, Wiley 1994.
2. B. Hammer, K. W. Jacobsen, and J. K. Nørskov, Phys. Rev. Lett. **70**, 3971 (1993); K. Gundersen, K. W. Jacobsen, J. K. Nørskov, and B. Hammer, Surf. Sci. **304**, 131 (1994).
3. B. Hammer, M. Scheffler, K. W. Jacobsen, and J. K. Nørskov, Phys. Rev. Lett. **73**, 1400 (1994).
4. A. Gross, B. Hammer, M. Scheffler, and W. Brenig, Phys. Rev. Lett. **73**, 3121 (1994);
5. P. Hu, D. A. King, S. Crampin, M.-H. Lee, and M. C. Payne, Chem. Phys. Lett. **230**, 501 (1994).
6. H. Yang and J. L. Whitten, J. Chem. Phys. **96**, 5529 (1992); O. Swang, J. K. Faegri, O. Gropen, U. Wahlgren, and P. E. M. Siegbahn, Chem. Phys. **156**, 379 (1991).
7. A. Zangwill, *Physics at surfaces*, Cambridge University Press, 1988.
8. R. I. Masel, *Principles of Adsorption and Reaction on Solid Surfaces*, Wiley, 1996.
9. S. Holloway and J. K. Nørskov, *Bonding at Surfaces*, Liverpool University Press, 1991.
10. R. Hoffmann, Rev. Mod. Phys. **60**, 601 (1988)
11. R. A. Van Santen and M. Neurock, Catal. Rev. Sci. Eng., **37**, 557 (1995).
12. For a review, see M. C. Payne, M. P. Teter, D. C. Allan, T. A. Arias, and J. D. Joannopoulos, Rev. Mod. Phys. **64**, 1045 (1992).
13. J. P. Perdew, J. A. Chevary, S. H. Vosko, K. A. Jackson, M. R. Pederson, D. J. Singh, and C. Fiolhais, Phys. Rev. B **46**, 6671 (1992).

14. M. Van Hove, Prog. in Surf. Sci. **46**, 305 (1994).
15. Y. Y. Yeo and D. A. King, to be published.
16. H Steininger, S. Lehwald, and H. Ibach, Surf. Sci. **123**, 1 (1982).
17. J. J. Mortensen, Y. Morikawa, B. Hammer, and J. K. Nørskov, Z. Phys. Chem., *in press.*
18. B. Hammer and J. K. Nørskov, Nature **376**, 238 (1995).
19. B. E. Hayden and C. L. A. Lamont, Phys. Rev. Lett. **63**, 1823 (1989).
20. A. Hodgson, J. Moryl, P. Traversaro and H. Zhao, Nature **356**, 501 (1992).
21. H. A. Michelsen, C. T. Rettner and D. J. Auerbach, Phys. Rev. Lett. **69**, 2678 (1992); H. A. Michelsen, C. T. Rettner, D. J. Auerbach and R. N. Zare, J. Chem. Phys. **98**, 8294 (1993).
22. H. F. Berger, M. Leisch, A. Winkler and K. D. Rendulic, Chem. Phys. Lett. **175**, 425 (1990).
23. H. J. Robota, W. Vielhaber, M. C. Lin, J. Segner, and G. Ertl, Surf. Sci. **155**, 101 (1985).
24. J. K. Brown, A. Luntz, and P. A. Schultz, J. Chem. Phys. **95**, 3767 (1991).
25. D. M. Jaffey, R. J. Madix, Surf. Sci. **311**, 159 (1994).
26. D. M. Ceperley and M. H. Kalos, in *Monte Carlo Methods in Statistical Physics*, edited by K. Binder, Springer, Berlin, 1979.
27. B. L. Hammond, W. A. Lester, Jr., and P. J. Reynolds, *Monte Carlo Methods in ab initio quantum chemistry*, World Scientific, Singapore, 1994.
28. L. Mitas and R. M. Martin, Phys. Rev. Lett. **72**, 2438 (1994).
29. J. Grossman and L. Mitas, Phys. Rev. Lett. **74**, 1323 (1995).
30. J. Grossman, L. Mitas, and K. Raghavachari Phys. Rev. Lett. **75**, 3870 (1995).
31. P. Hohenberg and W. Kohn, Phys. Rev. **136**, B864, (1964).
32. W. Kohn and L. Sham, Phys. Rev **140**, A1133, (1965).
33. D. M. Ceperley and B. J. Alder, Phys. Rev. Lett **45**, 566 (1980).
34. J. P. Perdew and A. Zunger, Phys. Rev. B **23**, 5048 (1981).
35. K. W. Jacobsen, J. K. Nørskov, and M. J. Puska, Phys. Rev. B **35**, 7423 (1987).
36. J. K. Nørskov, Rep. Prog. Phys. **53**, 1253 (1990).
37. A.R. Mackintosh and O.K. Andersen, in *Electrons at the Fermi surface*, edited by M. Springford (Cambridge University Press, Cambridge, England, 1980).
38. V. Heine, in *Solid State Physics*, edited by H. Ehrenreich, F. Seitz, and D. Turnbull (Academic, New York, 1980), vol. 35.
39. M. Methfessel, Phys. Rev. B **52**, 8074 (1995).
40. G. B. Bachelet, D. R. Hamann and M. Schlüter, Phys. Rev. B **26**, 4199 (1982).
41. N. Troullier and J. L. Martins, Phys. Rev. B **43**, 1993 (1991).
42. J. E. Inglesfield and G. A. Benesh, Phys. Rev. B **37**, 6682 (1988).
43. J. M. MacLaren, S. Crampin, D. D. Vvedensky and J. B. Pendry, Phys. Rev. B **40**, 12164 (1989).
44. H. L. Skriver and N. M. Rosengaard, Phys. Rev. B **43**, 9538 (1991).
45. S. H. Wosko, L. Wilk, and M. Nusair, Can. J. Phys. **58**, 1200 (1980).
46. J. P. Perdew, K. Burke, and Y. Wang, preprint.
47. L. Fan and T. Ziegler, J. Am. Chem. Soc. **114**, 10890 (1992).
48. M. A. van Daelen, Y. S. Li, J. M. Newsam, and R. A. van Santen, Chem. Phys. Lett. **226**, 100 (1994).
49. J. A. White, D. M. Bird, M. C. Payne, and I. Stich, Phys. Rev. Lett. **73**, 1404 (1994).
50. P. H. T. Philipsen, G. te Velde, and E. J. Baerends, Chem. Phys. Lett. **226**, 583 (1994).
51. H. Burghgraef, A. P. J. Jansen, and R. A. van Santen, J. Chem. Phys. **101**, 11012 (1994).
52. P. Kratzer, B. Hammer, and J. K. Nørskov, Surf. Sci. **359**, 45 (1996).
53. B. Hammer and M. Scheffler, Phys. Rev. Lett. **74**, 3487 (1995).
54. J. Jacobsen, B. Hammer. K. W. Jacobsen, and J. K. Nørskov, Phys Rev. B **52**,

14954 (1995).
55. B. Hammer and J. K. Nørskov, Surf. Sci. **343**, 211 (1995); *ibid.* **359**, 306 (1996) (E).
56. B. Hammer, Y. Morikawa, and J. K. Nørskov, Phys. Rev. Lett. **76**, 2141 (1996).
57. J. A. White, D. M. Bird, and M. C. Payne, Phys. Rev. B **53**, 1667 (1996).
58. P. A. Gravil, D. M. Bird and J. A. White, preprint.
59. S. Wilke and M. Scheffler, Phys. Rev. B **53**, 4926 (1996).
60. S. Wilke and M. Scheffler, Phys. Rev. Lett. **76**, 3380 (1996).
61. P. Kratzer, B. Hammer, and J.K. Nørskov, J. Chem. Phys, *in press*
62. H. Yang and J. L. Whitten, J. Chem. Phys **96**, 5529 (1992).
63. P. Nacthigall, K. D. Jordan, Smith and H. Jonsson, J. Chem. Phys. **104**, 148 (1996).
64. S. Holloway, B. I. Lundqvist, and J. K. Nørskov, Proc. Int. Congress on Catalysis, Berlin 1984, Vol. 4, p. 85.
65. P. Nordlander, S. Holloway and J.K. Nørskov, Surf. Sci. **136**, 59 (1984).
66. J. Phys.: Condens. Matter **6**, 9495 (1994).
67. E. M. Shustorovich, Surf. Sci. Rep. **6**, 1 (1986); Adv. Catal. **37**, 101 (1990); E. M. Shustorovich and A. T. Bell, Surf. Sci. **268**, 397 (1992).
68. P. J. Feibelman, Phys. Rev. Lett. **67**, 461 (1991).
69. D. Pettifor, *Bonding and Structure of Molecules and Solids*, Clarendon Press, Oxford, 1995.
70. J. Friedel in *The Physics of Metals*, edited by J. M. Ziman (Cambridge University Press, 1969), p 494.
71. H. L. Skriver, Phys. Rev. B **31**, 1909, (1985).
72. M. S. S. Brooks and B. Johansson, J. Phys. F. **13**, L197 (1983).
73. D. M. Newns, Phys. Rev. **178**, 1123 (1969).
74. P. W. Anderson, Phys. Rev. **124**, 41 (1961).
75. T. B. Grimley, J. Vac. Sci. Technol. **8**, 31 (1971) and **9**, 561 (1971).
76. J.W.Gadzuk, Surf. Sci. **43**, 44 (1974).
77. B. I. Lundqvist, O. Gunnarsson, H. Hjelmberg, and J. K. Nørskov, Surf. Sci. **89**, 196 (1979).
78. K. Stokbro and S. Baroni, Surf. Sci., *in press.*
79. O. K. Andersen, O. Jepsen, and D. Glötzel: *Highlights of Condensed Matter Theory* LXXXIX Corso Soc. Italiana di Fisica, Bologna, 1985, p. 59.
80. J. K. Nørskov, J. Chem. Phys. **90**, 7461 (1989).
81. A. Ruban, B. Hammer, P. Stoltze, H. L. Skriver, J. K. Nørskov, J. Mol. Catal. A, *in press.*
82. I. Toyoshima and G. A. Somorjai, Catal. Rev. Sci. Eng. **19**, 105 (1979).
83. K. Horn, J. DiNardo, W. Eberhardt, H.-J. Freund, and E. W. Plummer, Surf. Sci. **118**, 465 (1982).
84. J. C. Campuzano in 'The chemical physics of solid surfaces and heterogeneous catalysis', D. A. King and D. P. Woodruff, eds., Vol 3A, p 389 (Elsevier, 1990).
85. P. S. Bagus and G. Pacchioni, Surf. Sci. **278**, 427 (1992).
86. G. Blyholder, J. Phys. Chem. **68**, 2772 (1964).
87. A. M. Schoeb, T. J. Raeker, L. Yang, T. S. King, and A. E. DePristo, Surf. Sci. **278**, L125 (1992).
88. J. A. Rodriguez and D. W. Goodman, Science **257**, 897 (1992).
89. A. Nielsen (Ed.), *Ammonia, Catalysis and Manufacture*, Springer-Verlag, Berlin, 1995.
90. A. Gross, S. Wilke, and M. Scheffler, Phys. Rev. Lett. **75**, 2718 (1995).
91. M. Kay, G.R. Darling, S. Holloway, J.A. White and D.M. Bird, Chem. Phys. Lett. **245**, 311 (1995).
92. A.D. Kinnersley, G.R. Darling, S. Holloway, and B. Hammer, Surf. Sci. **364**, 219 (1996).
93. G. R. Darling and S. Holloway, Rep. Prog. Phys. **58**, 1595 (1995).

94. M. Holmblad, J. H. Larsen, I. Chorkendorff, L. Pleth Nielsen, F. Besenbacher, I. Stensgaard, E. Lægsgaard, P. Kratzer, B. Hammer, and J. K. Nørskov, Catal. Lett. **40**, 131 (1996).
95. T. P. Beebe Jr., D. W. Goodman, B. D. Kay, and J. T. Yates, J. Chem. Phys. **87**, 2305 (1987); A. Luntz and J. Harris, Surf. Sci. **258**, 397 (1991).
96. P. M. Holmblad, J. Wambach, and I. Chorkendorff, J. Chem. Phys. **102**, 8255 (1995)
97. M. B. Lee, Q. Y. Yang, and S. T. Ceyer, J. Chem. Phys. **87**, 2724 (1987).
98. P. Kratzer, B. Hammer, and J. K. Nørskov, *to be published.*
99. M. Boudart, Adv. Catal. Relat. Subj. **20**, 153 (1969).
100. P. J. Feibelman and D. Hamann, Phys. Rev. Lett. **52**, 61 (1984); Surf. Sci. **149**, 48 (1985).
101. L. M. Falicov and G. A. Somorjai, Proc. Natl. Acad. Sci. USA **82**, 2207 (1985).
102. W. Yang and R. G. Parr, Proc. Natl. Acad. Sci. USA **82**, 6723 (1985).
103. S. Wilke, M. H. Cohen, and M. Scheffler, Phys. Rev. Lett. **77**, 1560 (1996).
104. J. Harris and S. Andersson, Phys. Rev. Lett. **55**, 1583 (1985).
105. K. Fukui, T. Yonezawa, and H. Shingu, J. Chem. Phys. **20**, 722 (1952).
106. R. G. Parr and W. Yang, *Density Functional Theory of Atoms and Molecules* (Oxford University Press, N. Y., 1989).
107. M. H. Cohen, M. V. Ganduglia-Pirovano, and J. Kudrnovsky, Phys. Rev. Lett. **72**, 3222 (1994); M. H. Cohen, M. V. Ganduglia-Pirovano, and J. Kudrnovsky, J. Chem. Phys. **101**, 8988 (1994).
108. R. Linke, U. Schneider, H. Busse, C. Becker, U. Schröder, G. R. Castro, and K. Wandelt, Surf. Sci. **307-309**, 407 (1994).
109. J. H. Sinfelt: *Bimetallic Catalysts: Discoveries, Concepts, and Applications*, Wiley, N. Y. (1983).
110. V. Ponec, Adv. Catal. **32**, 149 (1983).
111. N. T. Andersen, F. Topsøe, I. Alstrup, and J. Rostrup-Nielsen, J. Catal. **104**, 454 (1987).
112. Y. L. Lam, J. Chiado, and M. Boudart, Nouv. J. Chim. **1**, 461 (1977).
113. M. Weinert and R. E. Watson, Phys. Rev. B **51**, 17168 (1995).
114. D. Hennig, M. V. Ganduglia-Pirovano, and M. Scheffler, Phys. Rev B, **53**, 10344 (1996).
115. M. V. Ganduglia-Pirovano, V. Natoli, M. H. Cohen, J. Kudrnovsky, and I. Turek, Phys. Rev. B **54**, 8892 (1996).
116. M. V. Ganduglia-Pirovano, J. Kudrnovsky and M. Scheffler, preprint.
117. I.Alstrup and M.T.Tavares, J.Catal. **139**, 513 (1993).

DENSITY FUNCTIONAL CLUSTER CALCULATIONS ON METAL DEPOSITION AT OXIDE SURFACES

N. RÖSCH and G. PACCHIONI[#]
Lehrstuhl für Theoretische Chemie
Technische Universität München
85747 Garching, Germany

1. Introduction

Metal-ceramic interaction is relevant in several areas of modern science such as corrosion, adhesion, microelectronic devices, photovoltaic cells, protective coating of metals, etc. In addition, highly dispersed metal particles supported on oxides make up an important class of heterogeneous catalysts. Oxide surfaces serve not only as inert support for the active component, but they are actually able to stabilise metal particles of a particular dispersion grade or to alter electronic and chemisorption properties of these species. Experimental findings of significant changes in the catalytic performance of supported transition metal species have stimulated a particularly high interest in the question of how the support affects the chemisorption and the catalytic properties of active sites. The complicated phenomenon of metal-support interaction comprises a variety of mechanisms, among them the direct local electronic effect of oxide surfaces on supported metal particles. Despite this considerable technological importance very little is known about the microscopic nature of the interface between the surface of an oxide substrate, like MgO, SiO_2, Al_2O_3, TiO_2, etc., and the contact metal atoms of a supported particle or of a metallic overlayer. This lack of information is both structural and theoretical. Most of the technologically interesting materials are based on amorphous components, but there are open structural questions even for single-crystal experiments, like the possibility of surface reconstruction, the presence of defects and dislocations, local disorder, etc. For these reasons the theoretical description of the metal-oxide interaction is especially challenging.

More recently, considerable experimental effort has been spent to better characterise the formation of metallic overlayers emphasising the very first stages of metal deposition [1–10]. This intense experimental activity is comple-

[#] Permanent address: Dipartimento di Chimica Inorganica, Metallorganica e Analitica, via Venezian 21, 20133 Milano, Italy

R.M. Lambert and G. Pacchioni (eds.),
Chemisorption and Reactivity on Supported Clusters and Thin Films, 353-370.

mented by a rather limited number of "first principles" theoretical studies dealing with the general problem of metal-ceramic interaction [11]. Density functional (DF) calculations have been reported on the adhesion of metal monolayers to the surface of a simple oxide like MgO [12–15]. Hartree-Fock (HF) and DF studies have also been carried out for single metal atoms interacting with this oxide [16–18], and recently the first attempts to theoretically investigate the cluster deposition on oxide surfaces [19,20] or the reactivity of surface vacancies [21] have been reported. However, systematic theoretical studies of the interaction of transition metal (TM) atoms and clusters with the surface of an oxide are still lacking.

For adsorbed metal atoms, two types of interactions are usually assumed, chemical bonds (mainly with the surface oxygen atoms) on the one hand, or van der Waals interaction and/or weak polarisation bonds with no metallization of the surface [22] on the other hand. In the initial step of forming a metallic film, metal atoms impinge on the substrate. These atoms can be reflected from the surface or they may stick to the surface, diffuse on it, and eventually reevaporate. Condensation can occur if the flux of adsorbed atoms is larger than the flux of reevaporated atoms and it is clear that the strength of the bond with the surface plays an essential role in this process.

In this chapter we review the results of DF calculations on single metal atoms [23] and on metal clusters interacting with oxide substrates, in particular MgO [19] and Al_2O_3 [20]. We will discuss the interaction of single, isolated, transition metal atoms with the oxide anions of the MgO surface to gain a more systematic understanding of the interface bond; we will consider small clusters of Ni and Cu atoms interacting with various adsorption sites of MgO to better understand the early stages of metal deposition. We will also study the extent of the perturbation induced by the substrate on the electronic properties of the supported metal cluster. In the last section we will consider some simple model calculations of the interaction of one- and two-layer Ni clusters with the more reactive surface of Al_2O_3.

2. The LCGTO-DF Method

The results discussed below were obtained in the model cluster approach with the help of the Munich version of the linear combination of Gaussian-type orbital density functional (LCGTO-DF) code [24]. The computationally economic local density approximation (LDA) to the exchange-correlation functional has successfully been used in chemical applications since the seventies. This functional (employed here in the parametrisation suggested by Vosko, Wilk and Nusair, VWN [25]) has been demonstrated to accurately describe important properties of adsorption complexes and other molecular species such as bond lengths and vibrational parameters [26,27]. However, the calculated LDA binding energies were often significantly overestimated. In the past few years it became apparent that density gradient corrections to LDA exchange-correlation functional considerably improve the calculated molecular energetics [28]. Several variants of such generalized gradient approximations (GGA) have been implemented in the LCGTO-DF program. The GGA expressions suggested by Becke [29] or, alternatively, by Perdew and Wang [30] were utilized to calculate density gradient corrections to the exchange energy functional. The corresponding correlation contribution was computed following the works of Lee et al. [31] or of Perdew [32]. The pertinent exchange-correlation contributions to the Kohn-Sham matrix are evaluated using a direct technique involving scalar products of the first order partial derivatives of the energy functionals and of the wave functions [33]. The following three GGA variants are in use: Becke-Lee-Yang-Parr (BLYP) [29,31], Becke-Perdew [29,32] and Perdew-Wang (PW) [30]. Recently, a self-consistent scalar-relativistic (SR) version of LCGTO-DF method has also been developed [34,35]. The SR variant employs a unitary second order Douglas-Kroll-Hess (DKH) transformation for decoupling large and small components of the full four-component spinor solutions to the Dirac-Kohn-Sham equation. The approximate DKH transformation most appropriate and most efficient for molecular calculations has been implemented; this variant utilises nuclear potential-based projectors and it leaves the electron-electron interaction untransformed.

Relatively large orbital Gaussian basis sets have been used in the LCGTO-DF all-electron calculations to be discussed below. The two auxiliary bases sets employed in the LCGTO-DF method to represent the electron charge density and, in some cases, the exchange-correlation potential were constructed from the orbital exponents as proposed in [24]. For weakly bound molecular systems the counterpoise technique was applied to correct interaction energies for the basis set superposition error (BSSE) [36]. Vibrational frequencies have been obtained either by diagonalising the mass-weighted force constant matrix which had been computed from finite differences of total energy gradients or by using a polynomial fit to a pointwise representation of the energy hypersurface. In the latter case both harmonic and anharmonic frequencies can be calculated by assuming a negligible coupling of the vibrational modes under consideration.

3. Adsorption of Transition Metal Atoms on Oxygen Ions of MgO(001)

Understanding of the bonding mechanims of individual metal atoms with the surface of a simple oxide like MgO is an essential prerequisite for the theoretical study of more complex systems, like clusters or organometallic fragments, supported on oxide substrates. The DF-GGA method is best suited for the description of this kind of interaction, although it should be noted that even this highly sophisticated approach presents some problems in the description of very weak and van der Waals interactions. In the following we describe the interaction of various TM atoms, Cr, Mo, W, Ni, Pd, Pt, Cu, Ag, and Au, with the

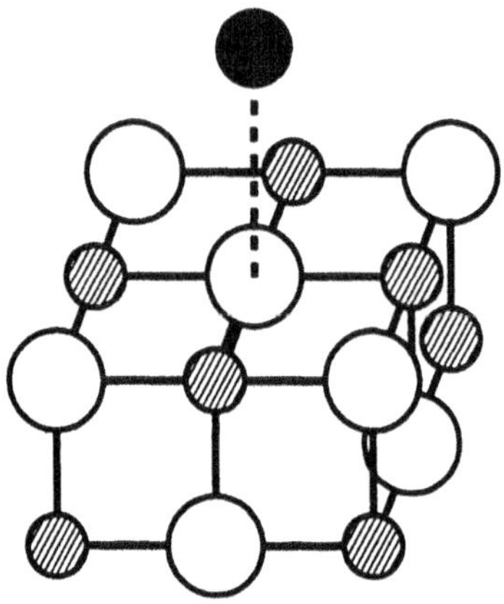

Figure 1. Schematic representation of a metal atom interacting with a O_9Mg_9 cluster model of an oxide anion on the MgO(001) surface. The cluster is embedded in a large array of point charges (not shown).

oxygen anions of the MgO(001) surface. The study is restricted to these sites as it is known from other theoretical investigations that the Mg cations of the surface are rather unreactive [16,20,21]. Indeed, metal atoms bind very weakly at these sites by polarisation-dispersion forces.

The calculations have been performed with the LCGTO-DF program employing a density gradient corrected energy functional [23]. These gradient corrections were introduced based on the self-consistent LDA density; all results presented in this section are at the BLYP GGA-DF level. The cluster used to simulate the MgO(001) surface is O_9Mg_9, Figure 1; it is embedded in an array of 17x17x6 point charges to represent the long-range Madelung potential.

3.1. Cu, Ag, Au

The presence of a filled d shell and a singly occupied s orbital prevents the Cu, Ag, and Au atoms from easily changing their configuration during bond formation. The lowest-lying excited state, d^9s^2, is in fact about 1.7 eV higher than the ground state for Cu and Au, and about 4 eV higher for Ag. On the other hand, Cu, Ag, and Au could, in principle, form relatively strong bonds with oxygen atoms of not fully reduced oxide surfaces by partial transfer of their outer electron to the substrate.

The ground state of the surface cluster plus adsorbate is $^2A_1$ for all atoms of the triad and the orbital analysis shows that the adsorbed atoms retain their $nd^{10}(n+1)s^1$ atomic configuration. However, a very small mixing of the metal σ orbitals with the O $2p_\sigma$ orbital is also found. All atoms are weakly bound to the surface, Table 1. The dissociation energy, corrected for the BSSE, goes from 0.30 eV for Cu, to 0.20 and 0.23 eV for Ag and Au, respectively. The weak

TABLE 1. Calculated adsorption properties of transition metal atoms on the oxide anions of the MgO(001) surface: adsorption height z_e(O-M), harmonic frequency ω_e(O-M) of the vibration perpendicular to the surface and BSSE corrected binding energy D_e (rel = relativistic, nrel = nonrelativistic calculation).

Atom	Method	Ground state	z_e(O-M), Å	ω_e(O-M), cm^{-1}	D_e, eV
Cu	nrel	$^2A_1$	2.18	98	0.30
Ag	rel	$^2A_1$	3.17	28	0.20
Au	rel	$^2A_1$	2.47	52	0.23
Ni	nrel	$^1A_1$	1.87	249	0.88
Pd	rel	$^1A_1$	2.15	134	0.81
Pt	rel	$^1A_1$	2.00	160	1.36
Cr	nrel	$^7A_1$	2.83	48	0.34
Mo	rel	$^7A_1$	3.14	29	0.33
W	rel	$^5A_1$	2.16	117	0.72

bonding is shown also by the very low values of the frustrated translation, perpendicular to the surface, which is characterised by a vibrational frequency of 100 cm^{-1} and smaller, Table 1.

The analysis of the bonding, and in particular of the dipole moment curve for the vertical displacement of the metal atom from the surface [37], excludes the occurrence of a substantial charge transfer from the metal atom to the substrate. The bonding can thus be described as mainly due to the polarisation of the metal electrons by the surface electric field, accompanied by a relatively modest chemical interaction. This result suggests that the oxide anions of the MgO surface are highly charged, with little tendency to ionise the adsorbed coinage atoms. The relatively weak bonding found for Cu, Ag, and Au, is therefore the consequence of the large adatom size (due to the singly occupied valence s orbital) which determines the long surface-adsorbate distance. Ag behaves somewhat differently from Cu and Au: the bond distance is even longer than for the other two members of the triad, Table 1. This difference may be related to the large energy separation between the Ag 4d and 5s levels compared to the d-s separation in Cu and Au which is reflected in the very different $s^1d^{10} \rightarrow s^2d^9$ transition energies. The smaller propensity for a metal s-d hybridisation and for a mixing with the oxygen levels is most likely the reason for the weaker bonding of Ag.

3.2. Ni, Pd, Pt

The interaction of Ni, Pd, and Pt atoms with MgO is considerably more complex than that of the coinage metal atoms because of the interplay between the d^{10}, the d^9s^1 and, at least for Ni, the d^8s^2 configurations of the atoms. The interaction of free atoms with the MgO cluster gives rise to several states; the lowest triplet and singlet states, $^3B_2$ and $^1A_1$, correlate at infinite adatom-surface separation with MgO + M(d^9s^1) and MgO + M(d^{10}), respectively. The state $^3B_2$ features the lowest energy for long surface-metal distances. At shorter distances, however, a crossing of the $^3B_2$ with the $^1A_1$ state occurs. Below about 2.2 Å (for Ni) or 2.5 Å (for Pt) the $^1A_1$ state becomes the ground state; the potential energy curve exhibits a deeper minimum and a rather high adsorption energy of about 1 eV or more, depending on the metal, see Table 1. The curve crossing implies that a magnetic quenching accompanies the formation of the bond. The key mechanism for the bonding is the formation of s-d hybrid orbitals which can conveniently mix with the O $2p_\sigma$ orbitals; in fact, in the $^1A_1$ configuration the metal s, d_σ and the O $2p_\sigma$ orbitals are strongly hybridised. The low-spin coupling of the metal open shell electrons seems to favour the bonding with the surface oxygen.

The dipole moment curves reveal a significant difference between Ni on one side and Pd and Pt on the other. Ni exhibits an almost linear curve with a relatively large slope of 0.48 au, indicative of a non-negligible polarity of the bond with MgO. Pd and Pt, on the other hand, feature dipole moment curves with small slopes and large curvatures, indicative of a more covalent character of the interaction between Pd and Pt orbitals and the oxygen 2p orbitals. A possible rationalisation of this difference may be gleaned from the values of the ionisation potential (configuration d^{10}) which in Ni is about 2 eV smaller than in Pd or Pt. The fact that Ni is easier to ionise is in agreement with the largest slope of the dipole moment curve and with a more polar bond. However, the bonding can by no means be viewed simply as a charge transfer from the metal atom to the oxide surface. The orbital analysis clearly shows a substantial mixing (hybridisation) of metal *n*d and O 2p orbitals. This mixing leads to a covalent bond with some polar character for Ni and little polarity for Pd and Pt. The fact that all three atoms in the group form relatively strong bonds with the surface, compared for instance to the coinage metals, can be explained by the more pronounced tendency of the group 10 atoms to form s-d hybrid orbitals and a more direct involvement of their valence d orbitals in the bond with the oxygen. This tendency can be generally related to the s→d (or d→s) transition energies which are considerably smaller than for Cu, Ag, and Au [23].

3.3. Cr, Mo, W

The interplay between high-spin and low-spin states is of great importance in the interaction of Cr, Mo, and W atoms with MgO. These atoms have high-spin atomic ground states $nd^5(n+1)s^1$ ($^7A_1$ in C_{4v} symmetry); their valence d shell is therefore half filled, and a comparison of their interaction with that of the late TM atoms will be particularly instructive. Low-spin, quintet, states have also been considered, but they are higher in energy for Cr and Mo. For Cr the ground state of the cluster is $^7A_1$ which correctly dissociates into Cr and MgO ground states. This state exhibits a very shallow potential energy curve, with a minimum around 2.8 Å and a binding energy of 0.34 eV (BSSE corrected), Table 1. Also for Mo the $^7A_1$ state is lowest in energy, with an equilibrium distance at large separation from the surface, close to 3 Å, and a binding energy of 0.33 eV; the force constant is even smaller than for Cr. At adsorption heights below 2.2-2.3 Å the low-spin $^5A_1$ state becomes energetically preferred, but again with a local energy minimum slightly higher than for the $^7A_1$ state. The two atoms, Cr and Mo, also have similar dipole moment curves, with very small slopes and pronounced curvatures. It is possible to conclude that the bonding is largely due to polarization and dispersion, with little, if any, chemical mixing of the metal orbitals with the surface electronic states.

A completely different bonding arises from the interaction of W with MgO. The interaction of W in the high-spin d^5s^1 configuration results in a flat curve, very similar to those computed for Cr and Mo. This state, however, is not the ground state, even for long surface-adsorbate distances. On the other hand, the low-spin $^5A_1$ state exhibits a deep minimum near 2.15 Å; in this minimum, W is bound to MgO by 0.72 eV (BSSE corrected value). This is a substantial chemical bond of similar strength than that calculated for the Ni triad. A population analysis shows that in this state the d^4s^2 atomic configuration is mixed not only with the d^5s^1 state, but also with the O 2p orbitals. This is consistent with the fact that the s-d transition energy is smallest in W where the s^1d^5 and the s^2d^4 states are almost isoenergetic, in contrast to the situation for Cr and Mo [23]. The facile s-d hybridization is actually the reason for the strong bonding of W. As for adsorbed Ni, the bond with W exhibits a considerable polar character, as shown by the rather linear dipole moment curve with its large slope, 0.71 au.

3.4. GENERAL TRENDS FOR TRANSITION METAL ATOMS ON MAGNESIUM OXIDE

The Ni triad is the only one for which strong interface bonds are formed. The adhesion energy of these atoms on top of surface oxygens is about 1 eV/atom or more, Table 1. The only other metal atom considered here which exhibits a tendency to form strong bonds with the surface is W. For W, in fact, we computed a bond strength and a force constant comparable to those of the Ni triad. This tendency to form strong bonds is connected to the fact that metal s and d orbitals hybridise and that these hybrid orbitals mix with the p orbitals of the surface oxygen. Pd is somewhat special in the Ni triad as it has the smallest binding energy and the longest distance from the surface. This reflects the general tendency toward a non-monotonous behaviour in many chemical properties as one moves down a TM group; in particular, second-row TM atoms often exhibit weaker interactions than the isovalent first- and third-row atoms [38,39]. To some extent, this trend is observed also for the other two triads considered here. The bonding of Cr, Mo, Cu, Ag, and Au to a MgO substrate, however, can be classified as weak, arising mainly from polarization and dispersion effects with only minor orbital mixing with the surface oxygen orbitals. This explains the very long bond distances found in some cases, and the flat potential energy curves which result in very small force constants.

Thus, the TM atoms considered can be classified into two groups, atoms which tend to form relatively strong chemical bonds with the surface oxygen anions of MgO (Ni, Pd, Pt, and W), and atoms which interact very weakly with the surface, with adsorption energies of the order of one third of an eV or less (Cr, Mo, Cu, Ag, and Au). In the first case the bonding originates from the direct mixing of the metal d orbitals with the surface O 2p band. This interaction is covalent in nature and *does not imply a significant charge transfer from the metal to the surface.* This is an important conclusion, connected to the highly ionic nature of the MgO surface where the surface oxygen atoms have their valence almost saturated. To a first-order approximation MgO can be described in terms of classical ionic model, Mg^{2+}-O^{2-}; a more refined estimate based on slab LDA calculations is that the Madelung field outside the surface is comparable to that of an array of point charges with ±1.8 au [40]. This means that the oxygen centers at the *regular surface sites of MgO(001) are almost completely reduced and are not able to oxidize adsorbed metal atoms.* Bearing in mind this important conclusion we consider now the interaction of small Ni and Cu clusters with the MgO(001) surface.

4. Ni and Cu Cluster Deposition on MgO

The deposition of Cu on MgO surfaces has been studied in detail experimentally by means of low energy electron diffraction, and by various electron spectroscopy methods, like Auger, electron energy loss (EELS), and X-ray photoemission [1–4]; the sticking probability of incident Cu atoms has also been measured [5,6]. Accurate studies of growth and morphology of Pd clusters on MgO have also been reported [8,9; see also the paper by C. Henry and coworkers in this volume]. This system is thus ideal for a theoretical investigation. The deposition of Al, Ag, Fe, and Ti metal overlayers on MgO(001) has been studied by means of band structure calculations at various level of theoretical sophistication [12–15].

4.1 Ni_4 AND Cu_4 ON MgO

Here we consider the deposition of Ni_4 and Cu_4 clusters on various sites of the MgO(001) surface and we shall discuss the nature of the cluster/substrate interface as well as the modifications induced on the cluster electronic structure [20]. The same approach described in the previous section has been followed here and the results described in the following are at the BLYP GGA-DF level. Two cluster models have been employed to simulate the MgO(001) surface, namely Mg_9O_9 and O_9Mg_9, Figure 2. These clusters model adsorption on a

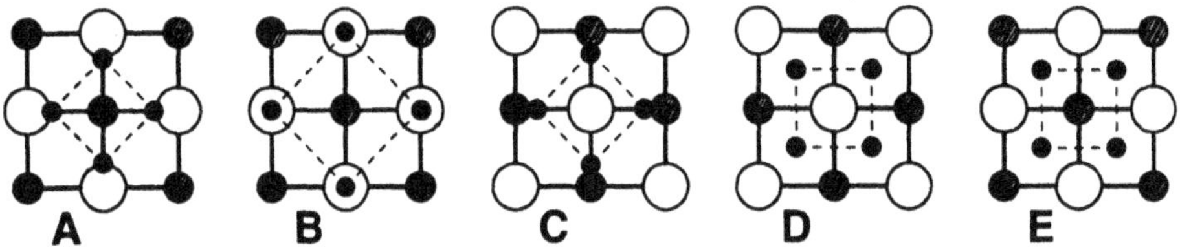

Figure 2. Schematic representation of the sites investigated for the adsorption of Ni_4 and Cu_4 clusters on the (001) surface of MgO. Small hatched circles represent Mg ions, large empty circles O ions, and small full circles Ni or Cu atoms. The cluster models Mg_9O_9 (A, B, E) and O_9Mg_9 (C, D) also contain the corresponding atoms from the "second" crystal layer, not shown here.

oxygen anion or a magnesium cation, respectively. The clusters are embedded in an array of 17x17x6 point charges. Further details can be found in [20].

To model the metal/oxide interface we have considered square planar (D_{4h}) clusters which may be viewed as precursors of an epitaxially growing metal film. An unsupported Ni_4 cluster is strongly spin-polarised, with about one unpaired electron per atom. The metal-metal distance (Ni-Ni) is 2.23 Å (2.49 Å is the bulk distance); the corresponding binding energy per atom is 1.76 eV. An unsupported square planar Cu_4 cluster, with a $^3A_2$ ground state, has an optimal Cu-Cu distance of 2.36 Å and a binding energy per atom of 1.3 eV. These two clusters have been deposited on the MgO(001) clusters in various orientations, see Figure 2.

A partial optimisation of the adsorbed cluster has been carried out. In two of the adsorption geometries, denoted as site A and site B, Figure 2, the interaction is primarily between the cluster metal atoms and the surface oxygen atoms. In site B each metal atom is placed on-top of the surface oxygen centres; this corresponds to pseudomorphic growth of a metal overlayer. In site C the metal atoms are closer to the surface cations; on-top adsorption has not been considered, given the weak binding of the single metal atoms. Sites D and E represent possible alternative orientations of the M_4 cluster with the metal atoms near the four-fold hollow sites of the (001) surface.

In general, adhesion of the metal clusters to the surface is moderately strong when a direct interaction with the oxygen cations is possible, while it is very weak when the interaction primarily occurs with the substrate cations or in the four-fold hollow sites. As already mentioned, the competition between the metal-metal and the metal-oxygen bonds is an important issue. If the adsorptive metal-oxygen bond strength is larger, the metal cluster may grow pseudomorphically until completion of the first monolayer (Frank-von der Merwe mechanism). If the intra-cluster metal-metal bond prevails, growth of three-dimensional metal particles is expected (Vollmer-Weber mechanism). When the square planar Ni_4 and Cu_4 clusters are adsorbed on site A, a bond of 0.62 and

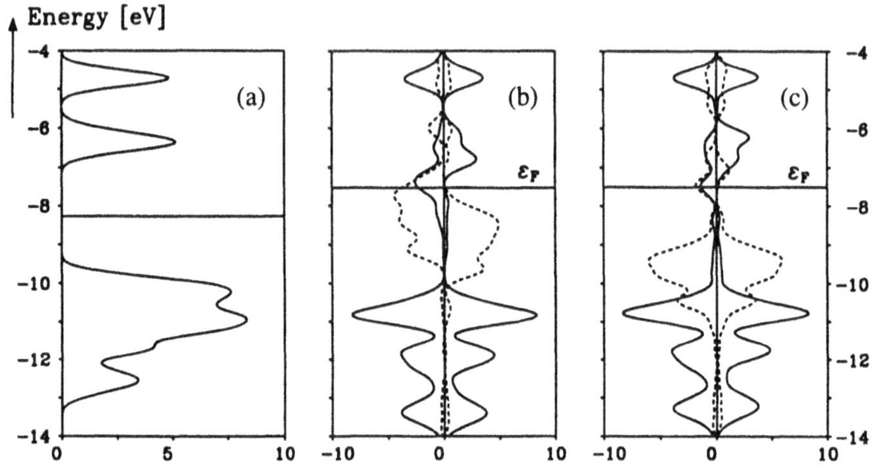

Figure 3. Density of states of (a) Mg_9O_9, (b) Ni_4/Mg_9O_9, (c) Cu_4/Mg_9O_9. Solid line MgO, dotted line M_4; left-hand side minority spin, right-hand side majority spin.

0.36 eV/atom is found. The surface-cluster distance is similar for the two metals, about 1.95-2.0 Å. Upon adsorption the metal-metal bond length expands from 2.2 to 2.5 Å for Ni and from 2.4 to 2.65 Å for Cu. This reflects a weakening of the metal-metal bond due to the interaction with the surface oxygen atoms. It is interesting to note that the metal-metal distance in the supported clusters is similar to the bulk metal values (2.49 Å for Ni and 2.56 Å for Cu). If the metal-oxygen bond were strong, the M_4 cluster epitaxially grown on-top of the surface oxygen atoms, site B (see Figure 2), should have a lower total energy; however, this is not the case. When the four metal atoms are placed directly above the four peripheral oxygen atoms of the cluster, site B, the bonding with respect to dissociation into free Ni_4 and Cu_4 clusters is weaker than for site A. It should be noted that in site B the metal-metal distance is 2.97 Å; at this distance, there is still considerable metal-metal bonding as shown by the atomisation energies of the free clusters Ni_4 and Cu_4 of this geometry which is at 2.97 Å 0.98 and 0.87 eV/atom, respectively. Thus, the calculations show that the metal-oxygen bonding is not strong enough to prevail over the metal-metal interaction and that the adsorbed clusters maintain to a large extent their identity. For adsorption on site C, with the metal atoms close to the Mg ions, or for sites D and E, where the metal atoms sit on four-fold hollow sites, the bonding is weak as expected.

The electronic interactions at the interface can be better visualised by means of density of states (DOS) plots, Figure 3. As usual, these plots have been generated by broadening the Kohn-Sham one-electron level spectrum using Gaussian functions of 0.2 eV half-width. The band structure of our MgO cluster exhibits a gap of about 3.5 eV, i. e. too small compared to the experimental

TABLE 2. Calculated properties of CO chemisorbed on free and MgO-supported Ni_4 and Cu_4 clusters (GGA calculations)

	Ni_4		Cu_4	
	supported	free	supported	free
z_e(M-C), Å	1.12	1.09	1.21	0.99
ω_e(M-C), cm^{-1}	205	276	228	275
ω_e(C-O), cm^{-1}	1625	1521	1585	1552
D_e, eV	1.12	1.22	-0.19	0.30

value, 7.8 eV. However, this result is not due to the cluster approximation, but rather is a general characteristic of many exchange-correlation functional approximations; one way to obtain a better representation of energy gaps in ionic substrates is to employ a self-interaction corrected functional. When a Ni_4 cluster is deposited on the surface the Fermi level ε_F moves to the edge of the Ni 3d band. The Ni cluster remains spin-polarised although the interaction with the substrate reduces the magnetisation from 4 unpaired electrons in unsupported Ni_4 to 1.75 unpaired electrons in the supported cluster. The formation of some covalent bonding between the O 2p band and the Ni 4s and 3d levels can be seen from the small tails of the Ni DOS into the O 2p band, Figure 3. The electronic structure of Cu_4 on MgO is similar, but here the 3d levels are completely filled and higher in energy than in Ni_4. The bonding involves primarily the 4sp Cu orbitals and not the 3d orbitals. For both Ni_4 and Cu_4, the 2s and 2p core levels are shifted to higher binding energies by 3-4 eV, a fact which is usually interpreted as an indication for some charge transfer from the cluster to the surface. However, given the existence of an electric field associated with the largely ionic MgO surface, some care is necessary in interpreting these shifts. They are not exclusively caused by charge transfer, but also by the interaction with the field.

4.2 CO ON Ni_4/MgO AND Cu_4/MgO

To characterise the electronic modifications induced by the oxide surface on the supported cluster, we have mimicked the strategy of probe molecules, often employed in experiments. To this end, we have adsorbed a CO molecule at the centre of the metal cluster. We shall compare the corresponding chemisorption properties with those of free Ni_4 and Cu_4 clusters, Table 2. We first determined the potential energy curve for the interaction of CO with the free and supported clusters, Figure 4. All energy curves exhibit a barrier followed by a deep minimum, indicative of an avoided crossing and of a configuration change as the CO molecule approaches the clusters. For instance, the magnetisation of Ni_4 disappears for short Ni-CO distances.

The physical reason for this quenching of the magnetisation and for the appearance of a barrier is a repulsive interaction at longer distances, occurring

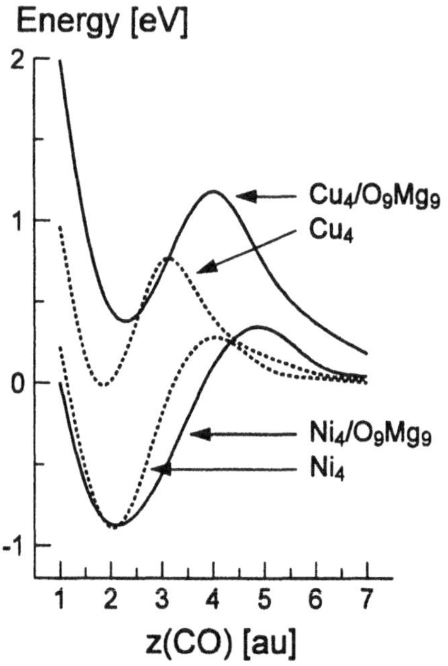

Figure 4. Potential energy curves for the interaction of a CO molecule with free and supported Ni_4 and Cu_4 clusters.

between the CO 5σ lone pair and the diffuse 4s-derived cluster levels [41]. The depth of the attractive well is much larger on Ni than on Cu; this reflects the different electronic structure of the two metals and the possibility of a configuration change in Ni (from $3d^9 4s^1$ to $3d^{10}$) to reduce the initial repulsion. As a result, CO is only weakly bound to free Cu_4, by about 0.3 eV, while it is bound by 1.2 eV to an unsupported Ni_4 cluster, Table 2.

On the supported clusters the height of the barrier is higher than for the free clusters, Figure 4, in particular for Cu; also, the position of the barrier moves somewhat. The shape of the attractive well changes as shown by the decrease of the M-C vibrational frequency, by 71 cm^{-1} for Ni and by 47 cm^{-1} for Cu. Similarly, the bond strength is reduced, by 0.1 eV for Ni and by 0.5 eV for Cu. The fact that the height of the barrier increases and that its position moves away from the surface suggests that the incoming CO molecule encounters a larger repulsion on the supported cluster. This is because the 4sp orbitals of the free metal cluster polarise away from the oxide surface, toward the vacuum, in order to reduce the repulsion with the substrate. On the other hand, a polarisation away from the incoming probe molecule is prevented through the presence of the rather rigid and unpolarisable MgO surface.

The reduction of the adsorption energy and the decrease of the vibrational frequency ω_e(M-C) suggest a weakening of the metal-CO bond in the supported cluster. To identify the origin of this weakening we have computed the vibrational frequency of the C-O stretching mode, Table 2. On free Ni_4 the CO vibrational frequency is 1521 cm^{-1}, i. e. 622 cm^{-1} lower than calculated for gas-phase CO; this is consistent with a substantial backdonation of charge from the Ni 3d orbitals to the CO $2\pi^*$ MO. The corresponding C-O vibrational frequency ω_e(C-O) on Cu_4 is 1552 cm^{-1}. On the supported clusters the red shift is reduced, by 104 cm^{-1} on Ni and by 33 cm^{-1} on Cu. Since the surface-CO distance used for the calculations is the same, 1.06 Å, the higher CO frequency in the supported clusters reflects a reduced back bonding. This is fully consistent with the picture of a weak oxidation of the metal atoms in contact with the surface oxygen atoms. The shifts, however, are not very large suggesting that there is only a moderate charge transfer from the cluster to the oxide substrate.

An important conclusion is that on MgO the metal-metal bond prevails over the metal-oxygen bond, in particular for Ni where the intra-cluster metallic bonds are stronger. As a consequence, both Ni_4 and Cu_4 moieties deposited on MgO keep their cluster identity. The metal-metal bonds are elongated upon adsorption, but the interaction with the substrate is not sufficient to stabilise a structure where the metal atoms are arranged directly on top of the substrate oxygen atoms. On an extended surface this latter structure would correspond to an pseudomorphic growth of the metal film. Thus, the calculations suggest that on MgO deposited Cu and, in particular, Ni particles should preferentially grow in a three-dimensional fashion.

5. Ni Clusters Deposited on Al_2O_3

A number of studies on the electronic structure of the metal/Al_2O_3 interaction have been reported based on both band structure calculations and cluster models and on various computational methodologies [42–52]. The somewhat confusing picture emerging from these different approaches is not so much a question of the level of theory used, but rather is connected to the assumptions introduced in the model chosen to represent the real situation. In fact, in the absence of reliable structural information, the conclusions about the nature of the metal/Al_2O_3 interaction vary dramatically depending on the assumptions made on the type of termination and oxidation state of the Al_2O_3 surface. Not surprisingly, an aluminium-terminated Al_2O_3 surface behaves differently from an oxygen-terminated one. Moreover, the chemical nature of the surface oxygen atoms, fully reduced, O^{2-}, partially reduced, O^-, or neutral, O, will result in a completely different interaction with the adsorbed metal.

If the theoretical situation looks rather complicated, the experimental side [53,54] is not more transparent since the deposition of different metals on the

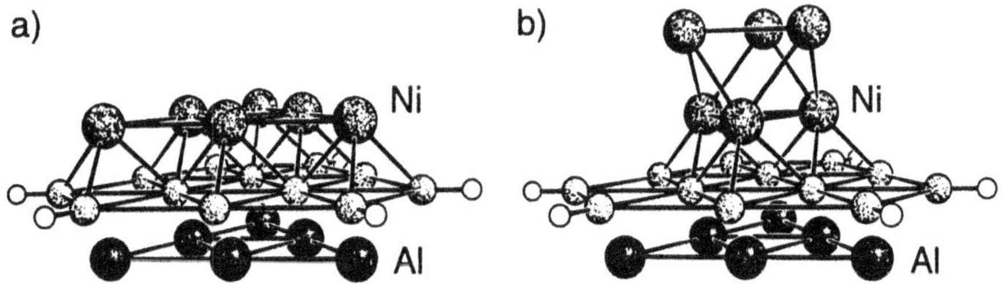

Figure 5. Structure of Ni_6 clusters supported on a $Al_6O_{12}H_6$ cluster model of an oxygen terminated alumina surface: (a) one-layer $Ni_6(6,0)$; (b) two-layer $Ni_6(3,3)$.

same substrate under similar conditions may result in completely different interfacial bonds. Of particular importance for the present work is the study of Chen et al. [55,56] on the $Ni/Al_2O_3/Al(111)$ interface. These authors found that the Ni atoms interact with the surface Al-O units of Al_2O_3 and that the d electron density on the Ni particles is decreased. This mechanism results in a reduced backdonation from the supported Ni particle to an adsorbed CO probe molecule.

5.1 Ni SPECIES ON Al_2O_3

The calculations on Ni clusters supported on Al_2O_3 have been performed at the LDA level [19], differently from the studies of metal deposition on MgO previously described. To represent the Al_2O_3 surface we have used a simplified model where 12 oxygen and 6 aluminium atoms are forming a two-layer cluster with overall C_{3v} symmetry, as shown in Figure 5. The Al-O distance was fixed at 1.89 Å so that the oxygen centres form triangles at the surface with edges of 2.52 Å. This may be considered as a model of an oxygen-covered Al(111) surface rather than as one of a basal (0001) plane of Al_2O_3. At the edge, the substrate model cluster was saturated with hydrogen atoms; thus the final substrate model is $Al_6O_{12}H_6$, see Figure 5. In this way, both the stoichiometric ratio, 2 Al : 3 O, and the number of valence electrons in the cluster are as close as possible to the real situation. We have deposited one- and two-layer Ni_6 clusters on this substrate. Except for the cluster "embedding", no a priori assumption was made concerning the oxidation state of the surface oxygen atoms; the geometry of the supported clusters has been partially optimised. Carbon monoxide was used as a probe molecule to monitor the modifications induced in the Ni_6 clusters by the interaction with the support.

A single Ni atom adsorbed on $Al_6O_{12}H_6$ gives rise to a significant charge transfer from the metal to the substrate, indicating a substantial oxidising power of the oxide surface at variance with the MgO(001) surface previously described. Of course, the character of the bonding may change considerably when a small Ni cluster interacts with the surface; in fact, in this case the formation of strong metal-oxygen bonds occurs at the expense of the metal-metal bonds within the metal cluster. We consider first the interaction of a $Ni_6(6,0)$ one-layer cluster of triangular shape, Figure 5 (the numbers in parenthesis indicate the Ni atoms in each layer). The optimal Ni-Ni distance for the free cluster, 2.22 Å, is much shorter than the Ni bulk distance, 2.49 Å, but for the adsorption on alumina we assumed a pseudomorphic growth of the metal film with the Ni atoms occupying the three-fold hollow positions, Figure 5; this corresponds to an intra-layer Ni-Ni distance of 2.52 Å, very close to the bulk value of Ni metal. The optimal Ni-O distances between Ni_6 and the oxide surface are about 2 Å.

The $Ni_6(6,0)$ cluster is strongly bound, by 17.7 eV, to the substrate. Of course, this is overestimated due to the LDA form of the functional. The calculated interaction energy corresponds to a binding energy of about 3 eV per Ni atom, much lower than that for a single adsorbed Ni atom, about 8 eV, and reflects the fact the Ni atoms are involved not only in the bonding with the oxygen centres, but also in the bonding among each other. Indeed, the energy required to atomise the adsorbed Ni_6 cluster, 31.1 eV, is 13.4 eV higher than the $Ni_6/Al_6O_{12}H_6$ dissociation energy, indicating the existence of considerable metal-metal bonds also in the adsorbed phase. Analysis of the Mulliken charges and of the core level shifts of the Ni 2p levels suggests the occurrence of a charge transfer from the cluster to the oxygen atoms, similarly as for a single Ni atom. The consequences of this moderate charge transfer on the CO adsorption properties will be discussed below.

Next, we consider an adsorbed two-layer Ni cluster, $Ni_6(3,3)$ of D_{3d} symmetry, Figure 5. We assume the same Ni-O contact distance as found for $Ni_6(6,0)$; the inter-layer Ni-Ni distance has been optimised with the constraint that the intra-layer Ni-Ni distances are fixed at 2.52 Å. $Ni_6(3,3)$ is bound to the $Al_6O_{12}H_6$ substrate cluster by 12.5 eV, or 4.2 eV per Ni-O bond. The estimated charge transfer from the first Ni layer to the oxygen centres is higher than in the case of $Ni_6(6,0)$, as shown by both the population analysis and the Ni 2p core level shift. On the other hand, we found a moderate perturbation of the Ni atoms in the second layer of the cluster; essentially, the electronic environment of the Ni atoms in the second layer is not too different in free and supported $Ni_6(3,3)$ moieties.

5.2 CO ON Ni/Al_2O_3

The vibrational modes of CO chemisorbed on the model catalyst $Ni/Al_2O_3/$ Al(111) [56] may be used to characterise the electronic modifications induced by the support on deposited Ni clusters and to probe the nature of the metal-oxide interface. As a reference, we have calculated the properties of CO chemisorbed on the three-fold hollow sites of free and supported Ni_6 clusters. CO is strongly chemisorbed at both free and supported Ni_6 clusters, Table 3;

TABLE 3 - Calculated properties of CO chemisorbed on free and supported Ni_6 clusters on $Al_6O_{12}H_6$ (LDA calculations)

	$Ni_6(6,0)$ one-layer		$Ni_6(3,3)$ two-layer	
	supported	free	supported	free
z_e(Ni-C), Å	1.28	1.16	1.13	1.16
ω_e(Ni-C), cm^{-1}	391	470	505	428
ω_e(C-O), cm^{-1}	1789	1689	1717	1757
D_e, eV	3.32	3.54	3.28	3.51

however, for the supported one-layer cluster the Ni-CO distance, vibrational frequency, and dissociation energy indicate a weakening of the adsorption bond, see Table 3. The most important property for the present analysis is the shift in the CO vibrational frequency on the supported cluster compared to that of CO on the free cluster. For the case of the one-layer $Ni_6(6,0)$ cluster, which gives rise to a moderate charge transfer to the surface,the CO vibrational frequency is 100 cm^{-1} higher than on free $Ni_6(6,0)$, consistent with the view of a partially oxidised monolayer. The vibrational blue shift is not found when CO is adsorbed on the two-layer $Ni_6(3,3)$ cluster, Table 3. In fact, the ω_e(C-O) shift between the supported and the free $Ni_6(3,3)$ cluster is negative, -40 cm^{-1}, thus providing no evidence of a reduced backdonation for this system. Clearly, a qualitatively different behaviour is observed for one-layer and two-layer Ni clusters.

These calculations allow us to draw some general conclusions concerning the nature of the interface between a small Ni particle and an oxidised Al surface. Starting from the assumption that the interaction occurs at an oxygen terminated surface, we found that the Ni cluster adheres quite strongly to the substrate. We found this bond to be 3-4 eV per adsorbed Ni atom, although this value is overestimated in LDA. This strong bond is connected to a partial oxidation of the Ni atoms which are in direct contact with the oxygen centres.

The temperature-programmed desorption (TPD) spectrum of CO adsorbed on the $Ni/Al_2O_3/Al(111)$ surface shows a peak at 375 K which has been attributed to the desorption of bridge bonded CO [56]; the corresponding peak on a single crystal Ni(111) surface is found at 410 K, indicating that a weaker interaction occurs between CO and the Ni particle when this is supported on the oxide substrate. Our model calculations, which show a moderate decrease of the binding energy for the supported compared to the free clusters (Table 3), are consistent with these measurements.

The adsorption of CO on a supported cluster provides a simple way to probe the reduced electron density at the metal overlayer. We find that when CO is adsorbed on a one-layer Ni cluster supported on the $Al_6O_{12}H_6$ substrate there is a positive shift, of 100 cm^{-1}, of the CO vibrational frequency. This is in qualitative agreement with the measured EEL spectra of Chen et al. [56] for CO adsorbed on $Ni/Al_2O_3/Al(111)$. The blue shift in the CO vibrational frequency

ω_e(C-O) is not found when a two-layer Ni cluster is considered. In this case, in fact, the calculated vibrational frequency shift is smaller and negative. This indicates that the partial oxidation at the interface is quite localised and probably restricted to the first layer of metal atoms.

6. Conclusions

The theoretical, quantum-mechanical, study of the metal/oxide interface is in its infancy. So far the few studies reported are dealing with only a limited number oxide materials, taken in their idealized structures, with no defects, hydroxyl groups, or other surface irregularities which, on the other hand, are known to play a very important role in the kinetics of metal deposition processes. The importance of this area of research from both a scientific and a technological point of view will stimulate additional efforts to investigate the mechanism of the metal/oxide interaction at the atomistic level. In this respect, a key issue in the near future will be the design of "realistic" models of oxide surfaces since both geometric and electronic structural experimental information will be necessary in order to calibrate the computational models and to verify the degree of reliability of their answers.

Acknowledgments

We would like to thank I. Yudanov, V.A. Nasluzov and K.M. Neyman for various contributions to the present work. This work has been supported by the Deutsche Forschungsgemeinschaft, by the Bayerischer Forschungsverbund Katalyse (FORKAT), by the European Community through the co-operation agreement INTAS-93-1876-ext and by the Fonds der Chemischen Industrie. G.P. thanks the "Vigoni Program" from CRUI and DAAD for supporting his visits at the Technical University of Munich.

References

1. He, J.-W. and Møller, P.J. (1987) *Surf. Sci.* **180**, 411.
2. He, J.-W. and Møller, P.J. (1986) *Surf. Sci.* **178**, 934.
3. He, J.-W. and Møller, P.J. (1986) *Chem. Phys. Lett.* **129**, 13.
4. Alstrup I. and Møller, P.J. (1988) *Appl. Surf. Sci.* **33/34**, 143.
5. Zhou, J.B., Lu, H.C., Gustafsson, T. and Garfunkel, E. (1993) *Surf. Sci.* **293**, L887.
6. Wu, M.-C., Oh, W.S. and Goodman, D.W. (1995) *Surf. Sci.* **330**, 61.
7. Didier, F. and Jupille, J. (1994) *Surf. Sci.* **307-309**, 587.
8. Meunier, M. and Henry, C.R. (1994) *Surf. Sci.* **307-309**, 514.
9. Henry, C.R., Meunier, M. and Morel, S. (1993) *J. Crystal Growth* **129**, 416.
10. Kizuka T., Kachi, T. and Tanaka, N. (1993) *Z. Phys. D* **26**, S58.
11. Noguera, C. and Bordier, G. (1994) *J. Phys. III France* **4**, 1851.
12. Schönberger, U., Andersen, O.K. and Methfessel, M. (1992) *Acta Metall. Mater.* **40**, S1.

13. Li, C., Freeman, A.J. (1991) *Phys. Rev. B* **43**, 780.
14. Li, C., Wu, R., Freeman, A.J. and Fu, C.L. (1993) *Phys. Rev. B* **48**, 8317.
15. Smith, J.R., Hong, T. and Srolovitz, D.J. (1994) *Phys. Rev. Lett.* **72**, 25.
16. Bacalis, N.C. and Kunz, A.B. (1985) *Phys. Rev. B* **32**, 4857.
17. Kunz, A.B. (1985) *Phylosoph. Magazine B* **51**, 209.
18. Li, Y., Langreth, D.C. and Pederson M.R. (1995) *Phys. Rev. B* **52**, 6067.
19. Pacchioni, G. and Rösch, N. (1994) *Surf. Sci.* **306**, 208.
20. Pacchioni, G. and Rösch, N. (1996) *J. Chem. Phys.* **104**, 7329.
21. Ferrari, A.M. and Pacchioni, G. (1996) *J. Phys. Chem.* **100**, 9032.
22. Gautier, M. and Duraud, J.P. (1994) *J. Phys. III France* **4**, 1779.
23. Yudanov, I., Pacchioni, G., Neyman, K.M. and Rösch, N., submitted.
24. Dunlap, B.I. and Rösch, N. (1990) *Adv. Quantum Chem.* **21**, 317.
25. Vosko, S.H., Wilk, L. and Nusair, M. (1980) *Can. J. Phys.* **58**, 1200.
26. Rösch, N. (1992) in: *Cluster Models for Surface and Bulk Phenomena*, Pacchioni, G., Bagus, P.S. and Parmigiani, F. (eds.), NATO Advanced Study Institute, series B; vol. 283, Plenum Press, New York, p. 251.
27. Rösch, N., Neyman, K.M., Birkenheuer, U., Krüger, S. and Nasluzov, V.A. (1996) *Kinetika i Katalis*, in print.
28. Ziegler, T. (1991) *Chem. Rev.* **91**, 651.
29. Becke, A.D. (1988) *Phys. Rev. A* **38**, 3098.
30. Perdew, J.P. and Wang, Y. (1992) *Phys. Rev. B* **45**, 13244.
31. Lee, C., Yang, W. and Parr, R.G. (1988) *Phys. Rev. B* **37**, 785.
32. Perdew, J.P. (1986) *Phys. Rev. B* **33**, 8822.
33. Johnson, B.G., Gill, P.M.W. and Pople, J.A. (1993) *J. Chem. Phys.* **98**, 5612.
34. Häberlen, O.D. and Rösch, N. (1992) *Chem. Phys. Lett.* **199**, 491.
35. Rösch, N., Krüger, S., Mayer, M. and Nasluzov, V.A. (1996) in: *Recent Development and Applications of Modern Density Functional Theory*, Seminario, J.M. (ed.), Elsevier, Amsterdam, in print.
36. Boys, S.F. and Bernardi, F. (1970) *Mol. Phys.* **19**, 553.
37. Pacchioni, G., Bagus, P.S., Nelin, C.J. and Philpott, M.R. (1990) *Int. J. Quantum Chem.* **38**, 675.
38. Pyykkö, P. *Chem. Rev.* **1988**, *88*, 563.
39. Chung, S.-C., Krüger, S., Pacchioni, G. and Rösch, N. (1995) *J. Chem. Phys.* **102**, 3695.
40. Birkenheuer, U., Rösch, N. and Boettger, J.C. (1994) *J. Chem. Phys.* **100**, 6826.
41. Pacchioni, G. and Rösch, N. (1995) *Acc. Chem. Res.* **28**, 390.
42. Johnson, K.H. and Pepper, S.V. (1982) *J. Appl. Phys.* **53**, 6634.
43. Zdetis, A. and Kunz, A.B. (1985) *Phys. Rev. B* **32**, 6358.
44. Raatz, F. and Salahub, D.R. (1985) *Surf. Sci.* **156**, 982.
45. Anderson, A.B., Mehandru, S.P. and Smialek, J. (1985) *J. Electrochem. Soc.* **132**, 1695.
46. Anderson, A.B., Ravimohan, C. and Mehandru, S.P. (1987) *Surf. Sci.* **183**, 438.
47. Nath, K. and Anderson, A.B. (1989) *Phys. Rev. B* **39**, 1013.
48. Ohuchi, F.S., French, R.H. and Kasowski, R.V. (1987) *J. Appl. Phys.* **62**, 2286.
49. Kasowski, R.V., Ohuchi, F.S. and French, R. (1988) *Physica B* **150**, 44.
50. Kohyama, M., Kose, S., Kinoshita, M. and Yamamoto, R. (1992) *J. Phys. Chem. Solids* **53**, 345.
51. Alemany, P., Boorse, R.S., Burlitch, J.M. and Hoffmann, R. (1993) *J. Phys. Chem.* **97**, 8464.
52. Ward, T.R., Alemany, P. and Hoffmann, R. (1993) *J. Phys. Chem.* **97**, 7691.
53. Sarapatka, T.J. (1993) *Chem. Phys. Lett.* **212**, 37.
54. Zhong, Q. and Ohuchi, F.S. (1990) *J. Vac. Sci. Technol. A* **8**, 2107.
55. Chen, J.G., Crowell, J.E. and Yates, J.T. (1987) *Surf. Sci.* **185**, 373.
56. Chen, J.G., Crowell, J.E. and Yates, J.T. (1987) *Surf. Sci.* **187**, 243.

THEORY OF HETEROGENEOUS CATALYTIC REACTIVITY USING THE CLUSTER APPROXIMATION*

R.A. VAN SANTEN
Schuit Institute of Catalysis, Laboratory of Inorganic Chemistry and Catalysis,
Eindhoven University of Technology,
PO Box 513, 5600 MB Eindhoven, The Netherlands.

Abstract
Density Functional Theory enables quantitative computational analysis of reaction intermediates. The cluster approach makes application to heterogeneous catalysis possible. Examples from transition metal catalysis and zeolite catalysis will be discussed. In the case of transition metals cluster choices will be analysed to simulate properties of the extended surface. The issue of coverage dependent reactivity and surface reconstruction will be discussed. The concept of adatom basicity in metal catalysis will be analysed. It will be shown that the information obtained on the elementary reaction steps from cluster can be used to predict the overall rate of a catalytic reaction.
In transition metal catalysis for an oscillatory reaction as the CO oxidation reaction catalysed by (100) it will be shown how knowledge on the elementary reaction rate constants of surface reactions can be used to predict the overall rate of the reaction with help of the time dependent Monte Carlo method. In zeolite catalysis the analysis of the reaction mechanism of methanol conversion will be used to show the strength of the cluster approximation for the case of a refractory oxide. The hydroisomerization reaction will be used to relate microscopic molecular information to the overall rate of a zeolite catalysed reaction.

1. Introduction

Heterogeneous catalytic reactions occur on solid surfaces. Hence it may seem that heterogeneous catalytic site models based on clusters cannot properly describe the surface chemistry that is the basis to heterogeneous catalysis. Clearly those aspects of the reaction that intrinsically relate to the presence of the extended lattice or micropore cannot be described by clustermodels but require full consideration of the two-dimensional or three-dimensional geometry. Surface reconstruction and island

---

*This text has also partially been presented at Francqui meeting, February 19-20, Bruxelles, 1996

R.M. Lambert and G. Pacchioni (eds.),
Chemisorption and Reactivity on Supported Clusters and Thin Films, 371-393.

formation belong to this category. In zeolites the shape and size of the micropore cavities may prevent particular reactionpaths when transition states are too bulky.
Hence it is important to discriminate between those aspects of the reaction that depend on local properties, that often can be treated using a cluster approach and those aspects of the reaction that are not.
Here we will discuss this for examples from transition metal catalysis and solid acid catalysis by zeolites. In both cases we will use clusters to also analyse the transition states of elementary reaction steps and we will demonstrate how this information can be used to predict the overall rate of a catalytic reaction. For the case of catalysis by a transition metal we will apply time dependent Monte Carlo methods to analyse the occurrence of oscillations in the rate of CO oxidation.
In zeolite catalysis we will simulate the turn over frequency (TOF) of hexane isomerization in Mordenite and ZSM-5. Here it appears essential to include the micropore size effects that determine the difference in the heats of adsorption of the reacting molecules.
Clusters are useful in those cases where the surface-chemical bond is localised and where boundary effects are well understood.
Quantum-chemical research on clusters has become fruitful especially due to advances in the development of Density Functional Theory techniques and the incorporation of gradient correction terms to the exchange-correlation energy. Several computer programs now have efficient procedures to find the geometries that correspond to stationary points on the potential energy surface. Within the harmonic approximation the vibrational normal modes can be found, so that ground and transition state energies as well as entropies can be computed.
This enables the application of Eyring's transition state reaction theory to predict reaction rate constants (1).
We will start with an analysis of cluster size dependence and relaxation effects on Pd clusters (2). A comparison will be made with slab calculations using the Car-Parinello method (3).
Then we will analyse dissociation paths of CH_3OH on Pt and Rh clusters. The aim is to study the role of coadsorbed oxygen atoms to promote dissociative adsorption of methanol. As we will also note later for the case of methanol activation by zeolitic protons, stabilisation of dissociation fragments is found to control the overall reactivity (4).
Differences in reactivity between surfaces are mainly determined by the degree of co-ordination of the surface atoms. Surface reconstruction effects can affect surface chemical reactivity by a change in the surface coordination number of adatoms. This local feature is also accessible to the cluster description.
After the discussion of transition metal catalysis, we will analyse methanol activation using protonic clusters. We will show how different reaction paths for dimethylether formation can be analysed.
Theory here appears to be essential for a proper understanding of carbenium and carbonium ion intermediates (5). The theoretical concepts will be used in the following analysis of hexane hydroisomerization.

2. Clusters in transition metal catalysis

In section 2a we will discuss cluster geometry relaxation of Pd clusters. Section 2b will discuss oxygen promoted methanol dissociation by Pt_{10} en Rh_{10} clusters and section 2c describes the difference in the stability of oxygen atoms adsorbed to clusters simulatory the reconstructed Cu (110) clusters.

2.1. SITE RELAXATION

Figure 1 (2,6) shows for a Pd_6 cluster the changes in geometry when an H atom or O atom becomes adsorbed to it. The calculated results presented here and in the next paragraphs have been obtained using Density Functional Theory including gradient correction terms.

The magnetic Pd_6 clusters become largely deformed by the coadsorbed atoms. The bonds next to the adsorbed atom change from 0.15 to 0.25 Å.

Large bond distance relaxation and adsorption induced reconstruction effects are common for small particles. However the changes tend to become less dramatic when the metal atom become more stabilised by coordination with more neighbours. This we have also found for the interaction of ethylene with very small Pd particles (7). For very large Pd_{18} particles the lattice expansions by coadsorption of H or O has reduced to only 0.07 Å (Fig. 1b), with excellent agreement of the computed adatoms bond energies with that to a Pd(111) surface, indicating that bonding has converged to its surface value. The increases in bond distances next to the adatoms are due to the decrease of the Pd-Pd bondorders due to the involvement of Pd with the additional adsorbing atom. At low surface coverage this leads to a large effective volume of the adsorption site compared to the free site inducing local lattice strain. STM data on C adsorbed to the Ni (100) surface at low surface coverage append to agree with this observation (8). However at high coverage the expansion in metal-metal bond distance next to the adsorbate has not been experimentally observed. Large surface reconstruction effects are usually reported (9) with displacement of surface atom positions. These changes must be the consequence of the local strain effects that tend to interfere at high surface coverage. Using the Car-Parinello method we have analysed the relaxation effect on a 3 layer Pt slab with 25% coverage of N (10). As shown in figure 2 very small changes are found in the surface layer, but a large expansion of the bond distances perpendicular to the surface. The expansion of the Pt-Pt bonds in the surface is now prevented because the displacements around different N atoms counteract. The bond weakening effect now only has consequences between the Pt-layers resulting in lattice expansions in close agreement with experiment (11).

One concludes that clusters that simulate surfaces can be used best with fixed bond distances and should be chosen such that boundary effects are minimised.

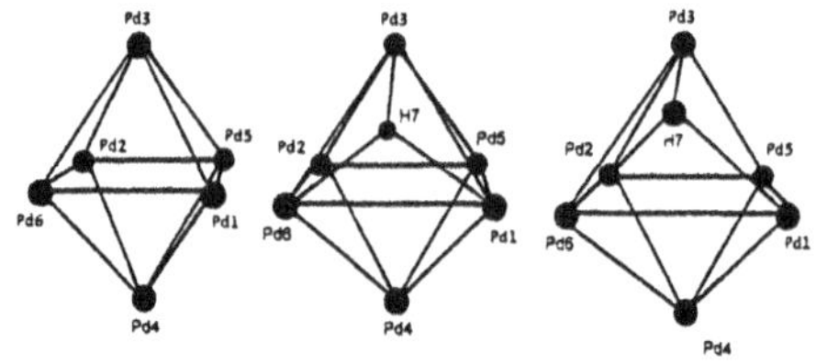

Pd6_Triplet		Pd6_Singlet			Pd6H_Doub		Pd6O_Trio	
1-3	2.756	1-3	2.697		1-3	2.918	1-3	3.131
1-6	2.631	1-6	2.698		1-6	2.898	1-6	3.074
3-6	2.756	3-6	2.697		3-6	2.918	3-6	2.918
2-4	2.756	2-4	2.697		2-4	2.667	2-4	2.651
2-6	2.632	2-6	2.697		2-6	2.756	2-6	2.757
4-5	2.757	4-5	2.697		4-5	2.661	4-5	2.628
					1-7	1.768	1-7	2.068
				Pd–X	3-7	1.768	3-7	2.068
					6-7	1.767	6-7	2.068

☐ Primary effects (local)

Figure 1a. Pd_6, Pd_6H, Pd_6O cluster geometries (2).

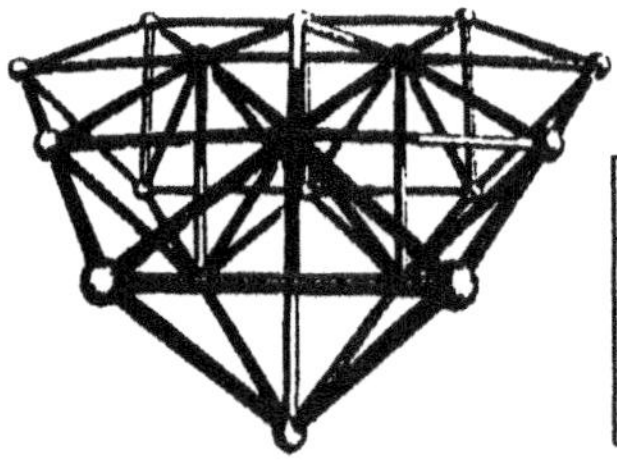

	Pd18 fixed	Pd18 Opt
Pd–H	1.78	1.75
Pd –Pd		
local-local	2.75	2.82
local-next neighbor	2.75	2.72

- ***Three local Pd surface atoms optimized with H.***
- ***Structural reconstruction shifts these atoms 0.15 ang. above the surface.***

Figure 1b. $Pd_{18}H$ cluster geometry (2).

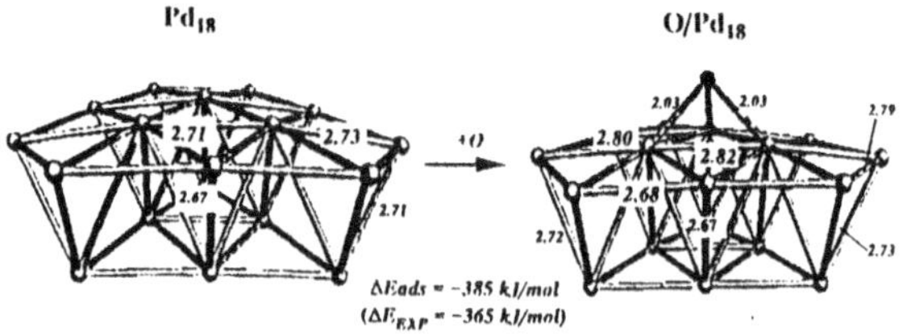

	Pd_{18}	$Pd_{18}O$
Pd*–Pd*	2.71	2.82
Pd–Pd* (neighbor)	2.73	2.80
Pd–Pd (next nearest)	2.70	2.69

Pd(ads) pulled up by ~0.09 Ang. from surface
Pd–Pd Elongated by 0.1 Ang.

Figure 1c. $Pd_{18}O$ cluster geometry (2).

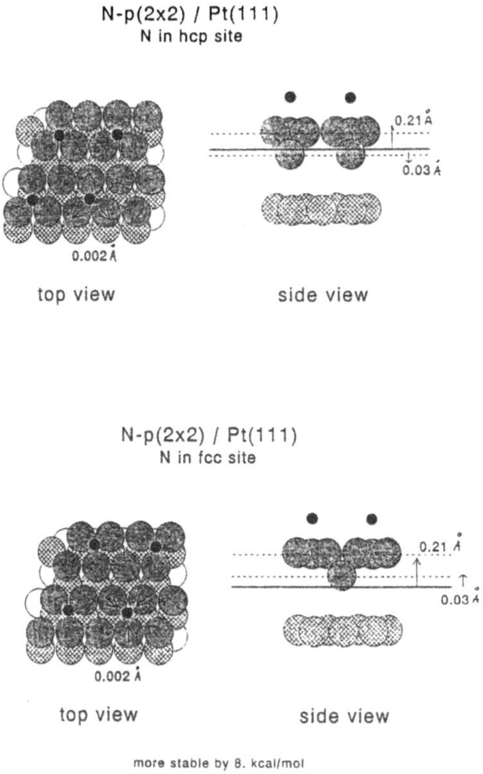

Figure 2. Car-Parinello predictions of the geometries of N overlayers adsorbed to the two different 3 coordinated positions of the (111) surfaces of Pt

2.2. DISSOCIATIVE ADSORPTION OF METHANOL. REACTION WITH COADSORBED OXYGEN ATOMS

Activation of OH, NH and CH on non-reactive metals is possible by reaction with coadsorbed oxygen.

$$RH + O_{ads} \rightarrow R_{ads} + Oh_{ads}$$

Such reactions have been extensively studied by Madix (12), Friend (13) and Roberts (14). On silver surfaces there appears to be a relation between the acidity of the substrate RH bond and surface reactivity and the reaction has been described as an Brønsted acid - Lewis base reaction. Interestingly this concept cannot describe the difference in reactivity of oxygen chemisorbed to the Rh (111) surface as compared to the reactivity of O adsorbed to the Pt (111) surface. Based on a comparison of the ionisation potentials or workfunctions one expects O adsorbed to Rh to be more negatively charged and hence the Lewis basicity to be larger on Rh compared to Pt.

The thermodynamics of the CH_3OH oxidation reaction has been analysed on M_{10} clusters by Bates e.a. (4) using the Amsterdam Density Functional code. For hydrogen adsorbed to Rh_{10} and Pt_{10} cluster (figure 3) one notes the expected result of a stronger bond of hydrogen to the Rh cluster than to the Pt clusters. This is a generally found result for adsorbates and is due to the reduction in the number of d-valence electrons on Rh compared to Pt (2).
Table 1 summarises the overall energy changes for the two reactions:

$$CH_3OH + M_{10} + M_{10}O \xrightarrow{E_2} CH_3OM_{10} + M_{10}OH$$
$$CH_3OH + M_{10} + M_{10} \xrightarrow{E_3} CH_3OM_{10} + M_{10}H$$

TABLE 1 Comparison of energy change (kcal/mol) in the two reactions: ADF results obtained with double zeta (dz) and triple zeta (tz) basis sets are compared.

	Rh		Pt	
	dz	tz	dz	tz
E_2	29.7	32.4	32.0	28.8
E_3	12.2	13.2	50.3	45.9

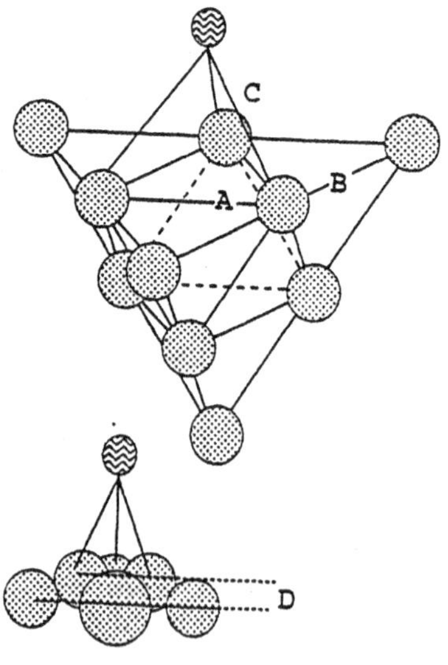

Figure 3. The M_{10} cluster used. Predicted geometries and interaction energies for H on Pt_{10} and Rh_{10} (kcal/mol). ADF results obtained with double zeta and triple zeta basis se's are compared.

One notes that according to the calculations both of the reaction are endothermic. Whereas on Rh dissociative adsorption of methanol to form adsorbed H and CH_3OH is significantly less endothermic than on Pt, the overall energy change for the reaction with oxygen shows a much smaller difference. Whereas the reaction with coadsorbed oxygen is more favourable on Pt, it is less favourable on Rh. The results can be understood when one realises that upon reaction two fragments are formed, hydrogen and methoxy. The reactivity of adsorbed oxygen with hydrogen is much less on Rh than on Pt because of the enhanced RhO bond energy compared to that with Pt (see table 2). In the overall methanol dissociation reaction this is counteracted by the enhanced CH_3O/Rh_{10} bond compared to the more weaker CH_3O/Pt_{10} bond. Clearly on the reactive transition metals the nature of the oxygen-adsorbate interaction has changed from the simple Lewis base picture to a covalent bonding scheme more in agreement with Bond Order Conservation rules (2).

TABLE 2 Energy changes (kcal/mol) of the reaction: $M_{10}O + H \rightarrow M_{10}OH$.

	Rh		Pt	
	dz	tz	dz	tz
E_1	-24	-29	-46	-55

2.3. DIFFERENCES IN REACTIVITY OF OXYGEN ON THE RECONSTRUCTED CU (110) SURFACE

The formation of surface overlayers, by lateral interaction between adsorbates or surface reconstruction, may change the reactivity of a surface dramatically (15). For adsorbed oxygen, Roberts (16) has proposed a difference between the reactivity of isolated oxygen atom and oxygen atoms that are part of an oxygen overlayer structure. Isolated oxygen atoms, with a short residence time and to be considered transient species, adsorbed to a Cu (111) surface have been found to be highly reactive and to promote dissociative adsorption of ammonia. However, oxygen atoms that are part of the oxidic overlayer are found to be non-reactive. A related observation has been made by Iwasawa e.a. (17). They studied the reactivity of O adsorbed to Cu (110) for the reaction with CO. They found a large reactivity for the oxygen as exposed surface at 100 K, whereas the reconstructed Cu (110) - (2x1) O surface has a very low reactivity. The reason for this decrease in reactivity relates to the large stabilisation of an oxygen atom in the reconstructed layer compared to the isolated atom. This can be studied using clusters that simulate the different local environments of the oxygen atoms in the two situations. Table (3) and figures (4) show some representative results (18) obtained with the ADF program. As figures 4 illustrate, oxygen coordination number to Cu is four on the reconstructed surface, but only two in the non-reconstructed surface. The atom in the oxide layer is considerably stabilised compared to the non-reconstructed surface.

TABLE 3 Computed heats of adsorption (kJ/mole) of O to Cu(111) cluster and reconstructed Cu(111) Cu2O2 cluster

Cu (6,6) + O	617.30
Cu (6,6) Cu_2O_2 + O	740.80

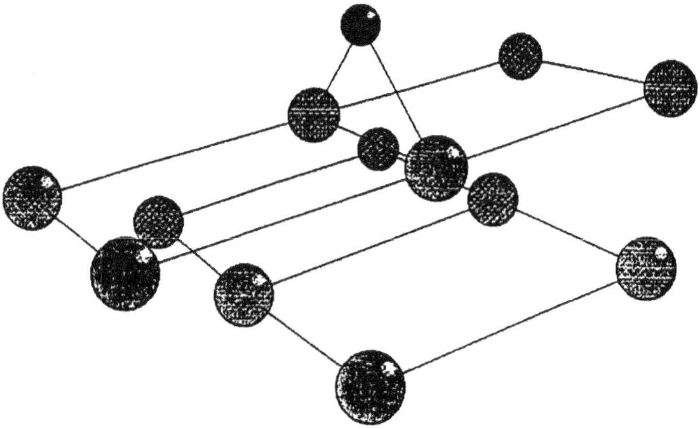

Figure 4a. Oxygen atom adsorbed to non-reconstructed Cu(110) surface cluster..

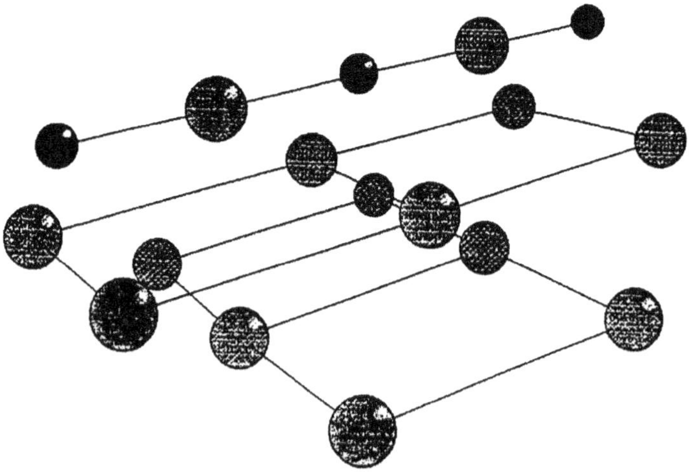

Figure 4b. Oxygen atom as part of an CuO chain on reconstructed Cu(110).

3. Surface reaction rates

In section 3a the computation of the elementary reaction rate constants using transition state reaction rate theory for CH activation are discussed. In section 3b the time-dependent Monte Carlo method is used to compute overall oscillatory behaviour on the basis of elementary surface reaction rate constants.

3.1. THE ACTIVATION ENERGY AND RATE CONSTANT

From a comparison (19) of chemisorption to many different clusters we have concluded that spherical cubo-octahedral clusters of thirteen atoms reproduce the reactivity of (111) surfaces reasonably well. The main advantage of these cluster is the absence of boundary effects, since all exposed atoms are boundary atoms. In figure 5 such a cluster is shown with the predicted transition state of CH_4 to produce adsorbed H and CH_3. Table 4 shows results for Ni as well as Co. Figure 6 illustrates the relation between trends in activation energies and reaction energy differences. Table 4 shows computed pre exponents as well as activation energies for the rate of dissociative adsorption and associative dissociation (20). The pre exponents to the reaction rate constants have been obtained applying Eyring's reaction rate theory expressions (1). They can be computed from the partition functions of ground- and transition state that depend on the computed vibrational frequencies. The pre-exponents contain information on the mobility in the transition state. In case there is no change in entropy between ground and transition state the pre-exponent is equal to ~ 1013 sec-1. This is typically found for dissociative or associative surface reactions. The computed

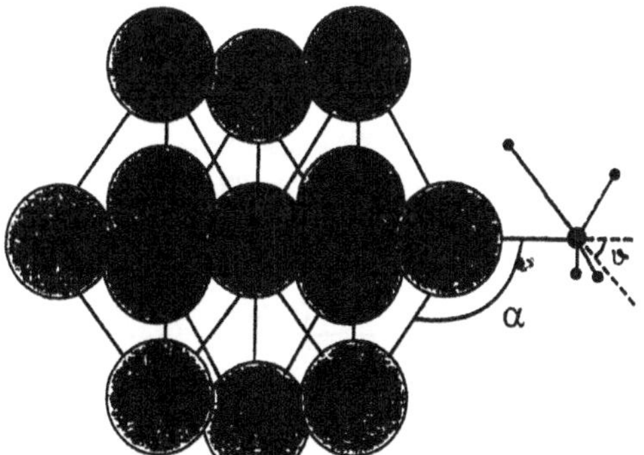

Figure 5. Transition state of CH_4 dissociation on a Ni_{13} cluster (20).

activation for dissociative adsorption is within 10 kJ/mole compared to experiment (22). The decrease in activation energy on Co follows from the larger bond energies of H en CH_3 to Co. According to the Polanyi-Brønsted relation, the change in the activation energy is approximately proportional to the change in reaction energy. This is illustrated in figure 5.

One of the most important reasons for the close correspondence between computed and experimental activation energies for dissociation is the proper modelling of the ensemble of surface atoms required for adsorption of the fragments after dissociation. Due to the high activation energy the contribution of tunnelling to the reaction rate is small. We have estimated this to be of the order of 2 (20). For a more extensive discussion of transition states of small molecules we refer to (2, 21).

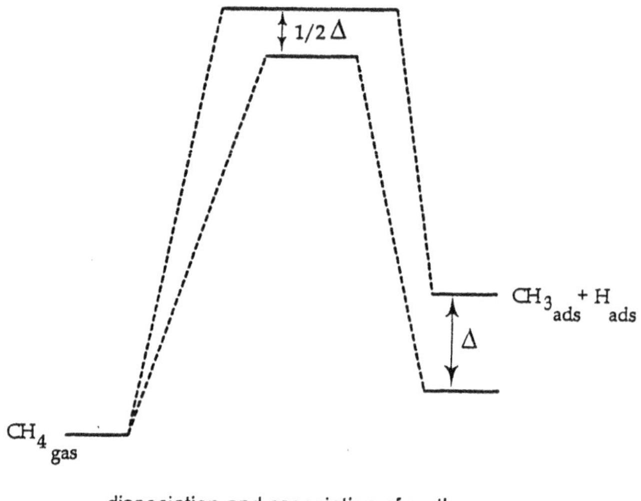

Figure 6. Brønsted-Polanyi reaction energy plot for dissociative adsorption of CH_4 to Ni and Co.

TABLE 4. Arrhenius rate parameters for computed rate of CH activation on M13 clusters

	Associatieve Desorption	
	pre-exponent (m^2g^{-1})	E_{act} (kJ/mole)
Co	2.53×10^{-5}	95
Ni	2.70×10^{-5}	86

	Dissociatieve Desorption	
	pre-exponent (m^2g^{-1})	E_{act} (kJ/mole)
Co	7.24×10^{3}	92
Ni	8.53×10^{3}	104

3.2. THE OVERALL REACTION RATE OF CO OXIDATION ON A PT(100) SURFACE (12)

Based on the assumption of ideal mixing of adsorbed surface species and concentration independence of the reaction rate constants surface kinetic equations can be formulated and readily solved (1). Agreement with experiment can be made by fitting rate parameters such that predicted and measured kinetics agrees. First principle prediction of these rate parameters requires an understanding how these "intrinsic" parameters, relate to the "apparent" rate constants and interaction energies of the actual adsorbates on the surface. Statistical methods as Monte Carlo techniques can be used to predict the overall kinetics of a reaction taking full account of non-ideal surface adsorbate mixing effects, that for instance are the consequence of adsorbate island formation. Here we will illustrate the results of the use of a time-dependent Monte-Carlo (24) approach to study the temperature dependence of the oxidation of carbonmonoxide catalysed by the Pt(100) surface in a pressure require where rate oscillations are observed (25). The oscillations are due to a surface phase transformation from the reconstructed low reactivity Pt (100)-hex phase to the reactive non-reconstructed bulk terminated Pt(100) phase.
The transformation between the two phases is modelled by a transition between a cubic and hexagonal network that occurs when 5 CO molecules become adsorbed next to each other on an ensemble of Pt atoms in the hex configuration. Dissociative adsorption of oxygen only occurs on the non-reconstructed (100) phase. When a surface ensemble of 2 Pt sites is free of adsorbates the (100) transfers back to the (100)-hex configuration.

Table 5 gives an overview of the reaction rate parameters used. Pre exponents and activation energies are typical for isolated surface reaction elementary steps. In the table a comparison is made between the predicted and experimental temperature interval (26) where rate oscillations occur and the computed and experimental oscillation frequencies.

Figure 7 gives the temperature dependence of the surface reaction rate. Initially the reaction is poişoned by adsorbed O_2. When the temperature increases reaction with CO is initiated and steady state reaction rate behaviour is observed. At higher temperatures more O_{ads} is reacted causing transformations to the cubic configuration so that the reaction rate starts to oscillate. At higher temperatures the reaction rate decreases because now CO adsorption becomes rate-limiting.
This approach incorporating phase transformation and incorporating island formation (figure 8 shows the presence of islands at oscillatory conditions) can be readily generalised to other systems, with very different energies and rate constants of elementary reaction steps within the islands

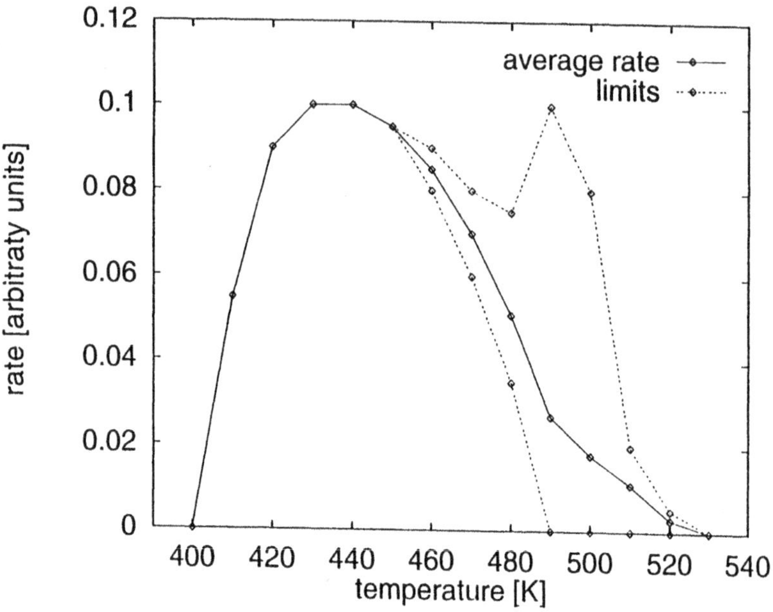

Figure 7. Temperature dependence of CO oxidation on Pt(100) according to Monte Carlo simulations (23). The upper and lower limits indicate the maximum and minimum between which the reaction rate oscillates.

TABLE 5. Pressures are given in Torr, prefactors in sec-1, activation energies in kcal/mol. S0 stands for initial sticking coefficient. For experimental values, see e.g. ref 18

reaction type	p^{sim}	p^{exp}	S_0^{sim}	S_0^{exp}
CO adsorption	$1\cdot10^{-5}$	$1\text{-}3\cdot10^{-5}$	0.0.1	0.1
O_2 adsorption	$4\cdot10^{-4}$	$2\cdot10^{-4}$		0.1
	ν^{sim}	ν^{exp}	E_{act}^{sim}	E_{act}^{exp}
CO desorpion	$1\cdot10^{15}$	$1\text{-}3\cdot10^{15}$	42	28-38
CO_2 production	$2\cdot10^{10}$	10^{10}	20	12-24
1x1 →hexagonal transformation	$1\cdot10^{9}$	-	25	25
Nucleation	0.03	-	0	0
Trapping	0.03	-	0	0

Oscillatory regime	
Simulation	Experiment
r (CO ads.) < r (O_2 ads.)	r (CO ads.) < r (O_2 ads.)
470 K < T < 510 K	460 K < T < 540 K
Period around 200 s	Period of 150 - 200 s

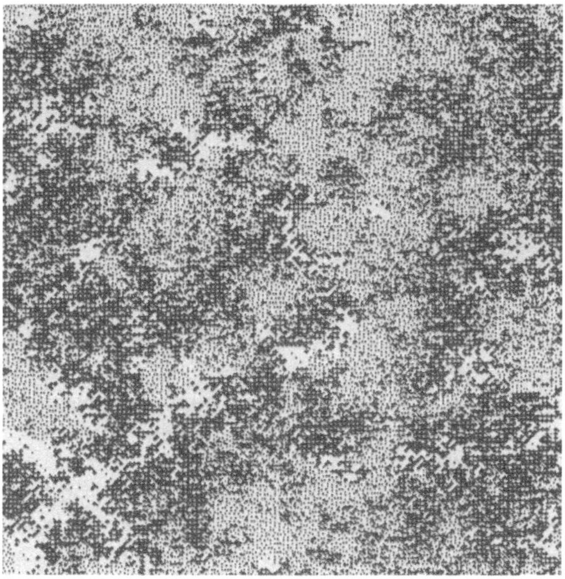

Figure 8. Island formation through oscillatory regime of CO oxidation (23). White areas indicate oxygen adsorbed at the (100) phase, grey areas are CO adsorbed on the (100) phase and black areas indicate (100)-hex phase, with low CO coverage.

4. Acid catalysis by zeolitic protons

In section 4a quantum-chemical cluster calculations on the activation of methanol to produce dimethylether are presented. Emphasis will be on the mechanism of the reaction. It will be shown that a bimolecular associative mechanism is to be preferred (27). In section 3b the concepts presented in section 3a will be extended and the modelling of the overall rate of a reaction is studied. The reaction chosen is the hydroisomerization of hexane catalysed by Pt promoted protonic Mordenite and ZSM-5 (28). The Al/Si ratios have been chosen such that the intrinsic acidities of the protons are the same. It will be shown that differences in performance relate to differences in the heats of adsorption of intermediate hexene, that are strongly dependent on the zeolite micropore diameter.

4.1. PROTON ACTIVATION OF METHANOL USING CLUSTER MODELS

On first sight it appears unlikely that protonation of a molecule by a zeolitic proton can be described well by cutting small cluster from the zeolite. The answer to this question has been extensively investigated (5).

The cluster approximation can be applied to reactions in which the transition state of the reaction or the reaction products are not controlled by the size of the zeolite micro cavities. The usefulness of the cluster approximation deduces from the covalent nature of the chemical bonds in a zeolite. Ninety percent of the bonding can be ascribed to covalent interactions, ten percent to electrostatic interactions.

The proton-oxygen bond in the ground state is highly covalent. Due to the covalent nature of the bonds proton addition to the lattice or protonation of a substrate is accompanied bond relaxation effects. This bond relaxation can be accommodated by the zeolite lattice by relaxation of the Si-O-Si bondangles. The bending energy of these angles is only of the order of a few kJ/mole varying the angle by 10 degrees. Finally it has been found (29) that three T-atom (three tetrahedral containing) clusters terminated by hydrogen bonds have a deprotonation energy close to that of a very large OH terminated cluster.

Lacking in small clusters is the possibility to study composition dependence or the consequences of hydrogen bonding. For the first however successfully larger ring models have been used (30).

The three T-atom clusters terminated by hydrogen atoms can be used to analyse the consequences of acid strength variation by exploiting the covalent nature of the chemical bond (31). Figure 9 illustrates this. By changing the bond length of the SiH bonds, the strength of the neighbouring SiO bonds is varied and hence the OH bond energy.

Here we will illustrate the use of the cluster approach to analyse proton activation of methanol.

It is now well understood (5) that carbenium ions, that are intermediates in proton catalysed reactions cannot freely exist in zeolites. The main reason for this is the absence of stabilising solvating interactions.

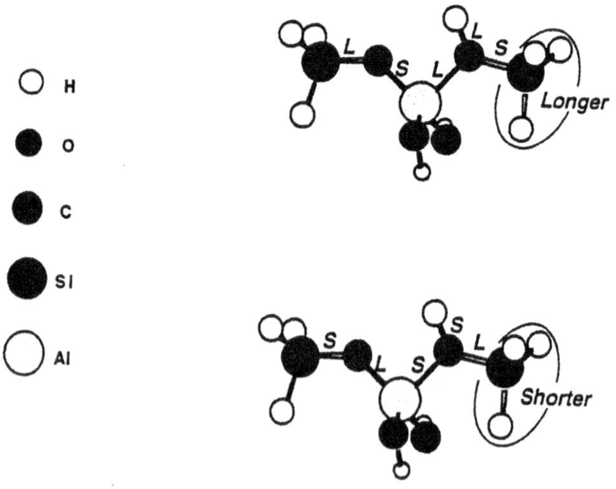

Figure 9. The zeolitic clusters with shortened or lengthened SiH bonds.

CH_3^+ ions in zeolites however are formed as part of the complex that is the transition state of a proton catalysed reaction (32). In the ground state the CH_3^+ ion will be bonded as a methoxy species to the basic oxygen atoms next to Al^{3+} located in the zeolite wall. These covalently bonded methoxy species have been observed by C13 NMR spectroscopy (33).

Figures 10a and 10b show respectively the energy-reaction diagram and the corresponding intermediate complexes for adsorbed methanol (two different modes), the dissociated product state and the transition states that are formed along the reaction path (32). One notes the carbenium type structure of the CH_3^+ intermediate in the transition state for C-O cleavage. Proton transfer from lattice to molecule is only observed for the transition state. The high energy with respect to the adsorbed ground state is due to the high energy cost of heterolytic OH splitting. The proton reacts with the OH^- from CH_3OH to form H_2O. The positive charge in the transition state complex resides on the CH_3^+ species and is stabilised by the electrostatic interaction with a basic oxygen atom around Al. This electrostatic stabilisation partially compensates for the cost of charge separation when the proton is transferred from the zeolite to methanol.

The overall reaction-energy diagram to dimethylether is shown in figure 11a (27). Dimethylether can be formed by reaction of the adsorbed methoxy species with a second methanol molecule. The overall activation energy with respect to the gas phase is seen to be 140 kJ/mole at low CH_3OH surface coverage and may increase to 215 kJ/mole at high surface coverage. The reaction order will vary between 2 and 0 dependent on the coverage.

The predicted activated energies are not consistent with experiment. Formation of dimethylether is an easy reaction that rapidly proceeds at mild conditions.

We will also analyse an associative reaction route, where reactions occur by direct recombination of methanol molecules. The corresponding reaction energy diagram is shown in figure 11b. The ground state is now an associative complex of two methanol molecules. One methanol molecule is bonded to the zeolite lattice via a zeolitic proton. The methanol proton converts to the oxygen atom of the second methanol molecule whose proton is bonded to a basic zeolitic oxygen. The computed infrared spectrum is consistent with experimental data by Mirth e.a. (34).

The transition state for methanol formation now corresponds to a Sn_2 nucleophilic substitution reaction. Again we observe the appearance of a carbenium ion intermediate as part of the transition state. Now the overall activation energy varies between 15 kJ/mole for low surface coverage and 145 kJ/mole at high surface coverage, consistent with experimental observations (35). Hence one concludes that the associative mechanism is the path for dimethylether formation.

We have shown that the methoxy species can react with methanol to methane and formaldehyde by hydride transfer (31). This process has a high activation-energy and therefore may become important at conditions where dimethylether is converted in consecutive reaction steps to ethylene and aromatics.

By varying the Si-H cluster bond length it has been found that hydride transfer is strongly dependent on cluster acidity (31).

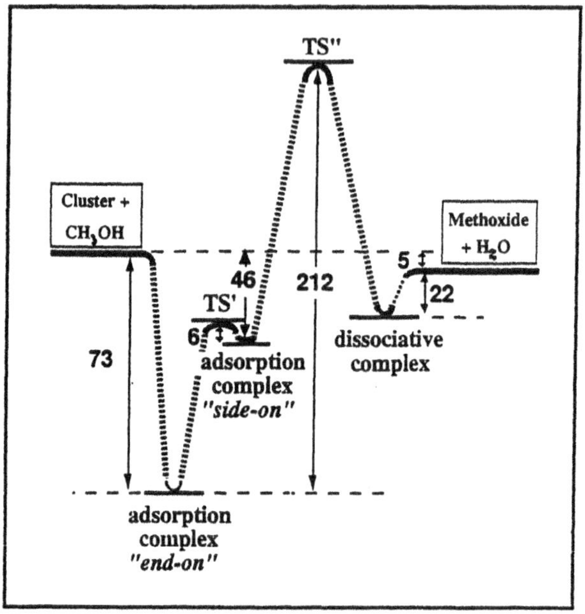

Figure 10a. Computed reaction energy diagram of CH_3OH protolysis (32).

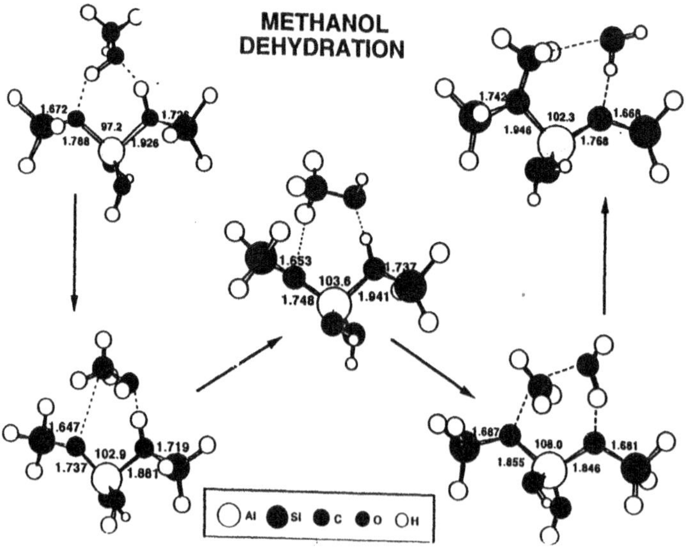

Figure 10b. The corresponding structures for ground- and transition states (32).

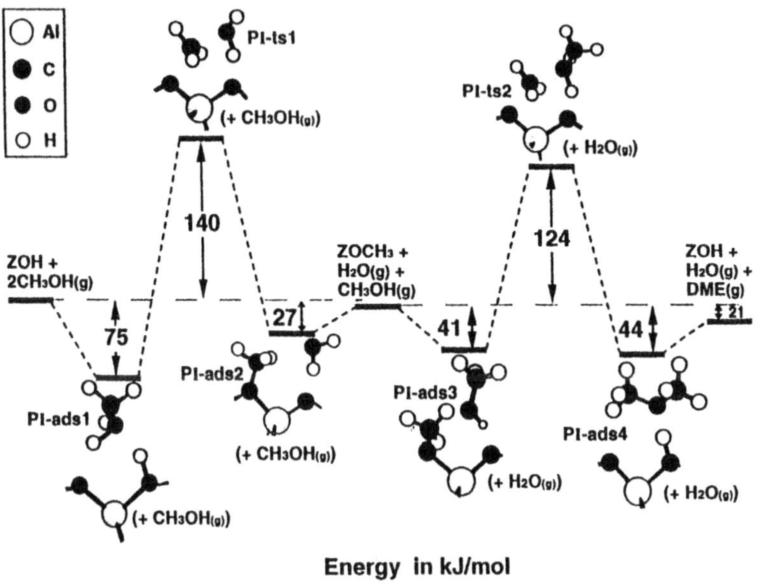

Figure 11a.. The Reaction energy diagram of dimethylether formation via adsorbed methoxy intermediate (27).

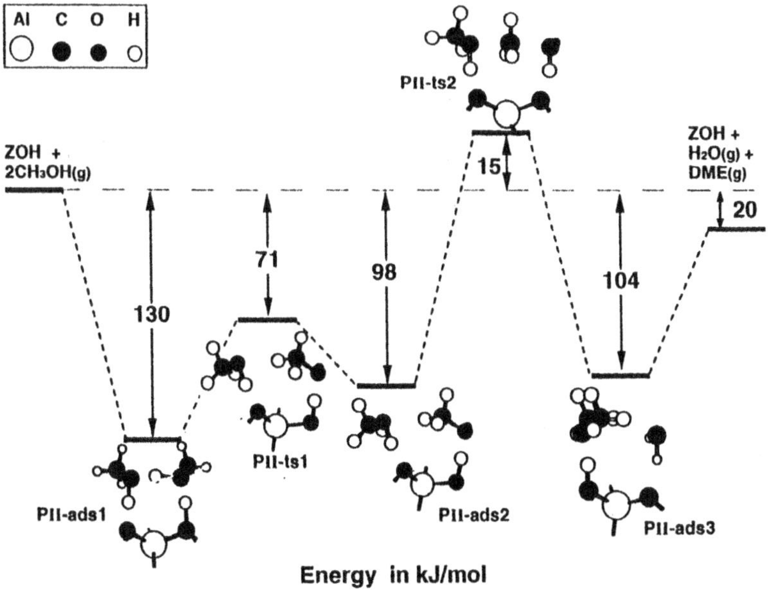

Figure 11b. The associative reaction mechanism of dimethylether formation (27).

The example of dimethylether formation discussed here illustrates the potential of current electronic structure computational methods. It is becoming a predictive tool to study reaction intermediates and transition states. The main current limitation is the small size of the clusters used so that zeolite cavity effects and composition dependence can be only indirectly deduced. The use of Car-Parinello techniques to the study of zeolite reactivity promises a major breakthrough in that extended systems can also be studied with the same accuracy as cluster models. The major issue to be resolved is the accuracy of pseudopotentials applied (36,37).

4.2. MODELLING THE HYDROISOMERIZATION OF HEXANE (20)

The hydroisomerization reaction was chosen to analyse kinetics of a zeolite catalysed reaction. The advantage of this reaction is its stability with time. The primary aim of this study was to demonstrate that the Turn Over Frequency (TOF) per proton of a zeolite catalysed reaction may differ significantly, even when the intrinsic acid strength of the zeolitic protons is the same. The experimental work was done with H-Mordenite (Si/Al = 10) and H-ZSM-5 (Si/Al = 28) (28). The proton chemical shifts (41 ppm with respect to TMS) and OH frequencies (~3610 cm-1) for these materials were the same. The mechanism of the hydroisomerization reaction is described by the kinetics scheme shown in figure 12. Platinum serves to equilibrate hexane and hexene (steps a and e). Conditions during the reaction are such that the partial pressure of hexene is so low that deactivating oligomerization reactions are suppressed. Once hexene is protonated (step b) it isomerizes (step c). After reaction the proton is backdonated and iso-hexene desorbs (step d), to be hydrogenated over Platinum. At high H_2/hexane ratio and large Pt/H^+ ratio equilibration by platinum is fast and the rate-limiting step of the reaction is the rate of isomerization of protonated hexene. Under those conditions one deduces the following expression for the rate of hexane isomerization:

$$\frac{d[\text{isohexane}]}{dt} = k_{iso} \bullet \frac{K^{p}_{ads} \bullet K_{eq} \bullet \frac{[\text{hexane}]}{[H_2]}}{1 + K_{ads} \bullet K_{eq} \bullet \frac{[\text{hexane}]}{[H_2]}} \qquad (1a)$$

$$= K_{iso} \bullet \left\{ K^{p}_{ads} \bullet K_{eq} \bullet \frac{[\text{hexane}]}{[H_2]} \right\}^{n} \qquad (1b)$$

$$0 < n < 1$$

k_{iso} is the rate of isomerization of protonated hexane, K^{p}_{ads} is the equilibrium constant of protonation of hexene and K_{eq} is the equilibrium constant of the hexane/hexene equilibrium. Figure 11 shows the large dependence of the heat of adsorption of alkanes to zeolite structure. Hence K^{p}_{ads} may be expected to be strongly zeolite

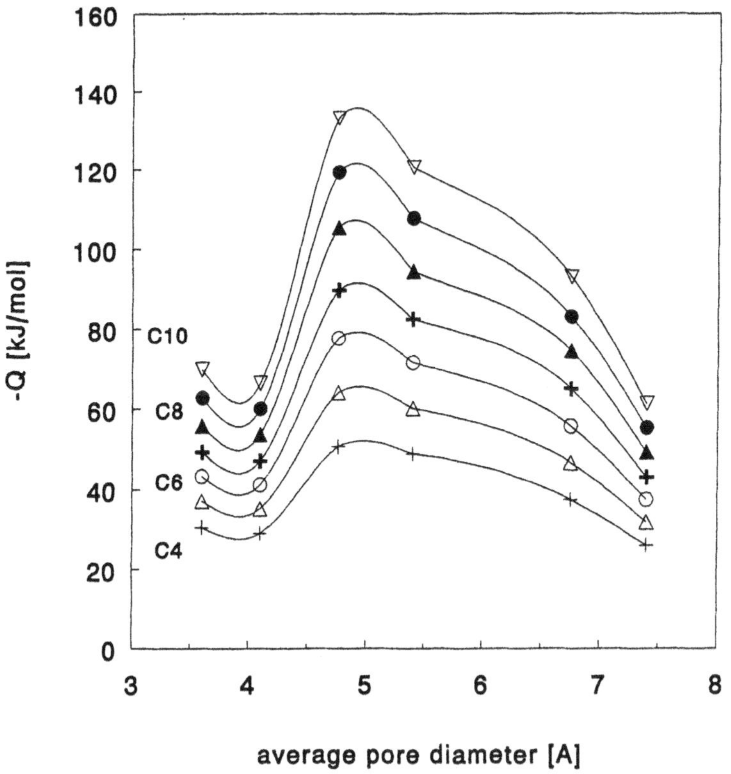

Figure 12. The bi-functional hydroisomerization reaction mechanism (28)

a.	hexane/hexene equilibration catalysed by Pt
b.	hexene protonation
c.	isomerization
d.	deprotonation
e.	iso-hexene/iso-hexane equilibration

structure dependent. The heats of adsorption of alkanes and alkenes do not differ more than 1 kJ/mole. One finds a correlation of the heat of adsorption with the micropore diameter (fig. 13). For zeolite micropore diameter in access of 5 Å, it decreases with micropore diameter for Mordenite and ZSM-5 the measured values are respectively 72 and 83 kJ/mole. The heat of protonation of hexene can be considered as the sum of two terms (see fig. 14):

$$\Delta H_{prot.gas} = \Delta H_{ads} + \Delta H_{prot.ads} (2)$$

DH_{ads} is the heat of adsorption of hexene with respect to the gas phase and $DH_{prot.ads}$ is the heat of protonation of adsorbed hexene. DH_{ads} is the heat of adsorption of hexene due to the dispersion interaction of hexene with oxygen atoms that form the wall of the zeolite micro cavities.

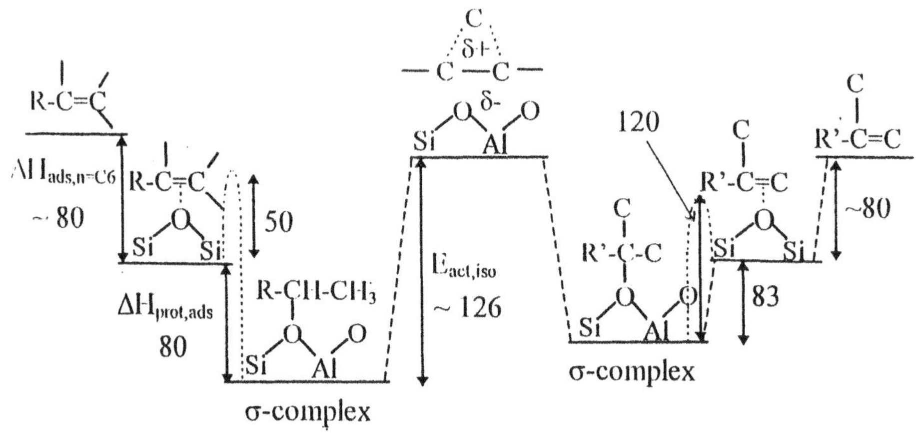

Figure 13. **Computed heats of adsorption of C_nH_{2n+2} hydrocarbons on zeolites with varying poresize dimensions (38).**

The energies of protonation of adsorbed alkenes can be considered independent of the zeolite structure. No measured values are available. However accurate theoretical estimates exist (39) and its value is around 80 kJ/mole. Hence for Mordenite and ZSM-5 the protonation energies of hexene with respect to the gas phase become 152 and 163 kJ/mole respectively. Note that this difference in protonation energy is due to the difference in the heats of adsorption of hexene.

ΔH_{hyd} the heat of hexane/hexene equilibration can be found from thermodynamics. This enables the determination of $E_{act,\ iso}$ from measured reaction orders and apparent activation energy:

$$E_{act,app} = E_{act,\_} + n(\Delta H_{prot,gas} + \Delta h_{hydr}) \quad (3)$$

The results are summarised in table 6.

The predicted values of $E_{act,iso}$ of 126 kJ/mole is considerably higher than found for isomerization reaction in superacidic homogeneous media. There its value is typically 30 kJ/mole (40). The main reason for the high value of the activation energy the elementary step of isomerization of the protonated hexene molecule in a zeolite is the strong covalent bond of the alkoxy complex of protonated hexene and the basic zeolite wall oxygen atoms locates around aluminium. The O_{zeol} - C_{hexyl} bond has to be broken so that isomerization can occur in the carbenium type transition state.

We have developed an adsorbate kinetic model (28) based on the concepts so far introduced and find that the differences in measured TOF's expressed per proton due differences in zeolite structure can be completely understood on the basis of the differences in the heats of adsorption of reacting molecules. This can be considered a microscopic basis to Derouane's confinement model (41).

TABLE 6 $E_{act,true} = E_{act,app} + n[\Delta H_{ads,n=C6} + \Delta H_{prot,ads}\ \Delta H_{dehydr}]$

Zeolite	$E_{ads,nC6}$ [kJ/mole]	$\Delta H_{prot,ads}$ [kJ/mole]	ΔH_{dehudr} [kJ/mole]	$E_{act,true}$ [kJ/mole]
MOR	71.9	80	-114.8	125.2
ZSM-5	82	80	-114.8	126.6

Mechanism of hydro-isomerization

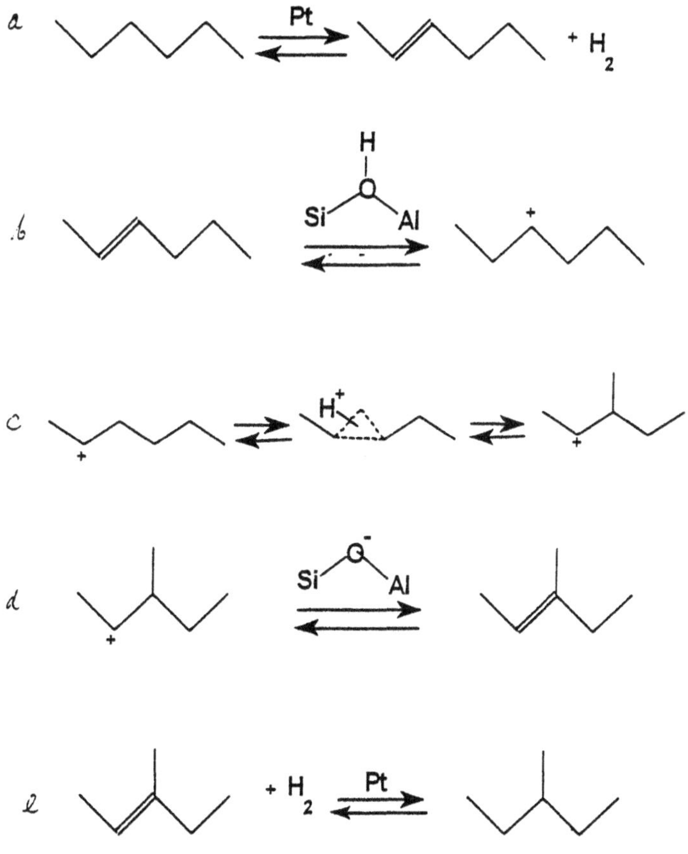

Figure 14. Reaction energy scheme of isomerization of an olefin by zeolitic proton (kJ/mole). The p-adsorbed state of olefin has not been included (28).

Summary

Reaction rate constants of surface chemical reactions can be calculated from first principles using clusters representative of the catalytically active site. Current quantum-chemical calculations enable the prediction of the geometry and energy of ground- and transition states of adsorbed reaction intermediates. Once the corresponding normal-mode frequencies have been computed Eyring's transition state reaction rate theory can be applied to compute the reaction rate constants from the ground state and transition state partition functions. To complete the overall rate of the reaction a kinetic scheme according to the reaction mechanism has to be composed. Statistical techniques as the Monte Carlo method can be used to incorporate non-ideal mixing behaviour of the adsorbed surface phase. In case lateral interaction effects can be ignored these equations can be directly integrated to predict reaction orders and apparent activation energies.

References

1. R.A. van Santen, J.W. Niemantsverdriet, Chemical kinetics and catalysis, Plenum, 1995.
2. R.A. van Santen, M. Neurock, Catal. Rev. -Sci. Eng., 37 (1995) 557.
3. M. Payne, M.P. Teter, D.C. Allan, T.A. Arias, J.P. Joannopoulos, Rev. Mod. Phys., 64 (1992) 1045.
4. S. Bates, M. Chen, C. Friend, R.A. van Santen, manuscript in preparation.
5. R.A. van Santen, G.J. Kramer, Chem. Rev., 95(3) (1995) 637.
6. M. Neurock, D. Dixon, to appear.
7. A. Fahmi, R.A. van Santen, J. Phys. Chem., 100 (14) (1996) 5676.
8. F. Besenbacher, P.T. Sprunger, L. Ruan, L. Olesen, I. Stensgaard, E. Loegsgaard, Topics Catal., 1 (1994) 325.
9. F. Besenbacher, J.K. Norskov, Progr. Surf. Sci., 44 (1993) 5.
10. A. Fahmi, D. Bird, R.A. van Santen, in preparation.
11. U. Starke, M.A. van Hove, G.A. Somorjai, Progr. Surf. Sci., 46 (1994) 305.
12. R.J. Madix, Adv. Catal., 29 (1980) 1.
13. X. Xu, C.M. Friend, J. Am. Chem. Soc., 112 (1990) 4571; J. Am. Chem. Soc., 113 (1991) 6779; J. Am. Chem. Soc., 113 (1991) 8571.
14. J.F. Roberts, R.J. Madix, J. Am. Chem. Soc., 110 (1988) 8540.
15. R.A. van Santen, Chem. Eng. Sci., 50 (1995) 4027.
16. M.W. Roberts, Surf. Sci., 299 (1994) 769.
17. T. Sasaki, T. Saeyoshi, Y. Iwasawa, Surf. Sci., 316, (1994) L1081.; T. Saeyoshi, T. Sasaki, Y. Iwasawa, Chem. Phys. Lett., 241 (1995) 189.
18. P. Davies, R.A. van Santen, unpublished results.
19. M.C. Zonnevylle, J.J.C. Geerlings, R.A. van Santen, J. Catal., 148 (1994) 417.
20. H. Burghgraef, A.P.J. Jansen, R.A. van Santen, J. Chem. Phys., 98 (1993) 8810; Surf. Sci. 324 (1995) 345.
21. M.A. van Daelen, Y.S. Li, J.M. Newsam, R.A. van Santen, Chem. Phys. Lett., 226 (1994) 100; Y.S. Li, M.A. van Daelen, J.M. Newsam, R.A. van Santen, J. Phys. Chem., 100 (1996) 2279.
22. P.M. Holmblad, J. Wambach, I. Chorkendorff, J. Chem. Phys., 102 (1995) 8255.
23. R.J. Gelten, A.P.J. Jansen and R.A. van Santen, J.J. Lukkien, J.P.L. Segers and P.A.J. Hilbers, in preparation.
24. A.P.J. Jansen, Comp. Phys. Comm., 86 (1995) 1; A.P.J. Jansen, Phys. Rev., B52, (1995) 5400.
25. G. Ertl, Adv. Catal., 37 (1990) 213.
26. R. Imbihl, M.P. Cox and G. Ertl, H. Müller and W. Brenig, J. Chem. Phys. 83 (1985) 1578.

27. S.R. Blaszkowski, R.A. van Santen, J. Am. Chem. Soc., to appear.
28. A. van de Runstraat, P.J. Stobbelaar, J. van Grondelle, B.G. Anderson, L.J. van IJzendoorn, R.A. van Santen, Proc. International Zeolite Conference, Seoul 1996, Stud. Surf. Sci. and Catal.
29. H.V. Brand, L.A. Curtiss, L.E. Iton, J. Phys. Chem., 96 (1992) 7725.
30. G.J. Kramer, R.A. van Santen, J. Am. Chem. Soc., 117 (1995) 1766.
31. S.R. Blaszkowski, M.A.C. Nascimento, R.A. van Santen, J. Phys. Chem., 100 (1996) 3463.
32. S.R. Blaszkowski, R.A. van Santen, J. Phys. Chem., 99 (1995) 11728.
33. F.G. Oliver, E.J. Munson, J.F. Haw, J. Am. Chem. Soc., 96 (1992) 8106.
34. G. Mirth, J.A. Lercher, M.W. Anderson, J. Klinowski, J. Chem. Soc. Far. Trans., 86 (1990) 3039.
35. J. Bandiera, C. Naccache, Appl. Catal., 69 (1991) 139.
36. E. Nusterer, P.E. Blochl, K. Schwarz, Ang. Chem., 108 (1996) 187; B.J. Hammer, J.K. Norskov, Nature, 376 (1995) 238; B.J. Hammer, M. Scheffler, K.W. Jacolsen, J.K. Norskov, Phys. Rev. Lett., 73 (1994) 1400; J.A. White, D.M. Bird, M.C. Payne, I. Stich, Phys. Rev. Lett., 73 (1994) 1404.
37. R. Shah, M.C. Payne, M.-H. Lee, J.O. Gale, Science, 271, (1996) 1395.
38 S. Bates, R.A. van Santen, B. Smit, J. Am. Chem. Soc., to appear.
39. V.B. Kazansky, M.B. Frash, R.A. van Santen, Appl. Catal., to appear.
40. D.M. Brouwer, H. Hogeveen, Progr. Phys. Org. Chem., 9 (1972) 179.
41. E.G. Derouane, J.M. Andre, A.A. Lucas, J. Catal., 110 (1988) 58.

CLUSTER MODELLING OF OXIDE SURFACES: STRUCTURE, ADSORPTION AND REACTIVITY

G. PACCHIONI
Dipartimento di Chimica Inorganica,
Metallorganica e Analitica
Università di Milano
via Venezian 21, 20133 Milano, Italy

1. Introduction

In the near future we will assist to the rapid development of an area of research which can be generally classified as "scientific computing". This relatively new field will complement the more classical experimental approach for the solution of scientific problems but should not be confused with the more general areas of theoretical chemistry and theoretical physics. Experiments are supposed to produce new facts and discoveries, and to disclose to humans secrets of nature; theory is supposed to provide a firm general framework for the explanation of the phenomena observed through a series of mathematical laws. Scientific computing will be something in between these two approaches: it will be based on well established theories and formalisms but it will be used to provide new facts, in a similar way experiments are performed. The reason why the importance of scientific computing will continuously and rapidly increase in the future, and why it will probably become one of the most intensive human activities in the next century, is quite clear. In the past 30 years we have assisted to an exponential increase of the computing power. This has stimulated the creation of an adequate software to solve problems in different scientific areas; indeed, the development of the software has been much slower than that of the computing power. Thanks to the combined use of advanced software and powerful computers, it will become possible to "simulate" an experiment on a computer before to perform it with a much lower cost and a much more rapid answer. Of course, to which extent this will become reality or will remain science fiction is a matter of the degree of reliability of the simulation. This will depend on the generality of the underlying theory, on the level of approximations introduced in the simulation, and on the accuracy of the software.

R.M. Lambert and G. Pacchioni (eds.),
Chemisorption and Reactivity on Supported Clusters and Thin Films, 395-424.

Coming to the more specific field of surface and material science, a lot of work has been done in this area in the last 20 years, but considerable more work is necessary. The fundamental theory of processes at surfaces or in solid materials is, of course, quantum mechanics. Different ways to implement and solve the Schrödinger equation have been proposed in the course of the years and through the intense application to concrete problems one has learned which are the best approaches for a given problem. Still, even when a first-principle, parameters free, approach is chosen, other approximations have to be done, mainly in the definition of the model used for the simulation which is usually much more simple than the real system. The knowledge of the errors introduced by the approximations in modelling a real situation is essential in order to assess the validity of the results.

Of course, modelling surfaces and new materials cannot be restricted to static electronic structure but has also to include the more complex area of dynamic phenomena. Also in this case considerable progress has been achieved in recent years through the generation of *ad-hoc* force fields or many-body potentials as well as *ab initio* molecular dynamics approaches, to describe in a classical or semi-classical way the dynamic properties of a system. This area of scientific computing, however, is quite recent and its validation through direct and close comparison with clean experimental data is still at the very beginning.

One of the central activities for the next years will be the description of oxide materials, of their surfaces and interfaces. The reasons are clear, since these materials have an extraordinary importance in many areas of science and technology [1]. The possibility to provide a global simulation of the properties of this class of materials is both scientifically appealing and economically useful. However, a significant amount of work is necessary before a sufficient level of reliability will be reached. The content of the present chapter is just a contribution to the understanding and the simulation, from a quantum-mechanical point of view, of phenomena occurring at the surface of an oxide.

We will present first a throughout comparison of local cluster models and periodic supercell approaches for the description of CO adsorption on $Cu_2O(111)$ and $TiO_2(110)$ non-polar surfaces. Different tools to analyse the surface-adsorbate bonding will be introduced and the analysis of the results will be extended to the surface of NiO(100), so that a comparison of transition metal oxides with very different d occupations will be possible. We will then consider the nature and structure of oxygen defects on the MgO surface and of their reactivity toward adsorbed metal atoms or small molecules. We will focus in particular on the nature of the electrons trapped on these surface vacancies and in their involvement in the bonding with adsorbates. In the last part of this chapter we will address the problem of the origin of electrochemical promotion in catalysis. We will propose a simple model based essentially on electrostatic concepts to explain the large changes in catalytic activity observed for metal particles supported on solid electrolytes under the effect of an external potential (also known as the NEMCA effect).

2. Electronic structure of transition metal oxide surfaces

2.1. CLUSTER VERSUS PERIODIC MODELS

Two basic approaches are usually employed to describe the electronic structure of surfaces and their interaction with adsorbed species: the cluster model [2,3], where the surface is described by a finite number of atoms and the surrounding is represented in some approximate way, and the periodic or supercell approach [4] where the periodic nature of the surface is correctly taken into account. Each approach presents advantages and disadvantages. Clearly, the simulation of the surface by a small number of atoms is appropriate only when the property under investigation is of very local nature; there is enough empirical evidence that this is the case in adsorption and bonding of atoms or molecules at surfaces [2,3]. However, a cluster is normally quite small and only one adsorbate is considered. In this respect, cluster models are representative of low-coverage situations. On the other extreme, in the slab or periodic models a unit cell is defined and repeated infinite times in two directions. In this case, an adsorbate is placed on each cell and different levels of coverage can be modelled, from very high to medium coverage. For obvious reasons, it is not possible to simulate the very low-coverage regime as very large supercells would be required. In general, however, it is possible to consider intermediate situations where the adsorbate-adsorbate repulsion is small enough to be considered negligible.

In this section we consider two examples of adsorption on transition metal surfaces described by cluster models or periodic slabs. The purpose is to compare, using as much as possible the same "ingredients" of the calculations, the two approaches. The examples considered are the adsorption of CO on the cation sites of the $Cu_2O(111)$ and $TiO_2(110)$ surfaces. All the calculations have been performed at the Hartree-Fock SCF level using the Crystal code for the periodic calculations [5] and the Hondo 8.5 code [6] for the cluster calculations. We will give here only a few technical details on the calculations; the reader is referred to refs. [7] and [8] for additional information.

To represent the $Cu_2O(111)$ and the $TiO_2(110)$ surfaces we used two relatively large clusters, $Cu_{28}O_{14}$, Fig. 1, and Ti_7O_{14}, Fig. 2. Notice that the number of atoms in the two clusters reflects the ratio of metal to oxygen atoms in the bulk materials. We refer to these clusters as to stoichiometric clusters. The importance of stoichiometric clusters is clear. Given the partially or largely ionic nature of the bonding in many oxides, the use of stoichiometric clusters does not require any *a priori* assumption about the cluster electron counting. For example, a TiO_4 cluster requires the addition of 4 e^- if one assumes a fully ionic model (Ti^{4+} and O^{2-}) and of 2 e^- is one assumes a lower ionicity (Ti^{2+} and O^-). The use of a stoichiometric cluster allows one to use neutral clusters; the actual charge distribution within the cluster will be determined by the self-consistent nature of the cluster wavefunction. The clusters must also reflect, as much as possible, the average co-ordination of the atoms in the real surface. The clusters have been surrounded by a large array of point charges, PC, to simulate the long-range effect of the Madelung Potential, MP. This technique is widely used to provide a sort of embedding of ionic crystals [9].

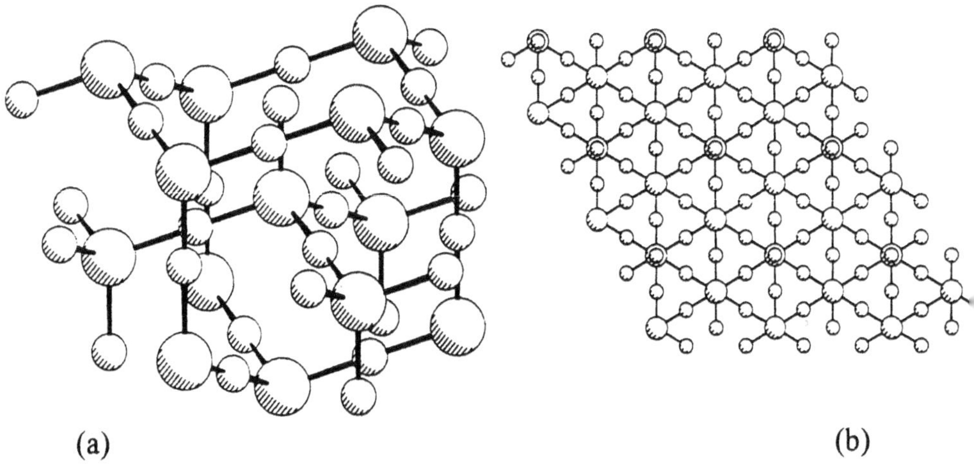

Figure 1. (a) The $Cu_{28}O_{14}$ cluster used to model the $Cu_2O(111)$ surface; the cluster is embedded in a large array of point charges (not shown). (b) Top view of the $Cu_2O(111)$ surface.

The use of the PC model, however, implies some assumption about the values of the PCs. Referring to the previous example of TiO_2, the charges could be set to +4/-2, to +2/-1, or to any other value. The importance of the surrounding PCs for the description of the adsorption properties will be discussed below.

The Cu ions of Cu_2O have been described by an Effective Core Potential (ECP); 11-electron ECP including explicitly in the valence the $3d^{10}4s^1$ electrons [10] were used for the Cu atoms closest to the adsorption site, and 1-electron ECP [11], treating explicitly only the $4s^1$ electron, have been used for the other Cu atoms [7]. For Ti in TiO_2 we used a 4-electron ECP which includes the $3d^24s^2$ electrons in the valence [10] except for the central atom of the TiO_2 cluster which was described at the all electron level [12]. Also the O atoms in the cluster and the CO molecule have been treated with all their electrons [13]. Double-zeta quality basis sets were used for the atoms closest to the adsorption site, while more contracted basis sets where used for the surrounding metal ions.

In order to evaluate the importance of correlation effects we also considered much smaller clusters. These clusters are subsections of the clusters described above, but, as it will be shown below, they provide qualitatively similar answers as the largest clusters.

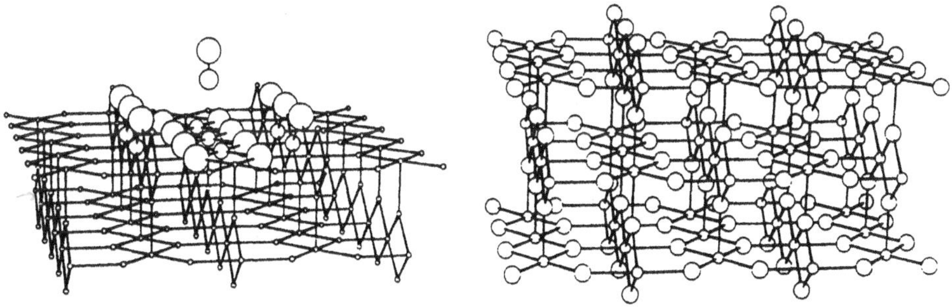

(a) (b)

Figure 2. (a) The Ti_7O_{14} cluster used to model the $TiO_2(110)$ surface; the cluster is embedded in a large array of point charges (not shown). (b) Side view of the three-layers slab used to model the $TiO_2(110)$ surface.

For the corresponding slab calculations we used three layers slabs. To maximise symmetry CO has been adsorbed on both sides of the slab. On Cu_2O we used a supercell so that low-coverage of CO is considered while for TiO_2 full monolayer coverage was studied. Exactly the same basis set as for the clusters was adopted in the slab calculations on Cu_2O (using ECP's on Cu) [7] while a slightly different all electron basis was adopted for the TiO_2 slab, see ref. [8]. The thresholds for integral truncation were set to values which can be regarded as rather strict (5 5 5 5 11); for the calculation of the vibrational frequencies we used better thresholds (5 5 5 6 12) and we selected comparable sets of integrals for all the points of the potential energy curve.

2.2. CO ON TRANSITION METAL OXIDES: A PROBE MOLECULE

We report in the following three properties of adsorbed CO which are quite critical for the description of the CO-oxide interaction: the CO-surface distance, $z_e(CO)$, the C-O vibrational shift, $\Delta\omega(CO)$, and the adsorption energy, D_e. The last two properties in particular, $\Delta\omega(CO)$, and D_e, are very delicate to compute as they easily oscillate as a function of the level of computational treatment. In all cases the CO molecule has been adsorbed C-down, perpendicular to the Cu or Ti ion.

For the Cu_2O case we compare first the slab results with those of a $Cu_{28}O_{14}$ cluster embedded in (+1/-2) PCs and with the same cluster without any PC embedding, Table 1. The use of (+1/-2) PCs is consistent with the data of Mulliken population for the slab and for the cluster with no embedding which show that the Cu_2O electronic structure is close to that of the full ionic model, Cu^+ and O^{2-} [7]. We first notice that the two clusters, with and without PC embedding, give very similar results, except for the CO vibrational shift which is somewhat smaller in the cluster without PCs, Table 1. When the periodic slab is considered, we found very similar z_e and $\Delta\omega$ as with the $Cu_{28}O_{14}$(+1/-2) cluster; only the D_e, corrected by the basis set superposition error (BSSE), a technical effect due to the incompleteness of the basis set [14], is different. In particular, we found a smaller adsorption energy in the periodic calculation. This is not due to the CO-CO repulsion which in this arrangement is 0.01 eV only. It is rather due to an overestimate of the BSSE in the periodic approach. Without this correction the difference between cluster and slabs is only 0.1 eV. These results are quite important because they have been obtained with exactly the same method, Hartree-Fock, and basis sets. The differences are thus entirely due to the model. The first conclusion is that a cluster embedded in PCs provides similar answers as the more rigorous periodic approach.

We have also considered a much smaller model, consisting of a $[CuO]^-$ unit embedded in (+1/-2) PCs; quite surprisingly, the results obtained with this cluster are very close to both the larger cluster and the slab results, Table 1. This provides a strong indication of the very localised nature of the interaction of CO with the substrate.

A similar comparison has been performed for TiO_2. Here the problem of the PC embedding is more severe than in Cu_2O because the analysis of the TiO_2

Table 1 - Adsorption properties of CO on $Cu_2O(111)$ and $TiO_2(110)$

Model	z_e(CO), bohr	$\Delta\omega$(CO) cm^{-1}	D_e eV
	$Cu_2O(111)$		
Cu_2O slab	3.91	+80	0.46
$Cu_{14}O_{28}$(+1/-2)	3.82	+79	0.76
$Cu_{14}O_{28}$	3.80	+53	0.75
$[CuO]^-$(+1/-2)	3.82	+93	0.89
	$TiO_2(110)$		
TiO_2 slab	4.52	+130	0.68
Ti_7O_{14}(+4/-2)	4.40	+155	0.84
Ti_7O_{14}(+2/-1)	4.35	+146	0.98
Ti_7O_{14}	4.23	+147	0.96
$[TiO_4]^{2-}$(+2/-1)	4.35	+128	0.93

electronic structure shows that the actual charges on the Ti and O ions are far from the nominal values of +4 and -2, respectively. In fact, from various measures one can conclude that charges of +2 and -1 are much more representative of the charge separation within the oxide [8]. Therefore, we have considered two Ti_7O_{14} clusters embedded in (+4/-2) and (+2/-1) PCs and a Ti_7O_{14} cluster without any embedding, Table 1. Depending on the external field provided by the PCs a moderate change in the surface-CO distance and in the binding energy is found, while the CO vibrational shifts remains constant within our error limits. The results can be compared with those of a $TiO_2(110)$ slab, but in this case slightly different basis sets where used [8] so that only a qualitative comparison is possible.

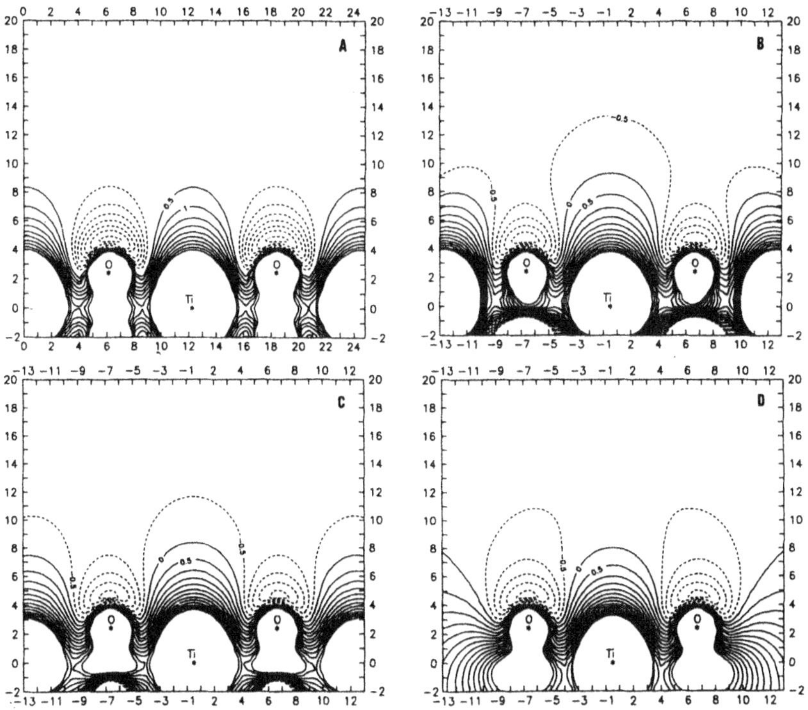

Figure 3. Electrostatic potential, in eV, of the $TiO_2(110)$ surface from various models: (A) TiO_2 slab; (B) $Ti_7O_{14}(+4/-2)$ cluster; (C) $Ti_7O_{14}(+2/-1)$ cluster; (D) Ti_7O_{14} cluster. Solid line = positive EP, dotted line = negative EP.

Nevertheless, we found also for TiO_2 that periodic slab and cluster calculations provide a similar picture of the bonding with CO: in particular, the adsorption is rather strong, 0.7-0.9 eV, and is accompanied by a substantial blue shift in the

CO ω_e. The most significant deviation between cluster and periodic results is in the adsorption energy, which is larger in the local model, as already found for $Cu_2O(111)$.

Also in this case we have considered a much smaller model, a $[TiO_4]^{2-}$ cluster embedded in (+2/-1) PCs, and we found a similar response as for the larger models. Again this is a sign of the localised nature of the interaction. The fact that small clusters provide similar answers as larger models is important for the inclusion of correlation effects (see below).

The reasons for the similar response of local cluster models and periodic slabs is that the charge separation naturally occurs within the clusters and resembles that of the infinite lattice. The role of the external potential through the addition of PCs is to favour the convergence of this charge separation to the "bulk" value, even for relatively small clusters. The consequence is that the electrostatic potential (EP) of the infinite surface is sufficiently well described, at least in the chemisorption region, by a cluster embedded in PCs [8,15]. This is shown in Fig. 3 where the EP for the periodic TiO_2 slab is compared to that of the Ti_7O_{14} clusters with various PCs embedding. In the region where the adsorption takes place the EP is similar. However, at long distances from the surface the EP is strongly affected by the presence of the PCs and behaves quite differently from that of the periodic calculation. Thus, some care is necessary in using clusters to compute properties at long distance from the oxide surface.

2.3. TOOLS FOR SURFACE BONDING ANALYSIS

Having established that cluster and periodic models give qualitatively similar results, at least for the case of adsorption on oxide cations, we consider now the nature of the bonding of CO with different transition metal oxides, namely $Cu_2O(111)$, $TiO_2(110)$, and NiO(100). For this latter system we used a minimal cluster, $[NiO_5]^{8-}$ embedded in (+2/-2) PCs [16]; it should be noted that NiO has a much more ionic character than TiO_2 and that in this case the use of nominal ionic values, +2/-2, for the surrounding charges is justified. While the ground state of the Cu_2O and TiO_2 clusters is $^1A_1$, a $^3A_2$ ground state was used for the calculations on $[NiO_5]^{8-}$ to take into account the high-spin coupling of the 3d electrons [16]. The comparison of these three oxides is particularly instructive since they are characterised by completely different occupations of the 3d shell: TiO_2 is formally a $3d^0$ oxide, Cu_2O has a full d shell, $3d^{10}$, while NiO has a partially filled d shell, $3d^8$. The calculation of the bonding properties of CO on the three TM oxides will provide information about the involvement of the 3d electrons in the bonding and in general on their participation to the chemistry of these oxide surfaces.

We first decompose the CO interaction energy into the sum of various contributions. This is done according the Constrained Space Orbital Variation technique, CSOV [17,18]. In this approach one constructs a zero-order SCF wavefunction as an antisymmetrized product of the occupied MO's on the two fragments A and B (CO and the cluster in this case): $\Psi_{AB} = A_{AB}\ \Psi_A\Psi_B$. At this initial step the orbitals on the two fragments are fixed as they are at infinite distance. The superposition of the two frozen orbital fragments at equilibrium bond distance is purely repulsive due to the Pauli or exchange repulsion, a bonding contribution whose importance is often underestimated in chemical bonding analyses. The energy change at the first Frozen Orbital CSOV step, E_{SCF}(CSOV step 1), with respect to the separated fragments, $E_{SCF}(A) + E_{SCF}(B)$, provides a measure of the Pauli repulsion. For ionic or partially ionic surfaces, however, there is another contribution included at this step, namely the electrostatic interaction between the surface electric field and the CO charge distribution; this term is attractive, and can partially compensate the Pauli repulsion.

Starting from the zero-order wavefunction, one can allow orbital mixing in subsequent steps. In particular, intra-unit polarisation can be determined by mixing the occupied with the virtual orbitals on the same fragment, $occ_A/virt_A$ or $occ_B/virt_B$; these steps measure the importance of the polarisation of the adsorbate or the surface when the two units are at equilibrium distance. When the mixing of the occupied orbitals on one fragment with the virtual orbitals on the other fragment is allowed, $occ_A/virt_B$ or $occ_B/virt_A$, charge transfer from A to B or *viceversa* is allowed. This is often referred to as the donation or back donation mechanism. The importance of the inter-unit charge transfer mechanisms can be deduced from the energy lowering at the corresponding CSOV steps or from the change in dipole moment of the system. Being the nuclei fixed during the CSOV analysis, a change in the dipole moment reflects the flow of electronic charge in a given direction.

Table 2 - Bonding contributions, in eV, of CO adsorbed on TM oxide surfaces; the analysis is done for a distance close to equilibrium: $z_e(TiO_2\text{-}CO)$ = 4.34 bohr, $z_e(Cu_2O\text{-}CO)$= 3.82 bohr, $z_e(NiO\text{-}CO)$= 4.70 bohr.

	TiO_2/CO	Cu_2O/CO	NiO/CO
Frozen orbital	-0.01	-0.07	-0.09
CO polarisation	0.45	0.23	0.06
σ donation (from CO)	0.43	0.31	0.07
π back donation (to CO)	0.15	0.24	0.11
D_e (total)	1.26	0.84	0.25

The most important "ingredients" of the bonding are therefore the electrostatic attraction and the Pauli repulsion measured at the Frozen Orbital step, the CO polarisation, the CO to the surface σ donation, and the surface to CO π back donation (where the σ and π notation refers to the symmetry of the orbitals involved). The contribution of each of these mechanisms for the three oxide surfaces considered is shown in Table 2. First, we observe that the sum of the electrostatic attraction and the Pauli repulsion almost cancels in the three systems; the net result is a very small repulsive interaction at equilibrium bond distance. The CO polarisation contributes almost 0.5 eV to the bonding in TiO_2, 0.2 eV in Cu_2O and almost nothing in NiO. A similar situation is found for the CO σ donation which is rather large, 0.4 eV in TiO_2, considerable in Cu_2O, 0.3 eV, and negligible in NiO, <0.1 eV. Thus, while the CO σ donation is an important bonding mechanism on TiO_2 and Cu_2O, it is not important for the stabilisation on NiO. The back donation from the surface to CO for metal oxides is a matter of debate. In fact, while there is strong evidence that this mechanism dominates on metal surfaces, there are doubts about its importance when a metal cation is involved [16]. Indeed, the CSOV data show that the back donation is not large; the largest contribution, 0.24 eV, is found for the Cu_2O surface, where Cu is in a $3d^{10}$ electronic configuration. This contribution is small in the other oxides. The final binding, see D_e in Table 2, takes into account also the polarisation of the substrate and the small couplings between the individual mechanisms. D_e is large for both TiO_2 and Cu_2O, while is very small for NiO. The conclusion is that while a real chemical bond forms between CO and the Ti and Cu oxides, the bonding on NiO is largely electrostatic, with very little charge transfer.

The situation can be rationalised as follows. On TiO_2 the metal cation is in a high oxidation state; this means that there are low-lying empty levels to accept charge from the CO donor. The 3d orbitals are only partially filled and mixed in with the O 2p band, so that very little back donation occurs from the surface to CO. Given the high formal charge of the exposed Ti cation, there is also a considerable polarisation of the adsorbed CO. On Cu_2O the accepting orbital is the Cu 4s but the 3d levels are filled; the donation from CO is less pronounced than on TiO_2. The low charge on the Cu ions, formally Cu(I), and the filled 3d shell favour the occurrence of some back donation to CO, at variance with other oxide surfaces like TiO_2. The regular NiO(100) surface behaves quite differently. The presence of the bulky O^{2-} ions around the adsorption site prevents the CO molecule from getting close to the surface, Table 2, and efficiently overlap with the Ni 3d levels. The bonding has only a minor covalent character, is very weak, and arises almost entirely from electrostatic and polarisation effects. From this point of view, the bonding of CO to NiO closely resembles that of CO to MgO, a non transition metal oxide [16].

Another property which turns out to be very useful for the characterisation of the bond of CO with oxide or metal surfaces is the CO dynamic dipole moment, $d\mu/dr$ or effective charge, e^*. At the Hartree-Fock level, $d\mu/dr$ for free CO is -1.03 a.u. [19], and it can be decomposed into a σ and a π component according to

$$\left(\frac{d\mu}{dr}\right)_{tot} = \left(\frac{d\mu}{dr}\right)_{\sigma} + \left(\frac{d\mu}{dr}\right)_{\pi} \qquad (1)$$

The σ and π contributions to dμ/r at the HF level are +0.46 and -1.49 a.u., respectively. When interacting with a surface, the CO dynamic dipole moment changes. The changes reflect the occurrence of σ donation or π back donation. In fact, a large π back donation, as for CO on metals, results in a substantially more negative π contribution to e^*; if, on the other hand, the σ donation dominates, dμ/dr will be less negative. It is therefore possible to use the relative change in dμ/dr of adsorbed compared to free CO to monitor the nature of the interaction with the substrate. In particular, three cases are possible:

$$\frac{(d\mu / dr)_{ads.CO}}{(d\mu / dr)_{freeCO}} > 1 : \text{the } \pi \text{ back donation dominates}$$

$$\frac{(d\mu / dr)_{ads.CO}}{(d\mu / dr)_{freeCO}} < 1 : \text{the } \sigma \text{ donation dominates}$$

$$\frac{(d\mu / dr)_{ads.CO}}{(d\mu / dr)_{freeCO}} \approx 1 : \text{no charge transfer or net cancellation}$$

The calculation of the dynamic dipole moment is very useful also because it is very sensitive to the inclusion of correlation effects. In general, electron correlation acts to increase the back donation from the surface to adsorbed molecules [19], and the changes in e^* reflect this enhancement. Correlation effects have been included through configuration interaction, CI, calculations, on small models of CO adsorbed on $TiO_2(110)$, $Cu_2O(111)$, and NiO(100) surfaces. In particular, single-double CI calculations were performed for CO adsorbed on $[CuO]^-(+1/-1)$, $[TiO_4]^{2-}(+2/-1)$, and $[NiO_5]^{8-}(+2/-2)$ clusters. The CO e^* has been compared to that of the free molecule computed at the same theoretical level, Table 3. For comparison, we report also the computed value in the case of a metal surface, Cu(100), represented by a Cu_9 cluster [19].

Two main conclusions can be drawn form the data of Table 3. First of all, three cases can be clearly distinguished: (a) oxide surfaces where π back donation occurs, like the $Cu_2O(111)$ surface; in this case dμ/dr increases upon adsorption (ratio adsorbed/free CO > 1), although much less than on a metal like Cu.

Table 3 - Ratio of adsorbed compared to free CO dynamic dipole moment, dμ/dr, for CO adsorbed on oxide and metal surfaces.

surface	$\frac{(d\mu/dr)_{ads.CO}}{(d\mu/dr)_{freeCO}}$		dominant bonding contribution
	SCF	CI	
Cu(100)	1.92	2.25	π back donation
$Cu_2O(111)$	1.20	1.32	π back donation
$TiO_2(110)$	0.82	0.80	σ donation
NiO(100)	1.01	1.02	electrostatic

(b) Oxide surfaces where there is little or no back donation but where the CO molecule donates charge to the substrate, like in $TiO_2(110)$; in this case dμ/dr is reduced upon adsorption because the σ component of dμ/dr, which is positive, becomes larger. (c) The third case is that of NiO(100) where no change is found in dμ/dr; this may indicate a net cancellation of the σ and π charge transfers or, as shown by the CSOV analysis for this system, a largely electrostatic bond with little or no charge transfer. In general, the conclusions from the analysis of the CO dynamic dipole moment are in perfect agreement with the CSOV results also for the other surfaces. The second observation is that the inclusion of correlation effects increases the back donation, as expected (the ratio of adsorbed to free CO dμ/dr increases when this mechanism is important); however, the changes are relatively small, and the conclusions obtained with uncorrelated wavefunctions are confirmed by the CI calculations.

2.4. REACTIVITY OF SURFACE OXIDE IONS

We have shown that clusters models and band structure calculations provide qualitatively similar results. Indeed, a considerable number of cluster studies of adsorption at oxide surfaces has been reported [20-37] and the method was proved to be quite successful for a number of problems. Very good review articles can be found in ref. [38-40]. However, a warning is necessary at this point. The fact that similar results are found with clusters and periodic calculations for the CO adsorption on cation sites can certainly be extrapolated to any other adsorbate. However, much more difficult can be the description of the adsorption and reaction at the oxygen surface sites. Very often this interaction involves charge transfer from the adsorbate to the surface (oxidising surfaces) or viceversa (reducing surfaces). The description of this interaction is much more delicate because when substantial charge transfer takes place, which is not the case of CO, the correct description of the "acid-base" properties of the surface becomes essential. In other words, the position of the acceptor or donor levels on both the adsorbate and the surface is a very important quantity which may be critically dependent on the size of the cluster, on the correct description of the long range coulombic potential, on the basis set, etc. A concrete example of this is provided by the different reactivity of low-co-ordinated oxygen anions on a surface like MgO or by the comparison of the reactivity of surface oxygens in different oxides like MgO or CaO [41-43].

Recent studies have shown that the tendency of gas-phase molecules like CO_2 or SO_2 to form surface carbonates or sulphites depends very strongly on the "basicity" of the surface [43]. This important property of oxide surfaces, which is simply related to the ability to transfer charge from the surface to an adsorbate, is critically dependent of the coulombic potential felt by the oxide anions and not, as often assumed, on the charge of the oxygen ion. The experimentally observed trend in basicity of the alkaline-earth oxides is BaO>SrO>CaO>MgO. This means that MgO is a weaker base because charge transfer from the surface oxygens to an adsorbed molecule occurs at an higher energy cost. It has been shown that a stronger potential stabilises the surface anions and reduces the surface basicity [43]. In MgO the Madelung potential is larger than in CaO because of the shorter cation-anion distance, and the oxide anion is more stable and less "basic"; on CaO, where the Madelung potential is lower, the surface oxide anions are much more reactive and form easily carbonates or sulphites without activation barriers. In a similar way, the low-co-ordinated oxide anions of the MgO surface (corners, steps, edges, etc.) where the Madelung potential is lower are the sites where the chemical reactivity is highest [41,42]. This shows that the origin of the surface basicity in these materials is primarily electrostatic. This can be generally understood by considering that the O^{2-} ion is unstable in the gas-phase and exists only in ionic crystals where is stabilised by the coulombic

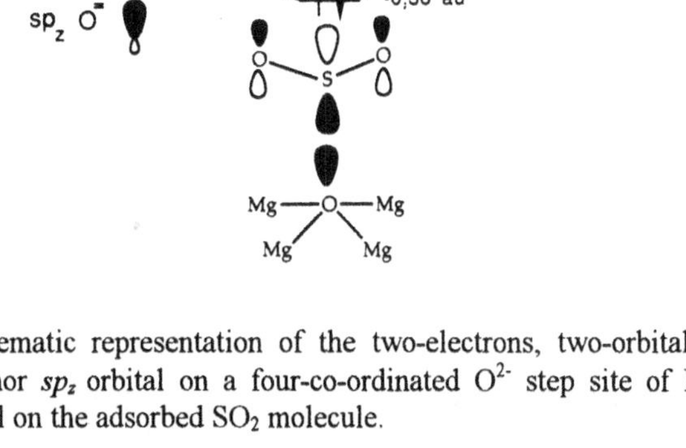

Figure 4. Schematic representation of the two-electrons, two-orbitals interaction between a donor sp_z orbital on a four-co-ordinated O^{2-} step site of MgO and an acceptor orbital on the adsorbed SO_2 molecule.

potential of the ionic lattice. Since the surface basicity is, in ultimate analysis, a case of HOMO-LUMO interaction where the donor orbital is an O $2p_\sigma$ level and the acceptor orbital is some level of proper symmetry on the adsorbed molecule, Fig. 4, it is clear that a less stable, energetically higher donor orbital will result in a larger charge transfer (or basicity). Thus, the description of the oxidising or reducing power of the surface is critically dependent on the representation of the long range coulombic potential of the ionic or semi-ionic oxide surface.

The oxygen sites are not only the sites where the chemical reactivity occurs but also the sites where the metal/oxide interface develops. The correct description of this kind of interaction is therefore quite challenging. It should also be noted that the comparison of cluster and slab calculations for charge-transfer reactions is no longer meaningful. Take the case of alkali adsorption on an oxidising surface. There is little doubt that at low coverage the metal atom will donate one electron to the substrate and become ionised. At high coverage, however, which is the situation usually considered in periodic calculations, the Coulomb repulsion among the positive ions will be too strong and will destabilise the system with consequent charge transfer back to the adsorbate and formation of a metallic overlayer. The situation is reminiscent of what occurs for alkali metals on transition metal surfaces where a ionic-metallic transition occurs as the coverage increases. Clearly, a cluster model, representing a zero-coverage limit, and a periodic calculation, with a finite, non small, coverage, cannot be compared in this case.

3. Oxygen vacancies on the MgO surface

In the previous sections we have analysed the bonding of a small molecule like CO with the regular sites of TM oxide surfaces. However, most of the chemical properties of oxides are connected with the low-co-ordinated defect sites or the cation or anion vacancies. While a considerable attention has been dedicated to the theoretical study of regular surfaces, very few investigations dealing with the electronic structure of defective surfaces have been reported. An ideal system to start looking at the problem is MgO. MgO is an highly ionic crystal, and the (100) face is unreconstructed with very little surface rumpling. Interaction of adsorbates on MgO(100) has been widely studied theoretically with both cluster and periodic approaches and a significant experience has been accumulated in the treatment of this system [15,16,20-37].

3.1. ELECTRONIC STRUCTURE OF F_s CENTRES

We have studied recently the nature and the electronic structure of both O and Mg vacancies at the MgO surface [44]. We report here only the major features related to the presence of O vacancies. These are naturally present in ionic crystals like MgO associated to a corresponding Mg vacancy (Frenkel or Schottky defects). The number of O vacancies, however, can be considerably increased by irradiating the surface by γ-rays or by bombarding it with Ar^+ ions. A simple, chemical, way to increase the concentration of O vacancies is through the dehydroxylation of the hydroxylated MgO surface at high temperature [45]:

$$\text{-Mg-}\overset{\text{H}}{\text{O}}\text{-Mg-}\overset{\text{H}}{\text{O}}\text{-Mg-} \xrightarrow{\Delta} \text{Mg-O-Mg-}|\underline{\text{F}}|\text{-Mg-} + H_2O \qquad (2)$$

A high concentration of O vacancies in MgO turns the colour of the sample to blue, and these defects are known as the colour F centres (from Farbe = colour in German). F_s indicates an O vacancy located at the surface of the material. F centres can be neutral or charged, depending on the atom or ion which is missing. In particular, three cases can be distinguished:

missing O^{2-}	charged F_s^{2+} centre
missing O^-	charged F_s^+ centre
missing O	neutral F_s centre

To study the properties of these centres we used rather large stoichiometric MgO clusters where one atom has been removed; an example is shown in Fig. 5 where a $O_{20}Mg_{21}$ cluster representative of an F_s centre is shown. Smaller clusters have also been used [44]. Double zeta basis sets where adopted for the atoms around the vacancy site, while smaller basis sets were used on the cluster borders.

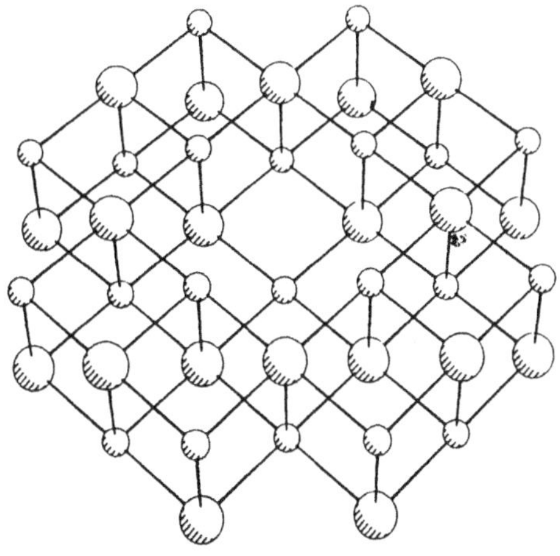

Figure 5. Structure of the $O_{20}Mg_{21}$ cluster used to model an oxygen vacancy on the MgO surface. The cluster is embedded in a large array of point charges (not shown).

All the clusters have been embedded in a large array of (+2/-2) PCs. Local relaxation of the ions around the cavity has been allowed. For more details see ref. [44].

Formation energies of O vacancies are not easy to determine [46]. In fact, while the calculation of the formation energy for the case of a neutral defect is relatively simple, two problems are connected to the formation energy of charged defects. (Here we define the formation energy of an F_s^{n+} centre simply as the energy required to remove and to bring to infinite distance an O^{n-} atom or ion (n=0,1,2) from the regular, non-defective, MgO(100) surface). First, charged defects induce long-range polarisation and relaxation of the lattice which are not easily included in a cluster calculation; second the energy of the missing ion, O^- or O^{2-}, is very critically dependent on the level of theoretical treatment. O^{2-} for instance does not exist as a free unit. By taking into account the experimental values for the O first and second electron affinity, EA, and by using a classical formula to estimate the contribution of the lattice relaxation due to the presence of a charged defect [44], the following formation energies have been computed: F_s = 7.7 eV, $F_s^+ \approx$ 11 eV, and $F_s^{2+} \approx$ 24 eV. These numbers, although purely qualitative, indicate that the order of stability of the centres is $F_s > F_s^+ > F_s^{2+}$. Furthermore, the energetics shows that F_s^+ and F_s^{2+} centres are strong electron-deficient species. This has consequences on their reactivity that will become clear in the following discussion.

As mentioned before, the removal of an atom or of an ion from the MgO surface results in a large relaxation of the lattice. In the calculations only the nearest neighbour ions to the vacancy were allowed to change their positions; for a neutral F_s centre there is an outward displacement of the Mg^{2+} ions and an inward movement of the second neighbour O^{2-} ions by 1-2% with respect to the unrelaxed surface. The effect is much more pronounced for charged vacancies, and changes of ≈5% and ≈10%, respectively, in the Mg-O distances around the cavity are found for F_s^+ and F_s^{2+} centres. This means that the relaxation is quite substantial and certainly not restricted to first and second neighbours. Much larger clusters are required in order to describe these geometric changes.

Surface O vacancies have an interesting electronic structure. MgO is a wide gap insulator, with a gap of 7.8 eV between the valence and the conduction band. The removal of an O atom or ion results in the appearance of impurity states in the band gap, Fig. 6. For an F_s^{2+} centre an empty level is found close to the top of the valence band. This is an acceptor level and indeed addition of electrons to this level results in a large stabilisation of the system. An F_s^+ centre is characterised by a singly occupied level well above the top of the valence band; an F_s centre has a doubly occupied level which is more or less half way between the valence and the conduction bands. This shows that the electrons occupying the cavity have a special character and are clearly different from the other electrons occupying the O 2p valence band. Therefore, the first conclusion from the analysis of the eigenvalues spectrum is that F_s^+ and F_s centres can be seen as very special "objects" and that their electrons have a very distinct character. Actually, the presence of impurity states in the band gap is a sign of the very localised nature of the electrons trapped in the cavity. This is particularly interesting for the case of F_s^+ paramagnetic centres.

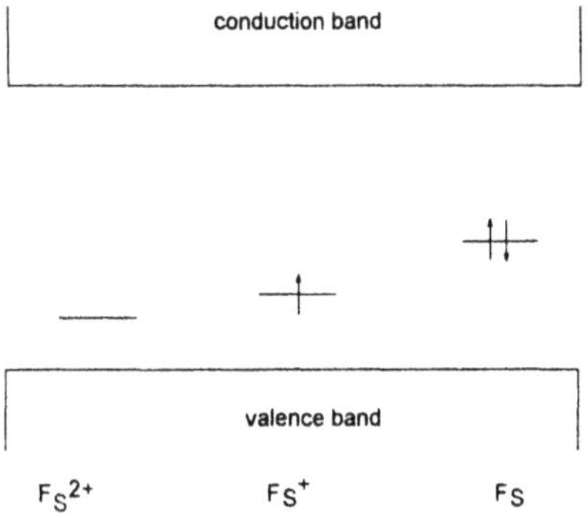

Figure 6. Schematic representation of electronic states in the MgO band gap for neutral and charged surface oxygen vacancies [44].

To better characterise the electron localisation at F_s^+ and F_s centres, we have determined spin or charge density plots. A question which has been matter of debate in the literature, at least for bulk F^+ centres, is in fact the degree of localisation of the unpaired electron. In principle, this electron could reside on the 3s Mg orbitals. However, the spin density plot in the MgO surface plane, Fig. 7, clearly shows the very localised nature of the unpaired electron, in agreement with the previous discussion of impurity states. In a similar way, charge density differences show that in neutral F_s centres the electronic charge is trapped in the centre of the vacancy and only to a very minor extent is distributed over the surrounding Mg cations [44]. The physical reason is that the electronic charge is confined at the centre of the vacancy by the Madelung potential of the ionic lattice. In fact, a completely different distribution of the electronic charge is found by using clusters without PC embedding (i.e. with no representation of the Madelung potential); when the external field is removed the electrons reside mostly on the Mg ions [9], at variance with the ESR data.

A direct, compelling, proof of the localised nature of the trapped electrons comes from the analysis of the EPR spectra through the determination of the hyperfine isotropic coupling constant, a_{iso}, of the electronic spin with the 5/2 nuclear spin of the ^{25}Mg nuclide (which has a natural abundance of 10.1%). a_{iso} is related to the spin density at the nucleus (the Fermi contact term) by the following equation:

$$a_{iso} = (8\pi/3)\, g_N \beta_N g_e \beta\, |\psi_s(0)|^2 \tag{3}$$

where g_N and g_e are the nuclear and electronic g-factors, and β_N and β are the nuclear and Bohr magnetons; $|\psi_s(0)|^2$ is the s spin density at the nucleus. The value of the hyperfine coupling constant for F^+ centres in bulk MgO is 3.94 G; using a Mg_6O_{12}+PCs cluster model of a bulk F^+ centre we compute a coupling constant of 4.5 G, in excellent agreement with the experiment. This makes us confident that the computed values for F centres at the surface of MgO are sufficiently accurate, although the inclusion of correlation effects is expected to change the values of the hyperfine coupling constants, in particular to lower them. For a five-co-ordinated surface vacancy, F_s^+, we found coupling constants in the range of 4 to 7 G, i.e. slightly larger than for the bulk, in close agreement with the experimental measures [45]. This is consistent with the fact that the "hole" is surrounded by only 5 instead of 6 Mg cations as in the bulk. The computed spin densities are about 10% of what one would expect if the unpaired electron would be entirely delocalized over the Mg 3s orbitals; this would correspond to hyperfine coupling constants of about 40 G, ten times larger than computed and experimentally observed. This shows unambiguously that the unpaired electron is largely localised at the centre of the vacancy, as shown by the spin density plot and indicated by the analysis of the orbital energies [44].

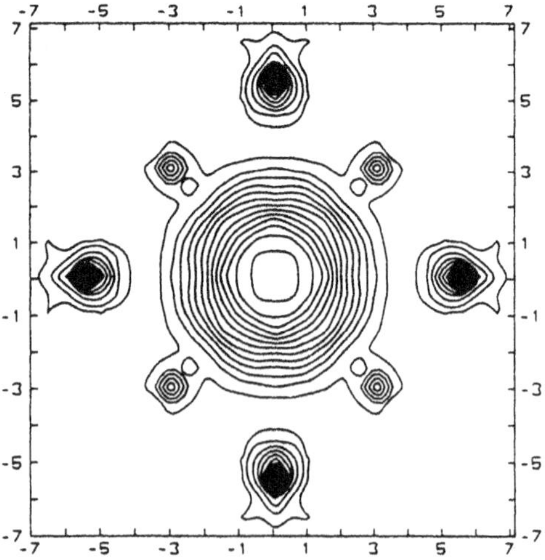

Figure 7. Spin density plot in the MgO surface plane for a paramagnetic F_s^+ centre [44].

3.2. ADSORPTION OF METAL ATOMS ON MgO SURFACE VACANCIES

Medium-energy ion-scattering experiments have shown that only 50% of the initially incident Cu atoms stick to the MgO (100) surface [47], indicating a weak adsorption. More recent experiments on the same system give a somewhat higher value of the sticking probability, which however remains < 1 [48]. Quantum-chemical calculations are in qualitative agreement with these results. Recent accurate density functional investigations have found a binding energy of less than 0.3 eV [49] for Cu atoms on MgO, in agreement with the weak nature of the interaction found experimentally. This does not apply to all metals; Ni, Pd, and Pt, for instance, seem to form relatively strong covalent bonds with the surface oxygens of MgO according to density functional theory (DFT) calculations [50]. On the other hand, the interaction is always very weak when the adsorption occurs at the surface Mg^{2+} cations [49,51].

In order to compare the adsorption at the surface vacancies with that on a regular surface, we have studied the bonding of neutral atoms and of positively charged ions with the O^{2-} sites of the regular surface by using a minimum, non-stoichiometric, $[OMg_5]^{10-}$ cluster. The interaction of Ag and Rb with a surface O^{2-} ion is purely repulsive, while a very small binding energy, 0.11 eV is found for Pd. These results are at the uncorrelated HF level. At the relativistic DFT level of theory, using gradient-corrected exchange-correlation potentials, Ag is very weakly bound, 0.2 eV, while Pd forms stronger bonds of about 0.8 eV [50]. The interaction energies are obviously larger even at the HF level for the cations, with Pd^+ being bound by 0.87 eV, followed by Ag^+, 0.58 eV, and by Rb^+, 0.22 eV [51]. Given the entirely electrostatic nature of the interaction, the stabilisation is a direct function of the distance from the surface, hence of the cation size; the trend in binding energy is inversely related to the cation-surface distance and the cation ionic radius.

These data, although qualitative, are in line with previous findings and with the general picture of the bonding of metal atoms and clusters on the MgO surface [49]. The regular (100) face is rather unreactive toward adsorbed species, including metal atoms. A larger binding energy is found for metal cations interacting with the surface oxide anions. The obvious conclusion is that regular terraces of the MgO surface are not the best candidates as the sites where a metal particle starts to grow during metal deposition. Steps, corners, and in general the morphological defects of the surface as well as the surface vacancies are much more likely the sites where the agglomeration process begins.

We have shown in the previous section that a neutral F_s defect is characterised by the presence of two electrons trapped in the centre of the cavity. In a sense, the two electrons "replace" the missing O^{2-} ion. For this reason, the reactivity of a neutral F_s centre is not very different from that of a regular surface site; in particular, the vacancy is not electron deficient and does exhibit a positive electron affinity, EA. In fact, the interaction of an F_s centre with Rb and Ag, two atoms with a singly occupied outer s orbital, is repulsive, as for the non-defective surface (for Rb more than for Ag because of the larger size of the alkaline atom). Pd, on the other hand, is bound by 1.6 eV. Compared to a regular MgO oxide anion, on a F_s site Pd is much more strongly bound. The special behaviour of Pd may be surprising at a first look. However, a closer analysis shows that the two electrons trapped in the vacancy (occupying a total symmetric a_1 level) can be partially delocalized over the empty 5s orbital of a ground state $4d^{10}5s^0$ Pd atom. The two electrons of the F_s vacancy belong to an impurity state in the band gap of the material; in a similar way, in F_s /Pd there is a molecular level well separated from the top of the valence band and the bottom of the conduction band with partial Pd 5s character. The nature of this orbital is shown in Fig. 8, where the delocalization of the electron pair over Pd is apparent. This is the origin of the stabilisation of the surface complex [51].

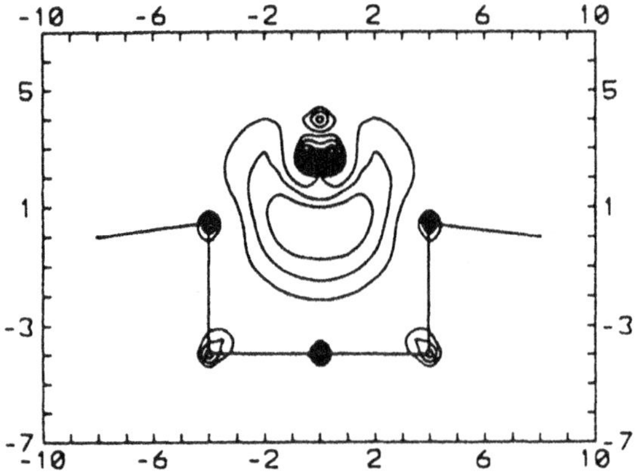

Figure 8. Orbital charge density plot of the electron pair localised between the vacancy and an adsorbed Pd atom in F_s/Pd. The plot is normal to the surface plane [51].

This stabilisation is obviously absent when the outer s orbital of the metal atom is partially or fully occupied because of the Pauli repulsion between the two fragments. In this respect, the behaviour of Pd is quite peculiar and is not expected for most of the other metal atoms of the periodic table. The different bond strength of a Pd atom on a F_s centre, 1.6 eV, compared to a regular surface O^{2-} ion, 0.1 eV, can be explained in terms of polarisability and donor ability of the two sites. The tendency to donate electrons of a regular O^{2-} site of MgO is very low (see section 2.4) as shown by the high IP of the $[OMg_5]^{10+}$ cluster, 13.4 eV. For comparison, the IP of the $[O_{12}Mg_{13}]$ model of an F_s site is 5.4 eV only.

Furthermore, the two electrons in the F_s cavity are much more polarisable than those of a surface O^{2-} ion.

The interaction of metal atoms with F_s^+ paramagnetic centres involves the electron localised at the vacancy. A F_s^+ centre has a negative EA, -5.4 eV; this means that a partial electron transfer from the metal to the surface is possible. Indeed, this is what happens with Rb which has a low IP, 3.2 eV. When Rb is adsorbed on a F_s^+ site looses its valence electron with formation of neutral F_s and Rb^+. The interaction is thus very similar to that between Rb^+ and a regular O^{2-} surface site where the bonding is very weak, about 0.2 eV. On a F_s centre the interaction is even weaker and the Rb^+ ion is unbound at the HF level.

A completely different interaction occurs with Pd. The interaction between of a closed shell Pd atom and the open shell vacancy leads to a doublet ground state. The EA of the F_s^+ site, -5.4 eV, is not large enough to compensate the cost of the ionisation of Pd (PI = 6.4 eV); nevertheless, a bonding of about 1 eV arises from the polarisation of the metal atom and from the partial delocalization of the unpaired electron on the empty 5s level of Pd, in analogy with the F_s/Pd interaction. In a similar way, on F_s^+ an Ag atom is bound by about 1 eV. The interaction with Ag involves the low-spin, singlet coupling of the two unpaired electrons with formation of a molecular orbital with bonding character. This orbital, formed from the Ag 5s and the vacancy a_1 orbitals is localised between the metal atom and the surface (Fig. 9). Since the IP of Ag is higher than the EA of F_s^+ there is no full ionisation of the Ag atom, differently from Rb, and the bonding can be described as a covalent polar bond.

F_s^{2+} centres are extremely reactive because of their very large EA, about 11 eV according to our calculations. Therefore, an F_s^{2+} site tends to extract electrons

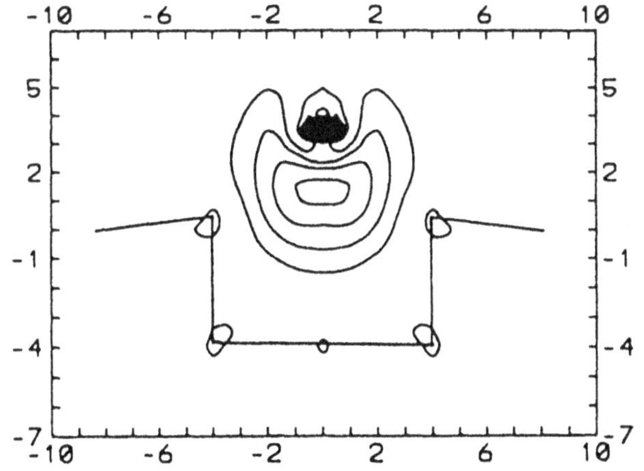

Figure 9. Orbital charge density of the electron pair resulting from the coupling of the F_s^+ and Ag 5s unpaired electrons. See Fig. 8 for definitions [51].

from adsorbed species to form the more stable F_s^+ centres. This is what happens when Rb, Pd or Ag atoms interact with these sites: one metal electron is removed (from the 5s orbitals in Rb and Ag or from a 4d orbital in Pd) and trapped into the cavity with formation of F_s^+ and M^+ (M=Rb, Pd or Ag). Clearly, the presence of two positively charged units leads to an electrostatic repulsion. The metal cations formed in this way are not stable over the resulting F_s^{2+} site. In other words, the F_s^{2+} sites on the MgO surface acts as very strong electron acceptors which are able to ionise virtually any metal atom deposited on the surface. The resulting cations however are unbound to these sites and diffuse on the surface where they become attracted by other sites, like a surface anion or a defect [51].

4. Catalysis by supported particles: electrochemical promotion

4.1. ELECTROCHEMICAL PROMOTION (NEMCA)

A new example of promotion in catalysis has been discovered a few years ago and named NEMCA, Non-Faradaic Electrochemical Modification of Catalytic Activity [52-54]. By varying in a controlled way the catalyst-electrode potential of metal particles supported on solid electrolytes, dramatic changes in the catalytic activity of oxidation reactions have been observed. Under the effect of an external potential there is a migration of ions through the supporting oxide (a solid electrolyte) to the surface of the metal catalyst. The solid electrolytes used are either anion (O^{2-} or F^-) conductors like Y_2O_3-stabilised ZrO_2, or cation (Na^+) conductors like β"Al_2O_3. The effect has been observed for various metals, Pt, Rh, Au, Ni, Ag [52-54]. Under these conditions the catalytic reaction rate of, for instance, ethylene oxidation on Pt increases by several orders of magnitude.

It has been observed that the spillover of ions on the catalyst surface is accompanied by a change in the metal work function (in particular an increase of $e\Phi$ for O^{2-} conductors [52]); recently, it has been shown that this increase in work function is linearly correlated with a decrease of the activation barrier for oxygen desorption [55]. Clearly, a less strongly bound oxygen will more easily oxidise another adsorbed molecule (e.g. ethylene) thus increasing the catalytic rate. An opposite effect has been observed for spillover of positive ions (Na^+) on the catalyst surface [52-54]. The observed correlation between oxygen bond strength and metal work function does not exclude that other possible effects take place and determine the origin of the electrochemical promotion. In particular, electrostatic effects or simply changes in the electrochemical potential of the adsorbed atoms can be of significant importance for cases where a sufficiently high concentration of adsorbed ions is present on the catalyst surface.

To better understand the origin of NEMCA we have performed Hartree-Fock calculations with the help of two clusters, Cu_{34} and Pt_{25}, see Fig. 10, to model oxygen adsorption on the open sites of the Cu(100) and Pt(111) surfaces, respectively [56]. Flexible basis sets of double-zeta plus polarisation quality (DZP) were used for the clusters; the central atoms of the cluster were treated with 11-electron ECP (Cu) [10] and relativistic 10-electron ECP (Pt) [57], respectively, while relativistic 1-electron ECPs were used for the surrounding Cu [11] or Pt [58,59] atoms. The calculations were performed with the Hondo 8.5 program package [6] except for Pt were the Hondo-Cipsi program package was used [60]. Coadsorbed ions have been modelled by placing point charges, PCs, above the cluster surface, and image charges of opposite sign below the cluster first layer. For symmetry reasons, 4+4 PCs have been used in the case of Cu and 3+3 PCs for Pt. For a more detailed description of the cluster electronic states, basis sets, etc. see ref. [56].

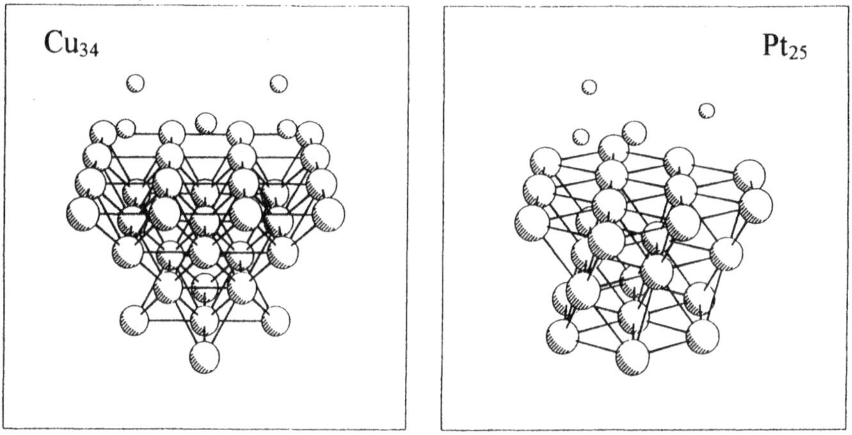

Figure 10. Cu_{34} and Pt_{25} clusters used to model coadsorption of atomic O and positive and negative ions on the Cu(100) and Pt(111) surfaces, respectively. Oxygen has been adsorbed on the central hollow sites. The position of the coadsorbed ions (or point charges) is also shown.

4.2. POINT CHARGE MODELS OF COADSORBED IONS

The binding energy of adsorbed oxygen has been computed with respect to the ionic dissociation limit, M_n^+ and O^-, a choice justified by the nature of the metal-oxygen bond. Of course correlation effects are essential for the determination of the metal-O bond strength [61], in particular for Pt [62]. However, the scope is not to obtain accurate values of the oxygen D_e but rather to reproduce the trends in D_e as a function of the presence of coadsorbates. In this respect it is important to correctly describe the nature of the Cu(100)-O and Pt(111)-O chemical bonds. A careful analysis of the bonding nature, based on various theoretical measures, has shown that O on Cu [61] and Pt [62] surfaces has a substantial ionic character. Given the high electronegativity of oxygen, it is not surprising that the charge associated to the adsorbed oxygen is not smaller than -1. The two metals, Cu and Pt, exhibit a slightly different tendency to transfer charge to the adsorbed oxygen; on Cu, in fact, the bonding has a more pronounced ionic character than on Pt where the covalent mixing seems to be more important. This is consistent with the higher experimental work function of Pt(111), 5.7 eV, compared to Cu(100), 4.6 eV. It is also important to stress that the largely ionic nature of the adsorbed oxygen is found with both correlated and uncorrelated wavefunctions [62]. This means that while electron correlation effects are essential for the correct reproduction of D_e, they do not change significantly the nature of the bond.

We first consider the PCs case. The PCs have been varied from q=+0.5 to q=−0.5 (these are the values of charges at +z above the surface; opposite charges were added at -z). The position of adsorbed O has been optimised in the presence of the PCs, Table 4.

Table 4 - Chemisorption properties of an O atom adsorbed on Cu_{34} and Pt_{25} clusters in the presence of point charges, PCs, +q/-q (and viceversa) above and below the surface.[a] The D_e values are computed with respect to the ionic asymptotic limit and corrected by the O EA.

PCs	Cu(100)			Pt(111)		
	eΦ,[b] eV	z_e, bohr	D_e, eV	eΦ,[b] eV	z_e, bohr	D_e, eV
+0.5/-0.5	3.56	1.79	6.60	3.82	2.93	3.45
+0.3/-0.3	3.89	1.72	6.63	4.13	2.91	3.43
+0.1/-0.1	4.00	1.64	6.44	4.49	2.90	3.34
no PCs	4.07	1.60	6.29	4.66	2.90	3.26
-0.1/+0.1	4.15	1.55	6.10	4.83	2.91	3.19
-0.3/+0.3	4.22	1.46	5.67	5.16	2.95	2.99
-0.5/+0.5	4.36	1.39	5.34	5.46	3.05	2.77

a) The PCs are placed at 4.5 bohr above and below the cluster first layer.
b) The work function eΦ is determined as -ε(HOMO)

With +q/-q PCs (positive adsorbed ions) a considerable increase in the D_e is found; the opposite is true for -q/+q PCs. This is the trend experimentally found, a decrease in oxygen D_e in correspondence with a spillover of negative ions. The change in binding energy is associated to a change in metal work function (as measured by the cluster HOMO according to the Koopmans' theorem [63]); in particular, the HOMO is shifted to lower energies by the negative ions, again consistent with the experimental observations. A simple interpretation is that the decrease of cluster HOMO (increase of eΦ) reduces the ability of the metal to transfer charge to oxygen thus decreasing the bond strength. This is the "chemical" mechanism originally proposed to explain the phenomenon [52]. However, we found that for Cu *the increase of D_e is associated with an* *elongation of the metal-O bond length!* This is not what is expected based on classical chemical bonding arguments: a stronger bonding should correspond to a shorter distance. The fact that the distance increases even if the bond becomes stronger suggests that other effects must take place, and in particular that the origin of the change in bond strength can be electrostatic, more than chemical.

4.3. ELECTRIC FIELDS AT SURFACES

To check this hypothesis we have considered the Cu_{34}-O and Pt_{25}-O clusters in the presence of an external uniform electric field, F. We used uniform electric fields normal to the surface to represent the effect of the ions. The applied fields are in the range of 0.005-0.025 a.u.; a field F = 0.010 a.u. corresponds to 5.7 x 10^7 V/cm. For comparison, in semiconductors fields of the order of 10^6 V/cm can be maintained, while fields within the double layer at the electrode-electrolyte interface can reach 10^7 V/cm. Exposed cations in zeolite cavities create fields of 10^8 V/cm (1 V/Å). The upper limit of electric field strength that can be maintained over microscopic distances before field emission of field evaporation takes place is of the order of ≈6 V/Å; this is also the field experienced by valence electrons in atoms and molecules [64].

The D_e of oxygen has then been computed at two levels. According to first order perturbation theory, for small, uniform, electric fields the total energy can be expresses as a function of F as

$$E_P\,(r,F) = E_{SCF}(r,0) - \mu\,(r,0) \cdot F \qquad (4)$$

where $E_{SCF}(r,0)$ and μ (r,0) are the energy and dipole moments of the system for F=0. E_P represents the first-order Stark effect [65,66] and the correction to the F=0 energy is simply due to the interaction between the dipole of the system (in this case the M_n^+-O^- polar bond) and the external field (representing the long-range effects of the coadsorbed ions). Changes in the electron distribution due to the field are not included at this level.

Thus, the Stark interaction energy does include only electrostatic effects with no change in the chemical bonding between the metal and oxygen. These changes are taken into account by relaxing the wavefunction for F≠0. The polarisation of cluster and adsorbate electrons and the modifications in chemical bonding mechanism are thus considered in the Full SCF wavefunction. This separation between Stark (electrostatic) and Full SCF (chemical) terms allows one to identify the importance of the two contributions.

A positive electric field (corresponding to negative adsorbed ions), causes in a decrease of D_e and a shortening of the surface-O distance, as for the PC models. For small fields, the dependence is linear for both Stark and Full SCF curves;

deviations from linearity are observed for larger fields, in particular for Cu, Fig. 11 and Tables 5-6.

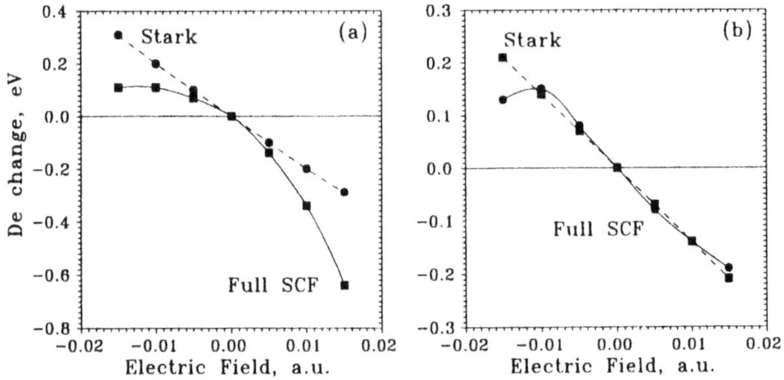

Figure 11. Stark and Full SCF curves showing the dependence of the oxygen adsorption energy, D_e, on the field strength. (a) Cu_{34}; (b) Pt_{25}.

The fact that the change in D_e found at the first-order perturbation theory level is virtually the same as for a Full SCF wavefunction provides a strong indication that the interaction between the adsorbed ions and oxygen is largely electrostatic and does involve only to a minor extent changes in the chemical bonding. The results show a more pronounced dependence of D_e on the PCs than on uniform fields. This shows the importance of local inhomogeneities in the field.

Table 5 - Chemisorption properties of an O atom adsorbed on a Cu_{34} cluster in the presence of a uniform electric field, F, normal to the surface plane. The D_e values are computed with respect to the ionic asymptotic limit, Cu_{34}^+ and O^-, and corrected by the Cu_{34} IP and O EA for F=0.

Field, a.u.	eΦ,[a] eV	z_e, bohr		D_e, eV	
		Stark	Full SCF	Stark	Full SCF
-0.015	2.96	1.73	1.72	6.60	6.40
-0.010	3.29	1.68	1.68	6.49	6.40
-0.005	3.68	1.64	1.64	6.39	6.36
no field	4.07	---	1.60	---	6.29
+0.005	4.48	1.55	1.55	6.19	6.15
+0.010	4.98	1.51	1.51	6.09	5.95
+0.015	5.56	1.46	1.47	6.00	5.65

a) The work function eΦ is determined as -ε(HOMO)

Table 6 - Chemisorption properties of an O atom adsorbed on a Pt_{25} cluster in the presence of a uniform electric field, F, normal to the surface plane. The D_e values are computed with respect to the ionic asymptotic limit, Pt_{25}^+ and O^-, corrected by the Pt_{25} IP and O EA for F=0.

Field, a.u.	eΦ,[a] eV	z_e, bohr		D_e, eV	
		Stark	Full SCF	Stark	Full SCF
-0.015	3.36	2.96	2.96	3.47	3.39
-0.010	3.65	2.94	2.94	3.40	3.41
-0.005	4.13	2.92	2.92	3.33	3.33
no field	4.66	---	2.90	---	3.26
+0.005	5.22	2.88	2.89	3.19	3.18
+0.010	5.81	2.86	2.87	3.12	3.12
+0.015	6.40	2.84	2.86	3.05	3.06

a) The work function eΦ is determined as -ε(HOMO)

In both the PC and field models we have determined the position of the metal work function for different values of F, see Tables 5 and 6. eΦ exhibits a similar dependence on uniform fields and PCs, see also Table 4. Since this is not the case for D_e which changes more rapidly with the PC models, it is interesting to verify whether a direct relationship exists between eΦ and the oxygen D_e. This is done in Fig. 12 by plotting D_e computed with the PC and the field models against the corresponding changes in eΦ, Δ(eΦ).

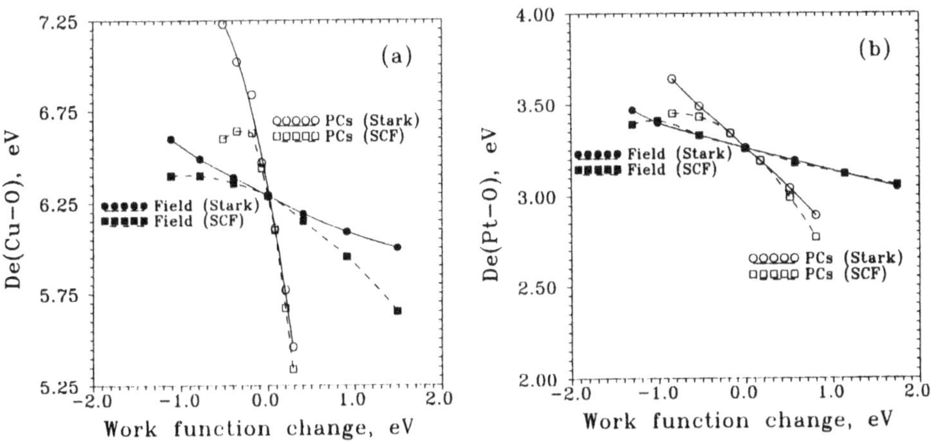

Figure 12. Work function change, −ε(HOMO), versus oxygen adsorption energy, D_e. The curves refer to the cluster with point charges (PCs), o and □, and uniform electric fields, • and ■. Both Stark (solid line) and Full SCF (dotted line) curves are shown. (a) Cu_{34}; (b) Pt_{25}.

We see that identical values of the change in work function, $\Delta(e\Phi)$, correspond to very different oxygen adsorption energies in the two models (PCs and electric field)! *Therefore, the oxygen bond strength is not directly related to* $e\Phi$. The first order effect is the electrostatic interaction between the electric field (uniform or non-uniform) and the surface dipole associated to the metal-oxygen bond. The chemical modifications connected to the change in metal work function is a second-order effect. The work function change is an observable consequence of the presence of ions on the surface, but is not directly responsible for the concomitant change in oxygen desorption barrier and catalytic rate.

5. Conclusions

Recent results of cluster model Hartree-Fock calculations on oxide surfaces have been reviewed. We have shown how the use of relatively simple models allows one to gain understanding of phenomena occurring at the surface of oxide materials. The chemistry of oxide surfaces is largely the chemistry of defects, vacancies, irregularities. This is why local models, like the cluster model, are particularly useful. Of course more sophisticated quantum-mechanical approaches are required for a more accurate evaluation of the energetics. The density functional approach is very promising in this respect. There is space for a considerable improvement of the theoretical approaches and the efficiency of the computing programs. On the other hand, this must be accompanied by a more detailed and accurate structural information from new experimental techniques. The design of "realistic" models of the surface of an oxide is in fact critically dependent on the structural information available. A considerable improvement in this direction will come from the dynamical simulation of these materials and their surfaces and the development of good force-fields will be of crucial importance in the near future.

To summarize, the future of scientific computing in the area of surfaces and heterogeneous catalysis, although at the very beginning, looks pretty brilliant. The potential use of different computational approaches (force-fields, cluster models, band structure calculations, etc.) and levels of theoretical treatment will make it possible to analyze and interpret a variety of experimental facts and measures. The combined use of computations and experiments will provide a solid basis not only for the understanding of the phenomena but also for the design of materials or catalysts with new, unprecedented, properties.

Acknowledgments
This work would have not been possible without the invaluable help of Anna Maria Ferrari (Milano), Thomas Bredow (Hannover), and Anna Clotet (Tarragona) and the cooperation with Francesc Illas (Barcelona), Jose Manel

Ricart (Tarragona) and Paul S. Bagus (San Jose, CA). Several useful discussions with experimentalists, in particular Elio Giamello (Turin) and Costas G. Vayenas (Patras) are gratefully acknowledged. I am indebted to Notker Rösch (Munich) for several very interesting discussions and a close collaboration on the subject of density functional calculations on oxide surfaces. Finally, I would like to thank NATO for the Collaborative Research Grant N. 941191, the Computational Center of Catalunya (CESCA) for supporting my visits to the University of Barcelona through the Human Capital and Mobility Program, contract CHGE-CT92-009, the "Vigoni Program" from CRUI and the Alexander von Humboldt Fundation for supporting my visits to the Technical University of Munich.

References

1. Heinrich, V.E., and Cox, P.A. (1994) *The Surface Science of Metal Oxides*, Cambridge University Press, Cambridge.
2. Pacchioni, G., Bagus, P.S., and Parmigiani, F. Eds., (1992) *Cluster Models for Surface and Bulk Phenomena*, NATO ASI Series B, Vol. 283, Plenum, New York.
3. Pacchioni, G. (1995) *Heterogeneous Chem. Rev.* **2**, 213.
4. Pisani, C., Dovesi, R., and Roetti, C. (1988) *Hartree-Fock Ab Initio Treatment of Crystalline Systems*, Lecture Notes in Chemistry, Vol. 48, Spinger, Berlin.
5. Dovesi, R., and Saunders, V.R. (1992) *Crystal92 User's Manual*, Gruppo di Chimica Teorica, Torino.
6. Dupuis, M., Johnston, F. and Márquez, A. (1994) *HONDO 8.5 for CHEMStation*, IBM Co. Kingston.
7. Bredow, T. and Pacchioni, G. submitted.
8. Pacchioni, G., Ferrari, A.M., and Bagus, P.S., (1996) *Surf. Sci.*, **350**, 159.
9. Pacchioni, G., Ferrari, A.M., Marquez, A., and Illas, F. (1996) *J. Comput. Chem.*, in press.
10. Hay, P.J., and Wadt, W.R. (1985) *J. Chem. Phys.* **82**, 270.
11. Bagus, P.S., Bauschlicher, C.W., Nelin, C.J., Laskowski, B.C., and Seel, M. (1984) *J.Chem. Phys.* **81**, 3594.
12. Faegri, K., and Speis, H.J. (1987) *J. Chem. Phys.* **86**, 7035.
13. van Duijneveldt, F.J., (1971) *IBM Res. Rep. N. RJ 945.*
14. Boys, S.F. and Bernardi, F. (1970) *Mol. Phys.* **19**, 553.
15. Ferrari, A.M. and Pacchioni, G. (1996) *Int. J. Quant. Chem.*, **58**, 241.
16. Pacchioni, G., Cogliandro, G., and Bagus, P.S. (1991) *Surf. Sci.*, **255**, 344.
17. Bagus, P.S., Hermann, K., and Bauschlicher, C.W. (1984) *J. Chem Phys.* **80**, 4378.
18. Bagus, P.S. and Illas, F. (1992) *J. Chem. Phys.* **96**, 8362.
19. Hermann, K., Bagus, P.S., and Bauschlicher, C.W. (1984) *Phys. Rev. B*, **30**, 7313.
20. Huzimura, R., Yanagisawa, Y., Matsumura, K., and Yamabe, S. (1990) *Phys. Rev. B*, **41**, 3786.
21. Borve, K.J., and Pettersson, L.G.M. (1991) *J. Phys. Chem.* **95**, 7401.
22. Pacchioni, G., Cogliandro, G., and Bagus, P.S. (1992) *Int. J. Quant. Chem.* **42**, 1115.
23. Borve, K.J. (1992) *J. Chem. Phys.* **96**, 6281.
24. Strömberg, D. (1992) *Surf. Sci.* **275**, 473.
25. Neyman, K. and Rösch, N. (1992) *Chem. Phys.* **168**, 267.
26. Pölchen, M. and Staemmler, V. (1992) *J. Chem. Phys.* **97**, 2583.
27. Pacchioni, G., Minerva, T., and Bagus, P.S. (1992) *Surf. Sci..* **275**, 450.

28. Freitag, A., Staemmler, V., Cappus, D., Ventrice, C.A., Shamery, K.A., Kuhlenbeck, H., and Freund, H.J. (1993) *Chem. Phys. Lett.* **210**, 10.
29. Nygren, M., Pettersson, L.G.M., Barandiaran, Z., and Seijo, L. (1994) *J. Chem. Phys.* **100** 2010.
30. Nakatsuji,K.,Yoshimoto, M., Hada, M., Domen, K.,and Hirose,C. (1995) *Surf. Sci.* **336**, 232.
31. Mejias, J.A., Marquez, A.M., Fernandez-Sanz, J., Fernandez-Garcia, M., Ricart, J.M., Sousa, C., and Illas, F. (1995) *Surf. Sci.* **327**, 59.
32. Neyman, K., Ruzankin, S.P., and Rösch, N. (1995) *Chem. Phys. Lett.* **246**, 546.
33. Ito, T., Tashiro, T., Kawasaki, M., Watanabe, T., Toi, K., and Kobayashi, H. (1991) *J. Phys. Chem.* **95**, 4476.
34. Sawabe, K., Morokuma, K., and Iwasawa, Y. (1994) *J. Chem. Phys.* **101**, 7095.
35. Yanagisawa,Y., Takaoka, K., and Yamabe, S.(1994) *J. Chem. Soc. Faraday Trans.* **90**, 2561.
36. Kobayashi, H., Salahub, D.R., and Ito, T. (1994) *J. Phys. Chem.* **98**, 5487.
37. Pelmenschikov, A.G., Morosi, G., Gamba, A., and Coluccia, S. (1995) *J. Phys. Chem.* **99**, 15018.
38. Sauer, J., Ugliengo, P., Garrone, E., and Saunders, V.R. (1994) *Chem. Rev.* **94**, 2095.
39. Sauer, J. (1989) *Chem. Rev.* **89**, 199.
40. Colbourn, E.A. (1992) *Surf. Sci. Reports* **15**, 281.
41. Pacchioni, G. (1993) *Surf. Sci.* **281**, 207.
42. Pacchioni, G., Ricart, J.M., and Clotet, A. (1994) *Surf. Sci.*, **315**, 337.
43. Pacchioni, G., Ricart, J.M., and Illas, F. (1994) *J. Am. Chem. Soc.*, **116**, 10152.
44. Ferrari, A.M. and Pacchioni, G. (1995) *J. Phys. Chem.*, **99**, 17010.
45. Murphy, D. and Giamello, E. (1995) *J. Phys. Chem.*, **99**, 15172 and references therein.
46. Pisani, C., Corà, F., Dovesi, R., and Orlando, R. (1994) *J. Electron Spectros. Relat. Phenom.* **96**, 1.
47. Zhou, J.B., Lu, H.C., Gustafsson, T., and Garfunkel, E. (1993) *Surf. Sci. Lett.* **293**, L887B.
48. Wu, M.C., Oh, W.S., and Goodman, D.W. (1995) *Surf. Sci.* **330**, 61.
49. Pacchioni, G., and Rösch, N. (1996) *J. Chem. Phys.* **104**, 7329.
50. Yudanov, I., Pacchioni, G., Neyman, K., and Rösch, N. to be published.
51. Ferrari, A.M. and Pacchioni, G. (1996) *J. Phys. Chem.* **100**, 9032.
52. Vayenas, C.G., Bebelis, S., and Ladas, N. (1990) *Nature (London)* **343**, 625.
53. Vayenas, C.G., Bebelis, S., Yentekakis, I.V., and Lintz, H.G. (1992) *Catal. Today* **11**, 303.
54. Vayenas, C.G., Bebelis, S., and Despotopoulou, M. (1991) *J. Catal.* **128**, 415.
55. Neophytides, S. and Vayenas, C.G. (1995) *J. Phys. Chem.* **99**, 17063.
56. Pacchioni, G., Illas, F., Vayenas, C.G., and Neophytides, S. (1996) *J. Phys. Chem.*. in press.
57. Durand, P. and Barthelat, J.C. (1975) *Theoret. Chim. Acta* **38**, 283.
58. Illas, F., Márquez, A., Zurita, S., and Rubio, J. (1995) *Phys. Rev. B* **52,** 12372.
59. Rubio, J., Zurita, S., Barthelat, J.C., and Illas, F. (1994) *Chem. Phys. Lett.* **217**, 283.
60. Dupuis, M., Rys., J. and King, H.F., *HONDO-76, program 338*; QCPE University of Indiana, Bloomington.
61. Ricart, J.M., Clotet, A., Illas, F., and Rubio, J. (1994) *J. Chem. Phys.* **100**, 1988.
62. Illas, F., Rubio, J., Ricart, J.M., and Pacchioni, G. submitted.
63. Koopmans, T.A. (1933) *Physica* **1**, 104.
64. Kreuzer, H.J. (1992), in: *Surface Science and Catalysis. In Situ probes and Reaction Kinetics*, Dwyer, D., Hoffmann, F. Eds.; Vol. 482, p. 269, ACS Symp. Series, Washington.
65. Lambert, D.K. (1983) *Phys. Rev. Lett.* **50**, 2106.
66. Bagus, P.S., Nelin, C.J., Müller, M., Philpott, M.R., and Seki, H. (1987) *Phys. Rev. Lett.* **58**, 559.

THEORETICAL MODELLING OF CHEMISORPTION AND REACTIONS ON METAL-OXIDE SURFACES

L.G.M. PETTERSSON, M. NYBERG AND J.-L. PASCUAL
Department of Physics, University of Stockholm
Box 6730, S-113 85 Stockholm, Sweden

AND

M.A. NYGREN
The Royal Institution, 21 Albemarle Street
London, UK, W1X 4BS

Abstract. Our approach to modelling metal-oxide surfaces and interactions of small molecules with these is described with particular emphasis on the determination of surface structure (using pair-potential simulations) and the embedding of the cluster-models. The electrostatic potentials obtained from different models are discussed and the computational strategy is described. Examples include the binding energy of CO to MgO(100), interaction of water with $Cu_2O(100)$, hydrogen dissociation over differently reconstructed ZnO surfaces. The static exchange approach to X-ray spectroscopies is described and applied to core excitons at alkaline-earth oxide surfaces.

1. Introduction

The present chapter will be mainly concerned with the modelling of metal-oxide systems that can be regarded as rather strongly ionic, *i.e.* in which the contributions of covalent bonding can be regarded as minor. In particular, we will discuss systems where a cluster model of an oxide surface or bulk may be constructed without having to consider how bonds should be cut between the central cluster and the rest of the crystal. Thus typical systems for the present modelling approach are *e.g.* alkali oxides, alkaline-earth oxides and many transition-metal oxides where there are no oxygen-oxygen bonds; systems excluded can be typefied by Al_2O_3 and other more cova-

R.M. Lambert and G. Pacchioni (eds.),
Chemisorption and Reactivity on Supported Clusters and Thin Films, 425-454.

lently bound systems. In the former case the interaction with the rest of the crystal may be given by some embedding potential while in the second case, termination of the cluster using some appropriate chemical group (often hydrogens) to saturate dangling bonds is a better approach.

Even with this choice of systems we find that modelling of chemisorption and reactions on strongly ionic metal-oxide surfaces, even "simple" oxides such as the alkaline-earths, may still in many respects be more complicated than the modelling of metal surfaces. There are several reasons for this: problems involving metals *can* concern mono-metallic (*i.e.* only one element) substrates while any metal-oxide *must* involve at least two different species with different charges. The metal surface is characterised by a high conductivity leading to short screening lengths, while the lower conductivity or insulating properties of the ionic oxides lead to long screening lengths and a relatively higher importance of long-range interactions. In particular, the electrostatic interactions with the rest of the ionic crystal will in many cases provide dominating contributions to the properties of a particular surface. This is especially true for cuts leading to a polar surface, *i.e.* with only one type of ion exposed at the surface; this is an electrostatically unstable situation and such a surface must somehow reconstruct in order to eliminate the dipole across the crystal. Even after reconstruction these polar surfaces can be expected to show a high reactivity for a variety of reactions; the reconstruction will certainly affect the properties of the surface and this is an effect that must be taken into account when modelling, or interpreting experiments on, polar surfaces.

The Madelung or crystal potential thus needs to be considered and may be expanded in different ways: a point-charge expansion may be used with a sufficiently large number of point-charges employed to give an accurate representation of the crystal potential in the volume of interest. For simple oxide structures the potential is fairly rapidly convergent, but when considering more complicated surfaces, such as reconstructed polar surfaces, the potential becomes slowly convergent and an excessively large number of point-charges may have to be employed. A simpler and more reliable way is then to perform the explicit Ewald [1, 2, 3] summation over the infinite crystal. Even in this case, one has to consider questions such as the value of the charge to be used and whether a point-charge is actually a good representation of the source of the local potential around the ion in question.

Surrounding a cluster model with point charges leads to further considerations: a $+2$ point charge may be considered as an α-particle and as such has a very high electron affinity and generates a (locally) strong electric field; since the point-charge is not associated with basis functions direct charge transfer from the cluster is not likely, but the anions will be strongly polarised towards the unscreened point-charges leading to artificial reductions in the charge density on the cluster [4, 5] and anisotropies in the electrostatic potential [6, 7]. Thus, embeddings including approximately quantum mechanical effects connected with the size of the surrounding ions must be considered [8, 9, 10].

In the following we will describe our approach to modelling of strongly ionic metal oxides beginning with the determination of the structure, the construction of an embedding potential and quantum chemical calculations on these embedded cluster models. Examples will be given from recent work on the chemisorption energy of CO on MgO(100) [11], from H [12] and H_2O [13] interaction with the polar (100) surface of Cu_2O and from H_2 interaction with reconstructed ZnO surfaces [14].

A goal with our work is to relate as closely as possible to experimental data as a check on the models used and as an aid in analysing the experimental results; in the field of surface chemistry and physics this implies analysis of X-ray absorption and emission spectroscopies. We will finish this chapter with a brief description of recent work on modelling complete NEXAFS and XES spectra for surface adsorbates and surface ions.

2. Modelling

2.1. STRUCTURES

For rock-salt structure materials, such as MgO and NiO, the perfect (100) surface is electrostatically stable and only exposes ions with high (*e.g.* five-fold) coordination. The surface layer may still relax inwards, and anions and cations may in addition move by unequal amounts leading to a rumpling of the surface [15]. For MgO(100) this effect is small [16] and will not be considered further here. From a reactivity point of view, sites like steps, corners or defects which have ions with lower coordination, are of more interest and the local structure around these has to be determined or guessed.

In the case of ionic crystals the particular facet that is exposed becomes particularly critical due to the long-range Coulomb interaction; only non-polar surfaces, *i.e.* surfaces without a dipole moment across the crystal, may be stable with only minor rearrangements of the ions. All surfaces, or bulk-terminations, which are built from alternating layers of positive and negative charges must reconstruct or be otherwise stabilised in order to exist [12]. This may be accomplished in several different ways: one possibility is for the surface to become chemically stabilised by decomposing *e.g.* water leading to $(M^{2+}OH^{-})^{+1}$ and OH^{-} occupancy at the surface, which gives the required reduction of the charges. This process is active in the stabilisation of the NiO(111) face [17]. A second possibility is to have the surface relax electronically, *i.e.* a change of charge state of the ions at the surface. A third way to remove the dipole across the crystal is by structural reconstruction such that (for a rock-salt structured crystal) half the ions are removed from the surface layer (see ref. [12] for a general discussion). This structural reconstruction may give rise to fairly large effects and we use atomistic modelling of the surface structure to find the resulting ion positions.

The atomistic modelling of surfaces and solids using two- (or more) body potential descriptions of the interactions between the ions has been reviewed by Catlow [18] and is also described in a separate chapter in the present volume. In our applications the potentials used for the inter-ionic interactions have been pair-potentials of Buckingham type with a shell-model description of the anion polarizability. The total energy is thus written as a sum of pair-energies

$$E = \sum_{i<j} V_{ij}(r_{ij}) \tag{1}$$

where r_{ij} is the distance between the ions and the potentials $V_{ij}(r_{ij})$ are given by

$$V_{ij}(r_{ij}) = \frac{Q_i Q_j}{r_{ij}} + A_{ij}\exp(-r_{ij}/\rho_{ij}) - \frac{C_{ij}}{r_{ij}^6} \tag{2}$$

Here Q_i is the charge of ion i determining the Coulomb interaction, the exponential term gives the short-range repulsion between ions i and j with the strength given by A_{ij} and the range determined by ρ_{ij}. The final term gives the van der Waals interaction. The parameters in the potentials are fitted either to experimental data or to results of quantum chemical calculations.

The polarisation of the ions (mostly anions) is described by dividing the charge into core and shell charges connected by a spring constant k (shell-model)

$$V_{ij}^{shell} = \frac{1}{2}kx^2 \tag{3}$$

In our determinations of surface structures and reconstructions we have used the MARVIN [19] program, in which the inter-atomic interactions are described through two- or three-body potentials with all interactions obtained through the proper Ewald summations over the infinite crystal. The program performs a minimisation of the energy of the system with respect to the ion positions in the surface unit cell and should thus provide the required reconstruction and relaxation effects. We have only performed static minimisations (no temperature effects included), but interesting applications of molecular dynamics simulations of oxide surface structures have been reported by other workers [20, 21].

The atomistic modelling of the surface generates relaxed ion positions within a surface unit cell. The main result of the relaxation of the surface is to make it more compact with a higher effective coordination of the exposed ions. This may or may not lead to important effects; for CO chemisorption on a Mg^{2+} at a step site of the MgO(100) surface, relaxation leads to a longer Mg-C distance (by 0.2 a_o), but has negligible effects on the binding energy (+0.01 eV) and vibrational frequency (-3 cm^{-1}) [11]. At a corner site larger effects may be expected: Pacchioni *et al* [22] found a decrease of the binding energy by 0.05 eV as the cation in a limited investigation of structural effects was relaxed inwards towards the fixed anions at the step; a full relaxation of the ions at the step should lead to somewhat larger effects.

For polar surfaces, which undergo reconstruction, the effects are larger and must be considered. As an example we discuss the Cu^+ terminated polar (100) surface of Cu_2O [23]. Cu_2O crystallizes in the cuprite structure with a lattice parameter of 4.267 Å [24]; the lattice parameter resulting from the simulation was 4.270 Å. The cubic unit cell has oxygen ions at the center and corners and four of the eight interstitial positions are occupied by the Cu^+ cations forming tetrahedra around the oxygens. The (100) surface, in a simple bulk-termination, would lead to alternating cation and anion layers with a dipole perpendicular to the surface and would thus be Madelung unstable. In the simulation of the surface structure we removed half the Cu^+ cations from the top layer and placed them at the bottom of the slab model. Several different possible reconstructions were investigated using the MARVIN program, with a (1×1) missing-row reconstruction giving the lowest surface energy. The ions in the first three unit cells were found to undergo substantial motion with a general downwards trend: the top-layer Cu^+ moved down by 1.36 Å, the second layer copper ions by 0.28 Å, while in the third layer one copper moves down by 0.32 Å and the other up by 0.44 Å. Similarly, the topmost oxygen moves down by 0.90 Å, while the second-layer oxygen moves up by 0.24 Å. There are also large lateral motions in the lattice.

The reconstructed surface is substantially more compact and the more densely packed environment around the top-layer Cu^+ cation has effects on the $Cu^+(d^{10} \rightarrow d^9s^1)$ excitation energy which increases from 1.38 to 1.89 eV. The computed atomic hydrogen chemisorption energy for the model of the reconstructed but unrelaxed surface was 2.27 eV, which would lead to the conclusion that dissociation of H_2 over $Cu_2O(100)$ would be energetically possible; this is contrary to what is observed experimentally[25].

The relaxed surface, on the other hand, gives a binding energy of 2.06 eV, which is sufficiently below half the binding energy of H_2 to allow the conclusion to be drawn that $Cu_2O(100)$ should not dissociate H_2 [23].

2.2. EMBEDDING

Several approaches exist in the literature for treating the interaction with the remaining crystal; a main subdivision may be made between periodic calculations using a supercell technique for the study of chemisorption and a cluster approach, where a small piece of the crystal is cut out to represent the local active site. In the latter case one may neglect the remaining crystal entirely and model the surface with a gas-phase cluster, which has the advantage that all theoretical techniques that have been developed for gas-phase systems are immediately available. The cluster in this approach may become too flexible allowing too large relaxations of the substrate, however, and the complete neglect of the interaction with the crystal is an approximation that needs to be investigated from case to case [26, 27]. In most studies some type of embedding scheme is thus employed where at least the Madelung potential from the remaining crystal is accounted for, either through a finite point-charge expansion or a complete Ewald summation [1, 2, 3, 28, 29].

In the case of chemisorption of CO on MgO [4] and NiO [5] it has been shown, however, that an embedding of the cluster that includes also approximately quantum mechanical effects through *e.g.* effective core potentials or by including the next neighbors leads to significant changes in the computed properties of the cluster: CO becomes unbound at the SCF-level and the small computed chemisorption energy is obtained only through inclusion of dynamical correlation. In [4] the origin of this effect was traced to an unphysical polarisation of the charge on the O^{2-} anions towards the surrounding $+2$ point-charges.

Several different embedding schemes exist; an embedding potential, including orthogonality effects on the cluster wave-function from the surrounding ions, may be extracted from the fully periodic solution for the bare surface [30, 31]. A simpler description is obtained by using effective core potentials based on the frozen ionic charge distribution [8, 9, 10] to describe the nearest neighbors to the cluster. These static potentials give an approximate description of the quantum mechanical effects, including orthogonality and exchange, from the frozen charge distributions on the cluster without explicitly including neither the electrons on the surrounding ions nor basis functions to describe them. This is the approach taken in the present work.

In our case we use the total ion ab initio model potential (AIMP) approach developed by Barandiarán and Seijo [8, 9], in addition to the expansion of the long-range crystal potential, to embed the central cluster. This provides an approximate representation, without basis functions, of all interactions between the cluster and the surrounding frozen charge distributions; thus, long-range (point-charge) and short-range (incomplete screening) Coulomb interactions, Pauli repulsion through level-shift operators as well as a spectral resolution of the exchange interaction are included.

The additional one-electron operator that is added to the Hamiltonian is thus

$$\hat{V}^{MP}(i) = \sum_I [\hat{V}^I_{lr-Coul}(i) + \hat{V}^I_{sr-Coul}(i) + \hat{V}^I_{exch}(i) + \hat{P}^I_{proj}(i)] \tag{4}$$

The summation runs over all model-potential centers, I. The long-range term gives the Coulomb potential from a point-charge with the appropriate ionic charge at the corresponding lattice position. The short-range Coulomb term gives the increased attraction felt by an electron that penetrates the charge-density around the model-potential ion. From a calculation on the ion (closed-shell) in the crystal environment, the charge density associated with that ion is obtained and the potential felt by an electron may be generated on a set of radial grid-points as

$$V^I(r) = -\frac{Z^I}{r} + \int \frac{\rho(\mathbf{r}')}{|\mathbf{r}-\mathbf{r}'|} dr' d\Omega \tag{5}$$

For convenience, a set of Gaussians is fitted $(\{A^I_j, \alpha^I_j\})$ to reproduce this potential out to a distance where the potential corresponds to that of a point-charge.

$$\hat{V}^I_{sr-Coul}(i) = \frac{1}{r_{Ii}} \sum_j A^I_j exp(-\alpha^I_j r^2_{Ii}) \tag{6}$$

The non-local exchange operator $\hat{V}^I_{exch}$ is spectrally represented:

$$\hat{V}^I_{exch}(i) = \hat{P}_I \hat{V}^I_{exch} \hat{P}_I \tag{7}$$

where $\hat{P}_I$ is the projection operator on the primitive basis set χ_I, with $\chi_I = \{|\chi_{I1}\rangle, |\chi_{I2}\rangle, ...\}$, used for ion I so that (S_I is the overlap matrix):

$$\hat{P}_I = \chi_I S_I^{-1} \chi_I^\dagger \tag{8}$$

Approximate orthogonality is maintained through the level-shift (projection) operator, $\hat{P}^I_{proj}$ defined by:

$$\hat{P}^I_{proj}(i) = \alpha \sum_{k \in I} (-2\epsilon^I_k) \mid \phi^I_{k,occ}\rangle\langle\phi^I_{k,occ} \mid \tag{9}$$

where ϵ_k^I is the orbital energy of the occupied atomic (ion) orbital $\phi_{k,occ}^I$. α is a constant, normally set to 1.0, which may be used to scale the level-shift operator (below).

Thus, to summarise: the total ion AIMP only requires the wave function of the ion computed in the proper environment. The Coulomb potential is then obtained from a straight-forward fitting of the potential given by the charge density and nuclear charge, the exchange and Pauli repulsion are described using the atomic orbitals and the orbital energies. To be explicit: calculations on *e.g.* Mg^{2+} and O^{2-} embedded in the MgO lattice, with some initial model potential description of the other lattice ions, give wave functions for the ions which are used to generate a new AIMP representation of the lattice ions. New calculations on Mg^{2+} and O^{2-} embedded in the new model potential representation of the MgO lattice are performed, and the process is repeated to self-consistency.

When the basis set is completely removed from the ions described by total ion AIMPs, the orthogonality between the cluster and the environment can no longer be strictly maintained; this leads to a spurious energy increase due to the level-shift operator. To reduce this effect the level-shift may be reduced by the scaling factor α (a value $\alpha = 0.65$ was used in ref. [4]) or totally contracted s and p functions (representing valence orbitals) may be added to the nearest AIMP sites to improve the orthogonality [32].

2.3. ELECTROSTATIC POTENTIAL

The Madelung potential is often included through expansion using a finite set of point-charges, but a direct evaluation of the Gaussian type integrals over the full infinite Ewald sum [1, 2, 3, 28, 29] is also possible and has the advantage that one does not have to worry about the convergence of the potential with respect to the number of charges included. The integrals of products of Gaussians over the electrostatic potential must thus be computed. The potential, $\hat{V}(\mathbf{r})$, is given by the sum over the entire crystal

$$\hat{V}(\mathbf{r}) = \sum_{\mathbf{R}} \sum_{i} \frac{q_i}{|\mathbf{R} + \mathbf{r}_i - \mathbf{r}|} \tag{10}$$

where $\mathbf{R}$ gives the position of the unit cell and $\mathbf{r}_i$ give the positions of the ions (charge q_i) in the unit cell. This summation is only conditionally convergent, but may be made absolutely convergent [1] by dividing the $1/r$ potential into a long-range and a short-range part. This is done by noting that

$$\begin{aligned} 1 &= \tfrac{2}{\sqrt{\pi}} \textstyle\int_0^\infty \exp[-t^2]dt = \\ &= \tfrac{2}{\sqrt{\pi}} \textstyle\int_0^x \exp[-t^2]dt + \tfrac{2}{\sqrt{\pi}} \int_x^\infty \exp[-t^2]dt = \mathrm{erf}(x) + \mathrm{erfc}(x) \end{aligned} \tag{11}$$

Applying this identity to each term in eq. 10 leads to a division of the crystal potential into a long-range and a short-range contribution (in terms of the distance $\mid \mathbf{R} + \mathbf{r}_i - \mathbf{r} \mid$) with relative weights determined by the parameter γ:

$$\begin{aligned}\hat{V}(\mathbf{r}) = \sum_{\mathbf{R},i} \frac{q_i}{\mid \mathbf{R} + \mathbf{r}_i - \mathbf{r} \mid} \mathrm{erfc}(\gamma \mid \mathbf{R} + \mathbf{r}_i - \mathbf{r} \mid) + \\ \sum_{\mathbf{R},i} \frac{q_i}{\mid \mathbf{R} + \mathbf{r}_i - \mathbf{r} \mid} \mathrm{erf}(\gamma \mid \mathbf{R} + \mathbf{r}_i - \mathbf{r} \mid) \\ \equiv \hat{V}_{sr}(\mathbf{r}) + \hat{V}_{lr}(\mathbf{r}) \end{aligned} \qquad (12)$$

The short-range contribution is absolutely convergent and may be summed directly, while the long-range contribution varies only slowly in direct space. Thus, a Fourier transformation of this part will only give a small number of significant (low-frequency) components that can be summed in the reciprocal space. Evaluation of integrals involving Gaussian basis functions over this potential has been implemented in the ECPAIMP one-electron integral program [33] that we use for the modelling.

Performing the summation in the integrals over the infinite crystal removes problems or questions associated with the convergence of the expansion of the crystal potential in terms of point-charges, but it does not provide the ultimate answer as to what the potential should be; the charges q_i assigned to each ion must still be determined. This is at the moment a somewhat open question. Rösch and coworkers [34, 35], based on fitting the potential obtained in local density slab calculations and on comparison with experiment for the shift in CO vibrational frequency, argue for a value lower than the nominal ± 2 for expanding the crystal potential of MgO. In our recent study of CO/MgO (section 3.1 and ref [11]) we obtain good agreement with experimental frequency shifts without reducing the charges giving the crystal potential, however.

It is clear that for cluster models exceeding a "certain size" the embedding electrostatic potential will be less important, but for small clusters the importance of the exact value of the crystal potential for the property of interest should be investigated. We have done this for a simple model reaction, where an atomic oxygen adsorbed on an oxygen anion at the MgO(100) surface reacts with CO to form CO_2 [36], by performing the calculations with an embedding potential based on charges ± 2 and ± 1, respectively. The geometry of the adsorbate and the location of the transition state in the reaction were not affected by the change in embedding potential. This was also true of the computed barrier (0.28 versus 0.29 eV), while the exothermicity and the atomic chemisorption energy were more strongly affected (by 0.4 eV). These results are not unexpected since the bonding of the oxygen atom to the MgO surface can be characterised in

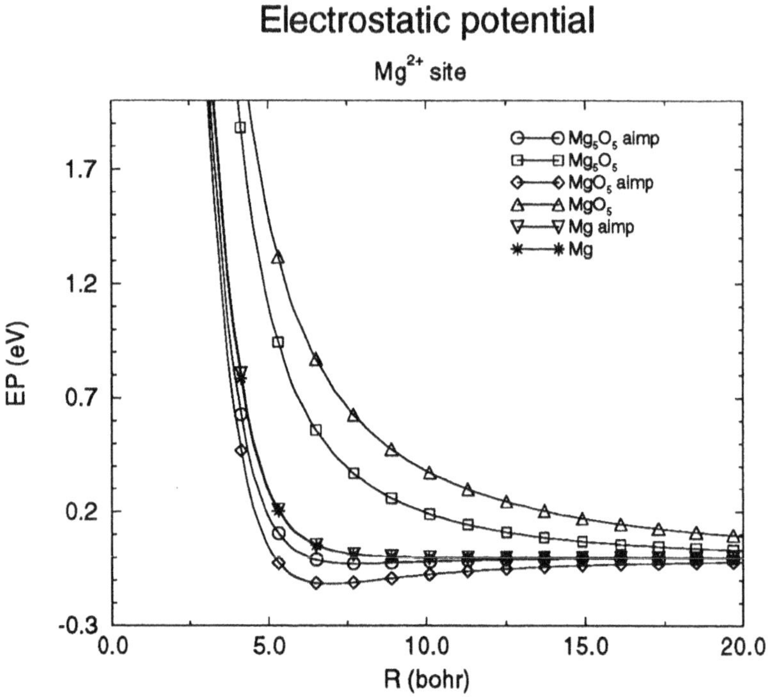

Figure 1. **Electrostatic potential felt by positive unit charge above Mg^{2+} site of MgO(100) surface from different cluster models with AIMP (aimp) or point-charge embedding.**

terms of the formation of an $[O_{surf}-O_{ads}]^{2-}$ species where the internal charge distribution is largely determined by the external potential.

However, irrespective of the value of the point-charges as discussed above, or explicit consideration of the multipole expansion of the actual, in-crystal charge density of the ions [37] it is clear that explicit consideration of the approximate quantum mechanical interactions between cluster and remaining crystal is of greater importance. Ferrari and Pacchioni [6] have recently discussed the electrostatic potentials obtained from different point-charge embedded cluster models and compared these with those obtained from periodic Hartree-Fock calculations using CRYSTAL [38]. In Figure 1 we show the corresponding radial electrostatic potential, using both point-charge and quantum mechanically embedded clusters, experienced by a positive point-charge above an Mg^{2+} cation in the MgO(100) surface [7].

With the AIMP embedding included, very stable results for the electrostatic potential are obtained and all curves are very close to the fully periodic, three-layer slab calculation reported in [6]. The electrostatic potentials given by the unembedded cluster models, both stoichiometric (*i.e.* containing cations and anions according to the chemical composition) and non-stoichiometric, show a substantially larger curvature at large distances, thus giving rise to spurious field-effects. The reason for this is the artificial polarisation of the anions at the cluster edges towards the unscreened positive point-charges which leads to a depletion of the charge density at the center of the cluster [4]. It is clear, that the stoichiometric model shows less of this artefact, but it is still far from the embedded cluster and periodic Hartree-Fock results. As a point of interest one may note the very rapid decrease in electrostatic potential with increasing height above the surface; this is true for these simple non-polar surfaces and shows that electrostatic effects will have a negligible contribution at distances longer than about 5 Bohr from the regular site.

Inclusion of the AIMP embedding, or equally well the next layer of cations, results in a potential surface much more similar to that of the extended surface. Some slight asymmetry persists, but at a much lower level. Indeed at this point, inclusion of the next four Mg^{2+} cations needed for a stoichiometric cluster has essentially no effect on computed properties in this highly ionic system. This is contrary to what has been reported for unembedded clusters where it has been argued that the cluster models must always be stoichiometric [39]; this is thus not a requirement from an electrostatic point of view once the cluster is properly embedded. In fact, in a recent study of CO/MgO(100) [11] the computed binding energy of CO to an embedded Mg_5O_5 cluster was found to differ by less than 0.01 eV from that computed using a non-stoichiometric MgO_5 embedded cluster [4]. The most important difference was actually found to be the reduction in basis set superposition error due to the presence of additional basis sets in the description of the cluster model [11]. We may furthermore note from the electrostatic potential above the anion in oxygen centered cluster models of the MgO(100) surface (Figure 2) that the stoichiometric Mg_5O_5 now shows a much poorer representation of the potential than the non-stoichiometric OMg_5 cluster. In the latter case the O^{2-} ion is fully embedded using a correct description (the actual ions) of the interaction with the nearest neighbors, while the Mg_5O_5 cluster model shows edge-effects due to the interaction of anions with the unscreened +2 point-charges. The properly embedded clusters again show a very consistent picture with electrostatic potentials very close to that of the infinite surface. Larger cluster models have been considered by Ferrari and Pacchioni [6], and it is clear that by increasing the cluster size the edge effects due to polarisation of anions at some point will no longer affect the electrostatic potential measured in the central parts of the cluster. It is also clear that, by including the embedding, the convergence is much more rapid and from the purely electrostatic point

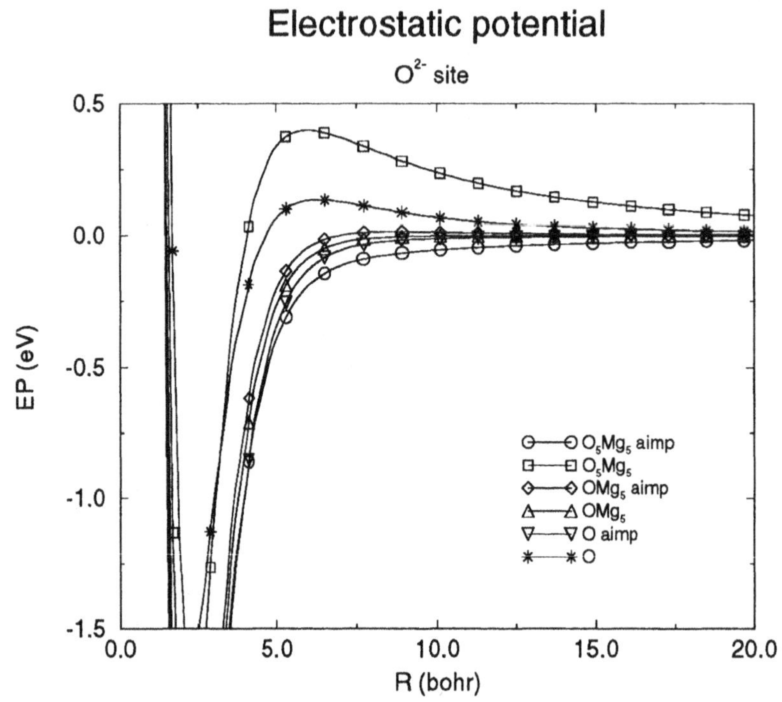

Figure 2. **Electrostatic potential felt by positive unit charge above O^{2-} site of MgO(100) surface from different cluster models with AIMP (aimp) or point-charge embedding.**

of view it is not necessary to go to larger cluster sizes.

A final note on the electrostatic properties of completely unembedded gas-phase clusters is in order. This is an attractive computational model since it allows for the use of all the different quantum chemical techniques that have been developed for gas-phase systems. In particular, geometry optimisations may be easily performed without having to develop and implement gradient techniques that involve complicated derivatives over the Madelung potential and different components of the embedding Hamiltonian. However, the electrostatic potential obtained from even relatively large gas-phase cluster models is, generally speaking, very different from the one obtained from embedded or periodic calculations. As pointed out by Ferrari and Pacchioni [6], who studied a stoichiometric $Mg_{13}O_{13}$ cluster, there is a small region localized around the center of the cluster in which the potential is similar to what is obtained from the slab-model or the embedded cluster calculations. The same behavior is found for an Mg_9O_9 cluster [7]: the presence of a net positive charge on the first layer (which for Mg_9O_9 contains five Mg^{2+} ions and four O^{2-} ions) makes the potential positive in a large region above the surface. The electric field above the central Mg^{2+} ion in the direction normal to the surface is small, however, allowing the calculation of adsorption properties. The anisotropy in the potential is large for neighboring sites which gives rise to spurious electric fields, that may or may not significantly affect the properties of interest. Thus, gas-phase clusters as models of the extended ionic substrate in adsorption studies should be applied with caution and checked for stability towards inclusion of the crystal embedding. However, this is certainly valid for any model and cases can easily be imagined where the full geometry optimisation, including the surface, is more important than the embedding, particularly if the latter restricts the situations that can be studied or the techniques that can be applied. Thus, a combination of modelling techniques, including both properly (AIMP or equivalent) embedded and gas-phase clusters, may be used to shed light on different aspects of a reaction on a metal-oxide surface.

2.4. COMPUTATIONAL STRATEGY

The application of the above considerations has become fairly straightforward with the combination of different approaches and recent program development. Thus, the structure is either taken from experiment or from an optimisation using the MARVIN program [19]. An appropriate cluster representation of the active site in question is selected; the number of atoms that should (or can) be included depends on the problem and the level of calculation that must be applied. In principle, one wishes to perform the calculations for a sequence of clusters to ensure that the results are not artefacts of a particular cluster size and to ensure convergence of effects of longer range (*e.g.* polarisation effects). For metal clusters oscillations in computed properties with cluster size has been a much-discussed problem (see *e.g.* ref. [40] and references therein), while for strongly ionic metal oxides fewer investigations have been performed; those that have been reported neglect the embedding potential and thus cannot exclude oscillations that have their origin in varying numbers of anions exposed to the unscreened point-charges [6, 41, 39] or multi-poles [37]. Thus, convergence of properties with cluster size is still largely an open question for metal-oxides, but proper embedding of the clusters gives, in our experience, a faster convergence (sections 2.3 and 3.4).

Having selected the central cluster and basis sets to describe these ions the (surface) unit cell (taken from experiment or as the output from MARVIN) is read into the quantum chemical program (ECPAIMP) [33]. This automatically generates the lattice and places AIMPs at the appropriate lattice points out to a given (input) distance from the ions in the cluster. The integrals over the Madelung potential and embedding are computed in addition to the standard one-electron integrals and these replace the one-electron integrals from the normal two-electron integral code.

The properties of the central, embedded cluster with or without adsorbed molecules, are now evaluated using any of the available standard techniques (Hartree-Fock, CI, MCPF, DFT etc). Geometry optimisation of adsorbates on embedded clusters is presently performed with the aid of numerical minimisers [13, 42].

3. Applications

3.1. CO/MGO(100)

The chemisorption energy of CO on a regular site of the MgO(100) surface has become an important test case for theoretical modelling of metal oxide surfaces. Early powder experiments [43, 44] gave low values, 0.15-0.17 eV, for the binding energy, while the first theoretical studies [45, 46] obtained a substantially higher value, 0.38 eV.

Recent experiments have resulted in values around 0.40-0.45 eV; Henry *et al* [47] gave a value of 0.3-0.4 eV from the kinetics of CO interacting with small Pd-clusters on an MgO film combined with an estimate of the barrier to diffusion. The binding energies of CO/MgO(100) [48] and CO/NiO(100) [49, 50] have furthermore been determined experimentally using both TPD (Temperature Programmed Desorption) and the Clausius-Clapeyron equation with very consistent results: for CO/MgO the TPD result for the activation energy for desorption is 0.45 eV, while the Clausius-Clapeyron result for the heat of adsorption is 0.43 eV [48]. Similarly, for CO/NiO(100) the TPD result is 0.30-0.37 eV [49], while application of the Clausius-Clapeyron relation gives 0.44 eV [50].

Meanwhile the computational and modelling techniques have been improved and more reliable results generated. Periodic Hartree-Fock calculations by Pisani and coworkers[30, 51] only gave a low binding energy: 0.17 eV for upright adsorption with the carbon end and 0.14 eV for adsorption with the oxygen end. Early cluster studies by Pacchioni and coworkers [22, 52, 53] used rather small cluster models, $[MgO_5]^{8-}$, embedded in a point-charge expansion of the Madelung potential. The SCF-level result for the binding energy of CO/MgO was reasonable, 0.25 eV, but no corrections for the basis set superposition errors (BSSE) were made. Pöhlchen and Staemmler [5] studied CO/NiO(100) and first pointed out the need to correct for the large effects of the BSSE. They furthermore found that when the cluster model was embedded in a surrounding of Mg^{2+} pseudopotentials, the CO became essentially unbound. The same effect was found for CO/MgO(100) by Nygren *et al* [4] who demonstrated the polarisation effects discussed in section 2.3 above. The charge depletion on the cluster lead to a reduced repulsion for the incoming CO and an artificially high binding energy; when the cluster was embedded, either in total ion model potentials (AIMP) or all-electron Mg^{2+} ions [4], the resulting BSSE-corrected SCF value for the binding energy was essentially zero. Only by including large-scale dynamical correlation was it possible to obtain a positive binding energy, and then only 0.07 eV [4]. In later work [54] application of gradient corrected density functionals and a BSSE correction lead to an estimated low (0.11 eV) binding energy, in agreement with the work of [4, 5]. We have recently addressed this problem again and by extending the treatment of the cluster model (see section 2.3 above) have succeeded in increasing our computed value by 0.01 eV to 0.08 eV [11]. Thus, an important discrepancy exists between theory and experiment.

We have reinvestigated the experimental data in an attempt to reconcile theory and experiment [11]. We begin by discussing the Clausius-Clapeyron experiment which is performed by measuring at fixed temperature the coverage as function of pressure; equilibrium between gas-phase and adsorbed phase is assumed and the coverage is obtained from the integrated intensity in the shifted (+35 cm^{-1}) vibrational frequency of the chemisorbed molecules. This is then converted to pressure as function of reciprocal temperature at fixed coverage and the Clausius-Clapeyron equation

$$\ln P = -\frac{q}{k_B T} + \text{constant} \tag{13}$$

is applied to determine the molecular heat of transition, $q = T\Delta s = T(s_{gas} - s_{ads})$. q is assumed independent of the temperature and s_{gas} and s_{ads} are the molecular entropies of the gas and adsorbed phases, respectively. Assuming non-interacting, non-diffusing CO molecules at the surface in equilibrium with the gas-phase we may use the grand canonical potential to construct the thermodynamic potentials required to compute the molecular entropies and the coverage:

$$\Omega = -k_B T \ln\Big\{\sum_N \sum_n \exp[(\mu N - E_{N,n})/k_B T]\Big\} \tag{14}$$

For non-interacting molecules the probability that there is a molecule at one site is independent of which other sites are occupied and, with either one or zero molecules adsorbed at a particular site, we obtain for Ω (with η the number of sites):

$$\Omega = -\eta k_B T \ln\Big\{1 + \sum_i \exp[(\mu - \epsilon_i)/k_B T]\Big\} \tag{15}$$

The coverage $\theta = N/\eta$ and the molecular entropy $s = S/N$ are now obtained from (assuming only one energy level contributing):

$$\theta = -\frac{1}{\eta}\frac{\partial\Omega}{\partial\mu} = \frac{1}{\exp[(\epsilon_0 - \mu)/k_BT] + 1} \tag{16}$$

$$s = -\frac{1}{N}\frac{\partial\Omega}{\partial T} = \frac{(1-\theta)(\mu - \epsilon_0) - k_BT\ln\theta}{\theta T} \tag{17}$$

This gives the molecular entropy and the coverage as functions of the temperature and the chemical potential. The latter is obtained by applying the equilibrium condition $\mu_{ads} = \mu_{gas}$ and computing the chemical potential for the gas

$$\mu = k_BT\ln P - 3.5k_BT\ln k_BT - 1.5k_BT\ln\frac{m}{2\pi\hbar^2} - k_BT\ln\frac{2I}{\hbar^2} \tag{18}$$

The molecular entropy for the gas phase is obtained from $s = -\partial\mu/\partial T$. The temperature range of the equilibrium regime is now obtained by plotting the coverage versus temperature for different pressures. For a low binding energy of 0.17 eV as deduced by Furuyama *et al* [44] *log P* must be between -2.0 and 2.0 Torr in order to maintain equilibrium at temperatures around 100 K, while under UHV conditions, with *log P* between -8 and -4.5 Torr as in the experiments by He *et al* [48], the temperature must be as low as between 50 and 70 K in order to maintain equilibrium. It is clear that under the UHV conditions in these experiments a high binding energy is required in order to maintain equilibrium between the adsorbed and gas phases. Furthermore, we find that for coverages $0.1 < \theta < 0.9$ the assumption of a linear relationship between temperature and coverage at fixed pressure is valid. The heat of transition (q), as opposed to the assumed binding energy (ϵ_0), is finally obtained by applying the Clausius-Clapeyron equation 13 and no significant discrepancy between the two quantities is found. These conclusions are not affected by introducing additional states in the description [11].

Thus, we must conclude that the analysis of the UHV experiments is correct and that the high (0.43 eV) binding energy coupled with the large (+35 cm^{-1}) vibrational frequency shift is indeed correct. Since theoretical model studies have failed to reproduce this for regular sites we will investigate possible alternative adsorption sites as an explanation of these experimental findings. Quantum chemical embedded cluster model calculations have thus been performed with the aim to find a chemisorption site that would match both the high frequency shift monitored in the Clausius-Clapeyron experiment and the high value for the binding energy that has been deduced from these experiments. The five-coordinated (Mg_{5C}) regular site, four-coordinated (Mg_{4C}) step site and a three-coordinated (Mg_{3C})

corner site were thus investigated using stoichiometric, embedded cluster models.

The adsorption geometry of CO on top of the Mg^{2+} ion was optimized at the MCPF [55] level for each of the cluster models with the internal CO geometry maintained at the gas phase value. In order to have a well-defined correlation space, the Mg^{2+} and the O1*s* orbitals in the cluster were localised and removed from the correlation treatment [56] and the remaining 40 (or 32) $O(2s, 2p)$ electrons on the Mg_5O_5 (or Mg_4O_4) cluster were correlated together with the 10 valence electrons on the CO molecule. Each point was BSSE corrected at the MCPF-level using the full counter-Poise correction [57]. The application of this procedure to the calculation of the CO/MgO binding energy has been discussed in ref [4]. The calculation of the vibrational frequencies was done keeping the center of mass of the CO unit fixed and varying the C-O distance by ± 0.05 a_o. For both the regular and the corner sites each point in the calculation of the frequency shift was furthermore BSSE corrected. This correction, however, turned out to be small (around 2-4 cm^{-1}).

For the cluster ions the same basis sets as in previous work [4] were used. This is a large [5*s*4*p*1*d*] basis for the oxygen anions and a [4*s*3*p*] contracted basis for the Mg^{2+} cations. In the calculations on the regular and corner sites the CO molecule was described using the ANO basis sets constructed by Widmark *et al* [58] in a [5*s*5*p*2*d*1*f*] or a smaller [4*s*3*p*1*d*] contraction. For the step site the [4*s*3*p*] contraction of Dunning [59] was used and extended by a diffuse *p* and a *d*-function in the optimisation of the geometry, while the final calculation of binding energy and vibrational frequency shift was done using the large [5*s*5*p*2*d*1*f*] basis set. The results are collected in Table 1.

At the regular site only a small, 0.08 eV, binding energy is found and this is coupled with a shift in the CO vibrational frequency of only 9 cm^{-1}. This is in good agreement with the 14 cm^{-1} shift that has been assigned to adsorption on five-coordinated Mg^{2+} from powder data [60]. The MCPF-level BSSE correction decreases the binding energy from 0.30 down to 0.08 eV, which is a somewhat smaller reduction than what was found for the MgO_5 embedded cluster model [4]; thus even with the full BSSE included in the computed binding energy the value would still be too low and it is clear that the frequency shift of 35 cm^{-1} cannot be reproduced in chemisorption at a regular site. Molecules chemisorbed at regular sites (as modelled here) thus do not correspond to the species that is monitored in the Clausius-Clapeyron experiments by Goodman and coworkers [50].

A somewhat stronger interaction is found for the four-coordinated step site, with and without atomistic relaxation of the step geometry. The computed chemisorption energy is 0.18-0.17 eV and the frequency shift is 24-

TABLE 1. MCPF results for geometries (in a_o), binding energies (D_e in eV) and vibrational frequencies and shifts (ω_e, $\Delta\omega_e$ in cm^{-1}) for CO on differently coordinated Mg^{2+} ions.

Site	CO-basis	R(Mg-C)	R(C-O)	D_e	ω_e	$\Delta\omega_e$
Gas phase	[4s3p1d]	—	2.159	—	2166	—
Gas phase	[5s5p2d1f]	—	2.140	—	2186	—
Mg^{2+}_{5C}	[5s5p2d1f]	5.11	2.139	0.08	2195	9
Mg^{2+a}_{4C}	[4s3p1d]	4.79	2.152	0.17	2193	27
Mg^{2+b}_{4C}	[4s3p1d]	4.99	2.153	0.18	2190	24
Mg^{2+}_{3C}	[4s3p1d]	4.60	2.144	0.44	2221	55
Mg^{2+}_{3C}	[5s5p2d1f]	4.60[c]	2.126	0.48	2242	56

[a] Unrelaxed step site. [b] Relaxed step site (see section 2.1). [c] Not optimized.

27 cm^{-1}, in very good agreement with the frequency shift (+25 cm^{-1}) assigned to step sites from powder data [60]; both binding energy and shift are thus computed to be lower for this site than what is observed in the Clausius-Clapeyron experiment. For the corner site, finally, attempts were made to optimise a (64 8 1) surface using the pair-potential techniques, but this was unsuccessful. Thus the three-coordinated corner site, Mg^{2+}_{3C}, was modelled by an embedded Mg_4O_4 cluster at the corner of a large cube. The geometry at this site has the CO molecule pointing out from the corner along the projected diagonal of the unit-cube [22] and for this structure the Mg-C distance was optimised at the MCPF-level as described above. The resulting bond distance of 4.6 a_o is the shortest found, and this is coupled with a quite high binding energy, 0.48 eV, which is quite close to the experimentally determined 0.43 eV; the shift, 55-56 cm^{-1}, in the vibrational energy is also quite consistent with the frequency shift, 58 cm^{-1}, that has been assigned to CO adsorbed at an Mg^{2+}_{3C} three-coordinated site [60]. Both binding energy and frequency shift are expected to be reduced if relaxation of the corner site is included in the calculations, but a reduction by 20 cm^{-1} of the shift is probably not likely (the neglected lower terrace, if included, would have been sufficiently far below the CO so as to be expected not to have a significant effect on either binding energy or shift). Thus, we are able to accurately reproduce the frequency shifts obtained from powder data for all three sites and it is clear that neither the regular nor the step site will fit the observations from the Clausius-Clapeyron experiment; the corner site modelled here gives the closest agreement, but the high value of the shift would suggest that some other low-coordinated site is what is actually observed.

Three different (regular, step and low-coordinated corner) sites at the surface have thus been characterised with respect to CO chemisorption: the highest binding energy, 0.4-0.5 eV, is computed for the corner site with the regular and step sites at substantially lower binding energies, 0.1 and 0.2 eV, respectively. In the TPD experiments only one main desorption peak is observed, however, rather than the three that would be indicated by the present calculations. Thus we need to set up theoretical TPD spectra based on the computed binding energies to see under what conditions desorption from the predicted regular and step sites should be observed.

In a typical TPD experiment the species to be studied is dosed onto the surface at low temperature. The surface is then heated and the number of molecules falling off per unit time is observed. Theoretical expressions for the rate of desorption are thus required for a simulation and such expressions have been developed for the desorption of both atoms and molecules from solid surfaces by Goddard and coworkers [61, 62]. Their treatment is based on classical stochastic diffusion theory and for molecular desorption they arrive at an expression for the rate which depends both on the temperature and the energies of vibration against the surface and the frustrated rotations; it is assumed that the rotational energy can be transferred to kinetic energy in the desorption process.

We have computed the frequencies of the frustrated rotational motion and find them to be extremely low, between 2 and 5 cm^{-1}, such that the contribution from these modes will be negligible in the desorption process and it is appropriate to instead use the simpler rate for atomic desorption given by [61]:

$$R = \frac{\Omega_o}{2\pi} \exp(-\frac{D_e}{k_B T}) \tag{19}$$

The value (80, 96 and 138 cm^{-1} for regular, step and corner sites, respectively) of the frequency, Ω_o, of the vibration against the surface are obtained from the cluster calculations and lead to preexponential factors of $1.5 \cdot 10^{14}$, $1.8 \cdot 10^{14}$ and $2.6 \cdot 10^{14}$ for the regular, step and corner sites, respectively. Taking the computed binding energies for the regular (0.10 eV) and step sites (0.20 eV) and the experimental value of 0.43 eV for the corner site gives the theoretical TPD spectra shown in Fig. 3. The rate maxima for both the regular (40 K) and the step sites (80 K) are predicted to occur at temperatures lower than (or at the initial temperature) what has been used in the experiments and these peaks would thus not have been observed in the experiment. It should be noted that in the TPD spectrum in Fig. 3 only the peak positions should be considered; since the coverages relating to the different adsorption sites are not known, the same coverage (0.5) was used for all sites.

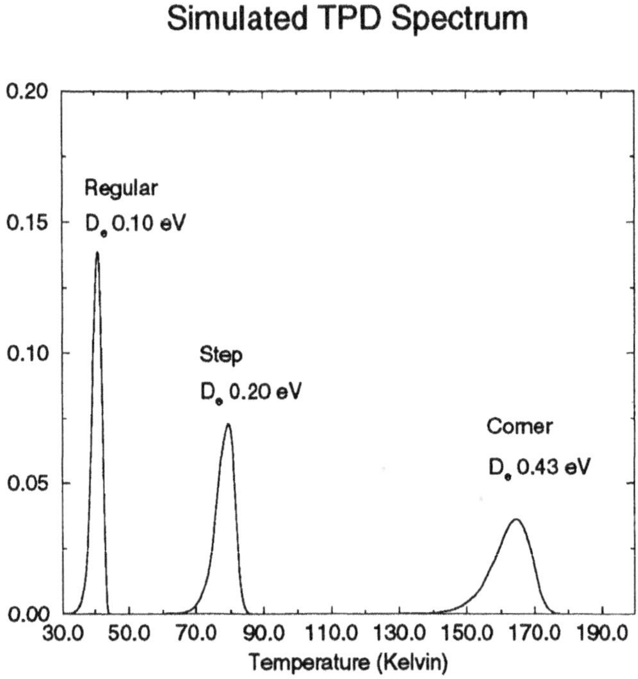

Figure 3. **Theoretical TPD spectrum for CO/MgO(001) including regular, step and corner sites.**

Adsorption at a low-coordinated corner site is thus predicted to be the most favorable and should correspond to what is actually observed in the experiments at the temperatures that have been used. The geometry of the CO molecule at such a site, at 45° against the surface normal and pointing out from the corner, would, however, seem to be in contradiction to the structure obtained from recent NEXAFS measurements performed on the similar system CO/NiO by Cappus *et al* [49]. The angular dependence of the $C(1s) \rightarrow \pi^*$ excitation was used to draw the conclusion that CO is standing up at the surface, *i.e.* aligned with the surface normal, but this is not necessarily in contradiction with the conclusion that CO chemisorbs at a low-coordinated corner site.

The computed vibrational frequency of CO for the frustrated rotations against the surface at the corner site is very low, of the order of a few wave numbers and thus this mode should be highly excited at the temperatures at which the experiments have been performed. Furthermore, the corner sites may be assumed to form part of terraces which would lead to an asymmetrical potential with larger repulsion against the terrace below the step. The highest probability of finding the molecule will then (for highly excited vibrational states) be towards the classical turning points and a probability function representing this is represented by *e.g.*

$$P(\theta) = C(\theta^2 + 0.1) \mid \theta - \frac{\pi}{2} \mid^3 (\theta + \frac{\pi}{2})^2 \tag{20}$$

for $-\pi/2 \leq \theta \leq \pi/2$. The constant C normalises the distribution and $\theta = 0$ corresponds to the molecule at 45° from the step. It should be understood that this probability function has not been obtained as a solution for the nuclear vibrational potential but constructed to fit this physical picture.

Assuming that the molecules are distributed according to this distribution function we may compute the averaged π^*-intensity for each photon incidence angle (assuming perfect polarisation of the synchrotron beam)

$$I(\phi) = I_z \int P(\theta) sin^2(\phi - \theta) d\theta + I_x \int P(\theta) cos^2(\phi - \theta) d\theta \tag{21}$$

Under the assumption of randomly distributed corner sites (in one dimension) the angular dependence of the π^* intensity is obtained in quite reasonable agreement with the experimental intensity distribution. The computed intensity distribution is rather sensitive to the probability distribution used in the averaging over angles. The probability distribution on the other hand should be expected to be strongly dependent on the excitation level of the frustrated vibrations such that at lower temperatures a higher probability for the 45° geometry should be found. The angular intensity distribution of the π^* excitation should thus be expected to exhibit a dependence on the temperature, which should be possible to investigate experimentally.

The computed binding energies and frequency shifts for CO adsorbed at different sites of the MgO(100) surface thus lead to agreement with the experimentally reported quantities only for adsorption at a low-coordinated Mg^{2+} ion at *e.g.* a corner site. For a regular site a low, 0.1 eV, binding energy combined with a frequency shift of <10 cm^{-1} is obtained. For the step site the interaction is somewhat stronger, but still giving a binding energy of only 0.2 eV combined with a frequency shift of 24 cm^{-1}; this is still too low to explain the observed 0.43 eV binding energy and 35 cm^{-1} frequency shift. Only small effects on the results are found from fully relaxing the step geometry using atomistic simulations of the surface.

Several different bands have been observed in IR spectroscopy on CO on MgO and a consistent picture is obtained if the 2178 cm^{-1} frequency observed by Goodman and coworkers [48] is interpreted as coming from CO adsorbed on corners or similar low-coordinated sites and the lower value around 2150 cm^{-1} reported by Zecchina [60] and Tashiro [63] as coming from adsorption on five-coordinated magnesium. The equilibrium regime for a CO binding energy of 0.15 eV is predicted to be around 50 K under UHV conditions, and such a species would thus not be present at the surface in the experiments reported in ref [48]. Furuyama *et al* [44] report the equilibrium regime to be between 90 K and 130 K for *log P* between -2.0 and 2.0 Torr. This is in agreement with the model of the Clausius-Clapeyron experiment using a binding energy of 0.15 eV.

The early determinations of the CO/MgO(100) binding energy were performed on powders, while the later experiments have been made either on thin MgO films deposited on a Mo substrate or on single crystals. In the powder work lower frequency shifts have been reported than what has been used to monitor the coverage in the work on thin films. Theoretical work is consistent in assigning higher frequency shifts to adsorption on less coordinated ions [11, 22, 39] and this is also the result of experimental studies on powders [60]; thus it would seem that adsorption on more highly coordinated ions would have gone undetected in the recent work. It should be noted that in the powder work higher pressures have been maintained which would be necessary in order to maintain a more weakly bound species in equilibrium with the gas phase at the temperatures used.

In order to conclusively resolve the issue of the CO/MgO binding energy the experiments should be redone at liquid helium temperature combined with a determination of absolute coverage; theory predicts the presence of additional states that should desorb at lower temperature and whatever the result of the experiment it will indicate an interesting area for further development: for theory if, in fact, low-binding energy sites are not found either on thin film, single crystal or powders and for the understanding of the properties of the thin films if, in fact, these states do turn up.

3.2. $H_2O/CU_2O(100)$

The results from our simulation of the surface reconstruction of the polar cation terminated $Cu_2O(100)$ surface were discussed in section 2.1 above. The lowest energy was found for a (1×1) missing-row reconstruction and we will now proceed to discuss the interaction of this surface with adsorbed water. Experimentally water adsorption on $Cu_2O(100)$ is dissociative at 300 K [64], while at 110 K both dissociative and molecular adsorption is observed at low coverages, about 10 % of a monolayer being dissociated. There is no dependence on the concentration of defects, thus the dissociation must be assumed to occur on regular sites of the reconstructed surface [64].

The cluster models consisted of a CuO, Cu_2O or Cu_3O unit embedded in a surrounding of total ion AIMPs out to a distance of 10.5 a_o from any ion in the cluster. The Madelung potential was generated directly from the optimised (using MARVIN [19]) surface unit cell and the integrals over the

Ewald sum and AIMPs were computed using the ECPAIMP program. The optimisation of the geometries was done at the SCF level using a numerical minimiser using Powell's method with a parabolic interpolation for the line-minimisations [42]. The substrate ions were kept fixed and all degrees of freedom varied for the two hydrogens and the oxygen in the approaching H_2O molecule in the optimisation; the final binding energies were computed at the MCPF-level [55]. The lowest energy dissociated structure has the OH group on-top of one surface Cu^+ cation with the dissociated proton forming a second OH group together with an O^{2-} anion at the surface; the computed binding energy relative to surface and H_2O is 0.6 eV. The dissociation thus leads to the formation of two OH^- groups at the surface and a concomitant redistribution of charge [13], which can be expected to affect the surface dipole in qualitative agreement with an observed reduction when the surface is covered by dissociated water [64]. Using the orbital energies obtained for the optimised structure, where a surface O^{2-} anion becomes protonated and an OH^- binds to a Cu^+ cation, we can furthermore generate theoretical UPS difference spectra in good agreement with the UPS data [64] obtained for the difference between the clean surface and the surface with dissociated water. Thus, the calculations give a resulting structure which provides a reasonable interpretation of the experimental data.

Two stable sites, with binding energies of 0.5 eV, were found for the adsorption of molecular H_2O [13]: on-top of a surface oxygen anion with the hydrogens pointing downwards and adsorption on a Cu^+-ion at the step in the surface formed by the missing-row reconstruction; here the hydrogens point away from the surface. The experimental TPD results as function of coverage show no indication of additional stability due to hydrogen bonding which again may be rationalised from the optimised structure; having alternating geometries for the water molecules with respect to the surface allows for at most pair-wise hydrogen bonding which should exclude the formation of larger areas or regions of hydrogen bonded water.

The binding energies obtained for the molecularly adsorbed waters are only slightly lower than the computed binding energy of 0.64 eV for the dissociation and so, from the present results, the driving force for dissociating the H_2O would be expected to be weak. All optimizations have been performed assuming a fixed cluster geometry, however, and it is very likely that adsorbate-induced reconstruction of the surface would favor the dissociation and lead to a larger difference in the binding energies. This is furthermore in agreement with the observation that heating to 300 K is required for all adsorption to be dissociative; a higher binding energy for the dissociated products than what has been computed here is also indicated by the high (465 K) desorption temperature found for the recombination peak in the TPD spectrum [64]. The recombination barrier should be lower than the binding energies so that, even if considering a barrier to the recombination, the computed binding energies are still too small to give the TPD high temperature peak.

Thus, inclusion of reconstruction of the surface as the water dissociates would be highly desirable, but is currently very difficult to include in the present approach. This is due to the combination of structures optimised at different levels of theory: using pair-potentials for the surface and quantum methods for the cluster with adsorbate. This is a point where additional work needs to be invested to couple the two descriptions together in a more consistent way.

3.3. H_2/ZNO

As a final example we will discuss the dissociation of H_2 over the non-polar $(10\bar{1}0)$ and $(11\bar{2}0)$ ZnO surfaces compared with the polar (0001) Zn^+ terminated surface [14]. The two non-polar surfaces are bulk-terminated and optimisation of the surface structures only produce an inwards relaxation of the surface of similar magnitude for both cations and anions. For the polar surface, however, the dipole across the crystal must be eliminated in order to produce a stable surface; we have considered the case where about 1/4 of the surface layer Zn^{2+} ions have been transferred to the opposite face of the crystal, which is sufficient to generate an electrostatically stable surface. This structure was then relaxed using the pair-potential techniques described above. Several different initial structures were studied, including small terraces and mesa-like structures [14].

Having obtained the relaxed structure, a quantum chemical model of a reaction site was constructed. It was assumed that H_2 will dissociate over a ZnO pair and the cluster models include either one, two or three ZnO units embedded in an environment of AIMPs and with the Madelung potential (based on ± 2 charges) included through a direct evaluation of the Gaussian type integrals over the full Ewald sum [1, 2, 3] generated by the ions in the repeated surface unit cell.

The results for the different surfaces and different sites are given in Table 2, where the exothermicities have been computed relative to the surface and an H_2 molecule at long distance from the surface.

Several conclusions may be drawn from the data in the Table. First, there is a strong correlation between the stability of the surface and its reactivity; for the stable, non-polar $(10\bar{1}0)$ and $(11\bar{2}0)$ surfaces none of the sites investigated shows a positive exothermicity in the reaction with H_2, while for the less stable, polar surface several reactive sites are found. Second, for the reactive, polar surface there is a strong correlation between the exothermicity of a particular site and its coordination: the lower the coor-

TABLE 2. Computed exothermicities (eV) for hydrogen dissociation at different sites of the surfaces studied. SCF and MCPF results.

Surface	Model	Coord Zn O	SCF	MCPF
		Non-Polar Surface		
$(10\bar{1}0)$	small	3 3	- 0.34	-0.42
$(10\bar{1}0)$	large	3 3	-1.0	
$(11\bar{2}0)$	parallell	3 3	- 0.03	-0.10
$(11\bar{2}0)$	orthogonal	3 3	- 0.19	-0.22
		Polar Surface		
(0001)	bulk	3 3	4.47	-
(0001)	2x2 near hole	3 2	0.63	0.32
(0001)	2x2 not near hole	3 3	-1	-0.93
(0001)	4x4 corner	3 2	1.07	0.85
(0001)	4x4 edge	3 3	0.57	0.26
(0001)	4x4 terrace	3 4	no bond.	-

dination the higher the reactivity. The third conclusion may be drawn from the calculation with a ZnO unit of a bulk-terminated polar surface. Here a strongly exaggerated reactivity is found due to the inherent instability of the unreconstructed surface. Small cluster models that do not include the embedding, *e.g.* a single gas-phase ZnO unit, or an insufficient treatment of the embedding, either through too few point-charges or through not taking charge reconstruction into account in the case of polar surfaces, may thus lead to erroneous results with strongly exaggerated binding energies.

The structural results for the two non-polar surfaces are interesting in themselves: LEED data [65] for the structure of the $(10\bar{1}0)$ surface have been interpreted as a relaxation inwards of the order of 0.3 Å and 0.1 Å for the zinc and the oxygen top layer atoms, respectively. For the $(11\bar{2}0)$ surface no relaxation was observed. Thus a differential effect of relaxation is found for the $(10\bar{1}0)$, but not for the $(11\bar{2}0)$ surface. Using the pair-potential techniques described in section 2.1 no differential movement of the Zn atoms compared to the O atoms was found for either surface [14]. In a tight-binding treatment (including Zn $4p$) of the surface relaxation, Wang and Duke [66] find a differential inwards relaxation of about 0.6 Å for both surfaces. More recently, Jaffe *et al* [67] have performed periodic Hartree-Fock and DFT calculations on the $(10\bar{1}0)$ surface using the CRYSTAL [38] program and report values of 0.075-0.115 Å for the differential relaxation. We have applied the same technique to calculate the relaxations of both surfaces, using a three-layer slab where one zinc and one oxygen atom in the top layer were allowed to move [68]. The result was for $(10\bar{1}0)$ a downward movement for zinc of 0.16 Å and for oxygen an upward movement of 0.07 Å. For the $(11\bar{2}0)$ surface the zinc moved upwards 0.02 Å and the oxygen moved downwards 0.04 Å. Thus a better agreement for the structures is obtained when using the full quantum mechanical potential than what was found either in the tight-binding treatment or in the pair-potential approach. This is perhaps not too surprising since the pair-potentials were fitted to bulk-data. It does indicate that potentials specifically developed for surface problems should be considered. However, since the reactivity is so strongly determined by the ion coordination, it would seem that reliable descriptions of differential effects between different types of sites should be obtainable also with the present approach.

3.4. X-RAY SPECTROSCOPIES

We employ the static exchange (STEX) formalism [69] to treat X-ray absorption and emission spectroscopies. This requires the wave functions of the initial (neutral in absorption, core hole state in emission) and final states (core hole state plus electron in absorption, valence ionised in emission) and the transition moments for the coupling between them. In the following we will make a brief discussion of the application to near-edge X-ray absorption fine structure spectroscopy (NEXAFS).

For a closed-shell ground state the static exchange hamiltonian corresponding to a doublet final, core hole state takes the form

$$\hat{H}^j = \hat{h} + \sum_{i \neq j}^{N/2} (2\hat{J}_i - \hat{K}_i) + \hat{J}_j + \hat{K}_j = \hat{F} - \hat{J}_j + \hat{K}_j \tag{22}$$

where the ionisation takes place from (spin-restricted) shell j. The Fock operator $\hat{F}$ is constructed from the final (relaxed core-hole) state orbitals, assuming double occupancy, so that the effects of relaxation are explicitly included. The relaxation of the core-hole state is performed in three steps by first relaxing the valence in the field of the frozen core hole, then relaxing the core hole in the field of the frozen valence and finally repeating the first step to check that the procedure has converged.

The wave functions of the initial and core hole state are generated in a good molecular basis, but the STEX hamiltonian, which is to provide the one-electron orbitals for the excited electron, must be constructed in a very large and diffuse basis set in order to describe accurately valence and Rydberg excitations and to provide a sufficiently dense representation of the continuum; for first-row systems we thus add nineteen functions of

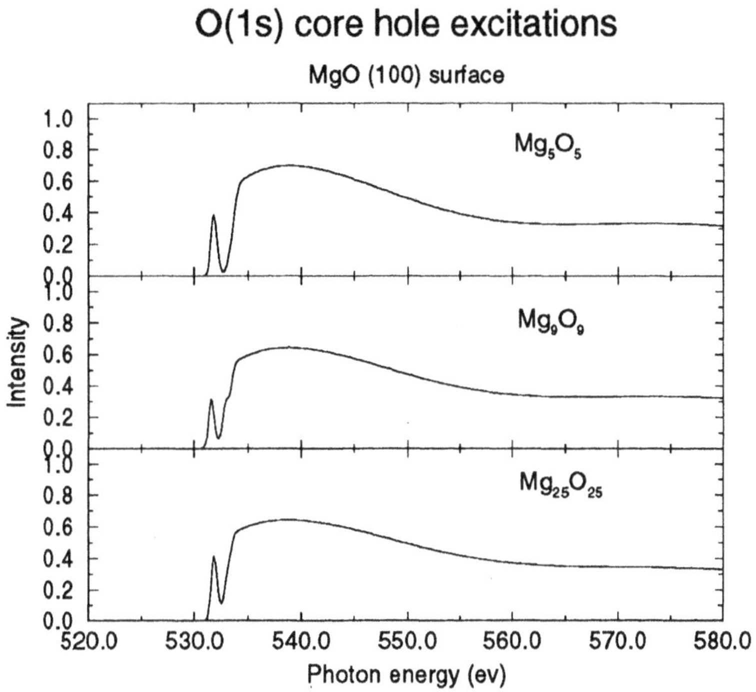

Figure 4. O(1*s*) X-ray absorption spectrum for cluster models of the MgO(100) surface.

s, *p* and *d*-type, respectively, to the basis set for the excitation center. The orbitals and eigenvalues are obtained by diagonalising the STEX matrix projected onto the unoccupied space; thus only the virtual orbitals of this Fock-like matrix are used. The NEXAFS excitation energies are obtained by adding the computed ΔSCF ionization potential to the eigenvalues of the STEX hamiltonian. Since the orbitals are obtained by diagonalisation of a Fock-like matrix, the (Hartree-Fock level) interaction of the excited electron with the relaxed molecular ion is fully described; the term *static exchange* is due to the neglect of relaxation of the molecular ion due to the excited electron.

The discrete levels, combined with the corresponding transition moments computed between the non-orthogonal sets of orbitals and broadened by a gaussian, give a true representation of the pre-edge structure. In the continuum the discrete eigenstates are used to construct a correctly normalized spectrum by means of the Stieltjes imaging procedure.

In order to treat large systems and basis sets the above formalism has been implemented in the direct SCF program DISCO [70]; the details of this implementation have been given in ref. [69] and some examples of applications may be found in *e.g.* ref. [71] (NEXAFS of OH/Cu(111), ref. [72] (NEXAFS of CO/Cu(100)), ref. [73] (orientational probing of polymeric thin poly-tetrafluoroethylene films). Here, considering the title of the contribution, we will show some results concerning the existence of O(1*s*) excitons at the (100) surface of MgO, CaO and SrO [74].

The calculations have been performed using properly embedded cluster models M_5O_5, M_9O_9 and $M_{25}O_{25}$ with an all-electron description of the central oxygen ion and core potentials for the inner shells of the remaining cluster ions. For the valence description of the cluster oxygens a [2*s* 3*p*] contracted basis set was used. The computed IPs show monotonic convergence with cluster-size, *i.e.* for MgO: 535.6, 535.2 and 535.1 eV including corrections for dynamical correlation (+1.2 eV), relativity (+0.4 eV) and inner-shell basis set extensions (-0.5 eV) computed for the oxygen ion. Extending the oxygen basis set in the case of M_9O_9 to [2*s* 3*p* 2*d*] to better represent the oxygen in-crystal polarisability reduces the computed IP by 0.5 eV (MgO). The trends in the O(1*s*) IPs for these systems have been discussed earlier by Pacchioni and Bagus [75].

For each of the systems we find a discrete level located at 2.0-1.5 eV before the ionisation edge (see Fig. 4). This corresponds to a $1s \rightarrow 3p_z$ excitation polarised out from the surface. In the case of MgO(100) an additional weaker excitation into a 4*p*-like level is found 0.9 eV before the edge. These states are only found at the surface; clusters simulating O^{2-} in the bulk show no discrete transitions before the edge, since the repulsion against the additional ions forces this level into the continuum. Thus, this surface exciton could possibly be used to fingerprint surface oxygens, but presence of other species such as OH^- at the surface may contribute to make unique identification difficult.

4. Acknowledgements

This work was carried out with support from the Swedish Natural Science Research Council (NFR) and by funds from the Swedish Consortium on Oxidic Overlayers.

References

1. P.P. Ewald. *Ann. Phys.*, 64:253, 1921.
2. D.E. Parry. *Surf. Sci.*, 49:433, 1975.
3. D.E. Parry. *Surf. Sci.*, 54:195, 1976.
4. M.A. Nygren, L.G.M. Pettersson, Z. Barandiarán, and L. Seijo. *J. Chem. Phys.*, 100:2010, 1994.
5. M. Pöhlchen and V. Staemmler. *J. Chem. Phys.*, 97:2583, 1993.
6. A.M. Ferrari and G. Pacchioni. *Int. J. Quantum Chem.*, 58:241, 1996.
7. J.-L. Pascual and L.G.M. Pettersson, submitted to Chem. Phys. Letters.
8. Z. Barandiarán and L. Seijo. *J. Chem. Phys.*, 89:5739, 1988.

9. Z. Barandiarán and L. Seijo. In S. Fraga, editor, *Studies in Physical and Theoretical Chemistry: Vol. 77(B), Computational Chemistry: Structure, Interactions and Reactivity*, page 435. Elsevier, Amsterdam, 1992.
10. J.A. Mejías and J. Fernández Sanz. *J. Chem. Phys.*, 102:327, 1995.
11. M.A. Nygren and L.G.M. Pettersson, J. Chem. Phys., in press.
12. J.G. Fripiat, A.A. Lucas, J.M. André, and E.G. Derouane. *Chem. Phys.*, 21:101, 1977.
13. M.A. Nygren and L.G.M. Pettersson. *J. Phys. Chem.*, 100:1874, 1996.
14. M. Nyberg, M. A. Nygren, L. G. M. Pettersson, D. H. Gay, and A. Rohl. *J. Phys. Chem.*, 100:9054, 1996.
15. H.H. Kung. *Transition Metal Oxides: Surface Chemistry and Catalysis*. Elsevier, Amsterdam, 1991.
16. S.A. Pope, M.F. Guest, I.H. Hillier, E.A. Colbourn, W.C. Mackrodt, and J. Kendrick. *Phys. Rev. B*, 28:2191, 1983.
17. D. Cappus, C. Xu, D. Ehrlich, B. Dillmann, C.A. Ventrice Jr, K. Al Shamery, H. Kuhlenbeck, and H.-J. Freund. *Chem. Phys.*, 177:533, 1993.
18. C.R.A. Catlow, R.G. Bell, and J.D. Gale. *J. Mater. Chem.*, 4:781, 1994.
19. D.H. Gay and A.L. Rohl. *J. Chem. Soc. Faraday Trans.*, 91:925, 1995.
20. L. Javier Alvarez, J. Fernández Sanz, M.J. Capitán, M.A. Centeno, and J.A. Odriozola. *J. Chem. Soc. Faraday Trans.*, 89:3623, 1993.
21. M. Baudin, M. Wojcik, and K. Hermansson. A molecular dynamics study of mgo slabs. Preprint.
22. G. Pacchioni, T. Minerva, and P.S. Bagus. *Surf. Sci.*, 275:450, 1992.
23. M.A. Nygren, L.G.M. Pettersson, A. Freitag, V. Staemmler, D.H. Gay, and A.L. Rohl. *J. Phys. Chem.*, 100:294, 1996.
24. R.W.G. Wyckoff. *Crystal Structures*. Wiley, New York, 1963.
25. K. H. Schulz and D. F. Cox. *Surf. Sci.*, 278:9, 1992.
26. K. Sawabe, N. Koga, K. Morokuma, and Y. Iwasawa. *J. Chem. Phys.*, 97:6871, 1992.
27. K. Sawabe, N. Koga, K. Morokuma, and Y. Iwasawa. *J. Chem. Phys.*, 101:4819, 1994.
28. V. R. Saunders, C. Freyria-Fava, R. Dovesi, L. Salasco, and C. Roetti. *Mol. Phys.*, 77:629, 1992.
29. M.A. Nygren. *Theoretical Studies of Metal Clusters and Oxidic Surfaces*. PhD thesis, Stockholm University, 1995.
30. C. Pisani, R. Dovesi, R. Nada, and S. Tamiro. *Surf. Sci.*, 216:489, 1989.
31. C. Pisani, R. Dovesi, R. Nada, and L.N. Kantorovich. *J. Chem. Phys.*, 92:7448, 1990.
32. J. L. Pascual, L. Seijo, and Z. Barandiarán. *J. Chem. Phys.*, 98:9715, 1993.
33. ECPAIMP is an integral program for ECP and AIMP calculations written by L.G.M. Pettersson, L. Seijo and M.A. Nygren.
34. N. Rösch, K.M. Neyman, and U. Birkenheuer. In H.-J. Freund and E. Umbach, editors, *Adsorption on Ordered Surfaces of Ionic Solids and Thin Films*. Springer, Berlin, 1993.
35. K.M. Neyman and N. Rösch. *Chem. Phys.*, 168:267, 1992.
36. M.A. Nygren and L.G.M. Pettersson. *Chem. Phys. Letters*, 230:456, 1994.
37. R. Orlando, R. Dovesi, C. Roetti, and V.R. Saunders. *Chem. Phys. Letters*, 228:225, 1994.
38. R. Dovesi, V.R. Saunders, C. Roetti, M. Causà, N.M. Harrison, R. Orlando and E. Aprà, *CRYSTAL95 user documentation*, Università di Torino, Torino 1995.
39. K.M. Neyman and N. Rösch. *Surf. Sci.*, 297:223, 1993.
40. U. Wahlgren and P.E.M. Siegbahn. In D. Salahub, editor, *Metal-ligand interactions: from atoms, to clusters, to surfaces*. Kluwer Academic Publishers, Dordrecht, Holland, 1991.
41. J. A. Mejías, A. M. Márquez, J. Fernández Sanz, M. Fernández-García, J. M. Ricart,

C. Sousa, and F. Illas. *Surf. Sci.*, 327:59, 1995.
42. W.H. Press, S.A. Teukolsky, W.T. Vetterling, and B.P. Flannery. *Numerical recipes in FORTRAN : the art of scientific computing, second edition.* Cambridge University Press, Cambridge, 1992.
43. E.A. Paukshtis, R.I. Soltanov, and N.E. Yurchenko. *Reaction Kinet. Catalysis Letters*, 16:93, 1981.
44. S. Furuyama, H. Fujii, M. Kawamura, and T. Morimoto. *J. Phys. Chem.*, 82:1028, 1978.
45. S.A. Pope, I.H. Hillier, M.F. Guest, E.A. Colbourn, and J. Kendrick. *Surf. Sci.*, 139:299, 1984.
46. E.A. Colbourn and W.C. Mackrodt. *Surf. Sci.*, 143:391, 1984.
47. C.R. Henry, C. Chapon, and C. Duriez. *J. Chem. Phys.*, 95:700, 1991.
48. J.-W. He, C.A. Estrada, J.S. Corneille, M.-C. Wu, and D.W. Goodman. *Surf. Sci.*, 261:164, 1992.
49. D. Cappus, J. Klinkmann, H. Kuhlenbeck, and H.-J. Freund. *Surf. Sci.*, 325:L421, 1995.
50. S.M. Vesecky, Xueping Xu, and D.W. Goodman. *J. Vac. Sci. Technol. A*, 12:2114, 1994.
51. R. Dovesi, R. Orlando, F. Ricca, and C. Roetti. *Surf. Sci.*, 186:267, 1987.
52. G. Pacchioni, G. Cogliandro, and P.S. Bagus. *Surf. Sci.*, 255:344, 1991.
53. G. Pacchioni, G. Cogliandro, and P.S. Bagus. *Int. J. Quantum Chem.*, 42:1115, 1992.
54. K.M. Neyman, S. Ph. Ruzankin, and N. Rösch. *Chem. Phys. Letters*, 246:546, 1995.
55. D. P. Chong and S. R. Langhoff. *J. Chem. Phys.*, 84:5606, 1986.
56. L.G.M. Pettersson and H. Åkeby. *J. Chem. Phys.*, 94:2968, 1991.
57. S.F. Boys and F. Bernardi. *Mol. Phys.*, 19:553, 1970.
58. P.O. Widmark, P.Å. Malmqvist, and B.O. Roos. *Theoret. Chim. Acta*, 77:291, 1990.
59. T.H. Dunning Jr. *J. Chem. Phys.*, 53:2823, 1970.
60. D. Scarano, G. Spoto, S. Bordiga, S. Coluccia, and A. Zecchina. *J. Chem. Soc. Faraday Trans.*, 88:291, 1992.
61. Y. Zeiri, A. Redondo, and W. A. Goddard III. *Surf. Sci.*, 131:221, 1983.
62. A. Redondo, Y. Zeiri, and W. A. Goddard III. *Surf. Sci.*, 136:41, 1984.
63. T. Tashiro, J. Ito, R.-B. Sim, K. Miyazawa, E. Hamada, K. Toi, H. Kobayashi, and T. Ito. *J. Phys. Chem.*, 99:6115, 1995.
64. K. H. Schulz and D. F. Cox. *Surf. Sci.*, 256:67, 1991.
65. C.B. Duke, A.R. Lubinsky, S.C. Chang, B.W. Lee, and P. Mark. *Phys. Rev. B*, 15:4865, 1977.
66. Y.R. Wang and C.B. Duke. *Surf. Sci.*, 192:309, 1987.
67. J. E. Jaffe, N. M. Harrison, and A. C. Hess. *Phys. Rev. B*, 49:11153, 1994.
68. M. Nyberg, L.G.M. Pettersson and M. Causà. To be published.
69. H. Ågren, V. Carravetta, O. Vahtras, and L.G.M. Pettersson. *Chem. Phys. Letters*, 222:75, 1994.
70. J. Almlöf, K. Faegri Jr., and K. Korsell. *J. Comput. Chem.*, 3:385, 1982.
71. L.G.M. Pettersson, H. Ågren, O. Vahtras, and V. Carravetta. *J. Chem. Phys.*, 103:8713, 1995.
72. L.G.M. Pettersson, H. Ågren, O. Vahtras, and V. Carravetta. Surf. Sci., in press.
73. H. Ågren, V. Carravetta, O. Vahtras, and L.G.M. Pettersson. *Phys. Rev. B*, 51:17848, 1995.
74. J.-L. Pascual, L.G.M. Pettersson, A. Nilsson and N. Mårtensson. To be published.
75. G. Pacchioni and P. S. Bagus. *Phys. Rev. B*, 50:2576, 1994.

STABILITY OF POLAR OXIDE SURFACES:

Oxygen vacancies and non-stoichiometric reconstructions

C. NOGUERA, A. POJANI, F. FINOCCHI, J. GONIAKOWSKI*

Laboratoire de Physique des Solides, associé au CNRS
Université Paris Sud, 91405 ORSAY (FRANCE)

**CRMC2, CNRS*
Campus de Luminy, 13288 MARSEILLE Cedex 9 (FRANCE)

Abstract

The stability of polar oxide surfaces is discussed on general grounds, and more specifically in the case of MgO(111), a prototype of standard polar surfaces, and in the case of $SrTiO_3(001)$, a prototype of 'weakly polar' surfaces. In both cases, on the unreconstructed stoichiometric surfaces, an electron redistribution takes place in the surface layers, which cancels the macroscopic electrostatic dipole. This process involves a metallization of the MgO(111) unreconstructed surfaces, and induces a high value of the surface energy. On $SrTiO_3(001)$, on the other hand, bond breaking at the surface suffices by itself to provide the compensating charges. A better stability of the MgO(111) surface may be achieved by the desorption of a macroscopic number of atoms in the outer layer(s). We study two surface configurations, associated with (2×1) and (2×2) unit cells, in which little electron redistribution is needed to cancel the macroscopic electrostatic dipole. We predict an even larger stability for $(2n \times 2n)$ reconstructed surfaces which present pyramid-like structures and $\{100\}$ facets of increasing sizes as n grows. In addition, we analyze the formation of oxygen vacancies and their interaction on the two terminations of $SrTiO_3(001)$. In this compound, oxygen vacancies do not provide a stabilization of the surface. However, their formation induces a whole set of reconstructed structures, on which we bring some elements of discussion.

1 Introduction

Surfaces of compounds which involve at least two types of atoms, bearing charges of opposite signs, have been classified in three families, on the basis of a simple electrostatic criterion [1]. In the first family, each layer parallel to the surface is neutral. However, in most compounds, due to the complexity of the atomic

R.M. Lambert and G. Pacchioni (eds.),
Chemisorption and Reactivity on Supported Clusters and Thin Films, 455-478.

structure, the atomic layers are not neutral. When the stacking sequence is such that the total dipole moment of the repeat unit vanishes, surfaces are called "type II surfaces". Finally, "type III surfaces", also called polar surfaces, are those whose layers are not neutral and whose repeat unit bears a non-zero dipole moment. For a cut of a crystal which produces a polar surface, non-zero dipole moments are present not only on the outer layers (which would not distinguish them from non-polar rumpled or reconstructed surfaces), but on *all* the repeat units throughout the material. This produces strong electrostatic effects which are not constrained within the surface layers.

Polar surfaces are generally unstable, because of a large electrostatic contribution to their surface energy. However, due to their potentially interesting applications — for example the high reactivity of their surface sites — , it is a challenge to produce and stabilize them, and get more information on their atomic and electronic structures. In rocksalt oxides, the {111} surface is polar, with alternating layers of anion and cation composition. In most experiments, this face turned out to be either facetted and/or non-stoichiometric. This was the case of NiO{111} [2], MgO{111} [3, 4] and CoO{111} [5]. Planar surfaces are more easily obtained when a thin oxide film is grown on a substrate [6, 7]. However, even so, nanofacets may be observed, as in the case of the p(2 × 2) reconstruction of NiO{111} [8, 9]. In the wurtzite structure, met in ZnO, the polar faces are the (0001) and $(000\bar{1})$ faces. It seems that they have a lower surface energy than the rocksalt {111} surfaces. The experiments performed on these materials do not mention impurity or facetting problems [10]. The $Cu_2O(100)$ polar surface also displays a $(3\sqrt{2} \times \sqrt{2})R45°$ reconstruction when it is terminated by copper atoms [11]. The (111) polar face of Fe_3O_4 was found to be non-stoichiometric [12].

Tasker's classification of surfaces relies on an ionic description, in which formal charge values are assigned to the ions: for example −2 for oxygen, +2 for strontium, +4 for titanium, etc. Under this assumption, the $SrTiO_3(001)$ surface is a "type I surface" with a stacking sequence: SrO/TiO_2 (Figure 1). However, in mixed iono-covalent materials, orbitals overlap, charge transfers take place and the estimation of the layer charge density is not a simple matter. In binary compounds, layers with anions and cations in the stoichiometric ratio are neutral, whatever the value of the anion-cation charge transfer: for example, a planar MgO(100) layer, with as many oxygens as magnesiums, is neutral, whether the ions bear ±2 or $\pm x$ charges. Nevertheless, in ternary compounds this is no longer the case. Let us take the example of $SrTiO_3$. In the bulk, the neutrality of the unit cell requires that $Q_{Sr} + Q_{Ti} + 3Q_O = 0$. If one assumes that the compound is fully ionic, the additional relationship $Q_{Ti} = 2Q_{Sr}$ is fulfilled, the (001) atomic layers in the bulk are neutral and the (001) surface can be classified as "type I". However, in a quantum description, regardless the approximation made, the relationship $Q_{Ti} = 2Q_{Sr}$ is no longer valid. The unit cell of the (001) layers bears charges equal to $\pm\sigma$, with $\sigma = Q_{Sr} + Q_O = -(Q_{Ti} + 2Q_O)$. A non-zero dipole moment is present in *all* the SrO-TiO_2 bilayers and the (001) surface turns out not to be of "type I" but

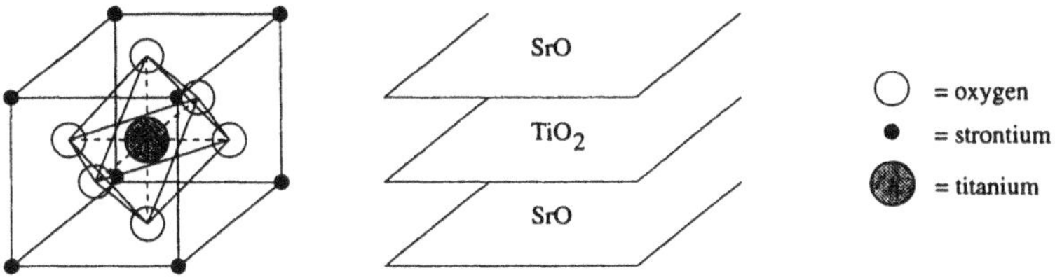

Figure 1: Non-equivalent (001) layers in $SrTiO_3$.

of "type III". We have recently proposed to call such surfaces: "weakly polar surfaces" [13, 14] because, as will become apparent in the following, the mechanism of charge redistribution turns out to be different from the one which takes place on standard polar surfaces. Planar $SrTiO_3(001)$ surfaces have been successfully prepared; however numerous reconstructions have been observed after annealing, which have been assigned to an ordering of oxygen vacancies in the TiO_2 outer layer [15, 16, 17, 18].

A discussion of the surface stability of polar and weakly polar surfaces is thus especially relevant and will be the subject of the present contribution. We will recall the electrostatic arguments which predict which charge distribution is required in the surface layers in order to cancel the macroscopic component of the dipole moment. Then, we will describe how such a charge distribution may be achieved on unreconstructed stoichiometric surfaces. For this purpose, we will focus on MgO(111) and $SrTiO_3(001)$ surfaces, and we will combine a numerical approach based on a semi-empirical Hartree-Fock total-energy code and analytical arguments for the interpretation of the charge density. In the two last sections, we will focus on non-stoichiometric reconstructions on these surfaces. On MgO(111), the reconstructions that we will consider provide a cancellation of the macroscopic dipole moment and lead to a surface stability which is higher than on the unreconstructed surface. On $SrTiO_3(001)$, we will analyze the energetics of neutral oxygen vacancies and we will propose some tracks for interpreting the observed reconstructions.

2 Electrostatic considerations

A simple electrostatic argument explains the instability of polar surfaces. It is illustrated in Figure 2a, in the case of a rocksalt {111} cut, whose layers bear alternating charge densities $\pm\sigma$ and are located at equal distances R. The electrostatic field is constant inside the bilayers, and zero between the bilayers. A monotonic increase of the electrostatic potential results, which amounts to $4\pi\sigma R$ per bilayer (several tens V for an assumed fully ionic MgO slab). The total dipole moment of a slab involving N bilayers is equal to $\mu = N\sigma R$ and its macroscopic electrostatic energy amounts to: $E = 4\pi N R\sigma^2$. It is infinite for macroscopic systems and very large even for thin films. This is the origin of the surface instability.

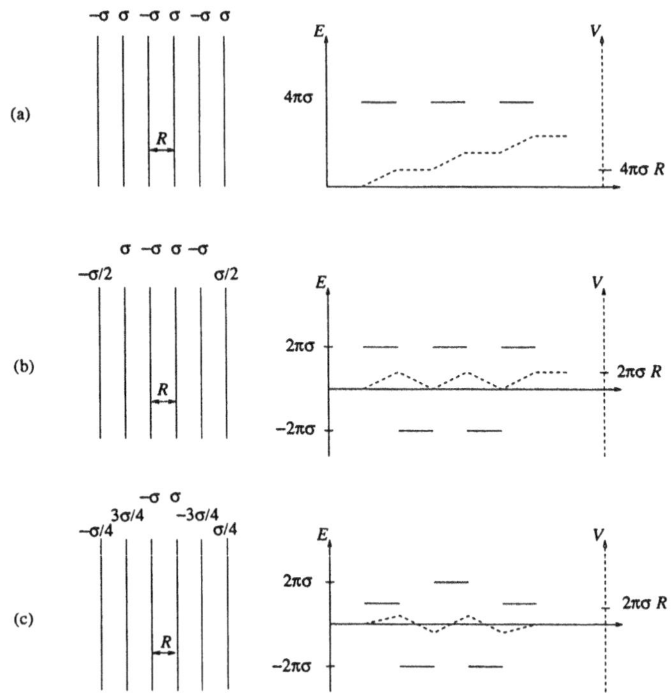

Figure 2: Spatial variations of the electric field $\mathcal{E}$ (full lines) and of the electrostatic potential V (dashed line) in a thin sample cut along a polar direction. (a) When the planes bear charge densities equal to $\pm\ \sigma$, the electrostatic potential increases monotonically through the sample. (b) The charge density on the outer planes is modified $\sigma_S = \sigma/2$: the electrostatic potential no longer increases proportionally to the sample thickness, but oscillates from one layer to another. (c) A dipole density $\sigma R/4$ is added on the outer faces in order to cancel the total dipole moment of the slab ($\sigma_S = \sigma/4$; $\sigma_{SS} = 3\sigma/4$).

The introduction of additional (compensating) charges on the outer layers of the slab may drastically change this picture. With $\pm\sigma_S$ the modified charge densities, the electric field is equal to zero outside the material, to $-4\pi\sigma_S$ inside each bilayer and to $4\pi(\sigma - \sigma_S)$ in the space region between two bilayers. Its mean value amounts to $\overline{\mathcal{E}} = 2\pi(\sigma - 2\sigma_S)$. A reduction of the surface charge by a factor of two with respect to the bulk: $\sigma_S = \sigma/2$, thus allows a cancellation of $\overline{\mathcal{E}}$ and transforms the monotonic increase of the electrostatic potential into small oscillations with maxima and minima on positively and negatively charged layers, respectively (Figure 2b). The dipole moment $\mu = \sigma R/2$ is no longer proportional to the slab thickness and the electrostatic energy has roughly decreased by a factor of 2: $E = 2\pi R\sigma^2(N - 1/2)$.

It is possible to further reduce the slab energy by adding a dipole on the two faces of the system. Compared with the preceding estimation, the electrostatic field is changed only inside the double-layer on which the dipole has been placed. A charge distribution $\pm\sigma_S = \pm\sigma/4$ and $\pm\sigma_{SS} = \pm 3\sigma/4$ on the surface and sub-surface layers, respectively (Figure 2c), yields a complete cancellation of the slab dipole moment and suppresses the potential discontinuity between the left and the right of the slab. The electrostatic energy now reads: $E = 2\pi R\sigma^2(N - 5/4)$.

More generally, a slab of N bilayers, whose m outer layers on each of its sides have charge densities $\pm\sigma_m$ different from σ, has a dipole moment equal to:

$$\mu = 2NR\left(\sum_{j=1}^{m}(-1)^{j+1}\sigma_j + (-1)^m\frac{\sigma}{2}\right) - R\left(\sum_{j=1}^{m}(2j-1)(-1)^{j+1}\sigma_j + (-1)^m m\sigma\right) \tag{2.1}$$

In this expression, the layers are numbered starting from the surface and are assumed to be equidistant. The condition of cancellation of the *macroscopic* component (the part of μ which is proportional to the slab thickness) of the dipole moment reads:

$$\sum_{j=1}^{m}(-1)^{j+m}\sigma_j = \frac{\sigma}{2} \tag{2.2}$$

In the following, we will refer to Equation (2.2) as the electrostatic condition for the stability of a polar surface. If $m = 2$, only the surface (S) and sub-surface (SS) layers are modified ($\sigma_1 = \sigma_S$ and $\sigma_2 = \sigma_{SS}$) and Equation (2.2) takes the simple form: $\sigma_{SS} - \sigma_S = \sigma/2$. The generalisation to a system in which the intra- and inter-bilayer distances are different is straightforward.

The electrostatic arguments which have just been given rely upon a macroscopic analysis. This means that (i) the discrete atomic structure inside each layer is neglected; only the uniform component of the electrostatic potential is taken into account (term $G_{||} = 0$ in the Fourier series); and (ii) the electron density is assumed to be localized in the planes parallel to the surface, with no inter-plane overlap. In addition, the compound is assumed to have no electronic degrees of freedom. Its charge distribution is not correlated to the electrostatic potential as it should be in a more correct self-consistent and quantum approach.

All the mechanisms for polar surface stabilization which have been hencefore proposed in the literature yield the charge modifications assumed in Figures 2b and 2c, either by internal electron redistribution, by the presence of charged adsorbates or impurities [2, 5] or by substoichiometry, as in compound semiconductors [19, 20]. The first of these mechanisms was discussed in a cluster approach of MgO(111) [21], and later generalized to periodic slabs [22]. A metallization of the surface layers and an electron exchange between the two opposite faces were predicted.

A stabilization of the surface due to the presence of charged adsorbates is exemplified by the hydroxylation of the rocksalt {111} surface, in which adsorbed protons or hydroxyl groups saturate all surface sites. The surface charges produced by this process exactly fulfil the condition $\sigma_{SS} - \sigma_S = \sigma/2$ (Equation 2.2). Relying on ab initio calculations of the surface energy, it was recently argued [23] that the most stable surface of MgO under ambient conditions is not the {100} surface but rather the hydroxylated {111} face. From an experimental point of view, the hydroxylation of NiO{111} was found to destroy the p(2 × 2) reconstruction and to restore the unreconstructed unit cell in a reversible manner [24].

Finally, a spontaneous desorption of a macroscopic number of surface atoms may yield the adequate charge densities. In some cases, facetted surfaces may be formed, and an ordering of vacancies may lead to surface reconstructions, as proposed long time ago [25]. In rocksalt oxides, the creation of {100} facets is the most likely, due to their high stability in this structure [26]. When only the two outermost layers are perturbed, as in Figure 2c, nanopyramids associated with nanofacets are formed [27].

3 Charge redistribution on stoichiometric polar and weakly polar surfaces

We now focus on the charge distribution on unreconstructed stoichiometric surfaces. We will analyse the MgO(111) [22] and $SrTiO_3$(001) [13] surfaces, by a combined use of numerical and analytical calculations.

3.1 STRATEGY

Surface and bulk properties of oxides have been studied numerically by various methods, ranging from ab initio (Density Functional or Hartree-Fock) to completely non-self-consistent approaches [28, 14]. Here, we present results obtained by a total energy self-consistent semi-empirical approach, working at the Hartree-Fock level, which has the advantages of (i) being fast enough to allow the consideration of large systems, and (ii) nevertheless being able to treat properly screening effects and the self-consistent interplay between charges and electrostatic potentials.

The Slater determinant is expanded on an atomic orbital basis set, assumed to be orthogonal, which includes the oxygens' $2s$ and $2p$ orbitals, the magnesiums' $3s$ and $3p$ orbitals, the strontiums' $5s$ and $5p$ orbitals and the titaniums' $3d$ and $4s$ orbitals. The resulting Hartree-Fock-Roothaan equations are solved self-consistently within a CNDO-like assumption (CNDO= Complete Neglect of Differential Overlap). Some parameters are required to calculate the matrix elements, such as intra-atomic electron-electron integrals, atomic energies of the neutral atoms, etc. They are adjusted to reproduce the bulk electronic and atomic characteristics of the compounds. In the particular case of $SrTiO_3$, the parameters have been taken from a previous investigation of the binary oxides TiO_2 and SrO [29] without any change. The details concerning the Hamiltonian matrix and the parametrization are presented elsewhere [30]. The calculations are performed in the slab geometry, with a minimal slab thickness determined by the requirement that the local electronic structure in the slab centre is identical to that in the bulk. Integration of the charge density is performed by means of a Brillouin Zone sampling using sets of special k-points [31]. For MgO, a geometry optimization is associated with the electronic structure calculation, in order to determine the atomic relaxations at 0 K. Starting from an assumed atomic configuration, the quantum ground state energy and the atomic forces are calculated and used to derive the new atomic positions, by a steepest descent method. The whole procedure is repeated until the energy minimum with respect to the atomic degrees of freedom is found. For $SrTiO_3$, results for an unrelaxed geometry will be given. The surface energies will thus be slightly overestimated but we have preliminary confirmation that the qualitative features of the surface stability are not soundly dependent upon the relaxation process.

A simple analytical model for the calculation of the charge values will be used [14, 13]. It starts from a fully ionic, tight-binding description of the oxide, in which the eigenstates are atomic orbitals localized on the ions, for example in the case of MgO: $|\Psi_{\mathrm{O}i\lambda}\rangle$ and $|\Psi_{\mathrm{Mg}j\mu}\rangle$ (the sites are denoted by i, j and the orbitals by λ, μ), with associated eigenvalues $\epsilon_{\mathrm{O}i\lambda}$ and $\epsilon_{\mathrm{Mg}j\mu}$. The valence band is constituted by oxygen states and the conduction band by magnesium states. The oxygens and magnesiums have respective charges $Q_{\mathrm{O}} = -2$ and $Q_{\mathrm{Mg}} = +2$. The wavefunctions are subsequently modified by introducing the oxygen-cation electron sharing (i.e. the resonance integrals β) as a first order perturbation. A mixing of anion and cation orbitals results for all the eigenstates; for example:

$$|\Psi'_{\mathrm{O}i\lambda}\rangle = |\Psi_{\mathrm{O}i\lambda}\rangle + \sum_{j\mu} \frac{\beta_{i\lambda,j\mu}}{\epsilon_{\mathrm{O}i\lambda} - \epsilon_{\mathrm{Mg}j\mu}} |\Psi_{\mathrm{Mg}j\mu}\rangle \tag{3.1}$$

We will assume that the basis set is orthogonal and that hybridization takes place only between orbitals localized on neighbouring sites. When only valence band states are filled, the charges of an oxygen i and of a magnesium j are equal to:

$$Q_{\mathrm{O}i} = -2 + \sum_{\lambda j\mu} 2 \left(\frac{\beta_{i\lambda,j\mu}}{\epsilon_{\mathrm{O}i\lambda} - \epsilon_{\mathrm{Mg}j\mu}} \right)^2 \tag{3.2}$$

and:

$$Q_{\mathrm{Mg}j} = +2 - \sum_{\mu i\lambda} 2 \left(\frac{\beta_{i\lambda,j\mu}}{\epsilon_{\mathrm{O}i\lambda} - \epsilon_{\mathrm{Mg}j\mu}} \right)^2 \tag{3.3}$$

respectively. With respect to the fully ionic limit, the charge reduction may be interpreted as arising from electron transfers $\Delta_{ij} = \sum_{\lambda\mu} 2(\beta_{i\lambda,j\mu}/\epsilon_{\mathrm{O}i\lambda} - \epsilon_{\mathrm{Mg}j\mu})^2$ along the ij bonds. In the case where only oxygen p orbitals and magnesium s orbitals are taken into account and in an isotropic environment (for example the bulk), these expressions simplify and give:

$$Q_{\mathrm{O}} = -2 + 2Z_{\mathrm{O}} \frac{\beta^2}{(\epsilon_{\mathrm{O}} - \epsilon_{\mathrm{Mg}})^2} \text{ and } Q_{\mathrm{Mg}} = +2 - 2Z_{\mathrm{Mg}} \frac{\beta^2}{(\epsilon_{\mathrm{O}} - \epsilon_{\mathrm{Mg}})^2} \tag{3.4}$$

with Z_{O} and Z_{Mg} the oxygen and magnesium coordination numbers. When, in addition, a conduction band state, localized on a given magnesium orbital $j\mu_0$, is partially filled with an occupation factor f, the charges of this magnesium and of the neighbouring oxygens are modified, in the following way:

$$Q_{\mathrm{Mgj}} = +2 - \sum_{\mu i \lambda} 2 \left(\frac{\beta_{i\lambda,j\mu}}{\epsilon_{\mathrm{O}i\lambda} - \epsilon_{\mathrm{Mgj}\mu}} \right)^2 + \sum_{i\lambda} 2f \left(\frac{\beta_{i\lambda,j\mu_0}}{\epsilon_{\mathrm{O}i\lambda} - \epsilon_{\mathrm{Mgj}\mu_0}} \right)^2 \tag{3.5}$$

and:

$$Q_{\mathrm{Oi}} = -2 + \sum_{\lambda j \mu} 2 \left(\frac{\beta_{i\lambda,j\mu}}{\epsilon_{\mathrm{O}i\lambda} - \epsilon_{\mathrm{Mgj}\mu}} \right)^2 - \sum_{\lambda} 2f \left(\frac{\beta_{i\lambda,j\mu_0}}{\epsilon_{\mathrm{O}i\lambda} - \epsilon_{\mathrm{Mgj}\mu_0}} \right)^2 \tag{3.6}$$

Under the same assumptions as in Equation (3.4), this leads to the simplified expressions:

$$Q_{\mathrm{Mg}} = +2 - 2(1-f) Z_{\mathrm{Mg}} \frac{\beta^2}{(\epsilon_{\mathrm{O}} - \epsilon_{\mathrm{Mg}})^2} \text{ and } Q_{\mathrm{O}} = -2 + 2(Z_{\mathrm{O}} - f) \frac{\beta^2}{(\epsilon_{\mathrm{O}} - \epsilon_{\mathrm{Mg}})^2} \tag{3.7}$$

Extension of these expressions to the cases of a partial filling of several conduction band states, partial filling of a valence band state, anisotropic environments, or ternary compounds is straightforward. This analytical approach will be useful for understanding the mechanisms of charge redistribution at the surfaces, and for proving that these mechanisms are not dependent upon the specific choice of the parameters made in the semi-empirical procedure.

3.2 THE UNRECONSTRUCTED (1×1) MgO(111) SURFACE

The unreconstructed (1×1) MgO(111) surface has an hexagonal symmetry. The outer layer is composed of atoms of a single type, which are three-fold coordinated to the atoms in the underlying layer.

Figure 3 shows the local density of states (LDOS) on surface and bulk atoms. The surface valence band, primarily of oxygen $2p$ character, is split into two peaks. The higher one is located in the bulk gap region, while the lower one lies in the lower part of the bulk valence band. The surface conduction band, mainly of magnesium character, is split as well, and the lower peak, located in the bulk gap, overlaps the higher surface oxygen peak. The Fermi level intersects these two surface bands, which gives a metallic character to the two faces of the slab.

As a result of metallicity, a strong electron redistribution takes place. In the surface layers, the charges are found equal to $Q_{\mathrm{S}} = \pm 0.59$. They amount to $Q_{\mathrm{SS}} = \pm\ 1.19$ in the sub-surface layers, while in all the inner layers, they are close to their bulk values $Q = \pm\ 1.22$. The charge reduction on the surface oxygens and magnesiums is assigned to the incomplete filling of the surface valence band and to the partial filling of the surface conduction band, respectively. We do not wish to emphasize too much on the precise charge values, but rather on the fact that $\sigma_{\mathrm{SS}} - \sigma_{\mathrm{S}}$ is close to one half the bulk charge density, as required from Equation (2.2).

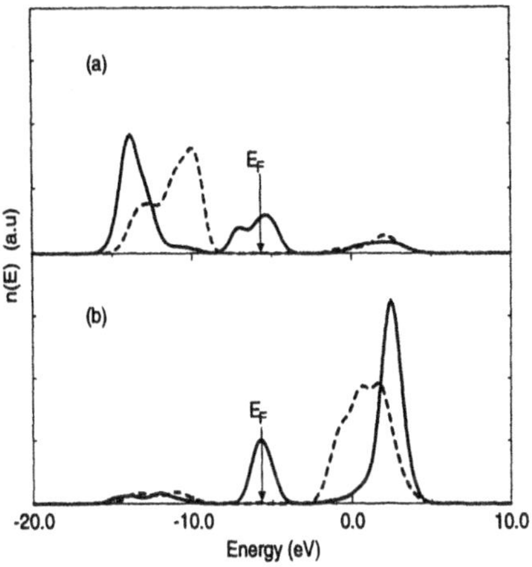

Figure 3: Local density of states (LDOS) on the unreconstructed (1 × 1) MgO(111) surface. (a): the oxygen LDOS; (b): the magnesium LDOS; the LDOS on surface and bulk atoms are respectively represented by plain and dashed lines. The calculated LDOS have been convoluted by a 0.7 eV wide Gaussian function.

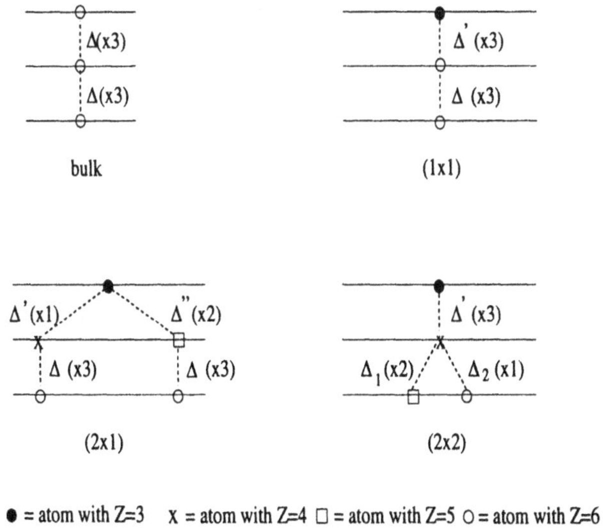

Figure 4: Schematic representation of the bond charge transfers, and their degeneracy, in bulk MgO and in the outer layers of the unreconstructed and reconstructed MgO(111) surfaces.

To understand the mechanism of charge redistribution, we apply the analytical model presented above. Bulk atoms are six-fold coordinated and the absolute value of their charge may be written: $Q = 2 - 6\Delta$. The atoms in the outer layers are three-fold coordinated ($Z = 3$) to atoms located in the underlying layers. The electron transfer Δ' between the surface and sub-surface layer is discriminated from the transfer Δ in the rest of the slab, to account for the shift of the effective atomic energies at the surface (Figure 4). First we wish to point out that, in an hypothetical insulating case, the surface charges would read:

$$\begin{aligned} Q_S &= 2 - 3\Delta' \\ Q_{SS} &= 2 - 3\Delta' - 3\Delta \end{aligned} \tag{3.8}$$

The condition $\sigma_{SS} - \sigma_S = \sigma/2$ would not be fulfilled, whatever the values of Δ and Δ'. Rather:

$$\sigma_{SS} - \sigma_S = \frac{\sigma}{2} - 1 \tag{3.9}$$

(hencefore we give the charge densities in electrons per unit cell).

However, a conduction band state in the surface magnesium termination is partly filled (occupation factor f) and the electron counting leads to surface and sub-surface charges equal to :

$$\begin{aligned} Q_S &= 2(1-f) - 3(1-f)\Delta' \\ Q_{SS} &= 2 - 3(1-f)\Delta' - 3\Delta \end{aligned} \tag{3.10}$$

A difference in the charge densities:

$$\sigma_{SS} - \sigma_S = 2f - 3\Delta \tag{3.11}$$

results, which fulfils the electrostatic condition Equation (2.2) provided that $f = 1/2$. For the oxygen termination, it is also found that a valence band state has to be half-filled. This is indeed what is found numerically (Figure 3). A metallization of the surfaces, associated with an overlap of the surface oxygen and magnesium bands, is thus a necessary condition for the cancellation of the macroscopic electric dipole. The numerical charge values are accounted for by the model with $\Delta = 0.13$, $\Delta' = 0.27$ and $f = 1/2$. Strictly speaking, the unreconstructed MgO(111) surface should not be called "polar": the oxide has enough electronic degrees of freedom to cancel the polarity. However this can only be achieved at the expense of a high surface energy (Section 4.3).

3.3 THE $SrTiO_3$(001) STOICHIOMETRIC SURFACE

An $SrTiO_3$ crystal, cut along the (001) orientation, has two inequivalent surfaces. Figure 5 shows the LDOS on surface and bulk atoms. On the SrO termination, the Sr LDOS is shifted to lower energies, compared to the bulk LDOS. The Sr-O

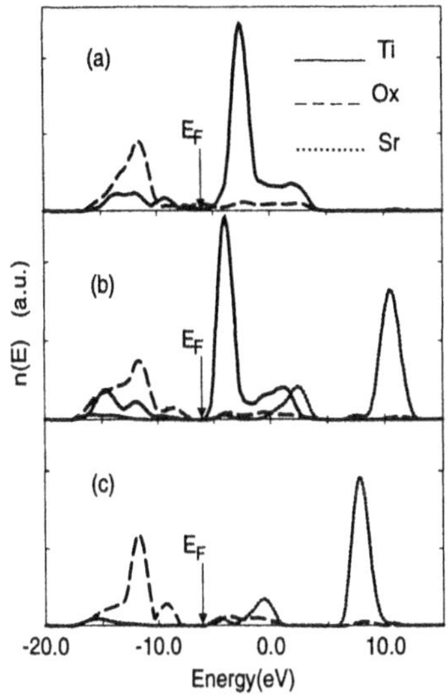

Figure 5: Local density of states (LDOS) on the unreconstructed $SrTiO_3(001)$ surface, as a function of energy (in eV). (a): the TiO_2 termination;(b): the bulk; (c): SrO termination. The calculated LDOS have been convoluted by a 0.7 eV wide Gaussian function.

bonds are thus slightly more covalent than in the bulk, but the local gap remains wide, of the order of 5 eV (peak to peak energy separation between the valence and conduction bands), in agreement with experimental findings [32]. Stictly speaking, the TiO_2 termination is semi-metallic but the density of states in the bulk gap region is very low. The energy separation between the higher valence band peak and the lower conduction band peak is of the order of 5 eV.

As expected for a compound known to present a large part of covalent character in the Ti-O bond, the bulk ionic charges in $SrTiO_3$ differ significantly from the formal values: Q_O = -0.97, Q_{Ti} = +1.47 and Q_{Sr} = +1.44. Again, we do not wish to focus on the absolute values of the charges which depend upon the chosen parameters, but rather on the fact that the (001) atomic layers have non-zero charge densities $\pm\sigma$, and that the bulk repeat units parallel to the (001) direction bear a non-zero dipole moment.

On the (001) surface, the main effect induced by the surface formation is a modification of the ionic charges on the outer layers and a weak electron transfer between the surface and sub-surface layers. On the TiO_2 termination:

$$Q_{Ti} = 1.29,\ Q_O = -0.78,\ \text{and}\ Q_O = -0.94,\ Q_{Sr} = 1.44 \tag{3.12}$$

respectively, on the surface and sub-surface layers. Similarly, on the SrO termination,

$$Q_O = -1.21,\ Q_{Sr} = 1.48,\ \text{and}\ Q_{Ti} = 1.46,\ Q_O = -0.98 \tag{3.13}$$

In the inner layers, the charges are identical to the bulk charges. The charge density on the surface layers is strongly reduced ($\sigma_S = 0.27$) while σ_{SS} remains close to the bulk value σ ($\sigma_{SS} = 0.50$, $\sigma = 0.47$). The electrostatic criterion for the stabilization of polar surfaces is thus fulfilled, without requiring any surface metallization.

According to the analytical model, the charges in the bulk of $SrTiO_3$ can be written:

$$Q_{Ti} = 4 - 6\Delta_{O-Ti},\ Q_{Sr} = 2 - 12\Delta_{O-Sr},\ \text{and}\ Q_O = -2 + 2\Delta_{O-Ti} + 4\Delta_{O-Sr} \tag{3.14}$$

since each titanium (strontium) atom has six (twelve) oxygens in its first coordination shell, and each oxygen is surrounded by two titaniums and four strontiums. The charge densities of the (001) atomic layers $\pm\sigma$ are thus equal to:

$$\sigma = |Q_{Ti} + 2Q_O| = Q_{Sr} + Q_O = 2\Delta_{O-Ti} - 8\Delta_{O-Sr}.$$

The difference of electronegativity between the Ti and Sr atoms and the difference between the Ti-O and Sr-O bond lengths reflect themselves in the much larger value of the electron transfer along the Ti-O bonds than along the Sr-O ones (the numerical results can be accounted for by taking $\Delta_{O-Ti} \sim 0.422$ and $\Delta_{O-Sr} \sim 0.045$). As a consequence, $\sigma \neq 0$ and since $\Delta_{O-Ti} > 4\Delta_{O-Sr}$, the TiO_2 layers bear a negative charge.

In the surface layer of TiO_2 stoichiometry, titaniums have lost one oxygen neighbour and oxygens have lost two strontium neighbours. In addition bonds inside the surface plane have a different covalency from the bulk, which gives rise to an electron transfer Δ'_{O-Ti} which differs from the bulk one. Consequently:

$$Q_{Ti} = 4 - 4\Delta'_{O-Ti} - \Delta_{O-Ti},\ Q_O = -2 + 2\Delta'_{O-Ti} + 2\Delta_{O-Sr}. \tag{3.15}$$

Similarly, on the other face of the slab, strontiums have lost four oxygen neighbours and oxygens have lost one titanium:

$$Q_{Sr} = 2 - 4\Delta_{O-Sr} - 4\Delta'_{O-Sr},\ Q_O = -2 + \Delta_{O-Ti} + 4\Delta'_{O-Sr}. \tag{3.16}$$

The values of Δ'_{O-Ti} and Δ'_{O-Sr} deduced from the numerical results ($\Delta'_{O-Ti} \sim 0.567$ and $\Delta'_{O-Sr} \sim 0.086$) show a strong increase of covalency in the surface layers. The absolute values of surface charge densities on both faces of the slab are equal and read:

$$\sigma_S = \Delta_{O-Ti} - 4\Delta_{O-Sr}, \tag{3.17}$$

while $\sigma_{SS} = \sigma$. The model predicts surface charge densities equal to $\pm\sigma/2$, whatever the specific values of Δ_{O-Sr}, Δ_{O-Ti}, Δ'_{O-Sr} and Δ'_{O-Ti}. On $SrTiO_3$(001), the bond breaking mechanism, by itself, yields the charge redistribution required for the surface stabilization (Equation 2.2). This conclusion is at variance with that obtained for the MgO(111) surface, and it is the reason why we proposed to give the qualificative "weakly polar" [13, 14] to such a surface.

4 Reconstructed MgO(111) surfaces

It has been shown in Section 3.1 that the MgO electronic structure is flexible enough to provide a cancellation of the (111) surface polarity, thanks to a metallization of the outer layers. However this process costs much energy, and better means can be found to stabilize the (111) surface. We will consider in the following two surface configurations in which a macroscopic number of atoms is missing in the outer layer(s), with a concentration such that little electron redistribution takes place.

4.1 THE RECONSTRUCTED (2 × 1) SURFACE

Starting from the stoichiometric MgO(111) surface and its hexagonal (1 × 1) unit cell (Figure 6a), a non-stoichiometric surface with half of the atoms missing in the outer layer and a doubled unit cell may be imagined as shown in Figure 6b. The reconstruction may be called (2 × 1), but the primitive cell that has the full symmetry of the array is the rectangular one, shown in Figure 2b, with $\vec{a'} = 2\vec{a} - \vec{b}$, and $\vec{b'} = \vec{b}$. The Wood notation does not apply to this reconstruction. We will keep the denomination (2 × 1), in the following, for the sake of conciseness. The surface displays parallel streaks, with {100} and {110} facets. The surface atoms are three-fold coordinated ($Z = 3$). In the sub-surface layer, the atoms belonging to the {110} facets have 4 first neighbours; those belonging to the {100} facets have 5 first neighbours. The interest of such a surface configuration lies in the fact that the surface stoichiometry automatically provides the charge compensation suggested in Figure 2b, without the need of a strong charge redistribution.

Indeed, we find that the surface electronic structure is not very different from the bulk one: the oxygen and magnesium bands are weakly shifted towards higher and lower energies, respectively. No band splitting takes place, and no metallization of the surfaces occurs. A finite gap is present between the surface valence and conduction bands, whose width is equal to 6.9 eV. This value is close to that found on the MgO(211) surface (6.5 eV) [29], a non-polar surface on which the atoms have a similar environment.

A weak charge redistribution takes place on the two outer planes on each side of the slab. We find atomic charges equal to $Q_S = \pm 1.14$ on the surface layers and $Q_{SS} = \pm\ 1.19$ on the sub-surface layers. After normalization to the surface unit cell of the unreconstructed surface, one obtains charge densities equal to $\sigma_S = \pm 0.57$, and $\sigma_{SS} = \pm\ 1.19$. The value of $\sigma_{SS} - \sigma_S$, which is close to half σ, fulfils the electrostatic criterion (Equation 2.2) for surface stability.

To apply the analytical model which describes the charge redistribution, we introduce two parameters Δ' and Δ'' to account for the electron transfer along the bonds linking the surface atoms to the {110} and {100} facet atoms, respectively (Figure 4). The absolute values of the charge Q_S of a surface atom and of the charges Q'_{SS} and Q''_{SS} on the two types of sub-surface atoms read:

$$Q_S \ = \ 2 - \Delta' - 2\Delta''$$

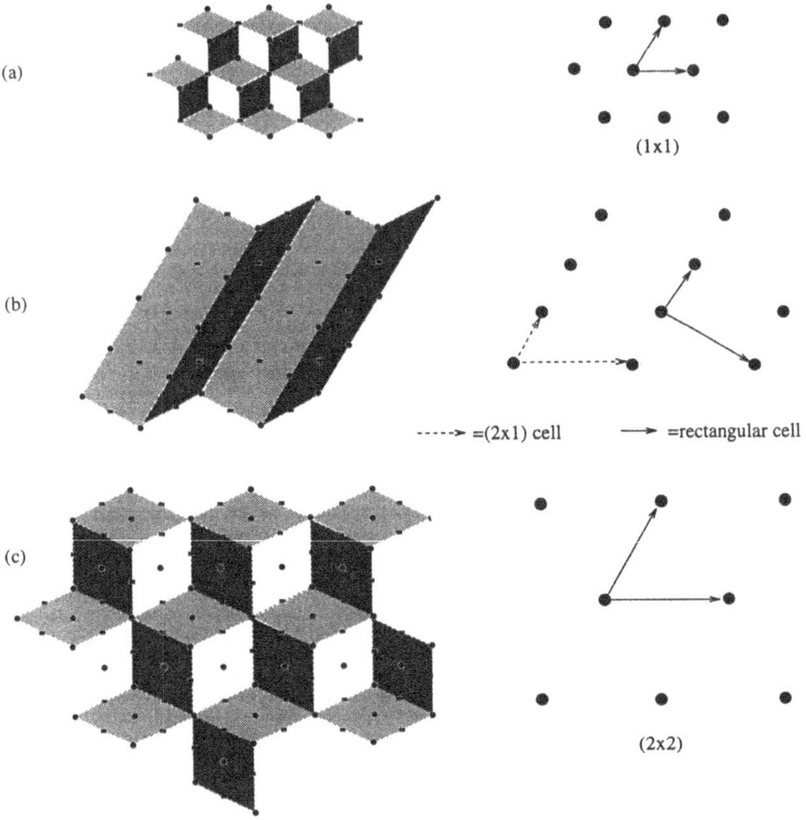

Figure 6: Structure of the MgO(111) surface. (a) the unreconstructed (1×1) surface; (b) the (2×1) surface; (c): the (2×2). Left: Relief view; Right: surface unit cells. On the (2×1) surface, $\{100\}$ and $\{110\}$ facets have been colored in dark and light grey, respectively.

$$\begin{aligned} Q'_{SS} &= 2 - \Delta' - 3\Delta \\ Q''_{SS} &= 2 - 2\Delta'' - 3\Delta \end{aligned} \tag{4.1}$$

The numerical results may be accounted for with $\Delta' = 0.43$ and $\Delta'' = 0.21$. The charge densities on the outer layers and on the sub-surface layers are equal to:

$$\begin{aligned} \sigma_{\mathrm{S}} &= 1 - \Delta'/2 - \Delta'' \\ \sigma_{\mathrm{SS}} &= 2 - \Delta'/2 - \Delta'' - 3\Delta \end{aligned} \tag{4.2}$$

in absolute value. As a consequence, the model predicts that, whatever the specific values of Δ, Δ' and Δ'':

$$\sigma_{\mathrm{SS}} - \sigma_{\mathrm{S}} = 1 - 3\Delta = \frac{\sigma}{2} \tag{4.3}$$

The electrostatic condition (Equation 2.2) is thus fulfilled without requiring a surface metallization.

4.2 THE RECONSTRUCTED (2 × 2) SURFACE

Figure 6c represents another non-stoichiometric surface structure in which three atoms over four have been removed in the outer layers, and one over four in each sub-surface layer. Square facets of {100} orientations involving four atoms, and pyramid-like structures around the three-fold coordinated surface atoms ($Z = 3$) are produced, a geometry already proposed by several authors [33, 27]. The surface cell is hexagonal and its area is four times larger than the unreconstructed (1×1) cell. The surface is p(2×2) reconstructed. The interest of such a surface configuration lies in the fact that the surface stoichiometry automatically provides the charge compensation suggested in Figure 2c, without the need of a strong charge redistribution.

The electronic structure found for this surface configuration is remarkably similar to that of the (2×1) surface. The shift of the surface valence and conduction bands with respect to the bulk bands is slightly smaller and a slightly wider gap results: 7.5 eV instead of 6.9 eV.

The surface charges on the outer planes of the (2×2) slab are close to the bulk values. We find $Q_S = \pm 1.20$ on the surface layers, $Q_{SS} = \pm$ 1.21 on the sub-surface layers and $Q_{SSS} = \pm 1.21$, $Q'_{SSS} = \pm 1.25$ respectively on the five- and six-fold coordinated atoms of the third layers. These values, once normalized to the unit cell of the unreconstructed surface, correspond to charge densities equal to $\sigma_S = \pm 0.3$, $\sigma_{SS} = \pm$ 0.91 and $\sigma_{SSS} = \pm$ 1.22. These charge densities are close to those assumed in Figure 2c and fulfil the condition (2.2): $\sigma_{SSS} - \sigma_{SS} + \sigma_S = \sigma/2$.

The analytical model applied to this surface configuration (Figure 4) yields charge values equal to:

$$\begin{aligned} Q_S &= 2 - 3\Delta' \\ Q_{SS} &= 2 - \Delta' - 2\Delta_1 - \Delta_2 \\ Q_{SSS} &= 2 - 2\Delta_1 - 3\Delta \\ Q'_{SSS} &= 2 - 3\Delta_2 - 3\Delta \end{aligned} \tag{4.4}$$

and charge densities equal to:

$$\begin{aligned} \sigma_S &= 1/2 - 3\Delta'/4 \\ \sigma_{SS} &= 3/2 - 3(\Delta' + 2\Delta_1 + \Delta_2)/4 \\ \sigma_{SSS} &= 2 - 3\Delta_1/2 - 3\Delta_2/4 - 3\Delta \end{aligned} \tag{4.5}$$

The numerical values may be accounted for with $\Delta' = 0.27$, $\Delta_1 = 0.20$ and $\Delta_2 = 0.12$. The model predicts that $\sigma_{SSS} - \sigma_{SS} + \sigma_S = \sigma/2$, which fulfils the criterion (2.2) for surface stability, whatever the specific values of Δ, Δ', Δ_1 and Δ_2.

4.3 DISCUSSION

The calculated value of the unreconstructed surface energy is very high: $E_{(1\times1)} = 7.3$ J/m^2. It is associated with the simultaneous formation of an oxygen and a magnesium termination. This high surface energy has various contributions, among which the most important is due to the filling of oxygen and magnesium states high in the gap region. It confirms previous speculations that this surface is highly unstable.

The energy required to form a (2×1) surface, on the other hand, amounts to $E_{(2\times1)}$=2.6 J/m^2. It is possible to obtain a good estimate of this energy by noting that the surface is actually made of {100} and {110} facets linked by edges. By counting the number of atoms involved in the facets and edges per unit cell, and by taking the energies calculated for these surface defects [34], one predicts a surface energy equal to 2.8 J/m^2, in agreement with the numerical value to within 10%.

The (2×2) surface energy amounts to $E_{(2\times2)}$=2.3 J/m^2. It is possible to account for it by summing the energies associated with the three {100} facets and the corner atom per unit cell. Using the values obtained in a preceding work [34], one obtains 2.8 J/m^2. The agreement is worse than on the (2×1) surface, because we have not taken into account the large relaxation of the corner atom, which induces a reduction of the facet area.

Nevertheless, this model allows to make predictions concerning the stability of surfaces in which a larger number of layers are under-stoichiometric. For example, $(2n \times 1)$ and $(2n \times 2n)$ reconstructed surfaces may be considered, as a mere generalisation of the (2×1) and (2×2) structures obtained by increasing the facet sizes. In an ionic picture — such as the one used in Figure 2 — their charge distributions (the layers are numbered starting from both surfaces) read:

$$\sigma_j = (2j-1)\frac{\sigma}{2n} \quad 1 \le j \le n \tag{4.6}$$

and:

$$\sigma_j = (2j-1)\frac{\sigma}{4n} \quad 1 \le j \le 2n \tag{4.7}$$

respectively; n layers are under-stoichiometric in the former case and $2n$ in the latter. In both cases, the criterion (2.2) for surface stability is fulfilled. The model developed above predicts a decrease of the surface energy, as n grows, due to the decreasing contribution of the edges and corners. Ultimately, for $(2n \times 1)$ and $(2n \times 2n)$ surfaces, when n goes to infinity, the surface energies tend to

$$\begin{aligned} E_{(2n\times1)} &= \frac{E_{100}}{\sqrt{3}} + \frac{E_{110}\sqrt{2}}{\sqrt{3}} \\ E_{(2n\times2n)} &= \sqrt{3}E_{100} \end{aligned} \tag{4.8}$$

with E_{100} and E_{110} the surface energies of the relaxed {100} and {110} planar surfaces. Since E_{110} is larger than $\sqrt{2}E_{100}$, this analysis predicts that the oxide will preferentially form reconstructed ($2n \times 2n$) surfaces. Thick samples should thus present a tendency to form large facets. In the case of thin films, the film thickness may control the size of the facets, as was likely the case in the experiments performed in NiO(111) [8].

5 Neutral oxygen vacancies on polar and weakly polar surfaces

It is well-known that, in most experiments performed on oxide surfaces, oxygen vacancies are present. They induce electronic structure modifications, structural distortions, as well as enhanced reactivity, and may in some cases drastically change the surface conductivity. In addition, the presence of a large density of vacancies may lead to changes in the size of the surface unit cells, recognized as reconstruction effects [28, 14]. In this context, the reconstructions which have been discussed for MgO(111) involve the desorption of a large number of surface atoms, oxygens or magnesiums. We will first reexamine the energetics of these surfaces, from the view point of the formation of oxygen vacancies. Then we will consider the (001) face of $SrTiO_3$ and we will propose some elements for understanding the reconstructions observed on this surface.

5.1 OXYGEN VACANCIES ON MgO(111)

We have not studied the formation of a single oxygen vacancy on the unreconstructed (111) face of MgO because this surface is unstable and has not been the subject of thorough investigations. However, we have modelled the desorption of a complete surface layer, either oxygen or magnesium, which gives information on the relative stability of the two slab terminations. In the previous study (Section 3.2), it was already possible to have some hints on this question, by analysing how the surface energy $E_{(1\times1)}$=7.3 J/m^2 is shared between the two terminations: the result is 13.6 J/m^2 for the oxygen termination and 1 J/m^2 for the magnesium termination. However, these partial energies are not measurable quantities, since the two terminations are formed at the same time when MgO is cut along the (111) direction. The energy needed to remove a complete (111) layer, on the other hand, is found equal to 3.1 J/m^2 and 5 J/m^2 for the oxygen and the magnesium layer, respectively. It is thus easier to remove a surface oxygen layer than a magnesium one on the MgO(111) surface. This shows that, in the absence of oxydizing atmosphere, the magnesium termination is more stable than the oxygen one.

It is interesting to note that the energy of oxygen vacancy formation E_{vf}=1.5 eV on this surface is much lower than on the (100) surface. On the latter surface, we had found [35] that E_{vf} varies from 10 eV for isolated vacancies to 3.8 eV when all the oxygens of the (100) top layer are removed. Here E_{vf} is low, because (i) the oxygen-terminated (111) surface has a high surface energy and (ii) the desorption of oxygens yields a charge redistribution constrained within the two outer layers. It does not give rise to a macroscopic dipole moment.

A comparison between the surface energies of the (1×1) and (2×1) MgO(111) surfaces allows to derive the cost in energy to remove half of the oxygen atoms from the unreconstructed surface. The result is -2.6eV per oxygen atom. The negative value tells that a desorption of a one-half density of surface oxygens is energetically favourable. This was expected from the discussion made in Section 4.1, which showed that it provides a cancellation of the macroscopic dipole without requiring charge redistributions. The energetics of vacancy formation — dependence of the energy upon the vacancy density, interactions between vacancies... — on a polar surface thus turns out to be quite different from that observed on a non-polar surface, e.g. the (001) face of MgO [35].

5.2 OXYGEN VACANCIES ON $SrTiO_3$(001)

We give here some results concerning neutral oxygen vacancies on the two terminations of $SrTiO_3$(001). These results have been obtained without taking into account the atomic distortions around the vacancies. In addition, the calculations have been performed using periodic boundary conditions and a (2×2) unit cell. While such a size turns out to be large enough on the SrO termination because the screening cloud is highly localized around the vacancy, this is no longer the case on the TiO_2 termination. The following discussion will thus remain at a qualitative level.

5.2.1 *Formation of a single vacancy*

On the SrO termination, the energy needed to remove a neutral oxygen is found equal to E_{vf}=4.1 eV. The electrons left behind by the missing oxygen are redistributed in the system, with a high localization on the titanium located just below the vacancy site. The LDOS on the atoms which surround the vacancy are shifted toward lower energies, as a result of an enhancement of their Madelung potential (associated with the loss of a negative charge in their neighbourhood). The lowest conduction band state is an hybridized d orbital on the titanium located below the vacancy. The electrons left behind by the departing oxygen thus mainly fill this orbital. The charge perturbation in the system is well represented by a vertical (i.e. perpendicular to the surface) dipole $(-Q,+Q)$, whose negative end is on the titanium below the vacancy, and whose positive end is on the vacancy site.

On the TiO_2 termination, the energy needed to create a neutral oxygen vacancy is equal to E_{vf}=3.2 eV. The electrons left by the missing atom are mainly redistributed on the two neighbouring surface titaniums. This can be explained with the same argument as on the SrO termination. However, on this surface, the screening cloud extends further from the vacancy site, along the Ti-vacancy-Ti axis. Calculations are under progress to get a more accurate description of the charge redistribution using

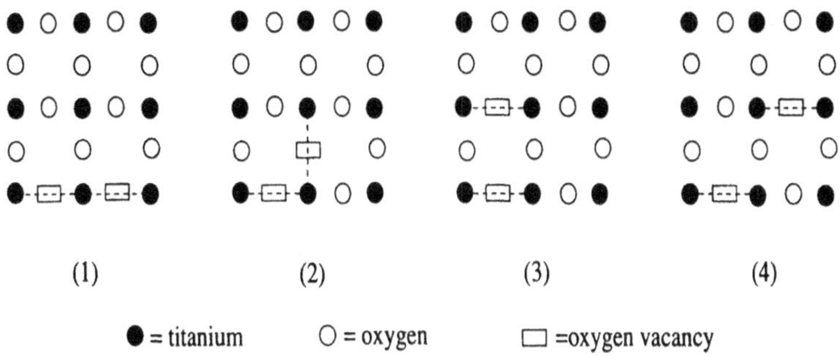

Figure 7: Four configurations of a (2 × 2) unit cell of the TiO_2 face of $SrTiO_3(001)$, with two vacancies. The dashed lines materialize the spatial extension of the charge redistribution.

(3 × 3) and (5 × 2) unit cells. However, no important correction to E_{vf} is expected; we have already checked that E_{vf} is changed by less than 0.05 eV when the unit cell is enlarged from (2 × 2) to (3 × 3). As a first approximation, the charge perturbation due to the oxygen vacancy may be represented by a linear horizontal (i.e. parallel to the surface) quadrupole ($-Q/2$,$+Q$,$-Q/2$), located on the Ti-vacancy-Ti unit, with two negative ends on the titaniums and a positive core on the vacancy.

The cost of energy to create an oxygen vacancy turns out to be larger on the SrO face than on the TiO_2 termination, in agreement with the electrostatic arguments given above (a dipole induces a stronger perturbation than a quadrupole). This result is consistent with experimental findings [36], which conclude that oxygen atoms in the topmost SrO layer are stable enough not to desorb even when annealing in UHV. However the lack of observation of gap states in the EELS experiment of Ref. 36 should not be taken as a conclusive proof of the absence of oxygen vacancies. On the oxygen deficient SrO termination the LDOS that we have calculated presents no well-defined gap states.

Considering the lack of experimental results on oxygen deficient SrO terminations, we have not tried to model several vacancies. Nevertheless we can argue that a repulsion between vacancies is likely to occur, by analogy with a previous study of vacancy-vacancy interactions on MgO(100) [35]. We had noted there that, despite other contributions (such as short range repulsion, covalent interaction...), the direct electrostatic interactions between the charge perturbations (the charge due to the missing lattice ion and the charge redistribution around) associated with the two vacancies accounted for all calculated features. We will see that this is also the case on the TiO_2 face. On the SrO termination, the interaction between two vacancies should be well described by the interaction between two parallel dipoles, i.e. by a repulsion. In the case where vacancies could actually be formed on this face, we would thus expect a tendency of the vacancies to maximize their spacing. Reconstructed structures should thus be observed, with unit cell sizes depending upon the vacancy density.

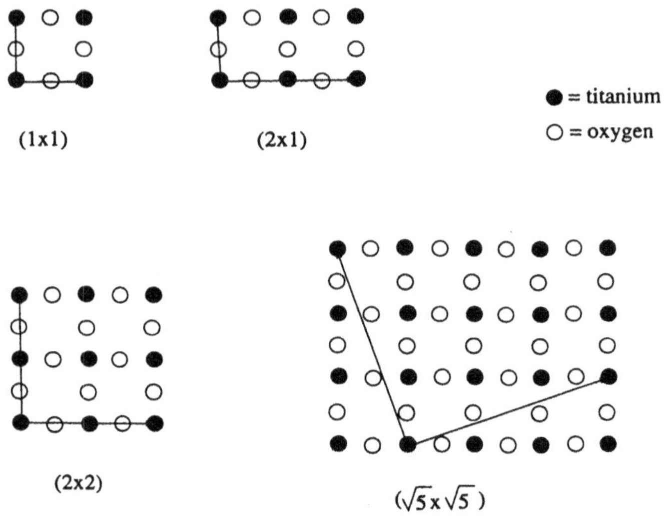

Figure 8: Unit cell vectors of the reconstructions observed on the TiO_2 termination of $SrTiO_3(001)$, superimposed on the ideal surface structure.

5.2.2 *Interactions between vacancies on the TiO_2 face*

We have studied four surface configurations with two neutral oxygen vacancies per (2×2) unit cell (vacancy density equal to 1/4). They are represented in Figure 7 and will be referred to C_1 ... C_4 in the following. The quadrupoles describing the charge redistribution around the vacancies are respectively aligned, perpendicular, parallel and shifted with respect to each other. From the energies of the unit cell in the presence and in the absence of the vacancies, we derive the total energy of the vacancies per unit cell. Substracting $2E_{vf}$ and counting the interactions between nearest vacancies, we obtain the following values for the interaction energy E_m^{int}:

$$E_1^{\text{int}} = -0.33 \text{ eV} \; ; \; E_2^{\text{int}} = 0.44 \text{ eV} \; ; \; E_3^{\text{int}} = 0.06 \text{ eV} \; ; \; E_4^{\text{int}} = -0.01 \text{ eV} \qquad (5.1)$$

The largest interactions thus take place between 'aligned' or 'perpendicular' vacancies. They are attractive in the former case, and repulsive in the latter.

As in the case of a single vacancy, a proper description of the charge redistribution requires a larger unit cell, and we postpone it to a forthcoming paper. However, the gross features can be understood by assuming that the charge perturbations due to each vacancy are additive. When two quadrupoles $(-Q/2,+Q,-Q/2)$ overlap, as in C_1 and C_2, the modification of charge on the atom belonging to the two quadrupoles is roughly equal to $-Q$. In C_1 for example, the Ti-vacancy-Ti-vacancy-Ti unit corresponds to a charge perturbation $(-Q/2,+Q,-Q,+Q,-Q/2)$. Electrostatic interactions are responsible for the respective values of the E_m^{int} found numerically.

This result suggests some comments on the possible orderings of vacancies on the TiO_2 termination of $SrTiO_3$ at 0K. Extending the results obtained for two vacancies

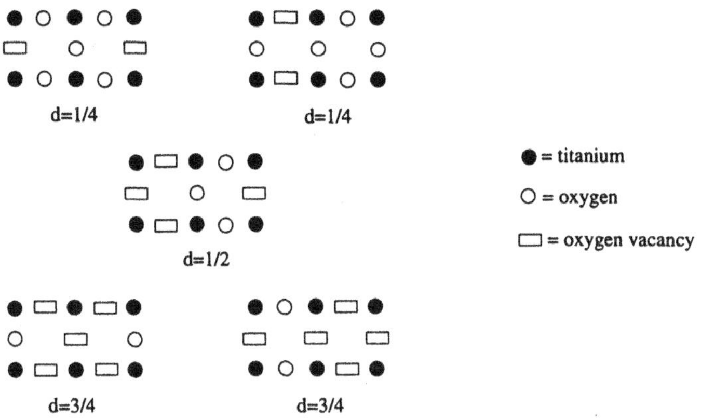

Figure 9: Examples of vacancy ordering in the (2×1) reconstructed unit cell of the TiO_2 termination of $SrTiO_3(001)$.

to a higher density of defects, it is expected that vacancies will preferentially form rows, in order to maximize the number of aligned quadrupoles, and that the rows will tend to be parallel, to minimize the number of intersecting points (where quadupoles are perpendicular). This prediction may be used to discuss the reconstructions which have been observed. Actually, planar, oxygen deficient surfaces, recognized to be the TiO_2 terminations of $SrTiO_3$ display (1×1), (1×2), (2×2) [15, 16], $(\sqrt{5}\times\sqrt{5})$R26.6° [18, 37] reconstructions, depending upon the annealing temperature of the surface. The unit cells are easily drawn (Figure 8), but, since in most experiments the density of vacancies is unknown, it is, at present, impossible to propose a structure for the reconstructed surface. For example, Figure 9 shows hypothetical vacancy configurations in the (1×2) unit cell. For a vacancy density $d = 3/4$, rows of vacancies are formed. In the case of the (2×2) reconstruction, there are eighteen inequivalent structures (one for a 1/8 density, 2 for a 1/4 density, 4 for the 3/8, 1/2 and 5/8 densities, 2 for a 3/4 density and one for the 7/8 density). The number is even higher for larger unit cells. Even a (1×1) unit cell is compatible with parallel rows of vacancies ($d = 1/2$). The understanding of the non-stoichiometric reconstructions on $SrTiO_3(001)$ may thus be considered as being at its starting point. From the experimental point of view, a thorough chemical characterization of the reconstructed surfaces is needed. From a theoretical point of view, work is currently under progress in our group (i) to obtain a more precise description of the charge redistribution around isolated vacancies, using larger unit cells, (ii) to take into account atomic relaxation, both on the stoichiometric and non-stoichiometric surfaces, and (iii) to actually compare the energy of various reconstructions, for a given density of vacancies, as we have done in the case of MgO(001).

6 Conclusions

We have discussed the stability of polar oxide surfaces, from different points of view. First we have considered unreconstructed stoichiometric surfaces, and we have distinguished the case of polar and 'weakly polar' surfaces. On standard polar surfaces (exemplified by MgO(111)), the dipole moment μ_0 of the repeat unit, in the direction perpendicular to the surface, is non-zero whatever charge values are assigned to the atoms. For weakly polar surfaces,(exemplified by $SrTiO_3(001)$), μ_0 is zero when calculated with formal charge values, but non-zero when the partial covalency of the anion-cation bonds is taken into account. We have shown, using numerical calculations and analytical arguments, that in both cases, an electron redistribution in the surface layers takes place in order to cancel the macroscopic electrostatic dipole. However, metallization of the surface layer is a necessary condition for its occurrence in MgO(111), while the bond breaking at the surface is sufficient to provide the compensating charges in $SrTiO_3(001)$.

The unreconstructed stoichiometric MgO(111) surface, although no longer polar once the metallization is achieved, remains unstable and has a high surface energy $E_{(1\times1)}$. We have studied two surface configurations in which a macroscopic number of atoms is missing in the outer layer(s), with a concentration such that no electron redistribution is needed to cancel the macroscopic electrostatic dipole. In the (2×1) reconstructed surface, a half of the atoms in the outermost layer are missing; in the (2×2) reconstruction, three quarters of atoms in the surface layer and one quarter in the sub-surface atoms are lacking. The surface energies of both surfaces are much lower than $E_{(1\times1)}$. Since the reconstructions are equivalent to nano-facetting, $E_{(2\times1)}$ and $E_{(2\times2)}$ can be estimated from the energies to form {001} and {110} facets, edges which delimit the facets and corner atoms. The most stable configuration is the (2×2) surface. It involves nanopyramids with {001} facets, which are the natural faces of the rocksalt structure. The transformation of the (1×1) into the (2×1) oxygen face may also be viewed as due to the creation of a 1/2 density of oxygen vacancies. We have shown that the energy associated with this process is negative, which is an indication of the instability of the unreconstructed oxygen termination.

We have analysed the formation of oxygen vacancies on the two terminations of $SrTiO_3(001)$, and sketched out a discussion of non-stoichiometric reconstructions on the TiO_2 termination. The two stoichiometric terminations are stable, and, at variance with MgO(111), the formation energy of oxygen vacancies E_{vf} is positive. We have shown that E_{vf} is higher on the SrO termination, and that the charge perturbation can be schematized as a dipole perpendicular to the surface. We thus predict that electrostatic interactions induce a repulsion between vacancies, which could possibly lead to ordered structures with unit cell sizes depending upon the vacancy density. On the TiO_2 termination, an oxygen vacancy induces a charge perturbation which can be represented as a quadrupole localized in the surface layer. Electrostatic considerations support the existence of an attraction or a repulsion between vacancies, depending upon the relative orientation of their associated quadrupoles. This study suggests that alignment of vacancies along the Ti-O rows might be energetically favourable. A thorough investigation of vacancy ordering on $SrTiO_3$ is currently under progress.

References

[1] P.W. Tasker, J. Phys. C: Solid State Physics 12 (1979) 4977.

[2] N. Floquet and L.C. Dufour, Surf. Sci. 126 (1983) 543.

[3] V.E. Henrich, Surf. Sci. 57 (1976) 385.

[4] H. Onishi, C. Egawa, T. Aruga and Y. Iwasawa, Surf. Sci. 191 (1987) 479.

[5] P.A. Cox and A.A. William, Surf. Sci. 152/153 (1985) 791.

[6] C.R. Henry and H. Poppa, Thin Solid Films 189 (1990) 303.

[7] O.L. Warren and P.A. Thiel, J. Chem. Phys. 100 (1994) 659.

[8] C.A. Ventrice Jr, T. Bertrams, H. Hannemann, A. Brodde and H. Neddermeyer, Phys. Rev. B49 (1994) 5773.

[9] H.J. Freund, H. Kuhlenbeck, V. Staemmler, Rep. Prog. Phys. 59 (1996) 283.

[10] M. Sambi, G. Granozzi, G.A. Rizzi, M. Casarin, E. Tondello, Surf. Sci. 319 (1994) 149.

[11] K.H. Schulz and D. F. Cox, Phys. Rev B43 (1991) 1610.

[12] A. Barbieri, W. Weiss, M.A. Van Hove and G.A. Somorjai, Surf. Sci. 302 (1994) 259.

[13] J. Goniakowski, C. Noguera, Surf. Sci. Lett. (1996) accepted

[14] C. Noguera, *Physics and Chemistry of Oxide Surfaces* (Cambridge University Press, 1996, under press); *Physique et Chimie des Surfaces d'Oxydes* (Eyrolles, 1995, collection Alea).

[15] B. Cord, R. Courths, Surf. Sci. 162 (1985) 34.

[16] T. Matsumoto, H.Tanaka, T. Kawai and S. Kawai, Surf. Sci. 278 (1992) L153.

[17] Y. Liang and D.A. Bonnell, Surf. Sci. 285 (1993) L510.

[18] H. Tanaka, T. Matsumoto, T. Kawai and S. Kawai, Surf. Sci. 318 (1994) 29.

[19] S.Y. Tong, G. Xu and W.N. Mei, Phys. Rev. Letters 52 (1984) 1963.

[20] M. Lannoo, Materials Sci. Ing. B 22 (1993) 1.

[21] M. Tsukada and T. Hoshino, J. Phys. Soc. Jpn. 51 (1982) 2562.

[22] E. Castanier, J. Goniakowski, A. Pojani, F. Finocchi and C. Noguera, accepted Il Vuoto (1996).

[23] K. Refson, R.A. Wogelius, D.G. Fraser, M.C. Payne, M.H. Lee and V. Milman, Phys. Rev. B52 (1995) 10823.

[24] F. Rohr, K. Wirth, J. Libuda, D. Cappus, M. Baümer and H.J. Freud, Surf. Sci. 315 (1994) L977.

[25] R.W. Nosker, P. Mark and J.D. Levine, Surf. Sci 19. (1970) 291.

[26] H. Onishi and Y. Iwasawa, Surf. Sci. 313 (1994) L783.

[27] D. Wolf, Phys. Rev. Lett. 68 (1992) 3315.

[28] V.E. Henrich and P.A. Cox, *The surface Science of Metal Oxides* (Cambridge University Press, 1994).

[29] J. Goniakowski, C. Noguera, Surf. Sci. 319 (1994) 68; *ibid.* 319 (1994) 81.

[30] J. Goniakowski, S. Bouette–Russo, C. Noguera, Surf. Sci. 284 (1993) 315.

[31] M.L. Cohen and D.L. Chadi, Phys. Rev. B8 (1973) 5747; S.L. Cunningham, Phys. Rev. B10 (1974) 4988.

[32] Y. Liang, D.A. Bonnell, Surf. Sci. 310 (1994) 128.

[33] R. Lacmann Colloq. Int. CNRS 152 (1965) 195

[34] J. Goniakowski, C. Noguera, Surf. Sci. 340 (1995) 191.

[35] E. Castanier, C. Noguera, Surf. Sci. 364 (1996) 1; ibid 364 (1996) 17.

[36] A. Hirata, K. Saiki, A. Koma and A. Ando, Surf. Sci. 319 (1994) 267.

[37] T. Matsumoto, H.Tanaka, K. Kouguchi, T. Kawai, S. Kawai, Surf. Sci. 312 (1992) 21.

COMPUTER SIMULATION OF STRUCTURAL, DEFECT AND SURFACE PROPERTIES OF SOLIDS

C.R.A. CATLOW, D.H. GAY, M.A. NYGREN AND D.C. SAYLE
The Royal Institution, 21 Albemarle Street
London W1X 4BS, United Kingdom

1. INTRODUCTION

A fundamental understanding of the chemistry of solids and their surfaces requires accurate models of structures and processes at the atomic level. Our knowledge of the physics and chemistry of solids is being advanced rapidly by the application of computer modelling techniques which are able, in general, to predict with growing accuracy and reliability the structural and dynamical properties of matter at the atomic level [1, 2, 3, 4]. This chapter will focus on the modelling of the structures of crystals, defects and surfaces, where we hope to show the detailed level of the information that can be generated by contemporary modelling methodologies.

In the next section, we outline the main methodologies used in contemporary computer modelling studies of bulk and surface structures of solids. Recent applications to the structures of a range of complex materials and oxide surfaces are then described. Later sections of the chapter will consider key problems relating to the properties of hydrolysed surfaces, to crystal growth inhibition, to the behaviour of oxide monolayers on oxide substrates and to the rôle of surface defects.

2. METHODOLOGIES: GENERAL ASPECTS

A powerful range of computational methodologies are now available to solid state science. There are two broad categories of technique: first, *electronic structure methods* in which we attempt to solve, at some level of approximation, the Schrödinger equation for the system under study; second, *simulation techniques* in which all electronic structure effects are subsumed into interatomic potential functions which describe the variations in the

R.M. Lambert and G. Pacchioni (eds.),
Chemisorption and Reactivity on Supported Clusters and Thin Films, 479-521.

energy of the system as a function of nuclear coordinates. Both methods have important rôles to play in modelling solids or their surfaces.

The electronic structure methodologies can themselves be divided into two general types of technique: the first, the *Hartree Fock methods* solve the Hartree Fock equations with explicit evaluation of the exchange energies; and in modern calculations, there is an increasing use of 'post' Hartree Fock techniques (configurational interaction or perturbation methods) to estimate correlation energies. A further subdivision of this class of technique is into *ab initio* and *semi-empirical approaches*: the former calculates all integrals directly, while the latter parameterises many of the important integrals and sets the less important ones to zero. Although semi-empirical methods will have a continuing rôle, the constant growth in computer power means that *ab initio* calculations can be performed on systems containing an increasingly large number of atoms.

An alternative approach of growing importance is the *local density functional* (LDF) method, the basis of which is the derivation by use of the variational principle of the *electron density* for the system, following a seminal study of Hohenberg and Kohn[5] which established that the energy is a unique functional of the density of a system. With this approach it is possible to calculate exactly the kinetic and 'Hartree' (electron ··· electron repulsion) energies. Exchange and correlation terms are evaluated in simpler approaches by using the *local density approximation*, *i. e.* these quantities are calculated as a function of the electron density using expressions appropriate for an electron gas of constant density. In practice, however, modern approaches add terms correcting for gradients in the density and for non-local effects. LDF have the advantage, over Hartree Fock methods, of being more economical in computer time, thus allowing larger systems to be investigated. It is difficult, however, to quantify precisely the errors introduced by the local density approximation, although the accuracy of the results achieved by many LDF calculations suggests that these are small

An important development in LDF theory was made by Car and Parrinello in 1985 [6] when they proposed a scheme whereby dynamics can be introduced in such calculations by employing a formalism based on Lagrangian multipliers. Their approach has been widely implemented in the context of calculations employing pseudopotentials (which represent core electrons in an approximate and economical manner) with plane-wave basis sets. LDF techniques are also being used increasingly with localised, atom centred bases.

Calculations employing both approaches may be performed either on clusters or periodic arrays of atoms. The latter is obviously more appropriate for modelling perfect crystal structures, while the former may be suitable for describing defects and localised states. However, considerable care must be exercised in the choice of cluster size and in the approach used to 'terminate' the cluster in particular to 'saturate' the dangling bonds of the peripheral atoms. If possible the cluster should be 'embedded' in an approximate representation of the surrounding lattice. The simplest procedure is to surround the cluster by point charges which at least describe the long range Coulomb field in the surrounding lattice. More sophisticated procedures represent the embedding matrix by using a 'simulation' approach (i.e. employing interatomic potentials) although there are difficulties associated with the linking of the quantum mechanical cluster and the embedding region in an appropriate way. Further discussions of the methods are given in the chapter of Pettersson in this book.

A vitally important aspect of all quantum mechanical investigations (indeed all modelling studies of solids) concerns the extent to which geometry optimisation is included. In cluster calculations on defects and active sites, it is essential that the site plus surrounding shells of atoms are relaxed to equilibrium. It may be desirable to fix the peripheral atoms of the cluster in order to impose the constraints arising from the surrounding crystal structure. It may still, however, be necessary to allow for polarisation of the distant region of the lattice. In periodic boundary conditions calculations, it is generally desirable to include a full geometry optimisation of the structure.

The account given here is, of course, the briefest of surveys of one of the most active fields of contemporary condensed matter science. Detailed discussions will be found in references [1, 2, 3, 7, 8] and the chapters of Pettersson, Norskov and van Santen in this book. In concluding our comments on these methods, we should note that for all the techniques outlined, good automated computer codes are available. 'Molecular', Hartree Fock codes, *e. g.* GAMESS [9] and CADPAC [10] may be used for HF calculations on clusters, while periodic boundary conditions HF studies can be performed with the CRYSTAL code [11]. Several LDF programs are available, *e. g.* CASTEP [12] (which uses a plane wave, pseudopotential approach), DSOLID [13] (employing localised basis sets for periodic systems) while cluster calculations can be carried out using e.g. DMOL [14] and DGAUSS [15] (both of which employ atomic centred basis sets). The availability of these good quality flexible codes greatly extends the range of applicability of these techniques.

The use of electronic structure methods is expanding rapidly with the growth in computer power. However, in many applications, simulations employing interatomic potentials, are the most appropriate and economical methods for modelling structural, elastic, dielectric and lattice dynamical properties of crystals. They are, moreover, able to handle highly complex systems containing a large number of atoms.

Simulation techniques comprise energy (and free energy) minimisation procedures, Monte Carlo and molecular dynamics methods. All have been fruitfully used in crystal structure modelling. In the first (and simplest) method, we start by calculating the lattice energy (U) which is defined as the energy of the crystal with respect to component ions at infinity (ions are a sensible reference point for ionic and semi-ionic solids):

$$U = \frac{1}{2}\sum_i V_i \, , \tag{1}$$

where the summation is taken over all ions in the unit cell and where V_i is the site potential of the ions which is evaluated by calculating its interaction energy with all other ions.

$$V_i = \sum_j^{\prime} U_{ij} + \sum_{jk}^{\prime\prime} U_{ijk} + \dots \, , \tag{2}$$

where the first term represents the summation over all ion pairs, the second all ion triplets; higher terms may also be included, but this is relatively uncommon. Indeed, the majority of calculations include only the first term the so-called pair potential approximation. This in turn may be usefully decomposed into Coulombic and non-Coulombic terms:

$$U_{ij}(\mathbf{r}_j, \mathbf{r}_\mathrm{j}) = \frac{q_i q_j}{r_{ij}} + \Phi_{ij}(r_{ij}) \, , \tag{3}$$

where q_i is the charge on the atom or ion. The Coulombic term which is purely two-body in nature is handled by standard procedures (in particular the Ewald summation technique) as discussed in references [1] and [2]. The non-Coulombic Φ_{ij} is then usually approximated by analytical functions which generally include both attractive and repulsive components; the latter describes the Pauli repulsion due to overlap of closed shell electron configurations, and the former attractive terms from dispersion (i.e. induced dipole-induced-dipole) and covalence effects. Several such functions are available, notably the Lennard-Jones potential:

$$V(r) = \frac{A}{r^{12}} - \frac{B}{r^6} \, , \tag{4}$$

the Morse potential:

$$V(r) = D\{1 - \exp[-\beta(|\mathbf{r} - \mathbf{r}_e|)]\}^2 \tag{5}$$

and the Buckingham potential:

$$V(r) = A\exp(-r/\rho) - \frac{C}{r^6} \tag{6}$$

Lennard-Jones potentials have been used widely in modelling rare-gas and molecular crystals. Morse potentials become more appropriate when covalent systems are being studied; D may then be interpreted as the covalent bond-dissociation energy and r_e the equilibrium bond length. Buckingham potentials have been very widely used in the study of ionic and semi-ionic solids.

Inclusion of many-body effects may be achieved by the use of angle dependent forces. A commonly used example is the simple 'bond-bending' terms of the type:

$$E(\theta) = \frac{1}{2}k_B(\theta - \theta_0)^2 , \tag{7}$$

where θ is the angle subtended by three atoms and θ_0 is an equilibrium value; k_B is the appropriate force constant. The use of such terms is most appropriate in systems with directional covalent bonding, e.g. silicates, where θ_0 will correspond to the tetrahedral angle subtended by O-Si-O bonds in the tetrahedral SiO_4 groups.

We note that atomic and ionic polarisation in solids is also essentially a many-body effect. Moreover in ionic materials, polarisation energies may be large, especially in the vicinity of charged defects and surfaces. Omission of polarisation while obviously leading to the failure to describe high frequency dielectric properties also results in marked inadequacies in calculated phonon dispersion curves. The simplest way of modelling polarisability is to use the point polarisable ion (P.P.I.) model which assigns a linear constant of proportionality i.e. the polarisability, α, between the dipole moment μ and the field E acting on the atoms, that is:

$$\mu = \alpha E . \tag{8}$$

However, as the dipole is a point entity, it cannot describe the physical basis of polarisability which is the displacement of the valence shell electron density in response to the applied field. Consequently, important physical effects, in particular the coupling of polarisability and short range repulsion, are omitted by these models which fail badly in describing both dielectric and lattice dynamical properties of crystals. These deficiencies are largely overcome by the shell model which describes the development of a dipole moment in terms of the displacement of a massless shell of charge Y (representing the valence shell electrons) from a core (representing the nucleus and core electrons), the core and the shell being connected by an harmonic restoring force for which the spring constant is k; the free atom polarisability, α, is then given by $\alpha = Y^2/k$. Shell model potentials have been very widely used in studies of lattice dynamical and defect properties of ionic solids. Further discussion is given in references [1] and [3].

Potentials of the type described above rest on the *Born model* description of ionic and semi-ionic solids. Such models, as is clear from the above discussion, are based on the idea of ions (which may be polarisable) interacting *via* Coulomb forces and short range pair potentials (which may be supplemented by three body or even by higher order terms). A contrasting approach is provided by molecular mechanics force fields, the conceptual basis of which is the covalently bonded network with energy terms associated with bond stretching and bending and with non-bonded interactions. For some systems, such as silica and silicates, both types of model may be used, and interestingly, similar results are obtained despite the very different approaches [16, 17, 18, 19].

Having established the type of potential model to be used, it is next necessary to fix the variable parameters, i.e. those used in the description of the short range potential, $V(r)$, the shell model constants Y and k and the effective atomic charges, q (although we note that in many modelling studies of ionic oxides, halides and even silicates, these have been fixed at integral, fully ionic values). This is undertaken by two procedures: first, *empirical methods*, in which variable parameters are adjusted, generally via a least squares fitting procedure to observed crystal properties. The latter must include the crystal structure (and the procedure of 'fitting' to the structure has normally been achieved by minimising the calculated forces acting on the atoms at their observed positions in the unit cell). Elastic constants should, where available, be included; and dielectric properties are required to parameterise the shell model constants. Phonon dispersion curves also provide valuable information on interatomic forces; and *force constant* models (in which the variable parameters are first and second derivatives of the potential) are commonly fitted to lattice dynamical data. This has been less common in the fitting of parameters in potential models which are our present concern as they are required for subsequent use in simulations. However, empirically derived potential models should always be tested against phonon dispersion curves when the latter are available. Again we refer to references [1] and [3] (and the references cited therein) for further discussion.

The second approach involves short range potentials being calculated directly by use of *ab initio* cluster or periodic boundary techniques, where the geometry is varied in a systematic manner with an effective potential being fitted to the resulting energy surface [17]. Such methods are being used increasingly in contemporary studies.

3. SIMULATION OF CRYSTAL STRUCTURES

The standard procedure used in modelling crystal structures, is first to calculate the lattice energy using our interatomic potentials. We then employ a minimisation algorithm which adjusts all structural variables (cell dimensions and atomic coordinates) until a structure of minimum energy has been generated. This straightforward procedure may be extended by including a calculation of the vibrational free energy using harmonic lattice dynamical theory[19, 20]. This method is now remarkably effective in simulating complex inorganic structures. It has, however, inherent weaknesses, first in its omission of explicit dynamic effects and second in the 'local minimum' problem, namely the fact that a minimisation procedure can only guarantee convergence to the nearest local minimum.

Fortunately an optimisation procedure which is much less susceptible to local convergence problems is available. This method, known as simulated annealing[21] introduces the possibility of accepting 'uphill' or energy increasing moves in the minimisation of a given function. To control the fraction of energy-increasing modifications which are made, an analogy with the process of annealing a physical system can be drawn. Here the relative probabilities of existence of two states are readily represented at a given temperature by Boltzmann statistics. The temperature of the systems governs the population of high energy states. On raising the temperature, states with widely differing energies will have increasingly similar probabilities while at low temperatures, low energy states will tend to dominate. In annealing a physical system, the elevated temperature permits a range of states to exist and the subsequent gradual reduction in temperature shifts the equilibrium until low energy states dominate. By equating the energy of the physical system to the object function to be minimised, with the temperature controlling the relative probabilities of two states, simulated annealing may be used as a general procedure for energy minimisation. Because the system is permitted to sample a wide range of states in the high temperature regime of the procedure, simulated annealing is less susceptible to becoming trapped in local minima than traditional gradient minimisation methods. It is this property which is particularly attractive in its application to the complex potential energy functions of the types found in crystal structure studies.

The searching algorithm used in the majority of simulated annealing studies is based on the Metropolis algorithm of the Monte Carlo scheme[22]. A particle is selected at random and its coordinates updated in a random fashion. The energy change associated with this perturbation in the structure is evaluated and the move either accepted or rejected on the basis of a comparison of the Boltzmann probability for the transition at the current temperature with a random number between 0 and 1. Application of this procedure over a sufficiently large number of moves leads to the generation of states which are representative of the equilibrium physical system at the specified temperature. A variety of schemes for governing the annealing of the system have been developed[23, 24, 25]. For example, the effective heat capacity can be monitored to detect the onset of melting and freezing, or alternatively, the fraction of successful Metropolis moves may be monitored to determine the rate of cooling applied to the system. In addition to the use of the Monte Carlo procedure in crystal structure simulation, methods based on 'genetic algorithms' may be employed in which, as discussed later, structures 'evolve' towards configurations with predefined structural characteristics. In addition, molecular dynamics procedures[1, 2, 4] can provide an effective means for exploring the physical structure of crystal systems. Such methods have proven extremely effective in the refinement of protein crystal structures[25] and have been used recently by Kaszkur *et al.* [26, 27] in assisting the refinement of data relating to sorbed molecules in zeolite pores, as discussed further below.

All the procedures described above are based on the idea of optimising the potential energy of the crystal calculated by an interatomic potential function. We emphasise that the quality of the models used in describing the interatomic force is, of course, vitally important in determining the reliability of the resulting prediction.

Optimisation procedures may use target functions other than the potential energy of the crystal. Indeed, traditional crystal structure refinement methods minimise the discrepancy between calculated and experimental diffraction intensity. This procedure may be blended with some of the approaches described above as in the powerful 'reverse Monte Carlo (MC)' technique which refines structural data using a MC algorithm in which the energy function is replaced by the deviation of calculated and experimental diffraction data[28]. Such methods have proved particularly effective in refining models of the structures of glasses and amorphous crystals[29]. Further developments of such methods can be expected in the future.

Application of minimisation procedures now cover a broad range of both organic and inorganic crystal structures. Modelling of silicates and aluminosilicates has been a particularly active field owing to the complexity and variety of the structures involved. Of special interest are the microporous zeolites and aluminophosphates, whose widespread use in catalysis is described in other chapters in this book. For these materials it has been shown that both the Born model[16, 17, 19] and molecular mechanics potentials [18] are able to model accurately their complex crystal structures.

Examples of the complexity and subtlety of the structural problems that can be handled include the recent work of Bell *et al.* [30] on the purely

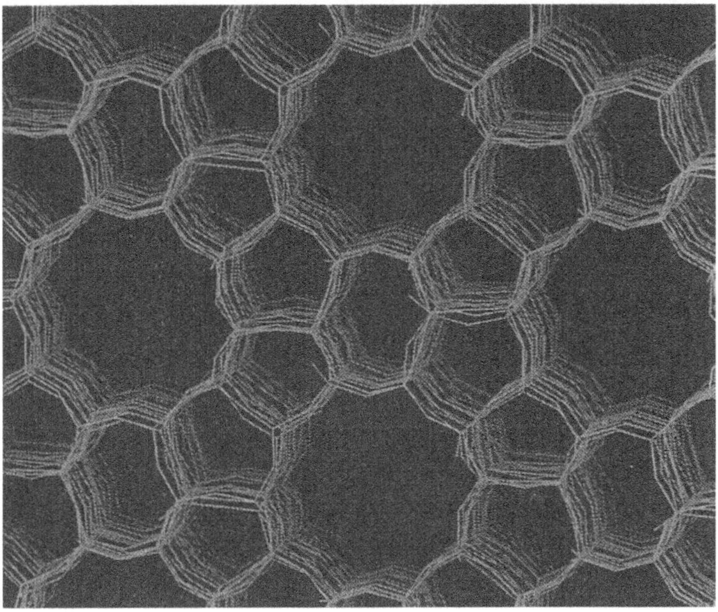

Figure 1. The energy minimised (dark) and experimental (light) crystal structures of silicalite.

siliceous zeolite silicalite which, below 330K, shows a small monoclinic distortion from a higher temperature orthorhombic structure. Energy minimisations accurately reproduced the distortion and gave an equilibrium structure that agrees very closely with experiment (see figure 1). A more complex problem in zeolite structural chemistry was investigated by Tomlinson et al.[31] who examined the difficult problem posed by zeolite β which had been shown by Newsam et al.[32] to comprise an intergrowth of different polymorphic structures. The two dominant polymorphs (A and B) were modelled by minimisation methods. The proposed structures were accurately reproduced and found to have identical lattice energies – a result which is obviously consistent with the formation of an intergrowth.

We should stress that the studies discussed above reproduce or confirm known structures. However, the accuracy that can now be achieved in modelling the most complex inorganic structures with straightforward interatomic potentials is remarkable, and lays the foundation for the more challenging problem of refining and predicting structures. A recent example where atomistic modelling played an important rôle in structure solution was that of the microporous metal-containing aluminophosphate MeAPO-36 (Wright *et al*.[33]). Structurally similar to zeolites, the (Me)AlPO family of molecular sieves are composed of three dimensional networks of corner sharing TO_4 tetrahedra, where T represents mainly aluminium and phos-

Figure 2. Crystal structure of MeAlPO-36 as revealed by high-resolution powder diffraction and computer simulation techniques.

phorus in the ratio 1:1 (although the aluminium can be substituted by minor amounts of metals (Mg, Zn, Co, Ni, Mn, *etc.*) which impart catalytic properties to the materials). A combination of adsorption and high resolution electron microscopy on the microcrystalline MeAPO-36 indicated that the solid possesses a one dimensional 12 ring channel system similar to that of $AlPO_4$-5 and consideration of the unit cell size and symmetry derived from electron diffraction suggested a model for the structure (figure 2). This model was then refined by the distance least square technique of Baerlocher *et al.*[34] using experimentally derived unit cell parameters and the maximum possible symmetry *Cmcm* which assumes disordered Al and P. The simulated XRD pattern closely resembles experimental patterns collected at high temperature (greater than 350°C) but did not explain peak splittings observed at lower temperatures and is indicative of lower symmetry (figure 3).

Lattice energy minimisation calculations were then carried out using the METAPOCS program [35] in which no symmetry constraints were applied, and all atomic positions and cell parameters were allowed to vary. The T atoms were replaced by Si in a first simulation, and by ordered Al and P in a second calculation. The 'SiO_2' simulation used silicate potentials and included the effect of polarisability. The '$AlPO_4$' simulations used the

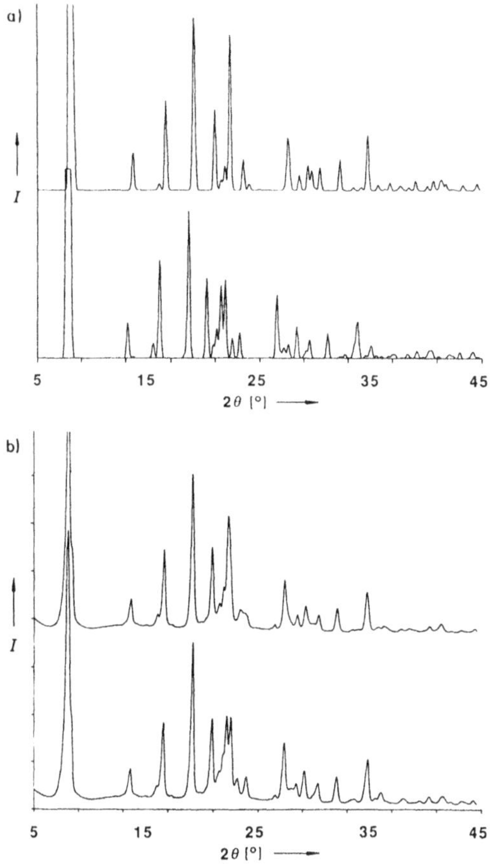

Figure 3. A. Simulated X-ray diffraction patterns for (a) ALPO and (b) SiO_2 energy minimised structures. B. Experimental X-ray diffraction patterns for MgALPSO-36 recorded at (a) 150 and (b) 375C. I is the intensity in arbitrary units.

partial charge potentials of van Beest *et al.*[17].

While the 'SiO_2' simulation retained Cmcm symmetry, the ordered '$AlPO_4$' simulation resulted in a reduction of space group symmetry to C_1 and a distortion of the unit cell. The simulated XRD pattern (see figure 3) closely resembles that observed at 150°C, although the simulated unit cell (a = 13.46 Å, b = 22.17 Å, c = 5.29 Å, α = 92.2° , β = 92.0°, γ = 90.0°) is about 2.5% larger than that observed experimentally (a = 13.16 Å, b = 21.65 Å, c = 5.19 Å, α = 90.2°, β = 92.0°, γ = 90.0°). Closer agreement still has recently been achieved using potentials assuming full formal charge models (Gale and Henson[36], Salvador-Ruiz *et al.*[37].

This computational approach then enables us not only to establish structural details in the absence of single crystal data, but also to predict them on the basis of a physically realistic model.

The most exciting and ambitious prospect in the field concerns the possibility of predicting crystal structures from a minimum of initial information. Useful 'demonstration' calculations have been carried out by Freeman and coworkers[38], who showed that it is possible to reproduce the crystal structures of the three phases of TiO_2 (rutile anatase and brookite) using a combined simulated annealing/energy minimisation methodology. The calculations included as input only the observed cell dimensions of each phase, and started their search for the low energy crystal structures from purely random distributions of the Ti and O ions of the appropriate density. A particularly interesting recent example of the use of these methods to a genuine problem in structure prediction is the recent work of Bush et. al.[39] who used evolutionary programming techniques employing 'genetic algorithms', referred to above, rather than simulated annealing to generate approximate structures which were subsequently refined using lattice energy minimisation techniques. We recall that genetic algorithm methods (which have been used in other optimisation problems in chemistry) involve 'breeding' successive configurations which then evolve towards a low specified 'cost factor'. In this case the cost function is designed to generate plausible structures with coordination numbers in accordance with Pauling's rules

In the present case the structure of a complex ternary oxide (Li_3RuO_4) was solved from knowledge only of its cell dimension; the resulting simulated structure is shown in figure 4. We note that it had previously proved impossible to obtain an adequate refinement of the powder diffraction data available for the crystal but that the computer generated model accurately refines the data. This success suggests that it may be possible to design increasingly general purpose automatic structure prediction packages for complex inorganic materials.

4. MODELLING DEFECTS

Accurate and reliable modelling of the structures and energies of defects in solids has been possible since the development in the early 1970s of automated computer codes implementing the Mott-Littleton [40] methodology. In essence, this procedure relaxes an atomistically modelled region surrounding the defects (containing usually several hundred ions) which is embedded in a quasi-dielectric continuum of the surrounding lattice. Further details of the methodologies and applications are given in references [41] and [42]. A large range of applications over the last 20 years has demonstrated the quantitative reliability of these methodologies, given the availability of reliable interatomic potential models.

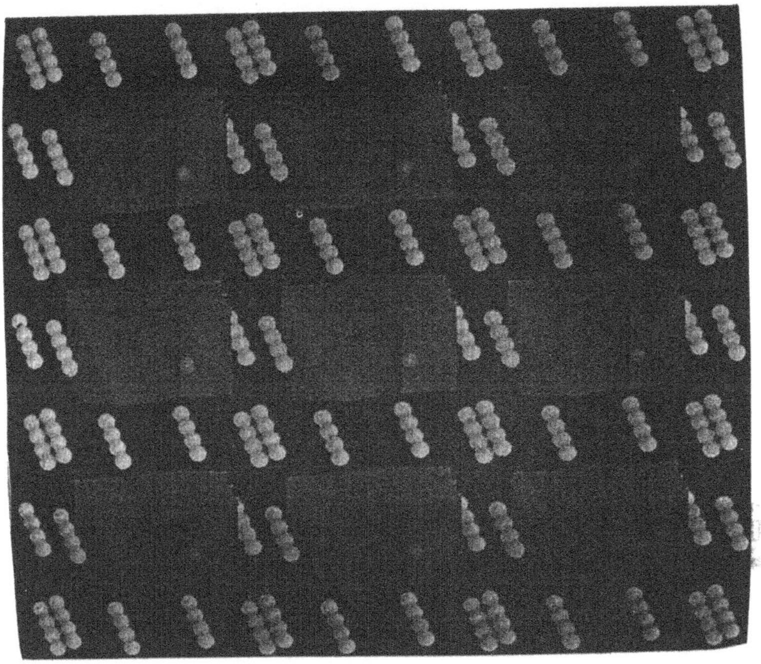

Figure 4. Structure of Li_3RuO_4 predicted by computer simulation techniques. Polyhedra represent RuO_6 octahedra; balls represent Li ions.

A field of considerable relevance to the theme of this book concerns catalytic materials where the local structures created by defects and impurities are often of crucial importance. There is an even greater incentive for developing procedures to model such states, as their investigation by experimental techniques is often difficult. As a first example of the accuracy which can be achieved in such studies, we will consider recent work on the topical Fe/silicalite material which has attracted attention owing to its use in a variety of shape selective isomerisation reactions.

Recent calculations of Lewis *et al.*[43] have provided detailed models for the localised states created by dissolving ferric iron into this important material. The simulations clearly establish that Fe^{3+} is accommodated within the framework at tetrahedral sites as shown in figure 5, which also shows that the neighbouring oxygen is protonated, thereby charge compensating the lower charge of Fe^{3+} compared with Si^{4+}. The calculated bond lengths

TABLE 1. Mean geometries of a Fe^{3+} substitutional with adjacent OH at all 96 unique sites.

	Calculated/Å	Experimental[b]/Å
Mean Fe-O(H)	2.02	
Mean Fe-O(over 4 Fe-O)	1.82	1.83(1.86,3.38)
Mean Si-O(H)	1.68	
Mean Si-Fe	3.12,3.40	3.2,3.38
Mean nn Si-Si	3.07	
Mean Si-Si	3.10	3.10
Mean Si-H	2.35	
Mean Fe-H	2.39	

[a] The Fe-Si distances are presented as the mean of the three shorter distances and the single longer distance. All sites exhibited this distribution of bond lengths. The experimental coordination number for the Fe-Si distances is two at both distances.
[b] Details of experimental data given in reference [43].

(which differ considerably from the perfect lattice T−O bond lengths) are given in Table 1, which also shows that they are in excellent agreement with those determined from EXAFS spectroscopic studies of the Fe K edge. (EXAFS, it will be recalled, provides a sensitive probe of the local atomic structure around the adsorbing atom.) The quality of the reproduction by the calculated structure of the EXAFS data is also clear in figure 6.

Subtle longer range structural changes accompany the substitution of Si by Fe. In particular, pore dimensions are modified significantly as shown in figure 7. Such changes would be expected to influence the shape selective catalytic activity of the material.

Extensive studies have also been reported on the Brønsted acid site in Al substituted zeolites, as it is well known that this is the active site for acid catalysed reactions in these materials. Detailed information is available from both quantum mechanical calculations [44, 45] and 'simulations' [46] on the structure, vibrational properties and proton affinities of these centres. For example, Table 2 reports results of calculations of Schroder *et al.* [46] on the Brønsted acid centre in zeolite Y where different configurations are possible as shown in figure 8. The simulation studies were able correctly to identify the low energy configurations, for which calculated vibrational frequencies were obtained that were in excellent agreement with experiment. Such calculations were also successful in modelling the differences in

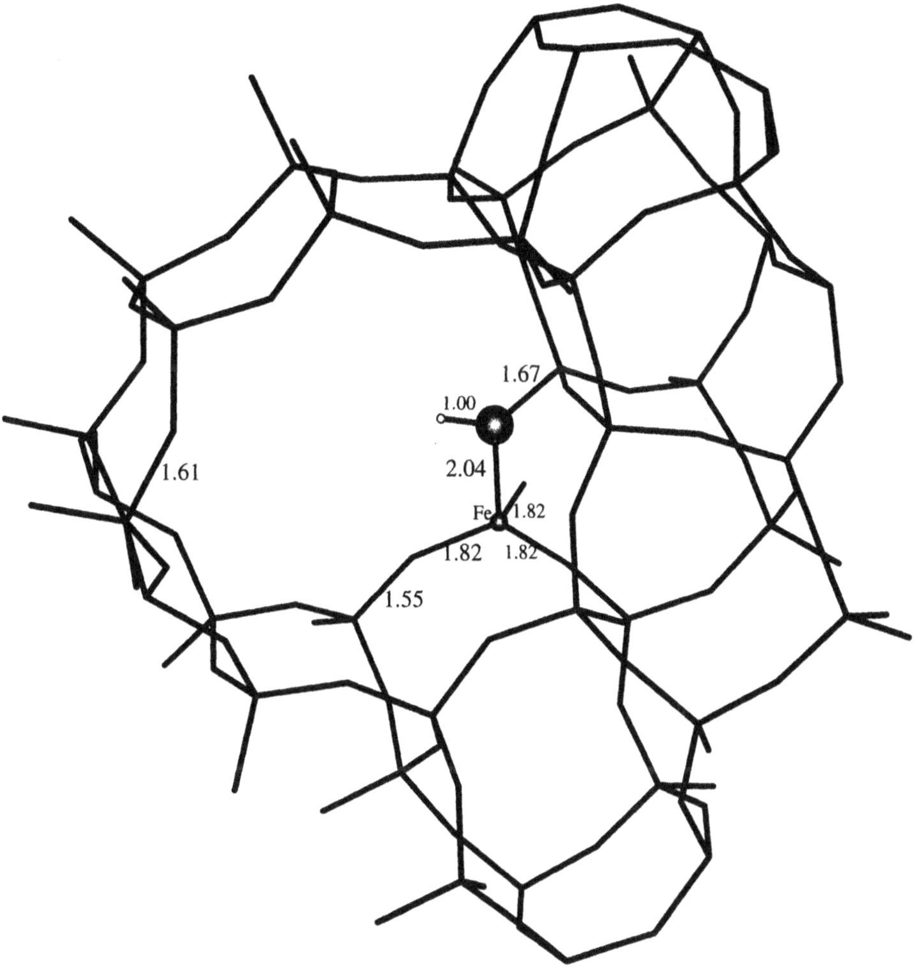

Figure 5. Local structure of Fe^{3+} in the framework of ZSM-5.

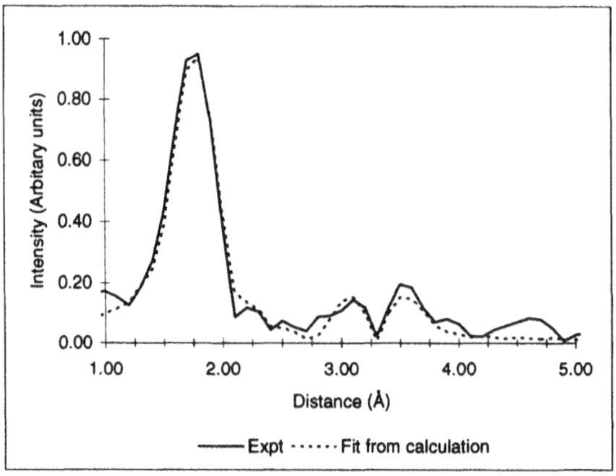

Figure 6. Fourier transform of the K-edge EXAFS for calcined Fe-ZSM-5. The best fit between the model derived from calculated distances and the experimental data for the calcined system is shown.

TABLE 2. Relative energies and harmonic OH stretching and SiOH bending frequencies of the four bridging hydroxyls in silicon-rich H-faujasite.

	E^{rel}(Kj mol^{-1})		ω_{OH}(cm^{-1})		δ_{SiOH}(cm^{-1})	
	Ref. [46]	CNDO/2[a]	Ref. [46]	obs.[b]	Ref. [46]	obs.[b]
O1H	5.3	7	3772	3770 or 3783	937	1015
O2H	19.8	39	3702	-	963	-
O3H	0.0	0	3736	3690	951	1055
O4H	23.7	58	3751	-	944	-

[a] ΔH^{rel} estimated on the basis of CNDO/2 calculations on molecular models.
[b] For references to experimental work see [46]

acidity between Al and Fe Brønsted acid centres. The study of Lewis *et al.* [43], referred to above, showed that the proton is more strongly bound to sites adjacent to Fe than to those adjacent to Al – a result which explains the observed reduced acidity of Fe/ZSM-5 compared with Al/ZSM-5.

Another important and topical field in which defect processes play a central rôle concerns ionic transport especially in materials with high conductivities, the so-called fast ion conductors. Simulation techniques using

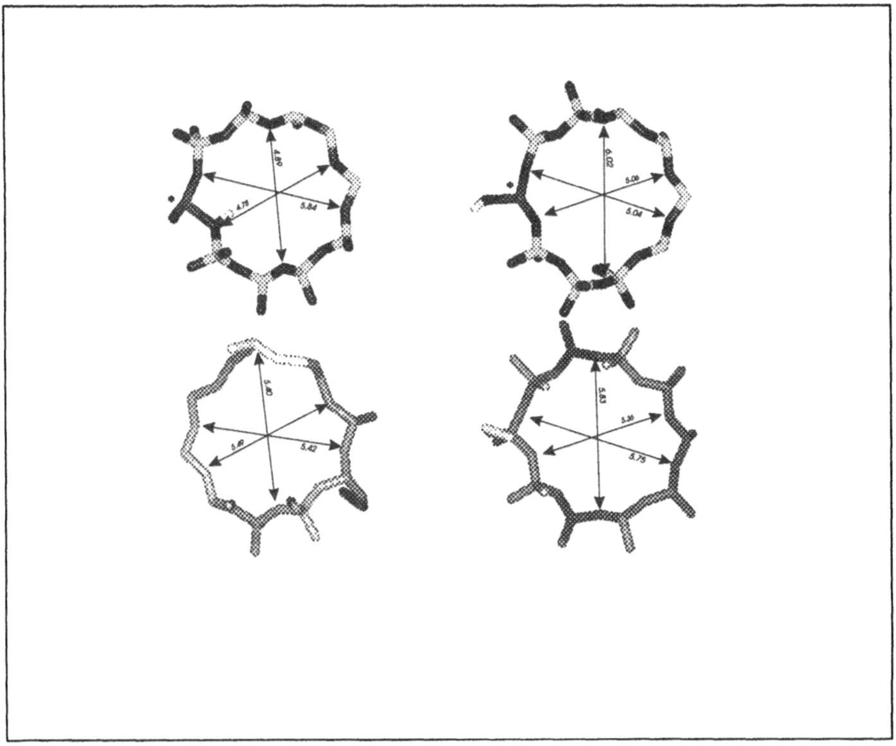

Figure 7. Effect of iron substitution on the pore dimension in Fe-ZSM-5. Upper figures show the sinusoidal and straight channels with the Fe-OH defect at the minimum energy site. The iron site is indicated by an asterisk(*). Lower figures show these pores in ZSM-5. Distances shown are measured using van der Waals radii of 1.35 Å for the oxygen.

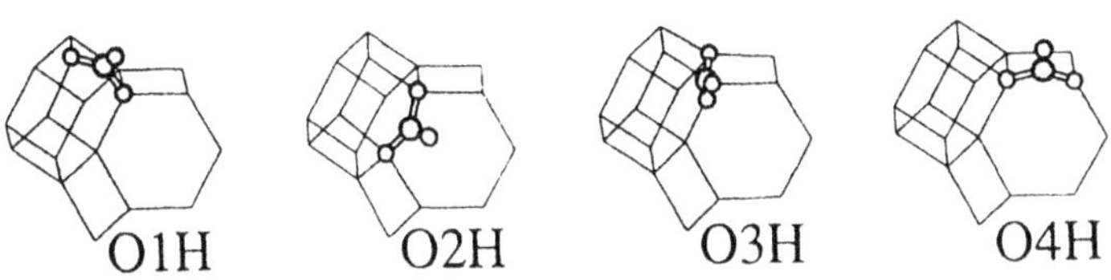

Figure 8. Different configurations for protons in zeolite Y.

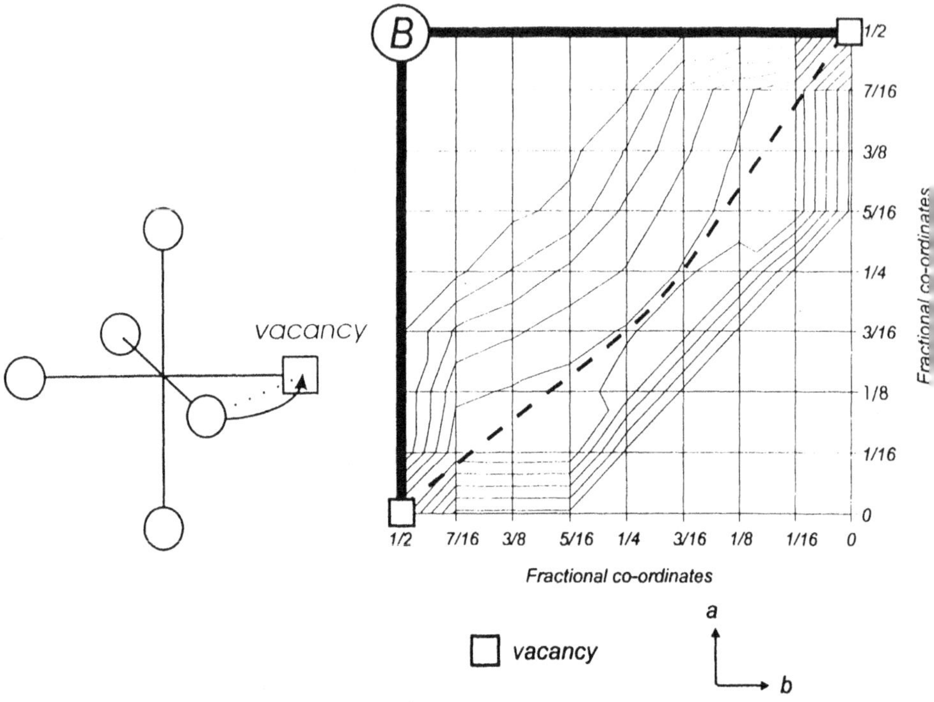

Figure 9. a) Schematic representation of the curved path for oxygen vacancy migration along the BO_6 octahedron edge;
b) Contour plot of the potential energy surface for oxygen vacancy migration, showing the curved path between adjacent anion sites of a BO_6 octahedron (in the ab plane).

both Mott-Littleton and molecular dynamics techniques have had a major impact on this field; reviews are available in references [42] and [41]. Of particular current interest are the perovskite structured oxides which show high oxygen ion conductivity when doped with low valence ions. The latter ions occupy host lattice sites with the resulting charge inbalance being compensated by oxygen vacancy formation. The oxygen vacancies are mobile (with both calculations and experiment giving activation energies of $\sim 0.5 - 1$ eV[47]). The recent work of Cherry *et al.*[47] also provided detailed information on the migration mechanism which, as shown in figure 9, involves migration between two oxygen ion sites *via* a slightly curved route.

Perhaps, however, the single most important aspect of this simulation study was the demonstration of the dependence with the vacancy activation energy on the radii of the A and B metal ions in the ABO_3 perovskite structured oxides. Figure 10 shows the calculated activation energy as a

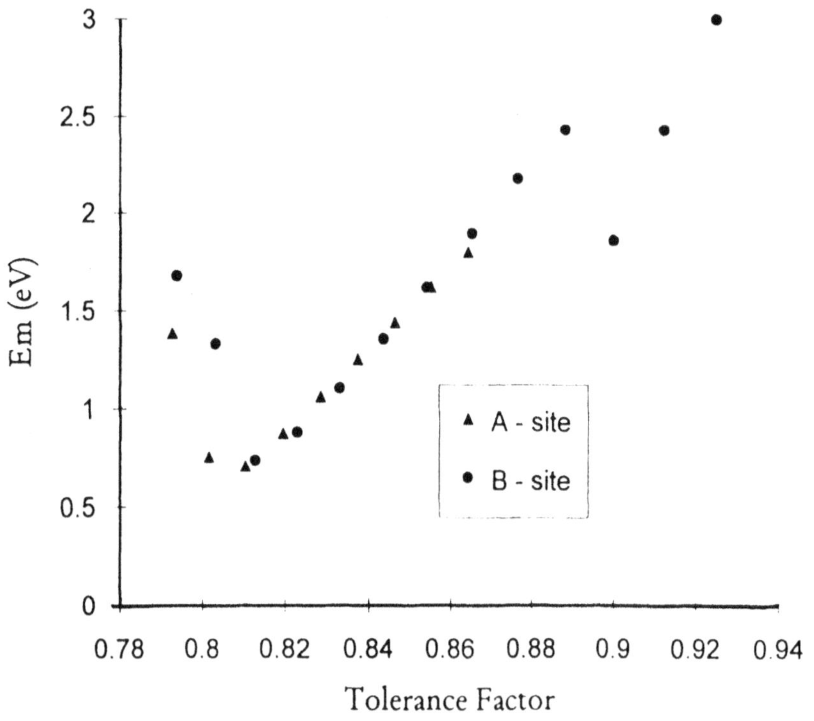

Figure 10. Calculated migration energy as a function of the tolerance factor (from both the A and B site simulations).

function of the tolerance factor 't' defined as

$$t = \frac{r_A + r_O}{\sqrt{2}\,(r_B + r_O)}\,, \tag{9}$$

where r_A and r_B are the radii of the A and B metal cations (which in the results shown all refer to trivalent ions) and r_O is the oxygen ion radius. This factor which is commonly used in discussing the stability of perovskite structured compounds must be unity for a stable perovskite structured compound given a hard sphere description of the component ions; in practice substantial deviations from unity are possible. The results, which clearly indicate a minimum in the vacancy migration energy as a function of t, accord with available experimental data and provide valuable guidance for future experimental work in this field.

The studies of defects in perovskite structured oxides have been extended to include the crucial issue of proton migration[48], in view of the important rôle of these materials as proton as well as oxygen ion conductors.

In particular, Cherry *et al.*[48] showed how the barrier to proton migration in these materials is low and is almost entirely attributable to the lattice energy change required to create an equivalent environment on both oxygen sites between which the proton may then tunnel. The resulting calculated activation energies of ~ 0.7 eV again accord well with experimental data. These and other recent studies on hydrogen containing defects (see e.g. the review in reference [49]) clearly show the continuing value of simulation studies using both interatomic potentials and quantum mechanical techniques in the field of the physics and chemistry of defects.

5. SURFACES.

5.1. METHODOLOGICAL ASPECTS

The methods used for modelling surfaces are similar to those used for modelling bulk systems. The main difference is how the simulation cell is chosen and what it represents. For bulk systems the assymetric unit or the P1 lattice cell is used. There is no information gained by including more cells than the fundamental unit. Likewise, for surfaces and interfaces, all the information is contained in a cell that is repeated in 2 dimensions, instead of three. The question of cell termination never enters into the bulk system, where the lattice is treated as infinite. For the 2 dimensional (2D) system, the lattice is infinite in two directions, but finite in the third. The question of how to treat the third direction makes surface simulations more demanding than bulk systems. The following description of surface modelling describes the techniques used by the program MARVIN[50], but could be used for any surface simulation technique.

The fundamental unit of a surface is a 2 dimensional (2D) cell, which has planar periodic boundary conditions parallel to the surface. The basic cell is shown in figure 11. The energy is computed with a 2D lattice summation using only the atoms contained within the cell. The cell consists of one or more blocks divided into 2 regions. The first region, which is near the surface or interface, is designated as region 1 (see figure 12). Atoms in region 1 are allowed to relax during energy minimisation. The other region is designated region 2 and these atoms are held in fixed locations. The purpose of region 2 is to reproduce the effect of the bulk crystal on the region 1 atoms.

For just the two–body interactions, $\phi_{ab}^{(2)}(\mathbf{r})$, the total energy of the system can be expressed as:

$$E_{\mathrm{Total}} = \sum_{a}^{N_1}\left[\frac{1}{2}\sum_{\mathbf{l}\neq 0}\phi_{aa}^{(2)}(\mathbf{l}) + \frac{1}{2}\sum_{b\neq a}^{N_1}\sum_{\mathbf{l}}\phi_{ab}^{(2)}(\mathbf{r}_{ab}+\mathbf{l}) + \sum_{b}^{N_2}\sum_{\mathbf{l}}\phi_{ab}^{(2)}(\mathbf{r}_{ab}+\mathbf{l})\right], \tag{10}$$

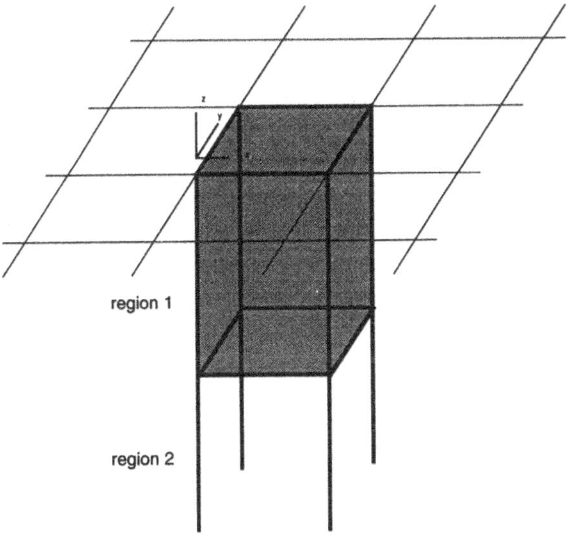

Figure 11. Schematic of the MARVIN simulation cell. The simulation cell is repeated infinitely in the x and y directions.

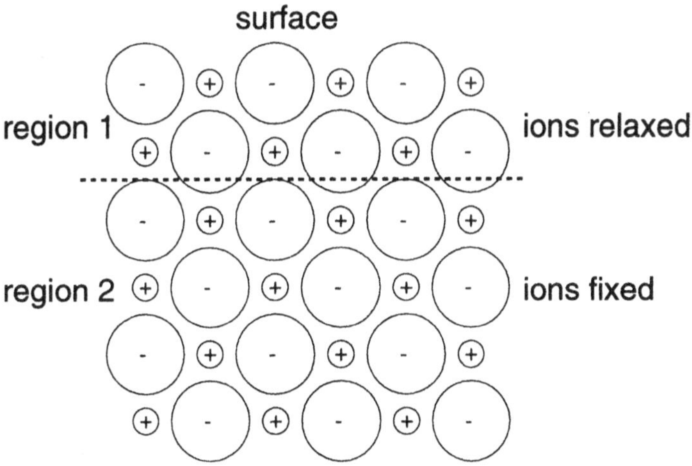

Figure 12. Schematic of the MARVIN surface simulation cell, emphasising the relationship between the relaxed and unrelaxed regions.

where N_1 is the total number of region 1 atoms, N_2 is the number of region 2 atoms, $\mathbf{l}$ are the 2D lattice vectors and $\mathbf{r}_{ab}$ is the vector from atom a and atom b. The first term inside the braces describes the interaction between an atom and its periodic images; the second term describes the interactions between all region 1 atoms and their images; while the last term describes the interaction between the atoms in region 1 and region 2.

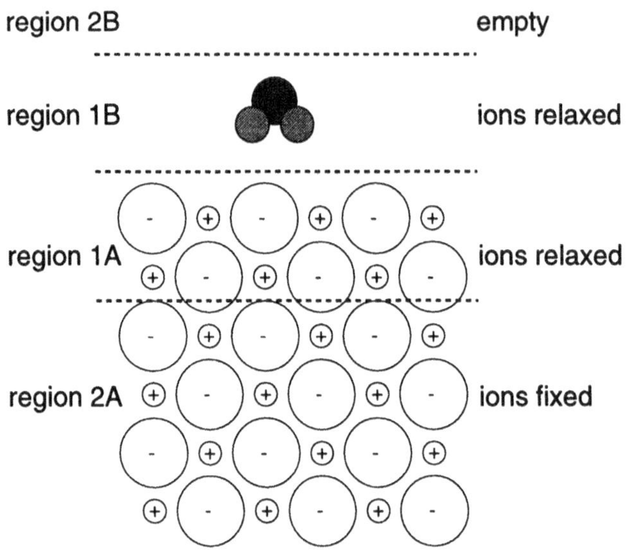

Figure 13. Schematic of the MARVIN surface simulation cell with a docked molecule

The main types of calculation that can be handled in this manner relate to free surfaces, adsorbed molecules or docking, and interfaces. For modelling a free surface one block is used. This block consists of the two regions, the sizes of which are determined by the energetics detailed in Equation 10. Adsorption or docking in this formalism can be described with 2 blocks, one containing the surface and the other the adsorbate (see figure 13), while an interface is two blocks with the two region 1s adjacent. The use of blocks and fixed regions is important in reducing the problem to a size that can be computed in a reasonable time.

Like the bulk solid, it is important to treat all the unique atoms individually and counter–productive to include symmetry replicated atoms. For surfaces, it is crucial that sufficient layers of atoms be included in region 1, so that the lowest layer of atoms does not move during minimisation. The depth of region 2 is important as well, since the lowest layer of atoms in region 1 should not see the vacuum beyond the bottom of region 2. In practice this can be done by increasing the size of region 2 until the total energy does not change (equation 10). Then the depth of region 1 can be increased until the surface energy does not change with increasing region 1 depth. The *relaxed* surface energy can not be calculated a priori, hence further calculations are necessary to determine if the depth is sufficient.

The region sizes depend upon the convergence properties of the potentials. Of the long–range interactions, it is the electrostatic that is the hardest to deal with. In 3D systems, the Ewald method is used to obtain a converged value for the electrostatic energy[51]. Parry produced a derivation appropriate for 2D slabs.[52][53] A more complete derivation is given by Heyes et.al.[54].

5.2. PREDICTING MORPHOLOGIES

Morphological prediction, based on the internal structure of a compound, has enjoyed many decades of active study. Initial approaches were based solely on the interplanar spacings, d_{hkl}, of the different crystal faces, with the lowest growth rates occuring at the faces with the greatest d_{hkl}[55][56]. Donnay and Harker improved this simple idea when they took account of the symmetry of a crystal by reducing the d_{hkl} growth slices at the faces containing a screw axis or glide plane, since the slice at these faces can be reduced into identical subslices[57]. The presence of these symmetry elements can be easily detected from the extinction conditions in the X–ray reflections, and is determined solely by the space group of the crystal. These arguments took no account of the detailed structure or energetics of the material.

Hartman and Perdok[58] revolutionised this field by considering the rôle of intermolecular forces in crystallisation. They classified crystal faces into three groups depending on the number of periodic bond chains (PBC) in the growth slice, where a PBC is an uninterrupted chain of nearest–neighbour particles. A slice with two or more PBCs is an F face, those with one PBC are S faces and the rest are K faces. They proposed that only F faces can grow according to a layer mechanism, with the other two having a continuous mode of growth. Thus the growth of F faces will be slow and therefore they will be the most important morphologically. They proposed that the attachment energy (the energy per molecule released when a new slice of depth d_{hkl} is attached to the crystal face, E_{attach}) be proportional to the growth rate and hence inversely proportional to its morphological importance[59]. Let $E_i(hkl)$ be the interaction energy per molecule between a slice of thickness d_{hkl} and the ith underlying slice. Then

$$E_{\text{attach}}(hkl) = \sum_{i=1}^{\infty} E_i(hkl). \tag{11}$$

Again the depth of the slice is reduced at the faces containing a screw axis or glide plane. The relationship between the attachment energy and the classification of faces by PBC analysis can be perceived in the relationship:

$$E_{\text{crystal}} = E_{\text{slice}}(hkl) + E_{\text{attach}}(hkl), \tag{12}$$

where $E_{\text{slice}}(hkl)$ is the energy between all the ions within the slice hkl (slice energy) and E_{crystal} is the energy of the crystal. Since E_{crystal} is constant for all faces, a large slice energy implies a face that will be morphologically important. A large slice energy in absolute value implies large interactions between ions in the slice which in turn means that there are at least two PBCs in the slice. This quantity has been used to predict the morphologies of a wide variety of materials[60][61][62][63]. A similar quantity has been defined by Dowty[64]. The morphology derived by this methodology is known as the growth morphology, since it is based on the idea of layers attaching themselves to a growing crystal.

Another approach to crystal morphology is *via* the surface energy i.e., the difference in energy of the surface ions compared with those in the bulk per unit surface area:

$$\gamma(hkl) = \frac{[E_{\text{Total}}(hkl) - E_{\text{boundary}}(hkl) - nE_{\text{crystal}}]}{Area(hkl)}, \tag{13}$$

where $E_{\text{Total}}(hkl)$ is the total energy of region 1, $E_{\text{boundary}}(hkl)$ is half of the interaction energy between region 1 and region 2, n is the number of unit cells in region 1 and $Area(hkl)$ is the surface area of the simulation cell. For a crystal in equilibrium with its surrounding, the surface energy must be a minimum for a given volume[65]. Hence the morphological importance of a face is inversely proportional to the surface energy. The morphology derived in this way is termed the equilibrium morphology. There has been some debate over which morphology is more likely to be observed[64][66]. The equilibrium morphology has been found to agree well with small experimentally grown crystals less than a micron in diameter[67][68]. For larger crystals and especially geological specimens, the growth morphology has been found to reproduce best the experimental results. The morphology derived by any of the methods discussed can be visualised using a Wulff plot[69] with the growth rate proportional to either d_{hkl}, the attachment energy or the surface energy.

These two methods of calculating the crystal morphology have traditionally ignored the effects of surface relaxation, *i.e.* the tendency of surface ions to move away from their ideal lattice sites. Although this seems a reasonable approximation for non-polar organic molecules, for inorganic systems it is know to be unrealistic.

An example of these different types of morphological prediction for zircon ($ZrSiO_4$) are illustrated in figure 14. The effect of relaxation is clearly evident by comparing (c) with (d) and (e) with (f) or (g). The observed morphology is more like (d) and (g) than the other forms shown here.

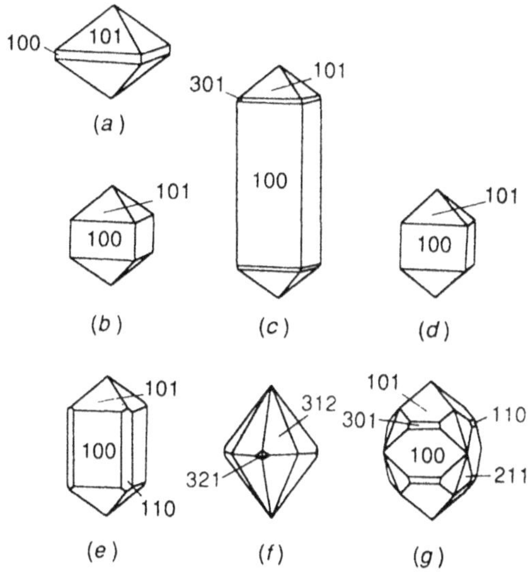

Figure 14. Zircon morphologies: (a) Donnay–Harker; (b) unrelaxed growth morphology; (c) unrelaxed growth morphology assuming (202) growth slices; (d) relaxed growth morphology; (e) unrelaxed equilibrium morphology; (f) relaxed equilibrium morphology; (g) relaxed equilibrium morphology with faces with invalid attachment energies removed.

5.3. DOCKING MOLECULES AND GROWTH INHIBITION

Molecular recognition at the interface between inorganic materials and organic templates is known to be a powerful force in the control of crystallization and mineralization in both synthetic and biological systems. Appropriately functionalized amphiphilic monolayers are able to nucleate a range of inorganic materials[70] and in biological systems, polysaccharidic and proteinaceous templates are responsible for skeletal development in a range of organisms[71]. Perhaps the simplest and best known example of this recognition process occurs during crystal growth in the presence of morphology modifying additives. In the past, morphological and structural data have been used to infer the essential elements of the interactions involved[72][73]. The control and inhibition of crystal growth is a major problem in several contemporary technologies. In many inorganic systems, the most efficacious growth modifiers are found to be polyanions, in which several charged groups are joined by a backbone. A particularly important case is provided by $BaSO_4$, whose precipitation in oil pipes is a signigicant problem in oil extraction and transport[74]. The most commonly used inhibitors for the growth of crystals of this compound are polyphosphonates of which the most widely studied are diphosphonates[75][76]. Growth inhibition of barite by these additives is thought to involve the incorporation of the inhibitor into a growing terrace, where it occupies the sites of two sulfate groups. Experimental evidence shows that diphosphonates are actually incorporated into the surface rather than just adsorbed onto the surface. In particular, it is found that similar dicarboxylates and disulfonates have no effect on crystallization[76]. The most obvious explanation of the latter is the mismatch between the single negative charge of the terminal groups and the double negative charge on the sulfate group. Moreover, scanning force microscope experiments[77] have shown that if commercially available polyphosphonate growth inhibitors are added to growing calcite crystals, they appear on the images at step edges. Once the diphosphonate is in place, it blocks the crystal growth when a step edge reaches it. Indeed, Monte Carlo simulations[78] have shown that the presence of a single ion impurity is enough to block step growth. But, a large impurity that binds simultaniously at two surface sites will be much more efficient at blocking step growth than a single ion impurity.

The docking of the diphosphonate involves the replacement of surface sulfates. Thus to determine which crystal faces will be affected by the additive, we must calculate the energetics of the process in which two surface sulfate ions are replaced by the additive. First, however, the surface structure of each growing face must be investigated. In general, faces corresponding to a given Miller index may have several different surface structures in which the crystal is cleaved in different places[79]. In barite, only the (010) face has a unique surface structure. For all the other faces considered, we calculated for each possible structure both the attachment energy and the surface energy. If the crystal growth is controlled by thermodynamic factors then the cut with the lowest surface energy will be dominant in the growing surface. If, on the other hand, the growth is kinetically controlled, the cut with the lowest absolute attachment energy will be most favoured. For the faces of barite that we have considered[80], the surface cut with the lowest surface and absolute attachment energies were the same, leading to an unambiguous surface configuration for each face, except for the (100) which has two possible cuts which have very similar unrelaxed attachment energies of -303 kJ/mol. The top layer of one of these is terminated with a sulfate layer, the other is terminated with a layer of barium atoms. Calculations have been performed on both configurations of this face. However, allowing the surface to relax produces a significant difference in the attachment energies (-274 kJ/mol vs -248 kJ/mol). Given the high quality of the potentials for $BaSO_4$, this difference is significant, suggesting that perhaps the barium terminated cut will be the favoured surface configuration.

For each surface, there are many combinations of pairs of sulfates which can be replaced by a diphosphonate. Since the P—P distance in the solution conformation of diphosphonate determined by MOPAC calculations is 6.17 Å, it seems reasonable to exclude those pairs of Sulfates with S—S distance greater than 7.5 Å, which results in the possible unrelaxed S—S distances 4.02, 4.52, 5.25, 5.47, 5.66 and 7.06 Å, although the list becomes more extensive after relaxation. In the first set of calculations, we used a simulation cell which was a 2 × 2 supercell of the surface of barite which was large enough to contain all S—S pairs less than 7.5 Å apart. For each pair of sulfates removed, a diphosphonate was inserted into the vacated area and both the surface ions and diphosphonate were allowed to relax to minimise the energy.

The most physically significant term is the replacement energy ΔE_{replac}, which is the energy change on replacement of two sulfates in a surface with a diphosphonate ion. This can be calculated as follows:

$$\begin{aligned}\Delta E_{replac} &= \left(E_{surface+diphosphonate} + 2E_{sulfate} + 2E^{Solvated}_{sulfate}\right) - \\ &\quad \left(E_{surface} + E_{diphosphonate} + E^{solvated}_{diphosphonate}\right), \qquad (14)\end{aligned}$$

where $E_{surface}$ is the total energy of the system containing just the relaxed surface; $E_{suface+diphosphonate}$ is the total energy of the system without two sulfates but with the docked diphosphonate; $E_{diphosphonate}$ is the internal energy of the isolated minimised diphosphonate ion (640 kJ/mol); $E_{sulfate}$ is the energy of the isolated minimised sulfate ion (-1,930 kJ/mol), and $E_{suface+diphosphonate}$ and $E^{Solvated}_{sulfate}$ are the the solvation energies of the diphosphonate (-2,734 kJ/mol) and sulfate (-1,019 kJ/mol), respectively. The solvation energies are calculated by inserting the target ion into a sphere of water with a radius of 15 Å and performing 350 steps of molecular dynamics at 270 K, followed by minimisation, resulting in an energy of the ion surrounded by its hydration sphere ($E_{water+ion}$). The target ion is then removed and the same steps are performed on the sphere of water only, yielding the energy of the hydration sphere (E_{water}). The solvation energies are then calculated by the expression:

$$\Delta E^{solvated}_{ion} = E_{water+ion} - (E_{water} + E_{ion}) \qquad (15)$$

where E_{ion} is the internal energy of the isolated ion. The procedure was repeated five times and the average value of the $\Delta E^{solvated}_{ion}$ calculated. Despite

a

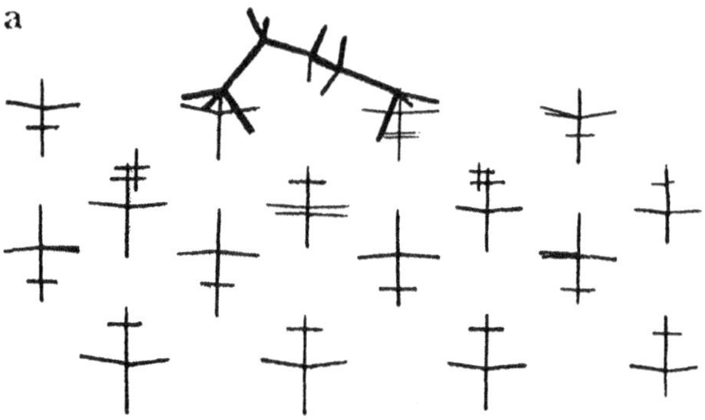

b

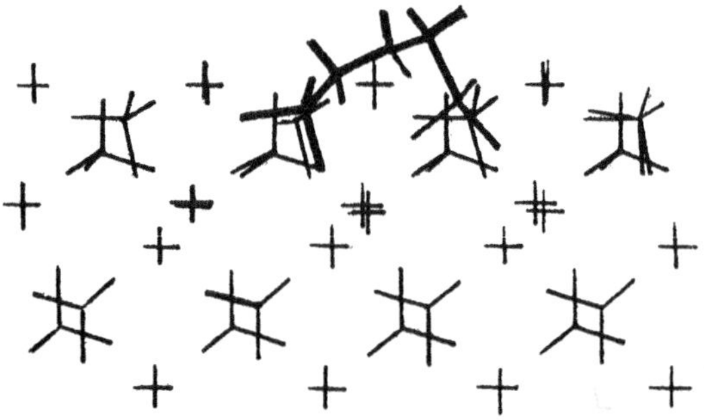

Figure 15. The two most stable docking sites for an alkyldiphosphonate on the surfaces of bartite (a) on the 5.47 Å site on the (100) surface and (b) on the 5.66 Å site on the (011) surface. The diphosphonate is coloured black and the surface barite ions gray.

the simplicity of this approach, it may be expected to yield a reasonable estimate of the hydration energy.

The two most stable configurations for the incorporation of an alkyldiphosphonate are shown in figure 15. These demonstrate clearly how the alkyldiphosphonate is included into the surface structure of the barite crystal. The ΔE_{replac} is -303 $KJ/mole$ for the (100) surface and -298 KJ/mol for the (011). The inclusion of the alkyldiphosphonate at these sites are not entire expected based upon simply comparing S—S (5.47Å and 5.66Å) and P—P (6.17Å) distances. There are sites that should provide a better fit based on distance measurements. However, these show how the alkyldiphosphonate controls the growth of the barite crystal.

5.4. SURFACES UNDERGOING CHEMICAL REACTIONS

In many cases the ions and molecules in the surface undergo chemical reactions. Not only are they exposed to the environment, but a clean surface will either expose dangling bonds or ions with incomplete coordination that may be stabilised chemically. It might also be that the chemically most stable cut of the crystal is for a surface which would otherwise be unstable due to a permanent dipole moment but for which the chemical reaction cancels the dipole.

In this section we will discuss two cases where reaction with water stabilises the surface. The first is α-quartz in which protonation of the oxygens saturates the dangling bonds and adsorption of hydroxide ions fills the coordination shell for silica ions. The second is α-Al_2O_3 where the oxygen terminated basal plane surface is a polar cut and halving the surface layer charge by hydroxylation cancels the dipole.

When cleaving quartz, the silicon exposed will always lack one of its four oxygens in the coordination sphere. For each silicon ion exposed there is one exposed oxygen. This means that by adsorbing protons and hydroxides in pairs, the hydroxide terminating the silicon and the proton the oxygen, the surface is saturated. This type of water surface reaction to saturate dangling bonds is schematically described in figure 16.

The basal plane surface of α-Al_2O_3 possesses one aluminium terminated electrostatically stable cut. For this cut, the surface aluminium ions lack three out of the six oxygens in its coordination sphere; it is thus highly coordinatively unsaturated. When relaxing the clean aluminium terminated surface the outermost planes moves inwards in a dramatic fashion, using both atomistic potentials [50, 81, 82] and periodic Hartree-Fock [83]. Large relaxations and high coordinative unsaturation are usually signs of instability and it is likely that a more stable situation may be found.

The only possible oxygen terminated cut results in a net dipole across the crystal. Moving half of the oxygens from the top to the bottom cancels the dipole giving an electrostatically stable situation. Reaction of the resulting surface with water results in a fully hydroxylated surface without any dangling bonds and coordinative unsaturation. This type of water surface reaction, depicted in figure 17, differs from the type occuring on quartz in that the reaction not only saturates the surface but also cancels the dipole removing the electrostatic instability.

In section (5.1) the surface energy was defined from the energy difference between the surface simulation cell and a part of the bulk of corresponding size. This is the work needed to cleave the crystal; but when the surface undergoes a chemical reaction the reaction enthalpy must be included into the surface energy. The reaction giving rise to the resulting surface is for both cases:

$$O^{2-}_{surf} + H_2O \rightarrow 2OH^- . \tag{16}$$

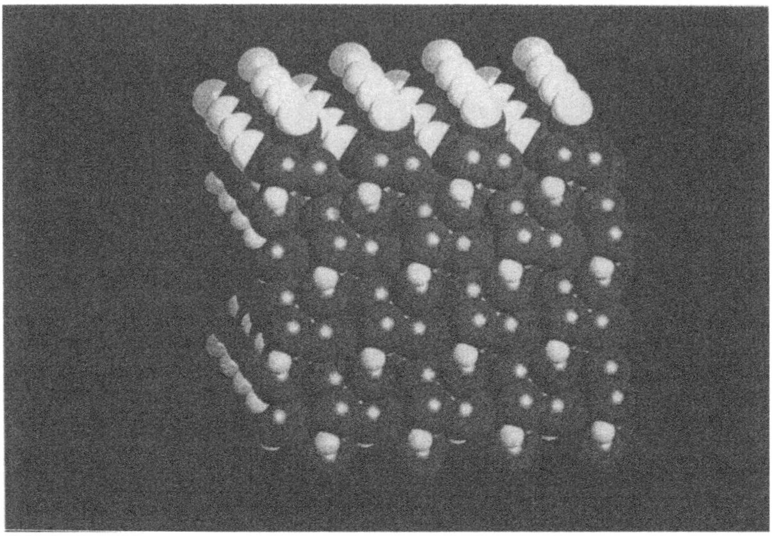

Figure 16. The surfaces of quartz are all exposing both silicon and oxygen ions. The ratio between oxygens having dangling bonds and the coordinative unsaturation of silicon is always such that protons and hydroxide ions are added in pairs. As can be seen the resulting surface has hydroxide ions in the two topmost layers.

To calculate reaction enthalpies involved in the formation of defects and surfaces one usually introduces a hypothetical gas phase reaction, via a Born-Haber cycle, for which the reaction enthalpy may be estimated. A suitable Born-Haber cycle for hydroxylation is

$$\begin{aligned} H_2O &\rightarrow OH_g^- + H^+ \\ O_{surf}^{2-} &\rightarrow O_g^{2-} \\ O_g^{2-} + H^+ &\rightarrow OH_g^- \\ 2OH_g^- &\rightarrow 2OH_{surf}^-. \end{aligned} \tag{17}$$

It should be noted that this Born-Haber cycle includes O_g^{2-} and therefore the second electron affinity of oxygen enters into the calculated energy. Since O_g^{2-} is an unstable species the second electron affinity of oxygen cannot be defined in a general manner. The second electron affinity of oxygen

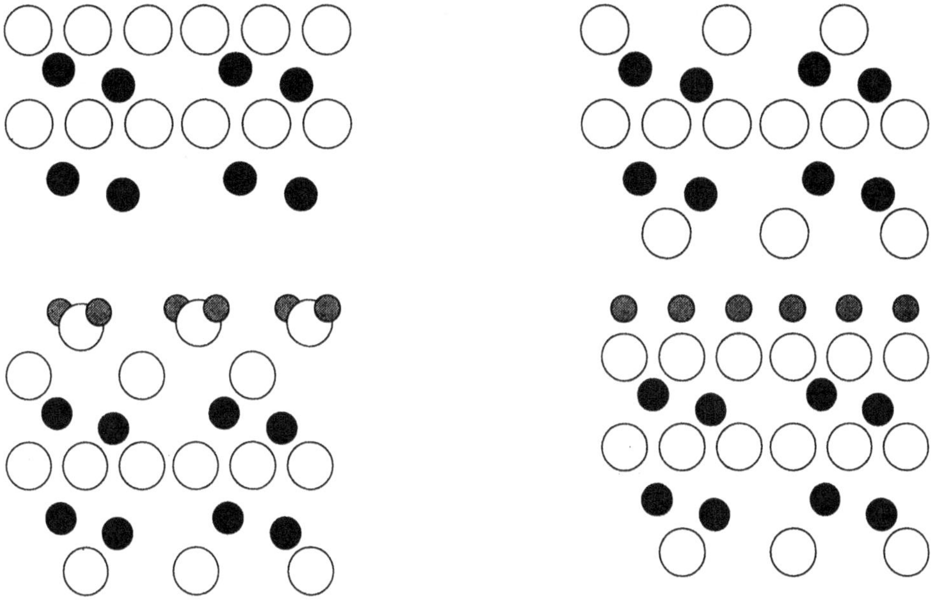

Figure 17. The oxygen-terminated basal plane has a dipole. This dipole is canceled when moving half of the oxygens to the other side of the crystal. Reacting the resulting halflayer oxygen terminated surface with water gives the hydroxilated surface.

is only meaningful when defined in connection with the formation of the bulk material and is thus structure dependent. Nevertheless, using.this procedure it is possible to calculate the reactive surface energy of $\alpha - Al_2O_3$.

For a clean surface the surface energy must be a positive quantity since it would otherwise be favourable for the crystal to maximise its surface area, *i.e.* disintegrate. When the surface undergoes a chemical reaction this is not necessarily true. A negative surface energy is possible if further reaction is either kinetically unfeasible or sterically inhibited. The most famous example of such a situation is the oxide coating covering aluminium metal. If we work out the corrections to the surface energy, using standard thermochemical data to estimate the reaction enthalpies [84], the surface energies stay positive for α-quartz but the α-Al_2O_3 basal plane surface energy is negative. It must be remembered that this negative surface energy show considerable uncertanity owing to the uncertanity in the definition of the oxygen second electron affinity.

5.5. THIN FILM INTERFACES

Thin film interfaces play a major rôle in many contemporary solid state applications such as catalysis, device applications (for example sensors) and printed circuitry. A detailed knowledge of the behaviour of these systems at the atomic level is therefore crucial to their further development and future applications. However, characterisation of thin film interfaces and in particular heteroepitaxial interfaces (between dissimilar materials) is difficult experimentally. For example, to examine these types of materials a surface sensitive technique is required to explore the thin film structure which is sensitive to a depth profile of as little as 10Å (or monolayer coverage) and furthermore may be required to detail a high degree of amorphisation of the supported material. Thin film heteroepitaxial interfaces are therefore ideally suited for study by simulation techniques.

The first theoretical studies of interfaces were by Duffy [85] who addressed defects in NiO grain boundaries. In this case the complexity of the problem is reduced as there is no question of misfit accommodation. Later, Cotter *et. al.* [86] rationalised, using simulation techniques, LEED patterns of BaO deposition on MgO and proposed that the misfit between the two materials is accommodated by a 45° rotation of the BaO monolayer with respect to the MgO substrate *i.e.* $a_1(BaO) \approx \sqrt{2}a_1(MgO)$.

In our work we have integrated a near coincidence site lattice theory (NCSL) in conjunction with simulation techniques [87, 88]; initial interfacial configurations are first constructed using NCSL theory to obtain configurations with low misfits and then energy minimisation techniques are applied. The driving force for minimisation is dominated by the elimination of the strain incurred as a result of the misfit, and maximising favourable (i.e. cation-anion) interactions across the interface region [89]. In contrast to the sometimes quite subtle local structural perturbations observed in, for example, defect calculations, interfacial relaxation can effect transformations of the whole thin film structure, occasionally resulting in complete amorphisation [90]. This method has now been sufficiently tested such that interfaces between two incommensurate oxide materials can now be routinely modelled. We show in the following sections some results of the various systems we have considered. In particular we focus on the effects of the interface on the thin film structure and properties and how this can (and has) been exploited in the various application of the materials such as catalysis.

We consider first a simple interface system between two rocksalt type structures, BaO and MgO [91],[92]. This system not only provides an ideal model system of a heteroepitaxial interface but also has implications for superconducting thin films. For example, to maintain the superconducting properties of $YBa_2Cu_3O_7$, the material must be deposited on a substrate such that little structural disruption is incurred. For this material the plane terminating the $YBa_2Cu_3O_7$ at the interface can be either BaO or CuO [89] and therefore the plane which provides the "cleanest" monolayer coverage would perhaps be the best choice. In this context small BaO clusters were constructed on an MgO(001) surface (with 1-8 BaO units) and each system energy minimised. Clustering of the BaO to form almost spherical BaO islands on the MgO surface (fig 18) is preferred rather than spreading of the BaO to form a thin film in accordance with the MBE results [93]. This behaviour was attributed, in part, to the instability of a BaO monolayer compared with a BaO bi-layer [89].

The BaO was also considered at higher coverages on MgO; one particular interface is illustrated in fig 19 which again suggests the BaO does not form a coherent monatomically flat film on the MgO. The corrugation of the BaO monolayer surface, is therefore very likely to perturb the structure of the $YBa_2Cu_3O_7$ and therefore one might expect a deterioration of the superconducting properties if the $YBa_2Cu_3O_7$ is terminated with BaO at the interface.

In a further study we considered the $YBa_2Cu_3O_7$/MgO interface with both CuO and BaO terminating the $YBa_2Cu_3O_7$ at the interface [94]. The calculations suggest the interface is more stable with a CuO termination. We also observe modifications in bond distances by as much as 10% within the $YBa_2Cu_3O_7$ structure as a result of the interface. Miceli et al [95] have postulated that the superconducting transition temperature, Tc, is a simple function of the Cu-O bond distance and therefore such changes in bond distances may influence the superconducting properties of the material. In the following sections we examine how the interface may influence the defect chemistry of the supported material.

5.5.1. *Interfacial Defects*

We have shown previously [96], that a reduced density of ions at the interfacial region of the MgO/MgO twist grain boundaries are essential to their stability. It is therefore pertinent to propose the same be true of hetero-interfaces. In this respect we considered again the BaO/MgO "model interface" and systematically removed stoichiometric BaO or MgO units from the interface [97]. The results show that removal of MgO units increased the stability of the interface. The reason for this behaviour is attributed to the removal of unfavourable cation-cation and anion-anion interactions across the interface. The term defect in this sense is therefore inappropriate as they are an intrinsic part of the low temperature interface configuration. This result also suggests that it is the local structure that governs the defect chemistry and clearly any continuum models of the interface employed to describe defect behaviour at these regions will fail [92].

In the next two sections we see how the modification in defect chemistry can be exploited in, for example, catalysis. We consider two aspects: catalytic activity and catalytic selectivity.

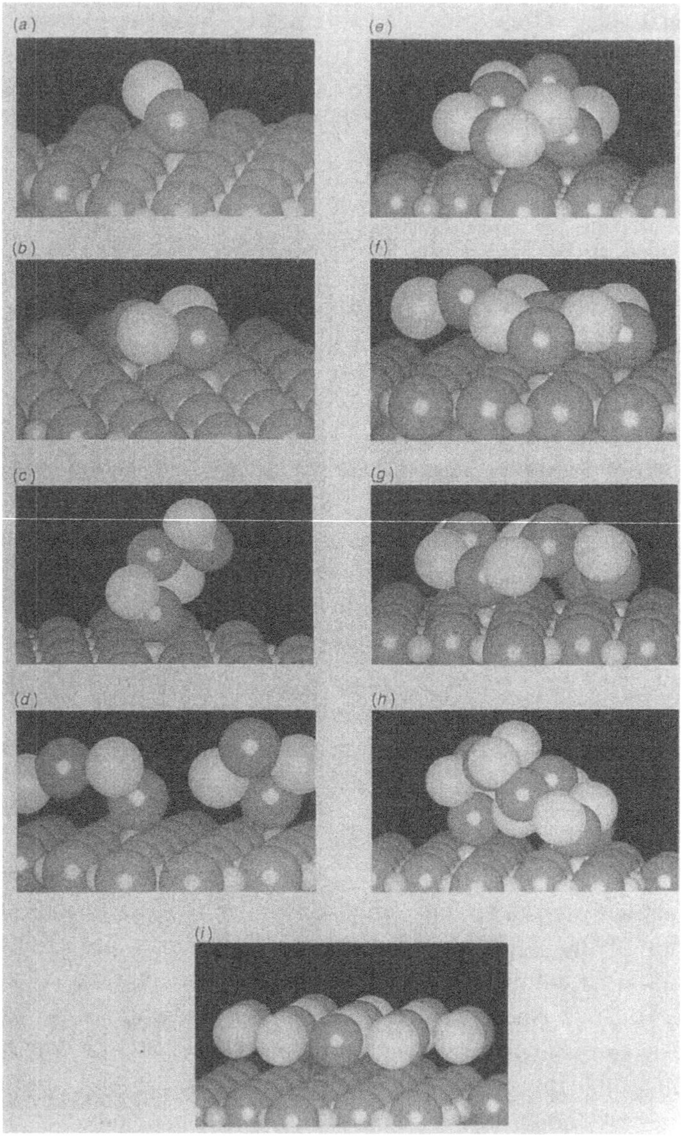

Figure 18. Representations of the relaxed configurations of from one to eight BaO units [(a)-(h)] placed on the MgO(001) surface. (i) illustrates the unrelaxed starting configuration of eight BaO units placed on the MgO(001) substrate. Barium ions are the larger light balls, magnesium ions are the smaller light balls and oxygen ions the dark coloured balls.

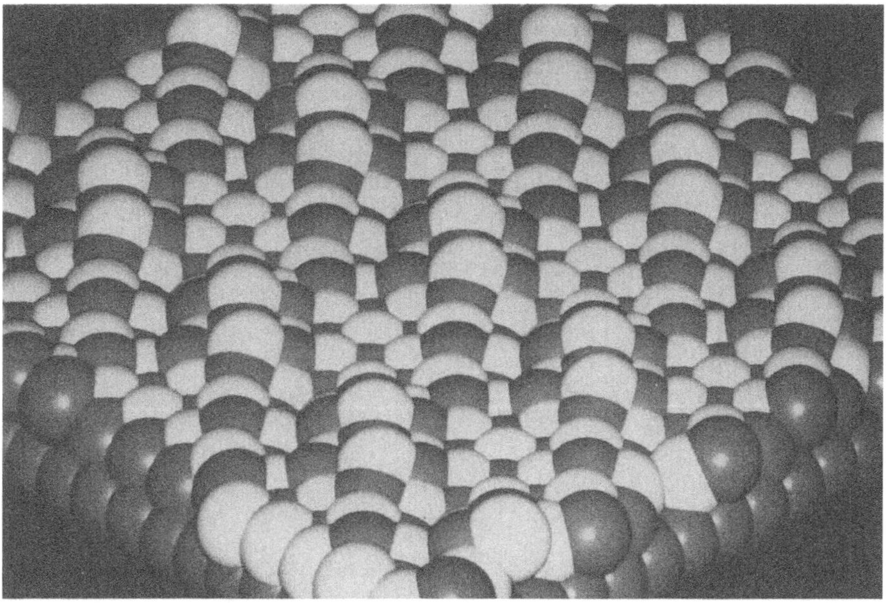

Figure 19. Relaxed configuration of a BaO monolayer on an MgO(001) surface. Barium ions are the lighter larger balls with oxygen the smaller darker balls. The MgO substrate is obscured by the monolayer.

5.5.2. *Catalytic activity*

In oxidation catalysts, the catalyst must have the ability to provide oxygen species (from the lattice) to participate in the reaction [98] which implies a link between the oxygen vacancy formation energy of the material and its catalytic activity. The sometimes quite considerable structural modification of a supported material is likely to influence its defect chemistry which in turn may affect its catalytic activity. Furthermore, by considering alternative support materials, one may be able to "fine tune" the catalysis. Clearly, detailed models of the defect chemistry at interfaces are of great importance within catalysis. In this respect we considered the CeO_2/Al_2O_3 [96] interface which is well known as an automobile exhaust catalyst. We

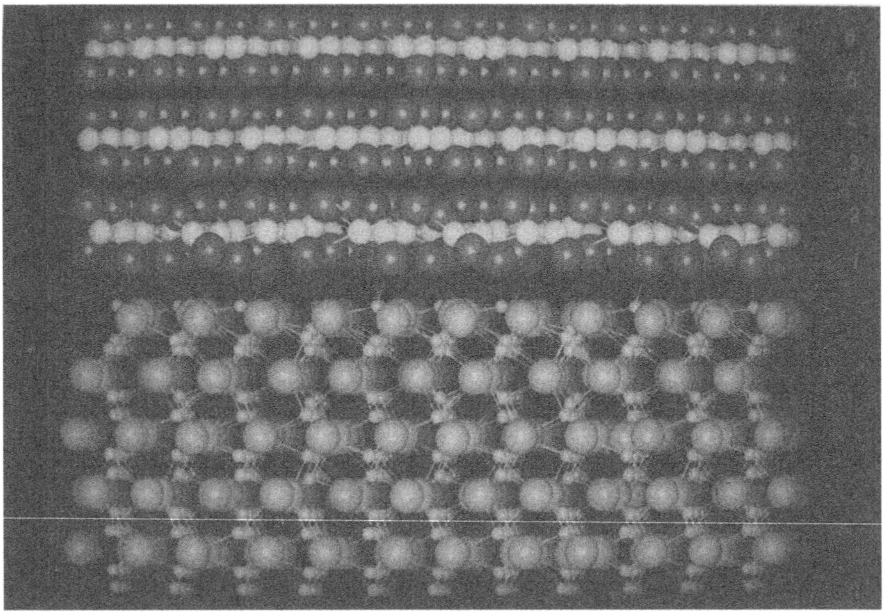

Figure 20. Diagrammatic representation of the $CeO_2(111)/Al_2O_3(001)$ interface after full energy minimisation illustrating oxygen plane notation. The CeO_2 is at the top of the figure with Al_2O_3 at the bottom. Oxygen ions are represented by the larger balls (dark in CeO_2 and light in Al_2O_3).

consider a CeO_2 thin film comprising 6 oxygen planes and calculated the energy to remove oxygen ions (14.3% coverage) from the CeO_2 from each of the six planes (fig 20). Charge compensation was facilitated by reducing two adjacent Ce(IV) species to Ce(III). The results, presented in fig 21, reveal that oxygen vacancy formation is favourable approaching the interface which implies that in this system, the presence of the interface will promote the migration of oxygen from the interface to the surface of the CeO_2 thin film, a process which may be of importance in catalytic reactions.

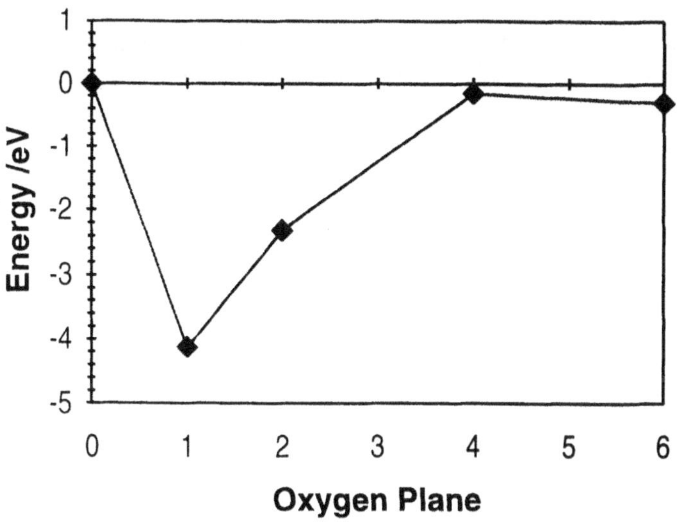

Figure 21. Variation of the oxygen vacancy formation energy in the $CeO_2(111)/Al_2O_3(001)$ interface as a function of oxygen plane (see figure 20). The energies are relative to oxygen vacancy formation at a pure CeO_2 surface (plane "0"). Negative values indicate the defect is more stable compared with a defect on the unsupported $CeO_2(111)$ surface. Plane 1 represents the interfacial plane and plane 6 the free surface of the CeO_2 thin film.

5.5.3. *Catalytic Selectivity*

The final area we consider is surface selectivity, in particular the way in which the modified surface structure of the supported thin film is likely to influence the sorption of adsorbed molecules. We examine the V_2O_5/TiO_2 interface [90, 99] which is an important partial oxidative catalyst. In this respect the low energy configurations of ethene on both the unsupported and supported V_2O_5 were calculated and are shown in figs 22a and 22b respectively. Clearly, there is a large difference in the sorption configuration; for ethene on the unsupported surface, the molecule lies flat, yet on the supported surface the modification in surface potential due to the structural change prevents the ethene from accommodating a flat configuration and instead lies at an angle. We also note that the low energy adsorption configurations for ethane are almost identical for the supported and unsupported V_2O_5. Such differences in the adsorption behaviour of ethene and ethane have implications for the selectivity of these materials.

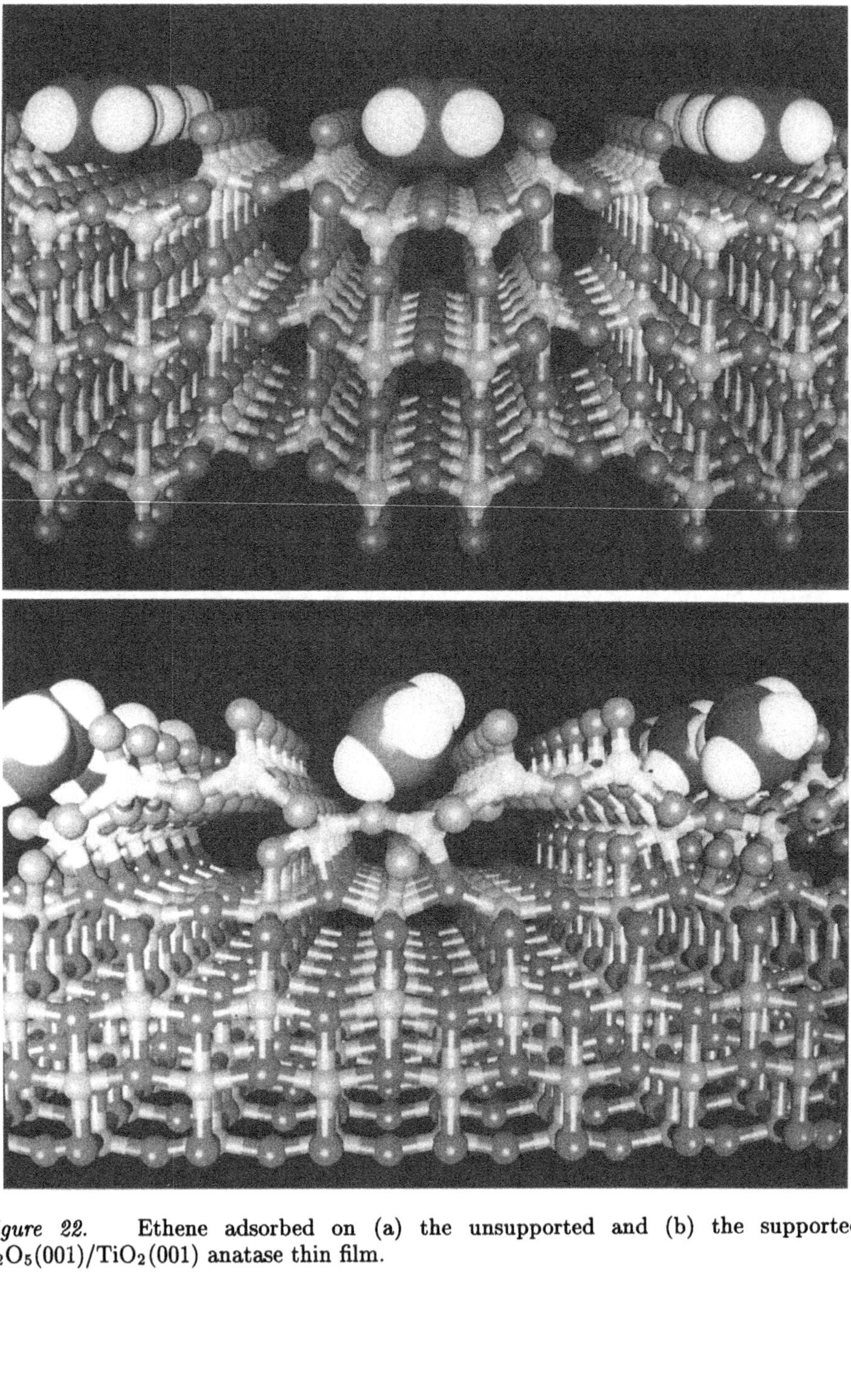

Figure 22. Ethene adsorbed on (a) the unsupported and (b) the supported $V_2O_5(001)/TiO_2(001)$ anatase thin film.

5.6. SURFACE DEFECTS

It has long been known that surface defects play a crucial rôle in many important phenomena, including catalysis. Perhaps the most widely studied case concerns the Li doped MgO catalyst [100], which functions as a partial oxidation catalyst, converting methane to C_2 and higher hydrocarbons. Such catalytic processes, although not commercially viable in the case of Li/MgO, are of great economic potential as they offer the possibility of converting natural gas into more readily transportable liquid fuels.

It is well known that doping of MgO with Li results in the creation of Li substitutionals (at Mg sites) which at normal oxygen partial pressures are compensated by 'hole' states, which to a first approximation can be described as substitutional O^- ions (see reference [101] and articles cited therein). In the bulk, the substitional defect traps the oppositely charged hole states forming $[Li]^\circ$ centres, i.e. states in which the hole is localised at the nearest neighbour site with respect to the Li. It is generally assumed that such species are present on the surface of the material where they provide the active sites for hydrogen abstraction *via* the reaction:

$$CH_4 + Li^+_{surf} + O^-_{surf} +_{-} \rightarrow Li^+_{surf} + CH_3 + OH_{surf} \ . \qquad (18)$$

The system has been studied in detail, using simulation techniques, by Lewis et al. [101, 102], which greatly extended our understanding of the surface defect chemistry of this material. Their work has established two crucial points:

i) There is a strong driving force for surface segregation of $[Li]^\circ$ centres, i.e. these species have substantially lower energies at surface sites.

ii) Hole states are stabilised at low coordination number sites (e.g. step sites) on the surface.

iii) Hole states may be displaced from sites adjacent to Li^+ and from low coordination sites by other monovalent ions, notably Cl^-, thereby rationalising the known effects of these species on catalysis (see reference [102] for a detailed discussion).

These and other structures clearly establish the rôle of simulation techniques in characterising surface defect structures, knowledge of which is increasingly important in understanding catalysis at the molecular level.

5.7. CONCLUSIONS

The applications discussed in this article have, we hope, demonstrated the increasingly central position of simulation techniques in solid state and surface chemistry. The horizons of the field continue to expand with the development in techniques and the constant growth of computer power.

Atomistic modelling techniques promise, therefore, to play a growing rôle in developing an understanding of key phenomena, e.g. catalysis, at a molecular level.

6. ACKNOWLEDGEMENTS

We are grateful to several colleagues for their contributions to the work summarised in the article, most notably A.L. Rohl, R.G. Bell, C.M. Freeman and D.W. Lewis. The work has been supported by research grants from EPSRC (including a ROPA grant). D.C.S. would also like to thank M.-A. Perrin and P. Nortier for useful discussions and Rhône-Poulenc for financial support.

References

1. C.R.A. Catlow, R.G. Bell, and J.D. Gale. *J. Mater. Chem.*, 4:781, 1994.
2. M.P. Allen and D.J. Tildesley. *Computer Simulation of Liquids*. Oxford University Press, ISBN 0-19-855645-4, 1987.
3. C.R.A. Catlow and G.D. Price. *Nature*, 347:243, 1990.
4. M.C. Payne, M.P. Teter, and D.C. Allan. *J. Chem. Soc. Faraday Trans.*, 86:1221, 1990.
5. P. Hohenberg and W. Kohn. *Phys Rev*, 140:1133, 1964.
6. R. Car and M. Parrinello. *Phys Rev Lett*, 55:2471, 1985.
7. D.E. Williams. *Top. Curr. Phys.*, 26:3, 1981.
8. E.A. Colbourn. In C.R.A Catlow, editor, *Advances in Solid State Chemistry*, volume I. JAI Press, 1990.
9. M. Guest. Gamess code. Daresbury Laboratory, UK, 1994.
10. R. Amos. Cadpac code. Chemistry Department, University of Cambridge, UK, 1994.
11. R. Dovesi, V.R. Saunders, and C. Roetti. Crystal92 user manual. Universita di Torino, 1992.
12. M. Payne. Castep code. Department of Physice, University of Cambridge, 1994.
13. DSOLID, Molecular Simulations Inc., 9685 Scranton Road, Suite A, San Diego, California 92121-2777, USA.
14. DMOL, Molecular Simulations Inc., 9685 Scranton Road, Suite A, San Diego, California 92121-2777, USA.
15. DGAUSS is available from the CRAY unichem project.
16. M.J. Sanders, M. Leslie, and C.R.A. Catlow. *J. Chem. Soc. Chem. Comm.*, page 1271, 1984.
17. B.W.H. van Beest, G.J. Kramer, and R.A. van Santen. *Phys. Rev. Lett.*, 60:1955, 1990.
18. J.R. Hill and J. Sauer. *J. Phys. Chem.*, 98:1233, 1994.
19. J. Purton, R. Jones, M. Heggie, S. Oberg, and C.R.A. Catlow. *Phys. Chem. Minerals*, 18:389, 1992.
20. S.C. Parker and G.D. Price. In C.R.A. Catlow, editor, *Adv. in Solid State Chem.*, volume 1. JAI Press, 1990.
21. S. Kirkpatrick, C.D. Gelatt, and M.P. Vecchi. *Science Wash.*, 220:671, 1983.
22. N. Metropolis, A.W. Rosenbluth, M.N. Rosenbluth, A.H. Teller, and G. Teller. *J. Chem. Phys.*, 21:1087, 1953.
23. M.W. Deem and J.M. Newsam. *Nature Lond.*, 342:260, 1989.
24. J.M. Newsam, M.W. Deem, and C.M. Freeman. National institute of standards and

technology special publication 846. In *Proc.Int.Conf.Accuracy in Powder Diffraction II*, page 80, 1992.
25. A.T. Bruger, D.J. Leahy, T.R Hynes, and R.O. Fox. *J.Mol.Biol.*, 221:239, 1991.
26. Z.A. Kaszkur, R.H. Jones, D. Waller, C.R.A. Catlow, and J.M. Thomas. *J. Phys. Chem.*, 97:426, 1993.
27. Z.A. Kaszkur, J.W. Couves, R.H. Jones, D. Waller, C.R.A. Catlow, and J.M. Thomas. In R. von Ballmoos, J.B. Higgins, and M.M.J. Treacy, editors, *Proceedings from the Ninth International Zeolite Conference*, volume 1, page 619, 1993.
28. R.L. McGreevy and L. Pusztai. *Mol. Simul.*, 1:359, 1988.
29. D.A. Keen and R.L. McGreevy. *Nature Lond*, 344:423, 1990.
30. R.G. Bell, R.A. Jackson, and C.R.A. Catlow. *J.Chem.Soc. Chem.Comm.*, page 782, 1990.
31. S.M. Tomlinson, R.A. Jackson, and C.R.A. Catlow. *J.C.S.Chem.Comm.*, (813), 1990.
32. J.M. Newsam, M.M.J. Treacy, W.T. Koetsier, and C.B. deGruyer. *Proc.R.Soc. Lond.A*, 420:375, 1988.
33. P.A. Wright, S. Natarajan, J.M. Thomas, R.G. Bell, P.L. Gai-Boyes, R.H. Jones, and J. Chen. *Angew.Chem.Int.Ed.Engl.*, 32:1472, 1992.
34. C. Baerlocher, A. Heppe, and W.M. Meier. *DLS-76*. Institute of Crystallography and Petrology, ETH Z73 Zurich Switzerland.
35. C.R.A. Catlow, A.N. Cormack, and F. Theobald. *Acta Crystallogr. Sect. B: Struct. Sci.*, B 40(3):195, 1984.
36. J.D. Gale and N. Henson. To be published.
37. R. Salvador-Ruiz, G. Sastre, D.W. Lewis, and C.R.A. Catlow. *J.Mater.Chem.*, in press.
38. C.M. Freeman, J.M. Newsam, S.M. Levine, and C.R.A. Catlow. *J.Mater.Chem.*, 3(5):531, 1993.
39. T.S. Bush, P.D. Battle, and C.R.A Catlow. *J.Mater.Chem*, 5:1269, 1995.
40. N.F. Mott and M.J. Littleton. *Trans. Farad. Soc.*, 34:389, 1938.
41. C.R.A. Catlow and W.C. Mackrodt, editors. *Computer simulations of solids, Lecture Notes in Physics*, volume 166. Springer-Verlag, 1995.
42. C.R.A. Catlow. *Ann. Rev. Mater. Sci.*, 16:517, 1986.
43. D.W. Lewis, S. Carr, G. Sankar, and C.R.A. Catlow. *J.Phys.Chem.*, 99:2377, 1995.
44. J. Sauer. In C.R.A. Catlow, editor, *Modelling of structure and reactivity in zeolites*, page 183. Academic Press Limited, London, 1992.
45. G.J. Kramer, A.J.M. de Man, and R.A. van Santen. *J.Amer.Chem.Soc.*, 113:6435, 1991.
46. K-P. Schroder, J. Sauer, M. Leslie, C.R.A. Catlow, and J.M. Thomas. *Chem.Phys.Lett.*, 188:320, 1992.
47. M. Cherry, M.S. Islam, and C.R.A. Catlow. *J. Solid State Chem.*, 118:125, 1995.
48. M. Cherry, M.S. Islam, J.D. Gale, and C.R.A. Catlow. *J.Phys.Chem.*, 99:14614, 1995.
49. C.R.A. Catlow, P.S. Baram, S.C. Parker, J. Purton, and K.V. Wright. *Phil.Trans.A*, 350(1963):265, 1995.
50. D.H. Gay and A.L. Rohl. *J. Chem. Soc. Faraday Trans.*, 91:925, 1995.
51. M.W. Deem, J.M. Newsam, and S.K. Sinha. *J. Phys. Chem.*, 94:8356, 1990.
52. D.E. Parry. *Surf. Sci.*, 49:433, 1975.
53. D.E. Parry. *Surf. Sci.*, 54:195, 1976.
54. D.M. Heyes, M. Varber, and J.H.R. Clarke. *J. Chem. Soc. Faraday Trans.*, 73:1485, 1977.
55. A. Bravais. *Etudes Crystallographiques*. Academie des Sciences, Paris, 1913.
56. G. Friedel. *Bull. Soc. Fr. Minér.*, 30:326, 1907.
57. J.D.H. Donnay and D. Harker. *Am. Mineral.*, 22:446, 1937.
58. P. Hartman and W. G. Perdok. *Acta Crystallogr.*, 8:49, 1955.
59. P. Hartman and P. Bennema. *J. Crystal Growth*, 49:145, 1980.

60. P. Hartman and W. G. Perdok. *Acta Crystallogr.*, 8:525, 1955.
61. P. Hartman. *J. Crystal Growth*, 49:157, 1980.
62. Z. Berkovitch-Yellin. *J. Am. Chem. Soc*, 107:8239, 1985.
63. R. Docherty and K. J. Roberts. *J. Cryst. Growth*, 88:159, 1988.
64. E. Dowty. *Am. Mineral.*, 61:448, 1976.
65. J.W. Gibbs. *Collected Works.* Longman, New York, 1928.
66. W.C. Mackrodt, R.J. Davey, S.N. Black, and R. Docherty. *J. Cryst. Growth*, 80:441, 1987.
67. P. Hartman. *Acta Crystallogr.*, 11:459, 1958.
68. P.J. Lawrence and S.C. Parker. In C.R.A. Catlow, S.C. Parker, and M.P. Allen, editors, *Computer Modelling of Fluids, Polymers and Solids*, page 219. Kluwer, Amsterdam, 1990.
69. G. Wulff. *Z. Krystallogr.*, 34:449, 1901.
70. B.R. Heywood and S. Mann. *Langmuir*, 4:756—768, 1992.
71. S. Weiner. *Crit. Rev. Biochem.*, 20:365–408, 1986.
72. L. Addadi, Z. Berkovitch-Yellin, I. Weissbuch, and M. Lahav. *Top. Stereochem.*, 16:1–85, 1986.
73. R.J. Davey, L.A. Polywka, and S.J. Maginn. In J. Garside, R.J. Davey, and A.G. Jones, editors, *Advances in Industrial Crystallisation*, pages 150–165. Butterworth-Heinemann, Oxford, 1991.
74. W.J. Benton, I.R. Collins, S.D. Cooper, I.M. Grimsey, G.M. Parkinson, and S.A.Rodger. *Faraday Discuss.*, 95:281–297, 1993.
75. S.N. Black and L.A. Bromley. *J. Chem. Soc., Faraday Trans.*, 87:3409–3414, 1991.
76. R.J. Davey, S.N. Black, L.A. Bromley, D. Cottier, and J.E. Rout. *Nature*, 353:549–550, 1991.
77. P.E. Hillner, S. Manne, P.K. Hansma, and A.J. Gratz. *Faraday Discuss.*, 95:191–197, 1993.
78. W.J.P. van Enckevort. In J.P. van der Eerden and O.S.L. Bruinsma, editors, *Science and Technology of Crystal Growth*, pages 355–366. Kluwer, Dordrecht, 1995.
79. P. Hartman and C.S. Strom. *J. Cryst. Growth*, 114:502–512, 1989.
80. N.L. Allan, A.L. Rohl, D.H. Gay, C.R.A. Catlow, R.J. Davey, and W.C. Mackrodt. *J. Chem. Soc, Faraday Discuss.*, 95:273–280, 1993.
81. W.C. Mackrodt, R.J. Davey, S.N. Black, and R. Docherty. *J. Cryst. Growth*, 80:441, 1987.
82. P. Hartman. *J. Cryst. Growth*, 9:721, 1987.
83. W. C. Mackrodt. *Phil. Trans. R. Soc. Lond. A*, 341:301, 1992.
84. C.R.A. Catlow. *J. Phys. Chem. Solids*, 38:1131, 1977.
85. D.M. Duffy. *J. Phys. C*, 19:4393, 1986.
86. M. Cotter, S. Campbell, L.L. Cao, R.G. Egdell, and W.C. Mackrodt. *Surf. Sci.*, 208:267, 1989.
87. H. Grimmer, W. Bollmann, and D.H. Warrington. *Acta Cryst.*, 30:197, 1974.
88. T.X.T. Sayle, C.R.A. Catlow, D.C. Sayle, S.C. Parker, and J.H. Harding. *Phil. Mag. A*, 68:565, 1993.
89. D.C. Sayle. *Computer simulation of metal oxide/metal-oxide and metal/metal-oxide interfaces.* PhD thesis, University of Bath, 1992.
90. D.C. Sayle, C.R.A. Catlow, M.-A. Perrin, and P. Nortier. *J. Phys. Chem.*, 100:8940, 1996.
91. D.C. Sayle, S.C. Parker, and J.H. Harding. *J. Mater. Chem.*, 4:1883, 1994.
92. D.C. Sayle, S.C. Parker, and J.H. Harding. *Phil. Mag.*, 69:787, 1994.
93. R.A. Mackee, F.J. Walker, E.D. Specht, L.A. Boatner G.E. Jellison, and J.H. Harding. *Phys. Rev. Lett.*, 72:2741, 1994.
94. D.C. Sayle, S.C. Parker, and J.H. Harding. *Mol. Simul.*, 12:127, 1994.
95. P.F. Miceli, J.M. Tarascon, L.H. Greene, P. Barboux, F.J. Rotella, and J.D. Jorgensen. *Phys. Rev. B*, 37:5932, 1988.

96. D.C. Sayle, T.X.T. Sayle, S.C. Parker, J.H. Harding, and C.R.A. Catlow. *Surf. Sci.*, 334:170, 1995.
97. D.C. Sayle, T.X.T. Sayle, S.C. Parker, C.R.A. Catlow, and J.H. Harding. *Phys Rev B*, 50:14498, 1994.
98. P. Mars and D.W. van Krevelen. *Spec, Suppl. Chem. Eng. Sci.*, 3:41, 1954.
99. D.C. Sayle, C.R.A. Catlow, M.-A. Perrin, and P. Nortier. *Catal. Lett.*, 38:203, 1996.
100. T. Ito, J.-X. Wang, C.H. Lin, and J.M. Lunsford. *J. Amer. Chem. Soc.*, 107:5065, 1985.
101. D.W. Lewis, R.W. Grimes, and C.R.A. Catlow. *J.Molecular Catalysis A: Chemical*, 100:103, 1995.
102. D.W. Lewis and C.R.A Catlow. *Topics in Catalysis*, 1:111, 1994.

INDEX

GPSR Compliance

The European Union's (EU) General Product Safety Regulation (GPSR) is a set of rules that requires consumer products to be safe and our obligations to ensure this.

If you have any concerns about our products, you can contact us on ProductSafety@springernature.com

In case Publisher is established outside the EU, the EU authorized representative is:

Springer Nature Customer Service Center GmbH
Europaplatz 3
69115 Heidelberg, Germany

Batch number: 10809436

Printed by Printforce, the Netherlands